Regreening the Bare Hills

WORLD FORESTS

Series Editors

World Forests Description

As forests stay high on the global political agenda, and forest-related industries diversify, cutting edge research into the issues facing forests has become more and more transdisciplinary. With this is mind, Springer's World Forests series has been established to provide a key forum for research-based syntheses of globally relevant issues on the interrelations between forests, society and the environment.

The series is intended for a wide range of readers including national and international entities concerned with forest, environmental and related policy issues; advanced students and researchers; business professionals, non-governmental organizations and the environmental and economic media.

Volumes published in the series will include both multidisciplinary studies with a broad range of coverage, as well as more focused in-depth analyses of a particular issue in the forest and related sectors. Themes range from globalization processes and international policies to comparative analyses of regions and countries.

David Lamb

Regreening the Bare Hills

Tropical Forest Restoration in the Asia-Pacific Region

Springer

David Lamb
Centre for Mined Land Rehabilitation
University of Queensland
Brisbane 4072, Australia
david.lamb@uq.edu.au

ISBN 978-90-481-9869-6 e-ISBN 978-90-481-9870-2
DOI 10.1007/978-90-481-9870-2
Springer Dordrecht Heidelberg London New York

Library of Congress Control Number: 2010937424

Cover illustration: The best place to discuss small-scale forestry issues is in the field. Photograph by David Lamb.

Printed on acid-free paper

Springer is part of Springer Science+Business Media (www.springer.com)

Preface

Large areas of the world's tropical forests have been cleared in the last 100 years. Much of this forest land has been transformed into productive agricultural land. However, many areas have not and are now marginal for agriculture or have been largely abandoned. The areas involved are huge. One estimate suggests there are 350 million ha of former tropical forest lands around the world that were once cleared but are now degraded in some way (ITTO 2002). This is equivalent to an area roughly double the size of Indonesia.

Of the forests remaining, extensive areas have been degraded by logging. This is not because logging necessarily causes such damage but simply because the logging was poorly managed. The damage they have suffered means many of these supposedly renewable resources are now incapable of supporting a second cutting cycle, at least within the foreseeable future. As a consequence, large areas of these badly logged forests will probably also be cleared for agricultural purposes. These changes have transformed tropical landscapes. In most countries the original forests were once a 'sea' that completely surrounded villages and farms. In less than a single human lifetime many of the residual forests have become small 'islands' in a 'sea' of agricultural land.

These events have affected livelihoods and changed environmental conditions. Some people have become very wealthy as a result of the changes. However, most have not and large populations of relatively poor people now live in these transformed landscapes. Many have been displaced from their traditional homelands and live on less than two dollars a day despite the wealth produced by logging. Deforestation and forest degradation have also caused massive environmental damage. Large numbers of species have become extinct and many areas are now largely devoid of native wildlife. In other areas wildlife persists but the surviving species have greatly diminished populations making them highly vulnerable to future changes. Erosion of topsoil has been common leading to changes in soil fertility. Water quality, and sometimes water availability, has been adversely affected.

There have been four broad policy responses to these events. One has been to try to increase the productivity of cleared agricultural lands to make better use of them and reduce the need to clear any more forest. A second has been to conserve more of the remaining forests in a network of National Parks and protected areas. A third has been to encourage the greater use of reduced-impact logging to decrease the damage caused by harvesting operations. The final response has been to reforest some of the land that has been cleared and then abandoned. It is this fourth response that is the focus of this book. Just how should such reforestation be carried out? In particular, how might it be done in a way that it overcomes degradation, restores some of the key ecological processes and functions of tropical forests and, at the same time, improves the livelihoods of local people?

Although many state agencies and large industrial corporations are engaged in tree-planting the present rate of tropical reforestation is much less than the present rate of tropical deforestation. Two things can be said about this reforestation. Firstly, it commonly involves a remarkably small number of species from an even smaller number of genera (e.g. *Pinus, Eucalyptus, Acacia, Tectona*) despite the extra-ordinary biological diversity of the region. Secondly, many of the large, homogenous plantations being established are good for producing goods such as pulpwood, particleboard or veneer but they are not necessarily able to provide many of the ecological services that disappeared when the first waves of deforestation swept across the tropics. These plantations are generating financial rewards to their owners but many are accentuating the process of landscape simplification and homogenisation that was begun with deforestation.

There is another group participating in reforestation whose role is often unrecognized and these are the small landholders. Some of these farmers are using the same species and reforestation methods being used by the bigger industrial growers because the silviculture of these species is well-known and they are profitable to grow. But many are interested in using a much broader range of species and reforestation methods because they want to produce a wider variety of goods and services. Their capacity to do this is hampered because much less is known about the silviculture of these other species, the site conditions they prefer or the ways they might be grown in plantations.

Reforestation to improve livelihoods involves more than simply planting trees. If landholders are to benefit from tree-planting then there must be markets for the goods and services they produce and these prospective growers must be aware of their opportunity costs. Reforestation may benefit the community as a whole but it may also be disadvantageous for certain people as well. How are these trade-offs to be made? The book attempts to address these types of questions while recognising that the value of generalisations are sometimes limited and that specific situations often require site-specific solutions.

Reforestation to improve conservation outcomes is equally difficult. Most forms of reforestation are likely to have some conservation benefits but there are very large differences in the capacity of different types of reforestation to generate these benefits. For example, large monocultures of exotic tree species will not be as effective as polycultures of native species in creating habitats for wildlife. Likewise, monocultural plantations may sometimes be less useful for watershed protection than a well-established grass cover. The task of reforesting degraded lands to conserve residual biodiversity is made especially difficult because so little is known of the habitat requirements of most tropical forest biota. Further research should help overcome this problem except that there is often a mismatch between the types of problems faced by field managers seeking to restore some kind of forest cover in degraded lands and those being addressed by conservation biologists. Sophisticated techniques are often being used by conservation biologists to investigate inconsequential or highly specialized problems but without any regard for the socio-economic or political context in which the problem is immersed. This book seeks to discuss some of the research priorities arising from deforestation and degradation although it deals more with ecosystems than with the conservation needs of particular species.

The geographic scope of the book needs some clarification. It covers the tropical and sub-tropical areas of Southeast Asia and the Southwest Pacific (subsequently referred to, rather loosely, as the Asia-Pacific region). This is a politico-geographical setting without any particular ecological rationale except that the dominant vegetation is tropical forest. On this basis, northern Australia is included but the temperate regions of that continent are not. This region is clearly diverse. It contains evergreen rainforests, deciduous forests and woodlands as well as large areas of grasslands, shrublands and secondary forests created by human activities. These various communities include some of the world's most biologically diverse ecosystems. In the opinion of the famous nineteenth century naturalist Alfred Russell Wallace '... no part of the world can offer a greater number of interesting facts for our contemplation, or furnish us with more extensive and varied materials for speculation in almost every great department of human knowledge' (Wallace 1863 p. 217).

The human societies present in the region are equally diverse. They include a variety of ethnic and religious groups which speak over 1,000 languages. Some of these people live on their traditional lands while others are recent migrants. Some have formal ownership of the land they are using while many others do not. Some people are comparatively wealthy, but large numbers of people remain living in circumstances below international poverty benchmarks. The states in which they live use a range of political, legal and economic systems with some governments being able to exert considerable control over land use practices while others are not. The region is one in which significant economic growth has taken place in recent years and includes some of the so-called 'Asian Tiger' economies but it also includes countries where economic development has been slow.

Despite this socio-economic and political diversity there appears to be growing recognition across the region that forest conservation is important and that some reforestation should take place. This does not mean that these views are held with equal enthusiasm or that all governments or communities are yet acting upon them. Nor does it mean there is agreement on how this reforestation should be done. In this respect it is interesting that one of the lessons emerging from recent experiences is that there is no single formula or recipe for reforestation. Rather, there are distinct benefits in exploring a variety of approaches.

To a person with a hammer every problem seems like a nail. The potential benefits of reforestation seem obvious to foresters and restoration ecologists. However, that should not blind us to the fact that reforestation may not always be the best way of dealing with degraded lands; nor might it be the most efficient way of improving livelihoods or overcoming poverty. Some degraded lands might be better used for cropping or other purposes that may have a far greater impact on poverty. Certainly some farmers will take this view. Nonetheless, reforestation may be the best way – and sometimes the only way – of rehabilitating some of the most degraded areas and, at the same time, generate improvements in livelihoods and conservation outcomes. This books aims to show that there are encouraging developments occurring across the region from which lessons can be drawn by anyone interested in rehabilitating other tropical forests.

This book has been written in four parts. The first section (Chapters 1 to 3) outlines the problems. It describes the scale of deforestation and degradation and some of the factors responsible and argues that those wishing to undertake reforestation for conservation reasons must also seek to improve rural livelihoods if there is to be any chance of success. The second section (Chapters 4 to 8) outlines some of the ways reforestation might be carried out. These include encouraging natural regeneration as well as various forms of plantation establishment and forest restoration. It considers the extent to which each of these different approaches is able to supply both goods and ecological services. The third section (Chapters 9 and 10) considers how these options might be implemented in practice. Implementation involves improving the financial returns from tree growing to land managers and determining the best ways of incorporating tree plantings into farms. The fourth section (Chapters 11 and 12) explores the problems of scaling-up and reforesting large areas. This section considers how the benefits from tree growing can be maximised at a landscape scale and the institutional settings needed to facilitate reforestation across larger areas. The book concludes by considering in Chapter 13 the likely opportunities and constraints on reforestation across the region in future.

Throughout the book there is an emphasis on creating new forests that are resilient rather than on simply improving productivity. This means there is a bias towards forms of reforestation that promote diversity and heterogeneity rather than just biomass. There is also a focus on silvicultural approaches that might appeal to smallholders rather than on techniques more appropriate for large industrial plantation owners. Overall, the intent is to show there are many silvicultural options open to those seeking to reforest degraded lands and that the technique currently most favoured – plantations of exotic species grown on short rotations – is only one of the many choices available.

Over the years I have been fortunate to have worked on tropical forest restoration and rehabilitation in various parts of Southeast Asia and the Pacific with a number of remarkable people. Much of what is within this book has been learnt from them or while in their company. They have included fellow academics and post graduate students, ecologists, field and research foresters and land owners interested in reforestation. I have also benefited from my association with the former Cooperative Research Centre for Tropical Ecology and Management in northern Australia (now sadly, closed), the Global Partnership on Forest Landscape Restoration, the Society for Ecological Restoration and the IUCN Commission on Ecosystem Management. The Australian Centre for International Agricultural Research funded research in Vietnam which was undertaken with my colleague Dr Huynh Duc Nhan (project FST 2000/003). Some of the as-yet unpublished results from that work are included in several chapters here.

I am especially grateful to a number of colleagues who have been kind enough to read sections of the manuscript and offer advice. They include Sai Bulai, Jim Carle, Margaret Chapman, Robin Chazdon, Richard Corlett, Paul Dargusch, Peter Dart, Peter Erskine, Jack Ewel, Bob Fisher, Christine Fung, Don Gilmour, Stephen Harrison, Andrew Ingles, Jiro Kikkawa, Dan Manson, David Pullar, Jeff Sayer and Doug Sheil. The feedback and advice they have given has been extremely useful. Needless to say any remaining errors of fact or interpretation are entirely mine.

I am also obliged to Paul Ryan for permission to use some of his unpublished data and to Simon Albert, Intu Boedhihartono, Mila Bristow, Sharon Brown, Jean-Christophe Castella, Peter Erskine, Jennifer Firn, Sean McNamara and Robert Ong for giving me permission to use their photographs or illustrations. These have been acknowledged in the text. All other photos are my own. I am thankful to Kate Moore for very ably preparing the figures and to Sue Jackson for editorial advice. I am grateful to the following for permission to reproduce previously published material: Ecological Society of Australia (Figs. 5.9, 5.10), Rural Industries research and Development Corporation of Australia (Figs. 6.8), Springer (Figs. 7.2, 7.4) and Elsevier (Figs. 7.3, 7.7).

Most of the manuscript has been complied while I have been a visiting fellow at the Center for Mined Land Rehabilitation at the University of Queensland and I am very grateful to the Director, Professor David Mulligan, for being such a generous host.

Lastly I must thank my family – Marg, Andrew and Kate – who have endured my physical and mental absences over the last few years. They have done so with remarkable tolerance and good humour and, as usual, I am indebted to them for this support.

Brisbane David Lamb

ITTO. 2002. ITTO Guidelines for the Restoration, Management and Rehabilitation of Degraded and Secondary Tropical Forests. ITTO Policy Development Series No 13. International Tropical Timbers Organization, Yokohama.

Wallace A.R. 1863. On the Physical Geography of the Malay Archipelago. *Journal of the Royal Geographical Society of London* 33:217–234.

Contents

Abbreviations

ADB	Asian Development Bank
m^3	cubic meter
cm	centimetre
CIFOR	Centre for International Forestry Research
dbh	diameter breast high
FAO	Food and Agriculture Organization of the United Nations
FSC	Forest Stewardship Council
gm	gram
ha	hectare
ICDP	Integrated Conservation and Development Project
IRR	internal rate of return
ITTO	International Tropical Timbers Organization
IUCN	International Union for the Conservation of Nature
kg	kilogram
MAI	mean annual increment
NPV	net present value
NGO	non government organization
NTFP	non timber forest product
PES	payment for environmental service
REDD	reducing emissions from deforestation and degradation
REDD+	reducing emissions from deforestation and degradation and reforestation
tph	trees per hectare

Chapter 1
Deforestation and Its Consequences in the Asia-Pacific Region

> At length, after existing thus for millions of years, the rain forest ecosystem has very recently – in most of its area only within the last 100 years or so – been rudely disturbed by the spread of western civilization to the tropics. This has involved not only the cultivation of rubber, coffee, cocoa and similar crops for export – but the increase of the native populations under settled government has given rise to even more widespread clearing of the forest in the interests of subsistence agriculture, mainly in the form of shifting agriculture. Thus, within a very short space of time the primeval climax forest communities have been replaced over immense areas by cultivation, ruderal communities and seral stages.
>
> (Richards 1952, p. 404)

Introduction

Large-scale deforestation and forest degradation have become common in most tropical areas in the 60 years since Richards first expressed his concerns. Part of the reason has been the need for more agricultural lands. The world's population in 1950 was 2.5 billion people but this had risen to 6.1 billion by 2000. Many of those carrying out deforestation did so to grow food and probably viewed the forests they were clearing as being endless. But concerns have grown about the transformation that is underway. Biologists, in particular, have worried about the dramatic impact tropical deforestation is having on global biodiversity. Because of this there have been increased efforts to save representative samples of the world's tropical forests in National Parks, protected areas or in some kind of forest reserves – before they have all been cleared.

The extent of tropical deforestation and reasons for concern about its impact on biodiversity conservation have been widely discussed (Aronson et al. 2007; Sodhi and Brook 2006; Sodhi et al. 2004). Apart from ethical or utilitarian concerns, ecologists believe biodiversity conservation is also important because of the role it has in regulating ecological processes. If biodiversity is lost from a landscape, a number of ecological processes are likely to be affected. This will have important functional consequences for other land use activities in that landscape, including agriculture (Foley et al. 2005;

D. Lamb, *Regreening the Bare Hills*, World Forests 8,
DOI 10.1007/978-90-481-9870-2_1, © Springer Science+Business Media B.V. 2011

Matson et al. 1997; Scherr and McNeely 2008). A network of protected areas is an important strategy for conserving biodiversity but these reserves alone will never be sufficient, because they will never be big enough to sustain it. Biodiversity must also be maintained in the broader landscape; otherwise it will slowly dwindle away.

One part of this broader landscape that deserves particular attention is the significant area of land classed as 'degraded'. This is land that may once have been used for agriculture but which has since been largely abandoned. One estimate suggests such lands are being created in the humid tropics at the rate of 2.3 million hectares each year (Mayaux et al. 2005). Reforesting such lands to restore biodiversity could generate significant benefits and, because these lands are often underused, the opportunity costs of doing so could be low.

Deforestation and biodiversity loss is taking place within a broader socio-economic context. The World Bank estimates 70 million people live in remote, closed tropical forests and another 735 million rural people live in, or near, tropical forests and savannas (Chomitz 2007). Many of these people are poor and their ability to escape from poverty in the future will depend on the ways that these landscapes are managed. What role might reforestation of degraded lands play in improving the livelihoods of these people? Might there be forms of reforestation that generate improved conservation and socio-economic outcomes?

This chapter begins the task of answering these questions by first examining more closely the patterns of deforestation and the types of landscapes being generated in place of the original forests. Not all of this degraded land is necessarily available for reforestation (not least because there are differences about the meaning of the term 'degradation'). However, it is important to know the scale of degradation to appreciate the impact it is likely to be having. For this reason, this chapter also considers the extent of the losses of biodiversity that are occurring and the changes in the supply of various environmental services. The chapter concludes by considering the capacity of the present protected area network to deal with these impacts.

Forests of the Asia-Pacific Region

The Asia-Pacific region is one of the most biological diverse parts of the globe. This region is where the Eurasian tectonic plate and its associated biota meet up with the Philippines, Australian, and Pacific plates, and their biota. The volcanoes and earthquakes associated with these boundaries have created and helped maintain some very fertile soils although less fertile soils are also present in other parts of the region. The high rainfall received in many locations has allowed forests to flourish. These forests vary enormously in structure, physiognomy and composition. Dense, evergreen rainforests with closed canopies are common and trees in many of these reach over 50 m in height. They contain a variety of plant life forms besides trees including vines and lianes, epiphytes, palms, bamboos, ferns and shrub species. The most structurally complex evergreen forests occur where the rainfall is well distributed and there is no strong dry season. In areas with an annual dry season (e.g. several months receiving less than, say, 50 mm rain) these closed canopy forests are usually replaced by seasonally deciduous forests, woodlands or savannas. The types of forests present are also affected by topography and some mountains in the region reach over 4,000 m (e.g. Puncak Jaya or Carstensz Mountains in New Guinea at 4,800 m). Lowland forests are replaced by montane forests above 1,000 m and, in New Guinea, tree lines occur around 3,900 m, above which alpine grasslands are more common (Paijmans 1976). In areas with impeded drainage or poorer soils, swamp forests, heaths and heath forests are often found.

It is convenient to recognise four broad floristic zones covering the Asia-Pacific area, although each of these contains evergreen as well as seasonally dry or deciduous forests and savannas. A substantial part of the area is included within the botanical region known as Malesia, which spreads from Malaysia across Indonesia and the Philippines to the Solomon Islands east of Papua New Guinea. Malesia is one of the richest botanical areas of the world and may contain up to 50,000 species of vascular plants. That is, it has around 20% of the world's flora on only 2% of the land area (Davis et al. 1995). Most of this flora is found in lowland rainforest. The dominant tree family in the Malesian region is the Dipterocarpaceae. This family has 386 species and is extremely abundant in the lowland forests of Malaysia, Indonesia and the Philippines it is much less dominant to the north and east of Malesia. Another 16 families of flowering plants in Malesia have more than 500 species.

To the north of Malesia is the Indo-China floristic zone covering Myanmar, Thailand, Laos, Vietnam and southern China. The flora of this region is botanically distinct from that of Malesia although it is not nearly as well documented. It probably includes 12–15,000 plant species and many of these are endemic (Davis et al. 1995). Dipterocarps are present but are much less common than in Malesia; a substantially temperate flora becomes more dominant with increasing latitudes and altitudes. Evergreen forests are present in southern parts and northern mountains but much of this very large area has a significant dry season and deciduous and semi-deciduous forests are common (Rundel and Boonpragob 1995). Teak (*Tectona grandis*) is especially common in these forests in Myanmar and Thailand. Unlike wetter parts of the region, the transition between different types of forest in drier areas can be quite sharp and follow changes in soil depth or topography. Fire regimes also affect boundaries between forest types.

To the east of Malesia are the relatively species-poor forests of the islands of the Western Pacific. The boundaries of this zone are somewhat artificial since the flora is largely derived from Malesia. There are probably around 11–12,000 native vascular plants present in the Pacific of which 7,000 are endemic. There is a distinct decrease in genera from west to east although the floristic diversity on any particular island is affected by the area, degree of isolation, island age and heterogeneity (Keppel et al. 2010; Mueller-Dombois and Fosberg 1998; van Balgoy 1971). Many of these taxa are now threatened by introduced species of plants and animals.

In the south, the tropical forests of Australia are distinctly different from Malesia with an especially sharp dividing line between Australia and New Guinea. Most species are representative of an ancient Gondwanan, rather than Asian flora, and many species are endemic. The tropical parts of the country contain closed rainforests as well as open woodlands and savannas. The rain forests of Australia are embedded within sclerophyll forests dominated by *Eucalyptus.* These sclerophyll forests extend into parts of New Guinea and the Lesser Sunda Islands of Indonesia that also have seasonal climates. More detailed descriptions of the many types of forest and some of the more common tree genera present across this huge region are given in Adam (1994); Gressitt (1982); Mueller-Dumbois and Fosberg (1998) and Whitmore (1984).

The plant species across the region are distributed along gradients of rainfall seasonality, soils, latitude and altitude. In some forests, the numbers of species are very high. For example, 240 canopy tree species per hectare have been recorded in lowland Borneo (Whitmore 1984) while Wright et al. (1997) recorded 228 species of tree and lianes (at 900 m elevation) in 1 ha of forest in Papua New Guinea. Much lower numbers are present in some of the more isolated Pacific Islands. In Tonga, Franklin (2003) found only 77 woody species across an area of around 3 ha (and these included trees down to a stem diameter of 5 cm dbh). Similarly, in Samoa, Webb and Fa'aumu (1999) found just 52 tree species with diameters >10 cm dbh in an area of a little over 1 ha. Although most forests in the region have a large number of canopy tree species, there are some unusual stands that are dominated by just one or two species. These so-called mono-dominant forests occur in mostly small patches surrounded by other species-rich forests throughout the region. A brief account of these and of their possible significance for reforestation methodologies is given in Box 1.1.

There are no commonly used methods of classifying the variety of different types of forest found across the region. Some studies have used terms such as lowland evergreen rainforest, semi-evergreen rainforest and seasonal or monsoonal forest. Others refer to closed, broad-leaved forests, humid tropical forest, moist deciduous forests and tropical dry forests. The problem is especially difficult in Australia where closed-canopy rainforests are found distributed over a large latitudinal and altitudinal range. One classification (Adam 1994; Webb 1959) uses several simple structural attributes including average leaf size and the presence or absence of lianas, palms, ferns and deciduous tree species. The classification recognises around 11 main forest types in north-eastern Australia. The distribution of these types is associated with differences in thermal and moisture conditions, the nutrient supplying capacity of soil parent materials and of upslope drainage conditions (Mackey 1993, 1994). The classification has been widely used for conservation planning in northern Australia.

Many forest species are economically valuable. Jansen et al. (1991) have identified 1,400 timber tree species, 1,100 medicinal plants, 170 rattan species and 390 species with edible nuts or fruits having some economic worth. Details of some of the more prominent timber trees are given in Soerianegara and Lemmens (1993) and Lemmens et al. (1995). Islands of the southwest Pacific were not included in this assessment but would bring additional species into each of these categories.

Box 1.1 Natural Forests with Limited Species Diversity

There are reports from across the region of patches of rain forest where the canopies are dominated by just one or two tree species. So, for example, Richards (1952); Whitmore (1984); Whitten et al. (1984) and Whitten (1989) all describe near mono-dominant stands of trees such as *Diospyros* spp., *Adina fagifolia, Elmerrilia ovalis, Eusideroxylon zwageri*, and *Drybalanops aromatica* occurring in various locations in Indonesia and Malaysia. Hart (1990) referred to these as mono-dominant forests. These stands are not obviously constrained by soil factors or poor drainage and the forests surrounding these patches are invariably rich in species meaning there is a ready supply of colonists from a much wider variety of taxa. One obvious explanation is that these are successional communities recovering from some kind of natural disturbance and the dominant species are those that had seed available at the time of the disturbance. This probably accounts for communities such as the even-aged stands dominated by *Eucalyptus deglupta* and *Anisoptera thurifera* in Papua New Guinea (Nir 2004), and the *Campnospermum brevipetiolatum* forests in Solomon Islands (Mueller-Dombois and Fosberg 1998; Whitmore 1984). In such cases, the understories in these forests invariably contain seedlings of other species that are rather more shade-tolerant and these eventually grow up and join the canopy.

This post-disturbance explanation is not sufficient in other situations where the stands have a variety of age classes (suggesting frequent reproductive events). Connel and Lowman (1989) argued that mono-dominance was related to the types of mycorrhizal associations developed in these more abundant species. Subsequent studies by Torti et al. (2001) were unable to find evidence supporting this hypothesis. Based on further work (Torti et al. 2001) they concluded that mono-dominance is maintained by the primary species having a large number of seedlings present in a seedling pool and that these are able to competitively exclude other species. The implications for reforestation practices are that species able to form these mono-dominant stands might not be the most appropriate to use if the goal is to create species-rich forests for conservation purposes, although they may be useful for plantations designed for timber production.

The forests of the region are also immensely rich in wildlife. Some of these, such as orang utans, elephants, rhinoceroses, tigers and the birds of paradise are very well-known but some fauna are much less known. Recent discoveries of several new species of large mammals in the Annamite Ranges between Vietnam and Laos PDR show how poorly understood the region's fauna are. These discoveries included a forest-dwelling bovine, the saola or Vu Quang ox (*Pseudoyrx nghetinhensis*), which represents not only a new species but a new genus (Kemp et al. 1997).

The distribution of fauna also shows strong regional patterns. One famous biogeographic boundary that defines the distribution of much wildlife is Wallace's Line. This lies between Borneo and Sulawesi and marks a sharp distinction between a Sunda or 'Asian' fauna (e.g. primates and ungulates) to the west, and a Sahul or 'Australian' fauna (e.g. marsupials and rattites) to the east (Whitten et al. 1989). The line is not an absolute boundary and a number of faunal groups are found on both sides, with invertebrates seemingly less affected than vertebrates. There are also differences at a species level between the avifauna of the Indo-China region and Malaysia and Indonesia. Likewise, there are latitudinal gradients in fauna in Vietnam (Sterling et al. 2006). In the Pacific, there is a west to east gradient in many fauna that parallels that observed in plants. For example, the number of forest bird species in Papua New Guinea is 445; this number decreases to 184 in Melanesia and only 40 species are found in southeastern Polynesia (Whiffin and Kikkawa 1992). Some of the geological and climatic events giving rise to these complex biogeographic patterns across both Asia and the Pacific region are described in more detail by (Sodhi et al. 2006; Sterling et al. 2006; Stoddard 1992; Turner et al. 2001; Whiffen and Kikkawa 1992; Whitmore 1984).

The biological richness of the Asia-Pacific region shows up in most assessments of global biodiversity. One of the first comprehensive global surveys to define the centres of plant diversity was that carried out by Davis et al. (1995). They identified many sites in the Asia-Pacific region with large numbers of plant species. They also identified many areas with high degrees of plant endemism including New Guinea with 80%, New Caledonia with 77%, Peninsular Malaysia with 50%, Fiji with 50%, and the Philippines with 39%. Other surveys have sought to identify the so-called global biological 'hotspots'. Some of these have focused on areas with large numbers of rare or taxonomically unusual species. Others focused on places with large numbers of endemic species. Mittermeier (1998) described 23 global hotspots based on the presence of high levels of vascular plant endemism and because they were also undergoing exceptional rates of habitat loss. Six of these hotspots were located in the Asia-Pacific region. They also identified New Guinea as being one of the world's three last remaining 'wilderness' regions. A subsequent update took account of animals (mammals, birds, reptiles and amphibians), as well as plants, and confirmed the region was of considerable biological significance (Myers et al. 2000). Three Southeast Asian hotspots were in the eight 'hottest' hotspots in the world (Philippines, Indo-Burma and Sundaland).

Such analyses are a function of the criteria used and don't necessarily take account of the full range of biological diversity that may be present or differences in habitat losses. This gap was partially filled by the work of Olson and Dinerstein (2002) who identified the world's 200 most biological important eco-regions. These were defined as areas of land (or water) harbouring a geographically distinct set of environmental conditions and species assemblages. One of the aims was to avoid a bias towards sites with high levels of species richness. Again, the Asia-Pacific region was well represented in this listing. While reflecting on these assessments, it is important to note that much of the region has still not been fully examined and more of the region's biological diversity probably remains to be documented.

Deforestation Rates

Much of the biodiversity described above is now threatened by forest clearing. There have been a number of attempts to quantify just how much has been cleared, as well as the rate at which clearing still continues. The Food and Agriculture Organisation of the United Nations (FAO) has taken the lead in this, but other groups have also been involved. These efforts have been hampered by a host of technical difficulties and there have also been problems in the comparisons between the various studies. These problems include inconsistent definitions of 'deforestation' (see Box 1.2), differences in the methodologies used, and differences in the types of forest recognised. The latter problem has been especially vexing. As noted above, there are a wide range of terms used to describe similar types of forest

Box 1.2 Definitions of Deforestation

Deforestation is a seemingly simple term, but in fact it is rather more difficult than it appears. Part of the problem concerns the definition of a forest. FAO (2000) defined a forest as vegetation with at least 10% crown cover and having woody plants >5 m tall. In tropical areas, this allows seriously degraded areas to be counted as being still intact forest. Some governments also include as land they declare to be state forest, even though these areas may be presently treeless. Another problem concerns changes over time. Some might simply say deforestation occurs when forests are cleared causing a complete loss of canopy cover. But does this include cases where the forest immediately regenerates (e.g. following shifting cultivation)? Or, does it include cases where plantations are established, either immediately or perhaps after several years of cropping? These questions suggest any definition requires distinguishing between temporary and permanent removal of forest cover and also between undisturbed natural forest, forest regrowth and tree plantations (although at some point old regrowth or secondary forest presumably ceases being distinguished from undisturbed forest). Most international statistics do not make these distinctions, but imply that deforestation occurs when forests are replaced by non-forest land uses and simply gather the various forest types together as 'forest'. The practical advantages of this are obvious but by doing so they hide the fact that natural forests might be being replaced by monoculture plantations of exotic species while, at the same time, the overall forest cover remains constant. A useful discussion of the complexities involved in defining and assessing deforestation is given in Watson et al. (2000).

(e.g. lowland evergreen rainforest, closed broad-leaved forests, humid tropical forest, etc.). Some of these terms are comparable, while others are not. Whitmore (1997) attempted to reconcile these terminological differences but there are still inconsistencies in the way secondary forests, degraded lands and plantation forests are dealt with in global forest assessments. For example, should these be classified with undisturbed forest, or included in a different category? These issues are discussed by Whitmore (1997) and Mayaux et al. (2005).

Two particularly useful recent analyses have been the Forest Resource Assessment study, carried out by FAO between 1990 and 2000 (FAO 2001), and the study known as the TREES project (Achard et al. 2002; Mayaux et al. 2005). These differ in the time periods covered, the methods of data collection and in the geographic areas represented but are still important. The FAO study reported on deforestation rates between 1990 and 2000 in the *dry and humid* tropical forests. By contrast, data from the TREES project reported by Achard et al. (2002) and Mayaux et al. (2005) covered the period 1990–1997 and was only carried out in *humid* tropical forests (i.e. evergreen and seasonal forests but not woodlands or dry tropical forests). The FAO study found a global annual net deforestation rate of 8.6 million hectares, or 0.52% loss. The TREES study found an annual net deforestation rate of 4.9 million hectares, corresponding to 0.43% loss (Table 1.1). Comparisons between the two studies were made by adjusting the data to include only humid sites and changes in the 1990–1997 period. This showed the net annual change in forest area was 2.0–2.5 million hectares in Southeast Asia and 4.9–6.4 million hectares globally (Achard et al. 2002). To place this in perspective, this annual global loss exceeds the land area of Switzerland (3.9 million hectares). The TREES study also sought to measure the amount of degraded forest (affected by logging, fragmentation and fire) and estimated this covered 2.3 million hectares, with half of this being located in Southeast Asia (Table 1.1).

A key question, of course, is whether this rate is increasing or decreasing? Determining longer term trends in tropical deforestation rates have been difficult because of the technical issues referred to earlier. Matthews (2001) used FAO data to argue that the loss of natural forest cover in the 1990s was greater than in the 1980s in tropical Africa, Asia and Oceania. Similarly, Mayaux et al. (2005) also noted a trend suggesting the rate was increasing over time, but the patterns were not clear-cut. In an attempt to assess long-term trends, FAO (2006) published a further update of its monitoring. This reviewed the cover of *forest and other wooded land* across the globe in 1990, 2000 and 2005. It also sought to differentiate primary forest from other forest

Table 1.1 Annual changes in forest cover in humid forests between 1990 and 1997 (Achard et al. 2002)

	Southeast Asia		Global	
	Million hectare	%	Million hectare	%
Deforestation	2.5	0.91	5.8	0.52
Regrowth	0.53	0.19	1.0	0.08
Net forest loss	2.0	0.71	4.9	0.43
Degraded forest	1.1	0.42	2.3	0.20

types. Primary forest cover was defined as forest composed solely of native species in which there was no visible evidence of human activity. The total forest cover category includes primary forest, disturbed forest (e.g. caused by logging), regrowth forest and plantations. Unlike the TREES study and the previous Forest Resource Assessment of 2000, this study relied heavily on information and expert opinion from individual countries. Because of the differences in the types of forest being assessed and in the methodologies used, the analysis is not directly comparable with the early FRA 2000 or TREES studies. It concluded that the annual global deforestation rate between 1990 and 2005 was 13 million hectares and, more to the point, that there were few signs of a significant decrease over this time period.

The results from this study of longer-term changes in forest cover in nine Southeast Asian countries are shown in Table 1.2. This showed the total forest cover declined from 244.7 million hectares in 1990 to 203.8 million hectares in 2005. Primary forest declined from 84.4 million hectares in 1990 to 62 million hectares in 2005. Primary forest represents about a third of the overall forest cover and this proportion declined from 34% in 1990 to 30% in 2005. The total forest cover and primary forest both fell over the period, with primary forest disappearing rather faster than overall forest cover. Perhaps most importantly, however, the annual rate of loss of both types of forests increased with time. In the case of primary forest, the loss accelerated from 1.78% during 1990–2000 to 2.15% between 2000 and 2005. In short, these data show there has been a steady loss of forests in Southeast Asia over the last few decades and that this loss seems destined to continue into the immediate future, despite efforts to halt it.

The quality of the available data means that the issue continues to be debated. In 2010, FAO revised its earlier estimates and concluded that global rates of deforestation were now beginning to decline, especially in Asia (FAO 2010). But Grainger (2008) cautions that the methods being by all of those seeking to monitor global changes are still insufficient to be able to track the fine-scale changes that are underway. He concludes that, despite the various revisions made, it is still impossible to say anything definitive about long-trends in overall forest area even after several decades of effort.

It is unclear what will happen in the next few years. Wright and Muller-Landau (2006) have argued that human population growth and a trend for people to move from the country to the city will soon begin reducing rates of tropical deforestation.

Table 1.2 Changes in forest cover in Southeast Asia[a] between 1990 and 2005 (Source: FAO 2006)

Forest type	1990	2000	2005	% Annual change 1990–2000	% Annual change 2000–2005
Total forest cover (000 ha)	244,728	216,846	203,847	1.13	1.19
Primary forest cover (000 ha)	84,472	69,462	61,977	1.78	2.15

[a]Countries include Brunei, Cambodia, Indonesia, Laos PDR, Malaysia, Myanmar, Philippines, Thailand and Vietnam

Their suggestions have been vigorously disputed by critics who point to the uncertainty about future population projections upon which Wright and Muller-Landau's argument is based and the assumptions about future per capita consumption of forest products (Brook et al. 2006; Gardner et al. 2007; Laurance 2007). They also highlight the ecological consequences that might flow from maintaining overall forest cover but losing primary (i.e. undisturbed) forest.

The New Landscapes

Most of the newly cleared forest land has been used for agriculture. But the process has often been inefficient and large areas of land have been used for a while and then abandoned. It is difficult to specify with any precision just how much of this 'degraded' land there is. There are three reasons for this. One is that degradation is a loaded term that means different things to different people (see Box 1.3). Some limit the term to soil degradation while others use it more broadly and refer to a wider range of ecosystem properties. Such definitional issues make cross comparisons difficult. Secondly, the situation is dynamic and changes are occurring rapidly. Forest can regenerate on abandoned lands when fires are excluded and seeds can reach the site from nearby intact forest. Under these circumstances, 'degraded land' can soon become 'forest'. Finally, the precision of any estimate depends on the scale used and many global or regional surveys are forced to adopt scales where land units containing several different types of vegetation (including degraded land) are mapped according to the dominant type. This will result in under-estimates of the other types. Despite these difficulties, there have been a number of attempts to quantify just how much land 'degradation' there is and to map where this is found. The issue has been tackled in a variety of ways.

General Assessment of Degraded Lands

Several global estimates of the amounts of degraded lands and forests have been made. One estimate done in the 1980s concluded there were then two billion hectares of tropical land in various states of degradation across the globe compared with about 650 million hectare of cropland (Office of Technical Assessment 1984). A later assessment by (ITTO 2002) arrived at a much lower figure and suggested there were 850 million hectares of degraded lands and forest (350 million hectares of degraded forest land and 500 million hectares of degraded primary and secondary forest) across the world's tropics. Mayaux et al. (2005) estimated an additional 2.3 million hectares was being added to these figures each year. Within Asia, the amount of degraded land was estimated to cover 125 million hectares, while there were a further 145 million hectares of degraded primary and secondary forest (ITTO 2002). Each year, an additional 1.1 million hectares of degraded land are thought to be added to these Asian totals (Mayaux et al. 2005).

Box 1.3 Definitions of Degradation

Like deforestation, degradation is an expression that is difficult to define with precision. It is a perceptual term and, being perceptual, it is open to a variety of interpretations. Some users may be primarily concerned about the loss of ecological productivity; some with the timber productivity and others with the agricultural productivity. So, for example, a grassland might be viewed as a degraded forest by some but as a grazing opportunity by others. According to them, it would be better described as productive agricultural land. Likewise, an area of regrowth forest area might be viewed as merely 'degraded forest' by some but, more optimistically, as 'regenerating forest' by others.

Certain things are obvious. There may have been a loss of native species, shifts in species dominance, changes in trophic interactions or changes in landscape connectivity. Likewise, there may be soil erosion, soil acidification, landslips, turbid rivers or weed invasions. All are indications of some degree of degradation because they indicate there may have been a reduction in the productivity, economic value or amenity value of a site. More particularly, they suggest human activities have caused a reduction in the capability of that site to satisfy a particular use and, importantly, to satisfy the capability of other alternative uses in the future. Note that deforestation does not necessarily lead to degradation in a financial or social sense; though it will certainly lead to changes in biodiversity, hydrology and micro-environmental conditions at that site.

It is useful to draw a distinction between degraded forest and degraded land. In this book, 'degraded forest' will be taken to refer to forest that has had its structure, biomass or species composition temporarily or permanently changed (e.g. by logging). These changes lower the quality of the forest and its capacity to supply goods and services. Under certain conditions, this forest may be capable of recovering much of its former ecological condition and timber productivity and cease being degraded (described further in Chapter 4).

'Degraded land' will usually be used to refer to land with a low agricultural productivity and capability because of soil infertility, erosion, weeds or recurrent fires. Such land is commonly land occupied by grasses or shrubs and many of these are likely to be exotic weeds. Biot et al. (1995) have distinguished between marginal, fragile and degraded lands. All have limitations on productivity with degraded lands being most adversely affected while marginal and fragile lands may still be used for agricultural purposes. Lands fall into one or other of these categories depending on their biophysical constraints, the current patterns of use, their location and the institutional and policy settings that apply to them. Degraded lands are more likely to have been abandoned and so be available for reforestation, but marginal and fragile lands may also be available for reforestation under certain circumstances. Perhaps the most unequivocal form of degradation is where the ecological system *and* the socio-economic systems it supports have both collapsed.

Another way of considering the problem is to assess the area of degraded soils in agricultural areas. Work by van Lynden and Oldeman (1997) suggested that 45% of land in South and Southeast Asia was affected by some form of human-induced soil degradation (mostly surface erosion). Of this area, 10–15% was likely to have degraded to the point where plant productivity is strongly affected and another 22–28% was land where productivity is likely to be moderately affected. The remaining land is negligibly or only lightly affected.

Given the imprecision of definitions of degradation and differences in what is being assessed, it is difficult to know how much faith to place in these various estimates. Perhaps the main conclusion to be drawn is simply that there are probably very large areas of degraded land and forest now present across the region.

Undisturbed or Human-Dominated Lands

A second approach sought to avoid the problem of defining 'degradation' and instead classified terrestrial lands as either, undisturbed, partially disturbed, or human-dominated (Hannah et al. 1994). It found that only 12% of lands in the Indo-Malayan biogeographic realm could be classified as undisturbed compared with a global average of 27%. In fact, this rating was skewed by high scores for Sumatra (at the time 27% undisturbed) and Thailand monsoonal forests (19% undisturbed); most countries in the region had no land that qualified as undisturbed. Sumatra had the lowest level of human dominated land (55%) while the Philippines had the highest (87%). Only two islands in the Pacific biogeographic realm were assessed. One of these, New Guinea, had 77% of the area classed as undisturbed but New Caledonia had none. Anecdotal evidence suggests that if the classification had been applied to the smaller islands in the Pacific they would all be rated as being heavily disturbed by human activity. In some extreme cases, the deforestation process is virtually complete (e.g. Easter Island and smaller islands like Nauru). In others, small fragments of forest remain in a largely agricultural landscape (e.g. Samoa, Cook Islands).

Mosaic lands

Measures of deforestation or disturbance are only partial descriptors of landscape changes. The degree of fragmentation is also important. Losses in forest cover are always spatially uneven, with some places being completely deforested and others being only patchily disturbed. Forests away from roads or on steeper lands are less likely to have been cleared for agriculture than more accessible forests. This means that some residual forest areas are still quite large while others only exist in small patches.

This unevenness in the spatial impact of deforestation led Chomitz (2007) to propose a tropical forest landscape classification based on three types:

1. Mosaic lands: areas with only small patches of forest remaining within a largely agricultural landscape.
2. Frontier and disputed areas at the forest edge: areas outside the mosaic lands but within 6 km of them. Some of these are lands where substantial clearing of intact forest for agricultural purposes is still taking place while others have rather more stable boundary zones.
3. Core forest: areas of mostly intact forest beyond the agricultural frontier.

The relative areas of forest in each of these three categories in 'Asia' (defined by Chomitz as the area from India and Sri Lanka in the east to southern China, Papua New Guinea and northern Australia in the east) are given in Table 1.3. Within the closed forest biome, the data show the area of forest now in mosaic lands is now similar to the areas of relatively undisturbed core forest. They also show that forest classified as being 'forest edge' was the largest category of all the remaining forest lands. In Asia, the savannah biome is much smaller than the closed forest biome and, within these lands, there was very little core forest left with most being in the mosaic forest or forest edge category. Chomitz (2007) estimated about 350 million people live in these forest edge and core forests in Asia and another 90 million live in the mosaic forest areas. These numbers are greater than for the Americas or Africa.

Frontier Forests

Another way of considering change is to assess the amount of forest still able to provide habitats for the most sensitive plant and wildlife species. Bryant et al. (1997) named these Frontier Forests and defined them as forests big enough to support viable populations of dependent species, as well as being large enough to withstand the effects of natural disturbances. The structure and composition of such forests are determined by natural events, but all are relatively undisturbed by human activities. They found Asia has lost 95% of its original Frontier Forests and only 20% of the remaining forests still qualify (the remainder being too fragmented or disturbed). The amount of Frontier Forest in some Southeast Asian and Pacific countries is shown in Table 1.4. This shows most countries have lost a large amount of their original Frontier Forests with the Philippines having lost the most (100%) while Papua New Guinea has lost the least (60%). Most of the residual Frontier Forests are now under some kind of threat.

Table 1.3 Areas of different types of forest in Asia (thousands of square kilometers) Source: Chomitz (2007)

Biome	Agricultural areas in mosaic lands	Forest in mosaic lands	Forest edge	Core forest
Closed forest	1,853	1,153	3,572	1,157
Savanna	16	14	20	3

Table 1.4 Frontier forests (FF) in Southeast Asia and the Pacific (Source: Bryant et al. 1997)

Country	Percent original FF lost (A)	Percent of current FF threatened (B)	Frontier forest index[a]
Australia	82	63	52
Cambodia	90	100	90
Indonesia	72	54	39
Laos PDR	98	100	98
Malaysia	85	48	41
Myanmar	94	56	52
Papua New Guinea	60	84	50
Philippines	100	–	99
Thailand	95	100	99
Vietnam	98	100	98

[a]The FF Index = A × B/100

An index was devised to take account of the amount lost and the proportion of the remaining Frontier Forest under threat. This was derived by multiplying column A in Table 1.4 by column B and dividing the result by 100. A score of 99 shows most Frontier Forests have been lost or are strongly threatened while a low score shows where many areas of Frontier Forests still remain. Most countries in the region have values exceeding 90 showing that there are only small areas of relatively undisturbed primary forest still present in the region.

Grasslands

Grasslands, especially those dominated by aggressive and fire tolerant species such as *Imperata cylindica,* are usually representative of the final stage of the degradation sequence and are often seen as wastelands. In fact, many grasslands are used for grazing and other purposes although *Imperata* is generally regarded as providing only poor quality pasture and is difficult to cultivate (Calub et al. 1996). Grasslands are not uniform but differ in their composition and structure. One simple classification simply divides them into tall and short grasslands. In Papua New Guinea, tall grasslands are often found not long after forests have been cleared and are viewed as the comparatively early stage of a deflected succession (Robbins 1960). These can contain species such as *Saccharum spontaneum, Ophiurus exattatus* and *Miscanthus floridus* and be more than three meters tall. With continued disturbance and fire, these tall grasslands are eventually replaced by shorter grasses including those dominated by *Imperata.* Different patterns are found in other locations.

Brown and Durst (2003) estimate there are 50 million hectares of *Imperata* grasslands in South Asia, China, the Pacific and Southeast Asia. Within Southeast Asia alone, Turvey (1994) quoted an estimate of 20 million hectares. This was supported by a subsequent more comprehensive regional survey by Garrity et al. (1996), which concluded grasslands covered 22 million hectares or around 4% of the land area. The countries with the largest area of *Imperata* grasslands were Indonesia (8.5 million hectare) and the Philippines (5 million hectare). Countries with the largest proportion of their land areas covered by *Imperata* were the Philippines (17%) and Vietnam (9%). Garrity et al. (1996) qualified the regional figure because of the mapping scales at which they worked and conceded the real value could easily be higher, perhaps reaching 34 million hectares.

In the Pacific, grassland areas dominated by *Imperata* are not as well defined, although large areas are found in Fiji and Papua New Guinea, as well as on many other smaller other islands (Gillison 1993; Mueller-Dombois and Fosberg 1998; Robbins 1960). An illustration of one of the large grassland areas in lowland Papua New Guinea is shown in Fig. 1.1. Extensive areas of a fern-grassland known locally as *talasiqu* occur on Fiji and are sometimes viewed as being the end stage of a long process of degradation (Mueller-Dombois and Fosberg 1998). Some of these tropical grasslands have been present for many years; Bennett (2000) notes some in the Solomon Islands were sufficiently extensive to have been recorded by passing Spanish sailors in the sixteenth century.

It might be expected that degraded former agricultural land now occupied by *Imperata* and other grasses would largely occur on less fertile soils. In fact, Garrity et al. (1996) concluded these occurred on both fertile and less-fertile soils and their distribution was related more to the occurrence of fires than the underlying soil fertility.

Fig. 1.1 Extensive grasslands in the Markham River valley in northern Papua New Guinea. Once grasslands like these are established recurrent fires usually prevent natural forest regeneration from occurring

Estimates of the Area of 'Degraded' Land Potentially Available for Reforestation

While it is difficult to say just how much deforested land is degraded, it is even more difficult to specify how much of this degraded land is likely to be available for reforestation. The amount will depend on the views of land owners or current users and what they perceive to be the opportunity costs of doing so. As noted earlier, even some seemingly worthless *Imperata* grasslands are sometimes used for cropping, grazing, and other purposes. Many other degraded lands are also used in various ways, even though they may be marginal for agriculture. Some of the people using these lands may be interested in reforestation provided they will benefit from doing so. On the other hand, all these people are likely to be opposed to reforestation if it means that they will lose access to the land and its resources.

An indication of the amount of land potentially available for reforestation in various countries in Southeast Asia is given in Table 1.5. This shows the total area of degraded lands (i.e. grasslands, shrublands and, sometimes, degraded secondary forest) as well as the amount of land identified as 'indicative areas to be reforested' (Nawir et al. 2007). The areas of degraded land are large, ranging from 1.2 million hectare in Malaysia (almost certainly an under-estimate given the way the figure was derived) to 57 million hectares in Indonesia. They represent around 4% of the land area in Malaysia and Thailand to over 30% of the land area in Indonesia, Lao PDR and Vietnam. The distribution of degraded lands potentially available for reforestation is not evenly distributed and some areas are likely to be more heavily degraded than others. An example of this is occurs in Vietnam where 60–65% of lands in the northern mountains were classed as 'barren', compared with a national figure of 35–42% (Sikor 1995). Indonesia, Philippines and Vietnam have identified reforestation targets that range from 5 million hectares in the Philippines and Vietnam to 47 million hectares in Indonesia. Although many islands in the Pacific

Table 1.5 Some recent estimates of areas of degraded lands (grasslands, shrublands and some secondary forests) and of land that might potentially be available for reforestation

Locality	Area of degraded land (000 ha)	Percent land area	Area to be reforested (000 ha)	Source
Cambodia	2,600	15		Gilmour et al. (2000)
Indonesia	56,900	30	47,000	Nawir et al. (2007)
Lao PDR	8,700	36		Gilmour et al. (2000)
Malaysia	1,200[a]	4		Chokkalingam et al. (2001)
Philippines	9,300	31	5,500	Chokkalingam et al. (2001)
	10,900	37		Kummer and Turner (1994)
Thailand	2,300	4		Gilmour et al. (2000)
Vietnam	9,700	30	5,000	Gilmour et al. (2000)
	13,000	40		Sikor (1995)

[a]Based solely on grasslands and old mine sites

are badly degraded, comparable data outlining the extent of degradation or areas deserving reforestation are lacking.

In summary, various assessments of deforestation and degradation in the Asia-Pacific region show that considerable forest has been lost and that only small areas of relatively undisturbed primary forest remain. At the same time, a large area of degraded land has accumulated. It is difficult to be precise about the extent of this area and how much might be available for reforestation. Deforestation seems certain to continue into the immediate future and it is likely, based on past experiences, that these areas of degraded land will also continue to increase.

Assessing the Extent of Biodiversity Losses

Forest loss and fragmentation is obviously a threat to regional biodiversity. But which species are most vulnerable? Some have sought to make predictions about the species most likely to be lost using current knowledge of the species' conservation status. Others have used theories concerning species-area relationships.

Predicting Future Extinctions

The species in a region most likely to be threatened by deforestation are those already known to be on the brink of extinction and the identity of these species is given in the IUCN Red Lists (Baillie et al. 2004). Table 1.6 shows the *proportion* of endemic mammals, birds and amphibians' species in the Indo-Malayan, Australasian and Oceania biogeographic regions that are now classed as threatened. Comparisons across geographic zones are complicated by differences in land areas or species richness but mammals are especially threatened in all three biogeographic regions. In fact, these three regions have the highest proportion of threatened mammals in all of the world's eight biogeographical regions with Oceania, in particular, having 70% of its endemic mammals listed (although this represents only seven species). There are also high proportions of threatened endemic birds and amphibians in the Indo-Malayan and Oceania regions though the proportions are less than those for mammals. Again, Oceania had very high proportions, reflecting the particular vulnerability of biodiversity on small, isolated islands. These data deal only with the

Table 1.6 Percentage of threatened endemic species from well-mapped taxonomic groups in the Asia-Pacific region (Source: Baillie et al. 2004)

Biogeographic region	Mammals	Birds	Amphibians
Indo-Malayan	36	22	36
Australasian	28	10	12
Oceania	70	54	33

better documented species in these three taxonomic groups and include species threatened by changes other than forest clearing and fragmentation. However, they point to a serious conservation problem being present across the whole Asia-Pacific region. This is one that can only be amplified by ongoing deforestation.

A second approach to predicting the consequences that deforestation might have for existing biodiversity is to use the well-known relationship between species numbers and area. The number of species present in a region increases as the area surveyed increases and, based on such relationships, it should be feasible to predict the proportion of species that will be lost as habitat areas decline because of deforestation. The approach assumes all species are equally vulnerable to being lost and so cannot predict which will be lost first. It also assumes no species can live outside the forest although many can, in fact, use multiple habitats, including some agricultural landscapes. Finally, it assumes that habitat loss is the primary cause of species extinctions, although other factors such as introduced predators or diseases can also be important, especially on some small islands. Nonetheless, it remains a useful first approximation of the changes that might occur when deforestation takes place.

The approach was used by Brooks et al. (1997) to assess the effect of deforestation on bird extinctions in islands in Indonesia and the Philippines. They counted the number of bird species only present on islands in the area. They then used the species-area relationship to predict the number of species that would become globally extinct following deforestation on each island. There was a strong relationship between the numbers of species that their modelling predicted would become extinct on individual islands and the number already listed as being threatened with extinction in the Red Data Book. The most severely impacted species were those with small ranges. Some local differences occurred but most of these could be linked to differences in habitat type or quality. A subsequent global analysis of the species lost from a number of biodiversity hotspots using species-area relationships also found that the numbers of birds and mammals lost because of habitat loss predicted by species-area relationships closely matched the numbers actually lost, or now classified as being threatened by extinction (Brooks et al. 2002).

The problem with using species-area relationships to predict extinctions, apart from the qualifications already noted, is that in smaller areas, estimates soon become entangled in the more complex effects of local fragmentation. Previous biogeographical research has provided some indication of the ecological processes at work at these scales (summarised in Table 1.7). Newly created fragments or patches of intact forest can be thought of as 'islands' left isolated in a 'sea' of agricultural land. Some may be close to the 'mainland' represented by large areas of undisturbed residual forest, while other fragments might be more isolated. Some species may be able to cross this sea if the gap is not too great, while others may find even a short gap is an uncrossable barrier. For example, Goosem (1997), working in rainforest in north Queensland, reported that even a narrow clearing made beneath an overhead power cable was sufficient to prevent the movement of a number of arboreal fauna from forest on one side of the clearing to forest on the other side. This means that the number of species able to survive on isolated patches (or 'islands') depends not only on the area of the islands, but also on their degree of isolation from mainland

Table 1.7 Possible mechanisms causing biodiversity losses in isolated forest fragments (After Turner and Corlett 1996; Laurance and Bierregaard 1997)

Factor	Reason
Further disturbances	Additional area is lost as fires burn into forest edges; over-hunting of wildlife by farmers and others living nearby
Restricted population sizes	Small forest patches will have only small populations of most species. Small populations may fluctuate below a demographically viable size
Reduction in immigration	Species normally found in undisturbed forests may be unable to cross open agricultural lands between forest patches. This will make the re-establishment of extinct populations unlikely and increase the genetic isolation of species present
Forest 'edge effects'	The edges of forest fragment represent major transition zones especially where they meet non-forest vegetation such as grasslands. Environmental conditions are commonly different at these edges. This 'edge effect' means the core area of small undisturbed forest patch may be significantly smaller than the actual area
Higher order effects	The loss of species able to form mutualistic relationships with other species can have significant knock-on effects. E.g. the loss of a specialist pollinator or seed disperser may affect plant regeneration. Likewise the loss of a predator may affect populations of prey species
Immigration of alien species	Alien species adapted to disturbed sites can invade the edges of small forest patches and may alter regeneration or reproductive patterns and hence the population dynamics of native species

and on the types of species present. This island-sea dichotomy is necessarily simplistic. The nature of the agricultural matrix often varies and parts of it may even form suitable habitat for some species, while other parts will not. Similarly, the matrix can change over time as agricultural crops or agroforestry plantings mature, or if agricultural lands are abandoned and natural regeneration of forest takes place. This means the analogy has limitations, but it remains a useful device by which to think about deforestation and species survival.

Monitoring Actual Species Losses Following Deforestation

These types of approaches are useful in predicting what might happen to species numbers if deforestation occurs but a better approach would be to monitor the actual species losses that do occur as deforestation takes place. By doing this, there is a better chance of identifying the factors behind the loss of different types of species. Despite the extent of deforestation, there have been surprisingly few studies that have studied the immediate impact of deforestation on species numbers. Instead, most studies have sought to understand the changes once they have occurred. One place where this has been done is Singapore (Turner et al. 1996; 1997; Turner and Corlett 1996). In this case, only small forest fragments still remain on the island and some of these have now been isolated for 150 years. Although these fragments represent less than 1% of the island's original lowland rainforest area, it appears they still contain more than 70% of the plant species thought to have been originally present (with about 50% remaining in one small, 4 ha fragment). Of these species, a higher proportion of trees and shrubs have persisted than epiphytes. This persistence may simply reflect the longevity of trees and some may now be functionally extinct because the reduced populations mean they are unable to reproduce. Nonetheless, these authors concluded that even small fragments can act to conserve biodiversity, perhaps for centuries, but that losses are inevitably after isolation.

The situation is different with wildlife and evidence from Singapore suggests wildlife have been more sensitive to fragmentation than plants. Brook et al. (2003) found losses of between 34% and 87% for the specialist forest taxa among butterflies, fish, birds and mammals over around 180 years. In the case of birds, the overall loss appears to be around 82%, with the larger species preferring undisturbed habitats in primary forest interiors being the most vulnerable species. Large frugivores that feed in the canopy layer appear to have disappeared at a relatively early stage of deforestation (Castelletta et al. 2000). Most of Singapore's remaining birds are now open habitat generalists, migrants and invasive species. The rapidity of the change is illustrated by the fate of tigers. These were still common in the nineteenth century and were claiming the equivalent of more than one human life a day in the 1850s, but the last was killed in the 1930s (with one being captured under the billiard room of the Raffles Hotel in 1902).

Another place where the consequences of past deforestation have been studied is Hong Kong. In this case, most deforestation probably took place more than 700 years ago. The forest patches now present are either regrowth forest or planted forests and none are remnants of the original forest. This means it is impossible to know the baseline status of biodiversity in Hong Kong. But despite this history, around 950 forest plant species have been recorded in these forest since 1841 (Corlett and Turner 1997). Around one third of these have not been observed in the last 30 years, although this may be more a function of search effort than evidence of extinction. It is difficult to quantify how this history has affected fauna, although but most of the forest passerines present are habitat generalists and all the remaining animals use secondary vegetation. In one sense, Hong Kong represents a worst-case scenario for the Asia-Pacific region. But some argue it provides grounds for optimism since there is considerable biodiversity still present and extinctions appear to have slowed or ceased in most taxonomic groups (Hau et al. 2005). On the other hand, Sodhi et al. (2005) argue that the past patterns of biodiversity loss in Hong Kong are not likely to be indicative of future changes in Southeast Asia.

Singapore and Hong Kong both represent situations where most of the original forests have disappeared and there have been only limited opportunities for recolonisation. The situation is quite different in more recently cleared landscapes that still retain large areas of nearby natural forest. A study carried out in northern Australia by Laurance (1991) explored such a situation. He examined forest patches left behind when agricultural clearing was carried out within the last 100 years.

Sixteen species of non-flying mammals in various feeding guilds were monitored to determine how these were able to persist in residual forest patches of different size and degree of isolation. The species least prone to local extinction were those most tolerant of a variety of habitats and able to move from the natural forests across the agricultural matrix (in this case pastures and forest regrowth). On the other hand, species with large body sizes, low fecundity and long longevities appeared to be more vulnerable to fragmentation even though they were still able to persist in the nearby intact forests.

The best way of studying the effects of deforestation and fragmentation is to follow changes in species populations from the time the forests are disturbed. This was done in a long-term study in Brazil. The study monitored changes taking place in plants and wildlife populations in several patches of forest that were deliberately left isolated when agricultural clearing took place in the early 1980s. These fragments ranged in size between 1 and 100 ha. After 22 years, the study showed that edge effects and area-related extinctions rapidly degrade most small fragments. The species most affected by fragmentation were larger mammals, primates, understorey birds and certain bee, ant, termite and butterfly species (Laurance et al. 2002). Not unexpectedly, some ecological generalists were able to expand their population sizes because of the changed habitat conditions. Wildlife populations in the patches tended to become dominated by species able to cross the agricultural matrix. Interestingly, experimental conditions at the sites changed over time. The pastures gradually lost productivity and were recolonised by secondary regrowth forest. This meant the forest fragments became less isolated, making it easier for some understorey hummingbirds and small frugivores to move from natural forests across the landscape and into the fragments once more (Bierregaard and Stouffer 1997). The study suggests strategically targeted reforestation that created habitat 'stepping stones' in the agricultural matrix of deforested landscapes could have a significant role in conserving regional biodiversity.

The impact of deforestation on small isolated islands such as those in the Pacific can be severe. The habitat areas are already small and the chances of recolonisation if species are lost from the island are extremely low. Of course, it is not just deforestation that causes problems but events such as storms or cyclones can also destroy forests. Some storms can damage the total forest cover on an island (e.g. Elmqvist et al. 1994; Pierson et al. 1996). However, there is evidence that many forest tree species appear to be surprisingly able to withstand such events. A study by Seamon et al. (2006) on a 12 hectare forest remnant on Samoa following storms found this had a unique range of native tree species and that all of these were reproducing well. There was some evidence of reduced regeneration near the edge of the remnant, but this only appeared to affect two of the 37 tree species present. Wiser et al. (2002) reached similar conclusions, following a study in Tonga. Many of the main non-successional tree species were able to regenerate within the fragment and the main threats to its future biological integrity were selective harvesting and several exotic vine species that had colonised some sites. The main reason why so many island plant species appear to be able to tolerate storms and clearing is because they can reproduce vegetatively. In the case of forest wildlife, their sensitivity to disturbances such as storms depends on their food requirements and those dependent on fruit rather than leaves are more likely to be affected than less specialised feeders (Pierson et al. 1996). A storm that damages food trees and limits fruiting for an extended period is likely to have a major impact on such species.

A general synthesis of the impact of logging, deforestation and fragmentation across the Southeast Asian part of the region has been carried out by Sodhi and Brook (2006). They found a strong and adverse effect of habitat degradation and deforestation on species diversity and abundance, as well as consistently negative impacts on a variety of different taxonomic groups. Perhaps the most severely affected were mammals (see also Table 1.6), while invertebrates appear to be the least affected. Overall, the most threatened biota were those with restricted ranges and specialised behaviours, particularly the species utilising undisturbed primary forests (also referred to earlier as the Frontier Forests; Table 1.4). Many species, especially plants, do appear to be able to persist in even badly fragmented landscapes, although some of these may be already functionally extinct because they are reproductively isolated. An overview of threats to biodiversity caused by deforestation has not been carried out in the Pacific, but studies by Clarke and Thaman (1993); Mueller-Dombois and Fosberg (1998); Rolett and Diamond (2004) and Wiser et al. (2002) provide ample evidence that essentially similar patterns of loss are being experienced there as well.

Consequences of Deforestation and Biodiversity Loss

The global loss of forests and the biodiversity they contain has given rise to considerable disquiet. Some people have an ethical concern about the species extinction that is underway. Others are concerned about the impact the loss of species might have on ecological processes in the residual forests. For example, Soule and Terborgh (1999) argue that the loss of top order predators will affect the populations of species in lower trophic levels and so alter rates of pollination, seed predation, seed dispersal and herbivory. This will cause a cascade of other effects that develop at different rates in the remaining forests. Another concern is that deforestation and these losses of species will lead to a reduction in the supply of a variety of goods, various non-material values and ecosystem services. Ecosystem services are the benefits received by humans from various ecosystem functions and processes. There is some differences in the terminology used. The Millenium Ecosystem Assessment (2003) recognised four categories of ecosystem services: Provisioning Services (e.g. ecosystem products such as timber, fuelwood, water, biochemicals), Regulating Services (e.g. disease regulation, water regulation, pollination), Cultural Services (e.g. non-material cultural, recreational and psychological values), and Supporting services necessary for the production of all other environmental services such as nutrient cycling, primary production and soil formation.

In this book, a rather simpler classification of the benefits provided by natural forests is used. This recognises three sets of benefits; namely, Goods, Non-material

Table 1.8 Goods and services provided by natural ecosystems that are likely to be affected by deforestation and biodiversity loss

Goods	Non-material values	Ecosystem services
Timber	Existence value	Major biogeochemical cycles (C, N, P), water cycle
Fuelwood	Spiritual	Natural pest control by predators in food web
Food (plant and animal)	Cultural	Pollination (by vertebrates and invertebrates)
Medicines	Historical	Seed dispersal
	Recreational	Decomposition of biomass, wastes, pollution
	Eco-tourism	Erosion prevention, management of soil fertility, soil formation
		Climate regulation

benefits, and Ecosystem Services (Table 1.8). Biodiversity is not an ecosystem service, but is a source of these because of its importance to various ecological functions. Some ecosystem services are provided by a range of species, but others are dependent on just one or two key species. Thus, watershed protection will be enhanced by the presence of a variety of species of trees and shrubs, while pollination of a particular plant species may depend on a single bird or insect. The loss of a specialised provider obviously has a major effect on the supply of that service. Likewise, the addition of that species in a reforestation program might also have a substantial effect.

Two of the most important ecosystem services affected by deforestation are changes to carbon sequestration and storage, and changes to hydrological cycles.

Deforestation and Greenhouse Gases

Forests sequester and store large amounts of carbon and one estimate suggests tropical forests contain about 25% of all carbon in the terrestrial biosphere (Bonan 2008). There is general agreement that deforestation has been a major contributor to increased greenhouse gas emissions (i.e. carbon dioxide but also methane and nitrous oxide) to the atmosphere. However, there is still considerable uncertainty over just how much has occurred as a consequence of past land use changes in the tropics. Changes involve not only the loss of forest biomass but related events such as the burning of peatlands. Ramankutty et al. (2007) summarise estimates ranging from 0.5 to 2.2 Pg C released per year in the 1990s (compared with 6.4 Pg C per year released from fossil fuels). They believed this large range was due to differences in the rates of deforestation used in calculations, the assumptions about the land use changes occurring after deforestation and because of the different carbon cycling models employed. Further work will refine the estimates, but it is already obvious that the amount is large and that it represents a significant proportion of total anthropogenic emissions.

Emissions could be reduced if deforestation was halted and there is now considerable interest in providing funds to countries still retaining large areas of forest for precisely this purpose (Miles and Kapos 2008). Priority areas for such REDD (Reduction in Deforestation and Forest Degradation) funds may not necessarily be sites where other forest values such as biodiversity or water movement are at risk. Hence these funds, if they eventuate, may help combat deforestation at some sites but not others. The other obvious response to greenhouse gas emissions is to reverse deforestation and use the capacity of new forests to sequester more atmospheric carbon. This has given rise to REDD+, which includes payments for reforestation, as well as compensation for owners not deforesting (RECOFTC 2009). The capacity of reforestation to do this will be discussed further in Chapter 9.

Deforestation and Watersheds

Deforestation also affects hydrological cycles and watersheds. The topic is one about which there are a number of strongly held views as well as some misconceptions (Bruijnzeel et al. 2005; Hamilton and King 1983; Sidle et al. 2006). This has made it more difficult to develop appropriate responses to the problem. Part of the reason for this confusion is that there have been surprisingly few long-term hydrological studies carried out in the tropics. However, there is widespread agreement among researchers that the deforestation of small catchments usually causes an increase in water yield when more than 20–30% of forest cover has been removed and that this increases as the proportion of forest lost increases (Bruijnzeel 2004; Bruijnzeel et al. 2005). The reason is because tree removal reduces evapo-transpiration and allows ground water stores to enlarge and produce additional streamflow. This enhanced water yield will eventually diminish if the forest is able to regenerate but the effect will persist if forest is permanently replaced by pastures or crops. These changes are scale-dependent and may be lost at scales >1 km^2 since only part of a large catchment may be affected by a particular rainfall event and because these larger areas tend to have a greater variety of land uses which dilute and modify the responses. The impact of deforestation may be less in drier regions since any reduction in evapo-transpiration may be balanced by an increase in evaporation from the soil surface (Sandström 1998).

There is rather less certainty about the impact of deforestation on dry season flows. A good deal of empirical evidence suggests deforestation is often followed by a reduction of water run-off in the dry season and a drying up of springs even though the overall amount of run-off may have initially increased (Bruijnzeel et al. 2005; Hamilton and King 1983). This may be caused by changes to the infiltration capacity of the topsoil caused by logging machinery or overgrazing. Topsoil erosion and surface sealing may have the same effect and can take place where rainfall intensities are high. Under these circumstances, less water is able to infiltrate into lower soil layers and more is quickly lost as overland surface flow. Such changes in infiltration capacity can take time to develop. If, on the other hand, land use changes are not accompanied by such changes in hydraulic conductivity, then the sub-surface soils are able to absorb rain and the baseflows, including dry season flows, may actually be enhanced. These processes will also depend on catchment geological characteristics, with modelling suggesting the difference is likely to be greatest in catchments with deep soils (i.e. having a large storage capacity), than in sites with shallow soils and limited storage capacities.

Many rural communities are likely to be more concerned about such modifications in the timing of water flow than they are about the overall quantity of water flowing from a particular catchment. There have often been disputes based on a perception that forest clearing in the uplands has been responsible for a decline in water resources for lowland farmers. In many cases, this has less to do with any change in *supply* and rather more to do with an increase in the *demands* for water (e.g. for irrigation) in the dry season (Walker 2003).

Despite popular opinion, deforestation has less dramatic effects on flooding. The loss of forest cover may increase the risk of flooding but the more important determinants of flooding are rainfall intensity and duration, slope morphologies and the hydrologic conductivity of the soils at different depths (especially the surface layers). Rainfall intensity is especially significant since it tends to be higher in tropical areas than in temperate regions (Bruijnzeel et al. 2005). However, variations in any one of these several factors can generate quite different degrees of stormflow. Deforestation will cause an increase in peakflow because of reduced evapo-transpiration, causing soils to be wetter and more responsive to rainfall. However, the effect is greatest for small rainfall events rather than larger, more intense storms because soil factors begin to override vegetation differences as soils get wetter. Again, scale is important and the impact of land use changes is thought to diminish in larger catchments. This view that forest cover is less influential at larger scales has been recently challenged by Bradshaw et al. (2007), although (van Dijk et al. 2009) argue that the evidence they produce is insufficient to overturn the generalization.

Deforestation can also affect erosion rates. The most common form of erosion is surface erosion in which sediment is eventually carried to rivers via overland flow. Most of this overland flow is concentrated into channels by gullies or roads. Logging or forest removal accelerates surface erosion by exposing areas of mineral soil, although the extent to which this occurs depends on the methods and machinery used to remove the forest, the road or track network that is created, and by the type of replacement vegetation (Fig. 1.2). Surface erosion may be limited if surface organic matter is retained and revegetation is rapid. On the other hand, surface erosion may be quite large if steep sites are cleared, burned and used for annual crops. Details of the erosion rates found with various land use practices are given by Sidle et al. (2006).

A second form of erosion is that generated by landslips. These areas are not as large as those affected by surface erosion, but each landslip can mobilise large volumes of soil. Forests reduce the risk of landslips by reducing the soil moisture content through evapo-transpiration, providing cohesion to the soil mantle through roots and by maintaining soil permeability. All of these are affected when forests

Fig. 1.2 River sedimentation caused by erosion from roads and logging tracks in the Solomon Islands (Photo: Simon Albert)

are removed. Landslips often increase several years after deforestation, corresponding with the time it takes for old root systems to decay. Again, much depends on the subsequent land use. If natural forest regrowth occurs, a new root network will eventually develop but regularly harvested and shallow-rooted crops will leave the site in a more fragile condition (Sidle et al. 2006).

A distinction needs to be made between local or on-site erosion and larger catchment-scale erosion with the former often being generally larger (per unit area) than the latter. This is because it may take some time for sediment to move from the site into permanent streams. Areas where there is continuity between the source of the erosion and the stream (i.e. deforestation is complete) are likely to generate more sediment than areas where the flow is broken by patches of residual vegetation or regrowth.

Reforestation has the capacity to help re-establish some of the former hydrological processes that operated in natural forests. As will be clear from the discussion above, its capacity to do this will depend on the scale at which it is done and the type of reforestation actually carried out. This, together with the role of biodiversity, will be discussed further in later chapters.

Is the Present Protected Area Network Able to Protect Regional Biodiversity?

The primary international model for conserving biodiversity is to create a network of protected areas that contain representative areas of all the key ecosystems within a region and that are managed by governments. This model has been widely used throughout the world and is being increasingly adopted by local bureaucracies. As originally conceived, the aim of these protected areas was to protect individual species. Over time, their purpose has broadened to also include fostering the sustainable use of biological resources. The International Union for the Conservation of Nature (IUCN) now recognises six options, including four categories (I–IV) that are strictly protected (Nature Reserves and Wilderness Areas, National Parks, National Monuments, Habitat/Species Management Areas) and two categories that allowed for some amount of regulated harvesting of natural resources across landscapes and seascapes. One of these (category V) is for situations where the interaction of people and nature has produced a landscape or seascape with a distinctive character and significant biological value. This category allows these interactions to continue provided the integrity of the landscape is not compromised. The other (category VI) allows for the sustainable harvesting of natural resources such as timber, while still protecting and maintaining biodiversity (Chape et al. 2003). Most national production forest estates would fit within this category.

Protected areas presently cover around 12.6% of land areas across the world (Chape et al. 2003). In Southeast Asia, they cover 14.8% of land area but, by contrast, only 2.1% of the land areas of the Pacific (excluding Australia and New Zealand). Much of the protected area network in Southeast Asia has been established since the 1990s and now includes 2,656 reserves covering around 760,000 km^2 (Chape et al. 2003). The coverage in particular countries and, more particularly, the extent to which actual forest areas in the region are now protected, is shown in Table 1.9. This suggests a surprisingly high proportion of existing forest area in most countries is now managed primarily for 'conservation' (although this includes areas of state forest also managed for timber production). But these reserves are under pressure. Table 1.9 shows that many of the countries with high proportions of their remaining forests conserved also have large numbers of poor people. For example, Laos PDR has 23% of its remaining forests largely managed for conservation purposes, while 74% of its people are living on less than US$2 per day and 45% have incomes below the nationally-defined poverty line. Similar patterns are evident for many other countries in Southeast Asia. Three of the larger counties in the Pacific also have low per capita incomes, but differ from countries in Southeast Asia in having only very small areas of land or forest permanently protected in reserves.

Within each country, large numbers of people often live in the very same places that most of the protected areas are located. Indeed, Colchester (2000) quotes one estimate suggesting that 85% of the world's protected areas might be occupied by local people. The actual numbers of people involved can be quite large and many of these people still regard themselves as being the customary landowners.

Table 1.9 The extent to which remaining forests are being managed for conservation purposes and several key economic welfare indicators

Country	Proportion of surface area protected[a]	Proportion of remaining forest primarily managed for conservation[b]	GDP per capita US$[c]	Percentage of population living on <US$ 2 per day[c]	Percentage of population below national poverty line[c]
Cambodia	22	21	2,432	78	36
Indonesia	9	19	3,609	52	16
Laos PDR	16	23	1,954	74	45
Malaysia	17	5	10,276	9	15
Myanmar	5	15	1,027	Na	Na
Philippines	6	12	4,614	47	41
Thailand	19	58	8,090	25	18
Vietnam	4	15	2,745	Na	29
Fiji	0.3	7	6,066	Na	Na
Papua New Guinea	4	5	2,543	Na	37
Solomon Islands	0.2	Na	1,814	Na	Na

[a] World Conservation Monitoring Center (www.unep-wcmc.org/wdpa/)
[b] FAO 2006
[c] World Bank, World Development Indicators 2006; Table 2.7

Threats to Asian Protected Areas

The success of the protected area model for conservation is predicated on there being a strong state with well-funded institutions able to manage and, where necessary, enforce boundaries and prescriptions. This does not always occur and some reserves have come to be regarded as 'paper parks' because they have been so badly degraded. There are a variety of causes.

Logging and Agricultural Clearing

Various forms of logging and agricultural clearing are still common in many protected areas. A survey by DeFries et al. (2005) of protected areas in South and Southeast Asia found 26% of the sites with moist forest and 24% of sites in dry forest areas had lost cover over the period between the early 1980s and 2001. A graphic account by Jepson et al. (2001) describes the scale of the problem in several Indonesian National Parks (Kerinci-Seblat and Gunung Leuser in Sumatra as well as Tanjung Puting and Gunung Palung in Kalimantan). In these cases, local officials, including the police, appeared to have been unwilling or unable to control what was happening, despite the scale of the damage being caused. The authors thought the stage had been reached where only extreme military action would be sufficient to prevent further damage. Several years after this report, Curran et al. (2004) measured the extent of illegal logging in West Kalimantan. They found 38% of lowland forest inside Gunung Palung National Park had been logged and deforested over the 14 year period prior to 2003. From addition surveys they concluded that more than 56% of Kalimantan's protected lowland forests were lost between 1985 and 2001. Other reports of illegal logging and agricultural encroachment have been reported from elsewhere across the region (Eaton 2005; Gaveau et al. 2007; van Schaik et al. 1997). ICEM (2003) notes that most of the protected areas in Vietnam, Cambodia, Laos and Thailand have been subjected to some form of illegal logging or clearing. Parks most at risk of being cleared are those with fertile soils or near good roads.

Gathering and Hunting

Hunting and gathering is common within many protected areas across the region (ICEM 2003; Nooren and Claridge 2001; Sodhi et al. 2004; van Schaik et al. 1997). Much of this has been carried out by people who gather materials such as medicinal plants or thatching and who live around the reserve boundaries. Park authorities often ignore small-scale harvesting because they know the people doing it are poor and have few alternatives. Some of these same people carry out sporadic hunting to supplement their domestic food supplies but others do so to sell meat to local restaurants. There are even reports of so-called bushmeat being exported to countries such as China, Vietnam, Thailand, Japan and South Korea (Nooren and Claridge 2001). Over time, the impact of such hunting can be significant and Redford (1992) describes how many forests in Amazonia are now 'empty' of wildlife because of hunting. The same has happened in some of the more densely populated areas of Asia. Donovan (2003, p. 91) quotes interviews with former hunters in Vietnam who 'readily admit that there are almost no animals left in the forest apart from rats and the passing bird'.

Lack of Commitment By Governments

Logging, clearing and hunting occurs because Park managers are unable to control it. This may be due to a lack of resources, or it may be because of a lack of political will. In some cases, governments have come to see protected areas as land banks which can be used for dams, oil exploration, industrial development, or that can be bisected by roads where this is necessary. In such cases, governments may believe that they have over-invested in the idea of protected areas and that biodiversity conservation does not seem to be contributing to either local or national development. Under these circumstances, any kind of degradation is not seen as an economic loss, nor is Park degradation factored into planning and development decisions as an economic cost. This situation is ironical given the finding by Balmford et al. (2003) that the benefit-to-cost ratio for conservation is far greater in most developing countries than it is in developed countries and that a small additional investment could go a long way to meeting unmet conservation needs. In reality, protected areas can contribute quite significantly to economic development and benefits can flow to local communities as well as to regional and national economies. Box 1.4 shows some of the ways in which managers might help enhance the contribution protected areas make to livelihoods and national develop and make their benefits more visible.

Size and Boundaries

The final threat to Asian protected areas is that many are too small to achieve their purpose in maintaining viable populations of plant and animal species. This is not just an Asian problem as the majority of protected areas around the world are less than 10,000 ha (Chape et al. 2003). The problem is sometimes accentuated by the fact that the land areas adjoining many protected areas have been cleared right up to the boundary and these reserves now exist as isolated patches in an otherwise totally cleared landscape. Such sharp boundaries represent a warning signal that land use practices outside the protected area may not be sustainable. DeFries et al. (2005) noted many Asian protected areas were especially sensitive in this respect.

Protected Areas in the Pacific

The situation in the smaller islands of the western Pacific is strikingly different to that in Southeast Asia. All communities in the Pacific have traditional methods of regulating resource use, but protected areas cover only 2% of the total land area compared with the 14.8% in Southeast Asia and it is clear that the global conservation blueprint has not been accepted. There are several reasons for this. One is that the amount of good agricultural land is in short supply. As populations grow, communities need to find increasing areas of land for subsistence farming. A second reason is that much of the land is owned by customary landowners and, with a few exceptions, these landowners have been unwilling to permanently cede significant areas of land to the state for conservation purposes. In any case, there are practical difficulties in doing so. If a proposed park is to be of any size, a large number of land-owning clans might be involved. This would require separate negotiations and agreements with each one.

Many people in the Pacific are sympathetic to the idea of 'conservation' but the narrow protectionist approach most usually promoted is not one that fits easily within common forms of land tenure. What is needed instead is a form of conservation that is community-based and which recognises people's need to actually use biological resources to sustain their livelihoods. This is, in fact, allowed for in categories V and VI in the IUCN Protected Area classification system, which permits the sustainable harvesting of resources. But there are relatively few management models that might permit this and which suit the rising populations and variety of social, economic and ecological situations present across the Pacific (Baines et al. 2002). Some promising schemes have been developed in Vanuatu

Box 1.4 Making Protected Areas Contribute to Development Needs

Many protected areas are islands of diminishing biodiversity in largely agricultural landscapes. People living in and around these areas often carry out unconstrained harvesting of plant and animal resources. Governments sometimes regard them as land banks. The key problem is that they are not seen to be contributing to national economic development or to improving the livelihoods of people living nearby. One review of the role of protected areas in promoting economic development investigated parks in Thailand, Laos PDR, Cambodia and Vietnam (ICEM 2003). Key findings were:

1. Managers of protected areas need to turn from being inward looking and concerned only with what happens within their lands to being more outward looking. They should promote their forests as being part of the development landscape and not as islands under siege.
2. Protected areas supply ecological and cultural services and, in some cases, goods. These are often regarded as being free. The market value of these services and goods need to be quantified so that protected areas can be recognised as economically beneficial uses of land, resources and investment funds. In some cases, it might be appropriate that a user-pays regime is developed. In such cases, appropriate economic policies and instruments will need to be created.
3. Protected areas can help maintain and boost the productivity of a number of economic sectors such as tourism and water resources. However, park managers need to express these linkages in economic terms so that managers in these other sectors will recognise the extent of the benefits they receive. Such financial flows should be recognised in the budgets of these other sectors.
4. Local people living in and around protected areas must benefit if these areas are to be effective over the longer term. These benefits may come from a share of the sale of ecological services such as clean water, tourism, or from the sustainable harvesting of resources such as medicinal plants.
5. Park managers should involve local people and other resource users in the management of their lands. This is especially important in cases where resources such as plants or wildlife within the protected area are being harvested. Collaborative management approaches are needed to involve local people in developing systems of self-regulation, monitoring and sustainable use of resources.
6. The success of protected areas is often dependent on the activities underway in the lands surrounding them. All protected areas should have specially managed buffer zones in which land uses compatible with conservation objectives are practised. Financial incentives may need to be offered to people living in these areas to achieve this.

where individual landowners have agreed to protect certain areas for periods up to ten years in exchange for cash payments from international donors to compensate for the logging income foregone (Tacconi and Bennett 1997). In other cases, land owners have not required cash payments but needed the local government to establish an appropriate legal framework. This had the side benefit of confirming the customary landownership rights of the current owners and preventing a drift to an open access system. In all these cases, the areas were mostly small and less than 5,000 ha. Another model providing rather longer term protection for larger areas has been developed on Western Samoa (see Box 1.5). But again, this scheme needed external funding to compensate landowners.

In summary, a significant protected area estate has been created in much of Southeast Asia and the areas involved are still increasing. But many of these parks

Box 1.5 A Pacific Conservation Reserve

An interesting example of how a conservation reserve was established in the Pacific is the Falealupo Rain Forest Preserve in Western Samoa (Cox 2000; Cox and Elmqvist 1991, 1997). In this case, western conservation ideals were successfully blended with traditional approaches to land use resulting in the development of a 5,000 ha forest reserve. The forest was to be logged but conservationists were able to devise a mutually satisfactory arrangement with the local community by bringing in external funds to compensate the community for not proceeding with the logging. In return, the community agreed to a 50 year conservation covenant that protects the forest although it allows traditional use of the forest such as harvesting of traditional medicinal plants and the felling of canoe trees.

The agreement has several important advantages. Firstly, from the point of view of international donors, only relatively small amounts of funds were needed. Secondly, the funds are being used to build a school, medical clinics and support the development of new ecotourism activities. In that sense, the funding achieves multiple goals. And, thirdly, there is unlikely to be any need to spend resources on protecting the reserve because the community will be in charge of its management. The covenant also provides the community with a share of any revenues arising from the commercialisation of biomedical discoveries arising from studies of the biodiversity now protected by the reserve. Without this covenant, the village would have been forced to allow logging to pay for the school.

Such seemingly simple arrangements are usually more complicated than they appear and depend on good faith and mutual respect between the parties. An attempt by another non-government organization to replicate the success of the Falealupo Covenant at another site on Samoa failed because of misunderstandings and the perception by the Samoans of a lack of respect and good faith on the part of the NGO (Cox 2000).

are threatened by logging, hunting and deforestation and some have lost much of their conservation value. The protected area model remains an important policy response to the threat of biodiversity loss but, by itself it may not be sufficient, especially in the Pacific.

Conclusions

The Asia-Pacific region contains a significant proportion of the world's tropical forests and biodiversity but it, like the rest of the tropical world, is being gradually deforested. Further losses are certain. These declines in forest cover are transforming the region from being one that is immensely biologically diverse to one with many biologically impoverished landscapes. Most countries in the Asia-Pacific region now have less than 20% of their original forests remaining in an undisturbed condition. Those natural forests that do remain are now much smaller and persist as patches scattered within a largely agricultural landscape. The ecological integrity of many of these patches is declining further because of logging and hunting.

Much of the deforested land has been used for agriculture to support a substantially increased population, but some has been used only briefly and has since been abandoned in a degraded state. Degradation is difficult to define and, consequently, hard to map. However, it is clear that very large areas of degraded or under-utilised lands are now found in most countries.

The boost to agriculture provided by deforestation has been accompanied by a loss in the supply of certain goods and a change in the ability of the remaining forests to maintain environmental services. Some goods can still be supplied by the remaining forests but the loss of biodiversity and environmental services is not as easily accommodated and alternatives may be expensive. Many of the people most affected by these changes are relatively poor.

The principal policy tool being used to counter biodiversity losses has been the establishment of a global protected area network. A substantial area of land is included within this network in Southeast Asia although the approach has been less successful in the countries of the Pacific. But many protected areas are not being effectively managed. This is because there are insufficient resources for managers and because in too many cases protected areas are seen as being irrelevant to the livelihoods of people living in and around them.

All of this suggests additional methods are needed to conserve biodiversity. The most obvious addition is to reforest some of the large areas of degraded land. This could help support some protected areas as well as restore some ecological services. However, the task faces ecological as well as socio-economic constraints and is easier said than done. As a first step in understanding just how it might be approached, it is useful to examine in more detail just how the present situation arose. The next chapter explores some of the recent events that have given rise to deforestation and land degradation across the Asia-Pacific region.

References

Achard F, Eva HD, Stibig HJ, Mayaux P, Gallego J, Richards T, Malingreau JP (2002) Determination of deforestation rates of the world's humid tropical forests. Science 297:999–1002

Adam P (1994) Australian rainforests. Oxford University Press, Oxford

Aronson J, Milton SJ, Blignaut JN (2007) Restoring Natural capital: Science, business, and practice. Island Press, Washington, DC

Baillie JEM, Hilton-Taylor C, Stuart S (eds) (2004) 2004 IUCN red list of threatened species: A global species assessment. IUCN, Gland

Baines G, Hunnam P, Rivers MJ, Watson B (2002) South Pacific Biodiversity Conservation Programme: Terminal Evaluation Mission: Final report. United Nations Development Program, New York

Balmford A, Gaston KJ, Blyth S, James A, Kapos V (2003) Global variation in terrestrial conservation costs, conservation benefits, and unmet conservation needs. Proceedings of the National Academy of Sciences 100:1046–1050

Bennett JA (2000) Pacific forest: A history of resource control and contest in Solomon Island, c. 1800–1997. Brill, Leiden

Bierregaard RO, Stouffer PC (1997) Understorey birds and dynamic habitat mosaics in Amazonian rainforests. In: Laurance WF, Bieeregaard RO (eds) Tropical forest remnants: Ecology, managment, and conservation of fragmented communities. University of Chicago Press, Chicago and London, pp 138–155

Biot Y, Blaikie PM, Jackson C, Palmer-Jones R (1995) Rethinking research on land degradation in developing countries. World Bank, Washington. DC

Bonan GB (2008) Forests and climate change: Forcings, feedbacks, and the climate benefits of forests. Science 320:1444–1449

Bradshaw CJA, Sodhi N, Peh KSH, Brook BW (2007) Global evidence that deforestation amplifies flood risk and severity in the developing world. Global Change Biol 13:2379–2395

Brook BW, Sodhi NS, Ng PKL (2003) Catastrophic extinctions follow deforestation in Singapore. Nature 424:420–423

Brook BW, Bradshaw CJA, Koh LP, Sodhi NS (2006) Momentum drives the crash: Mass extinction in the tropics. Biotropica 36:302–305

Brooks TM, Pimm SL, Collar NJ (1997) Deforestation predicts the number of threatened birds in insular southeast Asia. Conserv Biol 11:382–394

Brooks TM, Mittermeier RA, Mittermeier CG, da Fonseca GAB, Rylands AB, Konstant WR, Flick P, Pilgrim J, Oldfield S, Magin G, Hilton-Taylor C (2002) Habitat loss and extinction in the hotspots of biodiversity. Conserv Biol 16:909–923

Brown C, Durst P (2003) State of forestry in Asia and the Pacific - 2003. In: Status, change and trends. UN Food and Agriculture Organisation, Regional Office for Asia and the Pacific, Bangkok

Bruijnzeel LA (2004) Hydrological functions of tropical forests: not seeing the soil for the trees? Agricult Ecosyst Environ 104:185–228

Bruijnzeel LA, Bonell M, Gilmour DA, Lamb D (2005) Conclusion: Forest, water and people in the humid tropics: an emerging view. In: Bonell M, Bruijnzeel LA (eds) Forests, water and people in the humid tropics. Cambridge University Press and UNESCO, Cambridge, pp 906–925

Bryant D, Nielson D, Tangley L (1997) The Last Frontier Forests. World Resources Institute, Washington, DC

Calub AD, Anwarhan H, Roder W (1996) Livestock production systems for Imperata grasslands. Agroforestry Syst 36:121–128

Castelletta M, Sodhi NS, Subaraj R (2000) Heavy extinctions of forest avifauna in Singapore: Lessons for biodiversity conservation in Southeast Asia. Conserv Biol 14:1870–1880

Chape S, Blyth S, Fish L, Fox P, Spalding M (2003) United Nations List of Protected Areas. International Union for the Conservation of Nature and UNEP World Conservation Monitoring Center, Gland, Switzerland and Cambridge UK

Chokkalingam U, Bhat DM, von Gemmingen G (2001) Secondary forests associated with the rehabilitation of degraded land in tropical Asia: a synthesis. J Trop For Sci 13:816–831
Chomitz K (2007) At Loggerheads? Agricultural expansion, poverty reduction and environment in the tropical forest. World Bank, Washington, DC
Clarke WC, Thaman R (1993) Agroforestry in the Pacific Islands: Systems for sustainability. United Nations University Press, Tokyo, New York, Paris
Colchester M (2000) Self-determination or environmental determinism for indigenous peoples in tropical forest conservation. Conserv Biol 14:1365–1367
Connell J, Lowman M (1989) Low-diversity tropical forests: some possible mechanisms for their existence. Am Naturalist 134:88–119
Corlett RT, Turner IM (1997) Long-term srvival in tropical forest remnants in Singapore and Hong Kong. In: Laurance WF, Bierregaard RO (eds) Tropical forest remnants: Ecology, management, and conservation of fragmented communities. The University of Chicago Press, Chicago, pp 333–345
Cox PA (2000) A tale of two villages: Culture, conservation and ecocolonialism in Samoa. In: Zerner C (ed) People, plants and justice: the politics of nature conservation. Columbia University Press, New York, pp 330–334
Cox PA, Elmqvist T (1991) Indigenous control: An alternative strategy for the establishment of rainforest preserves. Ambio 20:317–321
Cox PA, Elmqvist T (1997) Eco-colonialism and indigenous controlled rainforest preserve in Samoa. Ambio 26:84–89
Curran LM, Trigg SN, McDonald AK, Astiani D, Hardiono YM, Siregar P, Caniago I, Kasischke E (2004) Lowland forest loss in protected areas of Indonesian Borneo. Science 303:1000–1003
Davis SD, Heywood VH, Hamilton AC (eds) (1995) Centers of plant diversity: A guide and strategy for their conservation. IUCN Publications Unit, Cambridge
DeFries R, Hansen A, Newton AC, Hansen MC (2005) Increasing isolation of tropical forests over the last twenty years. Ecol Appl 15:19–26
Donovan D (2003) Trading in the forest: Lessons from Lao history. In: Lye T-P, De Jong W, Abe K (eds) The political ecology of tropical forests in Southeast Asia: Historical perspectives. Kyoto University Press, Kyoto, pp 72–106
Eaton P (2005) Land tenure, conservation and development in Southeast Asia. Routledge Curzon, London and New York
Elmqvist T, Rainey WE, Pierson ED, Cox PA (1994) Effects of tropical cyclones Ofa and Val on the structure of a Samoan Lowland Rain-Forest. Biotropica 26:384–391
FAO (2000) FRA 2000 on definitions of forest and forest change, Forest Resources Assessment Program,Working Paper 33. Food and Agriculture Organisation of the United Nations, Rome
FAO (2001) Global Forest Resources Assessment 2000, Main Report. FAO Forestry Paper 140. Food and Agriculture Organization of the United Nations, Rome
FAO (2006) Global Forest Resources Assessment 2005: Progress towards sustained forest management. Food and Agricultural Organization of the United Nations, Rome
FAO (2010) The Global Forest Resource Assessment 2010. Food and Agriculture Organization of the United Nations, Rome
Foley JA, DeFries R, Asner GP, Barford C, Bonan G, Carpenter SR, Chapin FS, Coe MT, Daily GC, Gibbs HK, Helkowski JH, Holloway T, Howard EA, Kucharik CJ, Monfreda C, Patz JA, Prentice IC, Ramankutty N, Snyder PK (2005) Global consequences of land use. Science 309:570–574
Franklin J (2003) Regeneration and growth of pioneer and shade-tolerant rain forest trees in Tonga. NZ J Bot 41:669–684
Gardner TA, Barlow J, Parry LW, Peres CA (2007) Predicting the uncertain future of tropical forest species in a data vacuum. Biotropica 39:25–30
Garrity DP, Soekardi M, VanNoordwijk M, DelaCruz R, Pathak PS, Gunasena HPM, VanSo N, Huijun G, Majid NM (1996) The imperata grasslands of tropical Asia: Area, distribution, and typology. Agroforestry Syst 36:3–29

Gaveau DLA, Wandono H, Setiabudi F (2007) Three decades of deforestation in southwest Sumatra: have protected areas halted forest loss and logging, and promoted re-growth? Biol Conserv 134:495–504
Gillison A (1993) Grasslands of the South-West Pacific. In: Coupland RT (ed) Ecosystems of the World: Natural grasslands eastern hemisphere and resume. Elsevier, Amsterdam, pp 435–470
Gilmour DA, Nguyen VS, Tsechalicha X (2000) Rehabilitation of degraded forest ecosystems in Cambodia, Lao PDR, Thailand and Vietnam. IUCN Asia
Goosem M (1997) Internal fragmentation: The effects of roads, highways and poerline clearings on movements and mortaility of rainforest vertebrates. In: Laurance WF, Bierregaard RO (eds) Tropical forest remants: Ecology, management, and conservation of fragmented communities. University of Chicago Press, Chicago and New York, pp 241–255
Grainger A (2008) Difficulties in tracking the long-term global trend in tropical forest area. Proceedings of the National Academy of Science 105:818–823
Gressitt JL (1982) Biogeography and ecology of New Guinea. W. Junk, The Hague
Hamilton LS, King PN (1983) Tropical forested watersheds. In: Hydrologic and soils responses to major uses or conversions. Westview Press, Boulder, CO
Hannah L, Lohse D, Hutchinson C, Carr JL, Lankerani A (1994) A preliminary inventory of human disturbance of world ecosystems. Ambio 23:246–250
Hart TB (1990) Monospecific dominance in tropical rain forests. Trends in Ecology and Evolution 5:6–11
Hau BCH, Dudgeon D, Corlett RT (2005) Beyond Singapore: Hong Kong and Asian biodiversity. Trends Ecol Evol 20:281–282
ICEM (2003) Regional Report on Protected Areas and Development. Review of Protected Areas and Development in the Lower Mekong River Region. International Center for Environmental Management (www.mekong-protected-areas.org), Indooroopilly, Australia
ITTO (2002) ITTO Guidelines for the Restoration, Management and Rehabilitation of Degraded and Secondary Tropical Forests, vol No 13, ITTO Policy Development Series. International Tropical Timbers Organization, Yokohama
Jansen PCM, Lemmens RHMJ, Oyen LPA, Siemonsma JS, Stavast FM, van Valkenburg JLCH (1991) Plant resources of South-East Asia. In: Basic list of species and commodity grouping. Final version. Pudoc, Wageningen
Jepson P, Jarvie JK, MacKinnon K, Monk KA (2001) The end for Indonesia's lowland forests? Science 292:859–861
Kemp N, Dilger M, Burgess N, Chu VD (1997) The saola *Pseudoyrx nghetinhensis* new information on distribution and habitat prefernces, and conservation needs. Oryx 31:37–44
Keppel G, Buckley Y, Possingham H (2010) Drivers of lowland rainforest community assembly, species diversity and forest structure in islands in the tropical South Pacific. J Ecol 98:87–95
Kummer DM, Turner BL (1994) The human causes of deforestation in Southeast Asia. Bioscience 44:323–328
Laurance WF (1991) Ecological correlates of extinction proneness in Australian tropical rain-forest mammals. Conserv Biol 5:79–89
Laurance WF (2007) Have we overstated the tropical biodiversity crisis? Trends Ecol Evol 22:65–70
Laurance WF, Bierregaard RP (eds) (1997) Tropical forest remnants: ecology, management and conservation of fragmented communities. University of Chicago Press, Chicago
Laurance WF, Lovejoy TE, Didham RK, Stouffer P, Gascon C, Bierregaard RO, Laurance S, Sampaio E (2002) Ecosystrem decay of Amazonian forest fragments: A 22 year investigation. Conserv Biol 16:605–618
Lemmens RHMJ, Soeriannegara I, Wong WC (eds) (1995) Plant resources of South-East Asia No 5(2) timber trees: Minor commercial timbers. Backhuys Publishers, Leiden
Mackey B (1993) A spatial analysis of the environmental relations of rainforest structural types. J Biogeograp 20:303–336
Mackey B (1994) Predicting the potential distribution of rain-forest structural characteristics. J Vegetat Sci 5:43–54

Matson P, Parton WJ, Power AG, Swift MJ (1997) Agricultural intensification and ecosystem properties. Science 277:504–509

Matthews E (2001) Understanding the FRA 2000, vol No 1, Forest Briefing. World Resources Institute, Washington, DC

Mayaux P, Holmgren P, Achard F, Eva H, Stibig H, Branthomme A (2005) Tropical forest cover change in the 1990s and options for future monitoring. Phil Trans R Soc B Biol Sci 360:373–384

Miles L, Kapos V (2008) Reducing greenhouse gas emissions from deforestation and forest degradation: global land use implications. Science 320:1554–1455

Millenium Ecosystem Assessment (2003) Ecosystems and Human Well-being: A Framework for Assessment. Island Press, Washington

Mittermeier RA, Myers N, Thomsen JB, da Fonseca GAB, Olivieri S (1998) Biodiversity hotspots and major tropical wilderness areas: approaches to setting conservation priorities. Conserv Biol 12:516–520

Mueller-Dombois D, Fosberg FR (1998) Vegetation of the Pacific. Springer-Verlag, New York

Myers N, Mittermeier RA, Mittermeier CG, da Fonseca GAB, Kent J (2000) Biodiversity hotspots for conservation priorities. Nature 403:853–858

Nawir AA, Murniat, Rumboko L (2007) Forest Rehabilitation in Indonesia: Where to after Three Decades? Center Int Forestry Res, Bogor

Nir E (2004) The stands of *Anisoptera thurifera* ssp. *polyandra* and their management in Papua New Guinea, PhD Thesis, School of Integrative Biology. The University of Queensland, Brisbane

Nooren H, Claridge G (2001) Wildlife trade in Laos: the end of the game. Netherlands Committee for IUCN, Amsterdam

Office of Technical Assessment (1984) Technologies to Sustain Tropical Forest Resources. Office of Technical Assessment, US Congress, Washington, DC

Olson DM, Dinerstein E (2002) The Global 200: Priority ecoregions for global conservation. Ann Missouri Bot Garden 89:199–224

Paijmans K (ed) (1976) New guinea vegetation. Australian National University Press, Canberra

Pierson ED, Elmqvist T, Rainey WE, Cox PA (1996) Effects of tropical cyclonic storms on flying fox populations on the South Pacific Islands of Samoa. Conserv Biol 10:438–451

Ramankutty N, Gibbs HK, Achard F, Defries R, Foley JA, Houghton RA (2007) Challenges to estimating carbon emissions from tropical deforestation. Global Change Biol 13:51–66

RECOFTC (2009) Decoding REDD: Restoration in REDD+. In: Forest restoration for enhancing carbon stocks an Asia-Pacific perspective. Regional Community Forestry Training Center, Bangkok

Redford KH (1992) The empty forest. Bioscience 42:412–422

Richards PW (1952) The Tropical Rain Forest. Cambridge University Press, Cambridge

Robbins RG (1960) The anthropogenic grasslands of Papua and New Guinea. Symposium on the Impact of Man on Humid Tropics. Vegetation Administration of the Territory of Papua and New Guinea and UNESCO Science Co-operation Office for South East Asia., Goroka, Territory of Papua and New Guinea, pp 313–329

Rolett B, Diamond J (2004) Environmental predictors of pre-European deforestation on Pacific Islands. Nature 431:443–446

Rundel PW, Boonpragob K (1995) Dry forest ecosystems of Thailand. In: Bullock SK, Mooney HA, Medina E (eds) Seasonally Dry Forests. Cambridge University Press, Cambridge

Sandström K (1998) Can forests'provide' water: widespread myth or scientific reality. Ambio 27:132–138

Scherr S, McNeely JA (2008) Biodiversity conservation and agricultural sustainability: towards a new paradigm of 'ecoagriculture' landscape. Phil Trans R Soc B Biol Sci 363:477–494

Seamon JO, Mann SS, Steele OC, Utzurrum RCB (2006) Conservation value of remnant forest patches: Tree composition, spatial patterns, and recruitment in the Ottoville Lowland Forest, American Samoa. Pacific Sci 60:319–332

Sidle RC, Ziegler AD, Negishi JN, Nik AR, Siew R, Turkelboom F (2006) Erosion processes in steep terrain – truths, myths, and uncertainties related to forest management in Southeast Asia. Forest Ecol Manage 224:199–225

Sikor T (1995) Decree 327 and the restoration of barren land in the Vietnamese highlands. In: ATR Reed R.R., Le TC, Di Gregorio MR (eds) The challenges of highland development in Vietnam. East-West Center, Honolulu

Sodhi NS, Brook BW (2006) Southeast Asian birds in peril. The Auk 123:275–277

Sodhi NS, Koh LP, Brook BW, Ng PKL (2004) Southeast Asian biodiversity: an impending disaster. Trends Ecol Evol 19:654–660

Sodhi NS, Koh LP, Brook BW, Ng PKL (2005) Response to Hau et al: Beyond Singapore: Hong Kong and Asian biodiversity. Trends Ecol Evol 20:282–283

Sodhi NS, Brooks TM, Koh LP, Acciaioli G, Erb M, Tan AKJ, Curran LM, Brosius P, Lee TM, Patlis JM, Gumal M, Lee RJ (2006) Biodiversity and human livelihood crises in the Malay Archipelago. Conserv Biol 20:1811–1813

Soerianegara I, Lemmens RHMJ (eds) (1993) Plant resources of South-East Asia 5 (1): Timber Trees; Major commercial timbers. Pudoc Scientific Publishers, Wageningen

Soule ME, Terborgh J (1999) The policy and science of regional conservation. In: Soule ME, Terborgh J (eds) Continental conservation: Scientific foundations of regional reserve networks. Island Press, Washington, DC, pp 1–17

Sterling EJ, Hurley MM, Le DM (2006) Vietnam: A natural history. Yale University Press, New Haven

Stoddard DR (1992) Biogeography of the tropical Pacific. Pacific Sci 46:276–293

Tacconi L, Bennett J (1997) Protected area assessment and establishment in Vanuatu: A Socio-economic Approach. Australian Center for International Agricultural Research, Canberra

Torti SD, Coley PD, Kursar TA (2001) Causes and consequences of monodominance in tropical lowland forests. Am Naturalist 157:141–153

Turner IM, Corlett RT (1996) The conservation value of small, isolated fragments of lowland tropical rain forest. Trends Ecol Evol 11:330–333

Turner IM, Chua KS, Ong JSY, Soong BC, Tan HTW (1996) A century of plant species loss from an isolated fragment of lowland tropical rain forest. Conserv Biol 10:1229–1244

Turner IM, Wong YK, Chew PT, bin Ibrahim A (1997) Tree species richness in primary and old secondary tropical forest in Singapore. Biodiv Conserv 6:537–543

Turner H, Hovenkamp P, van Welzen PC (2001) Biogeography of Southeast Asia and the West Pacific. J Biogeograp 28:217–230

Turvey N (1994) Afforestation and rehabilitation of imperata graslands in Southeast Asia. Australian Center for International Agricultural Research, Canberra

van Balgoy MM (1971) Plant -geography of the Pacific. Blumea (Suppl) 6:1–222

van Dijk AIJM, van Noordwijk M, CaldeR IR, Bruijnzeel SLA, Schellekens J, Chappell NA (2009) Forest-flood relation still tenuous – comment on 'Global evidence that deforestation amplifies flood risk and severity in the developing world' by C. J. A. Bradshaw, NS Sodi, K. S.-H. Peh and BW Brook. Global Change Biol 15:110–115

van Lynden GWJ, Oldeman LR (1997) The assessment of the state of human induced soil degradation in South and Southeast Asia. UNEP/FAO/ISRIC, Nairobi, Rome, Wageningen

van Schaik C, Terborgh J, Dugelby B (1997) The silent crisis: The state of rain forest nature preserves. In: Kramer R, van Schaik C, Johnson J (eds) Last stand: Protected areas and the defense of tropical biodiversity. Oxford University Press, Oxford, pp 64–89

Walker A (2003) Agricultural transformation and the politics of hydrology in Northern Thailand. Dev Change 34:941–964

Watson RT, Noble IR, Bolin B, Ravindranath NH, Verardo DJ, Dokken DJ (eds) (2000) Land use, land-use change, and forestry. Cambridge University Press, Cambridge

Webb LJ (1959) A physiognomic classification of Australian rain forests. J Ecol 47:551–570

Webb EL, Fa'aumu S (1999) Diversity and structure of tropical rain forest of Tutuila, American Samoa: effects of site age and substrate. Plant Ecol 144:257–274

Whiffin T, Kikkawa J (1992) The status of forest biodiversity in Oceania. J Trop Forest Sci 5:155–172

Whitmore TC (1984) Tropical rain forests of the far East. Clarendon Press, Oxford

Whitmore T (1997) Tropical forest disturbance, disappearance, and species loss. In: Laurance WF, Bierregaard RO (eds) Tropical forest remnants: Ecology, management, and conservation of fragmented communities. University of Chicago Press, Chicago and London, pp 3–12

Whitten AJ, Damanil SJ, Anwar J, Hisyam N (1984) The ecology of Sumatra. Gadjah Madah University Press, Yogyakarta

Whitten AJ, Mustafa M, Henderson GS (1989) The ecology of Sulawesi. Gadjah Madah University Press, Yogyakarta

Wiser SK, Drake DR, Burrows LE, Sykes WR (2002) The potential for long-term persistence of forest fragments on Tongatapu, a large island in western Polynesia. J Biogeograp 29:767–787

Wright SJ, Muller-Landau HC (2006) The uncertain future of tropical forest species. Biotropica 38:443–445

Wright DD, Jessen JH, Burke P, Garza HGD (1997) Tree and liana enumeration and diversity on a one-hectare plot in Papua New Guinea. Biotropica 29:250–260

Chapter 2
Forest and Land Degradation in the Asia-Pacific Region

If it should happen that the forests of the mountains are exhausted, and that of the forests of the piedmont only scattered remnants remain, or that they have disappeared, and that the thickets and marshes have come to the last days of their existence, the forces of the people will be weakened, the farmlands where the cereals and hemp grow will be uncultivated, and they will lack resources. The superior man should be concerned about this problem in a spirit of altruistic urgency, and without relaxation. How could it ever delight him?
Duke Mu of Shan, a minister in the court of the Zhou high king, China, 524 BC.

Elvin (2001, p. 17)

The snapshot of the history of forest management in the Asia-Pacific over the last three decades is one of steady destruction. Scorched earth is increasingly the final picture.

Dauvergne (2001, p. 27)

Introduction

The first chapter described how large amounts of deforestation and degradation have occurred in tropical forests across the Asia-Pacific region and how this is threatening biodiversity, while still leaving many people poor. This discussion covered the problem from a global and regional perspective but did not explore the reasons why these changes are occurring. Why should societies destroy their forests for what are sometimes only modest benefits? And why should different countries repeat the mistakes made previously by others? Is there no capacity to learn from neighbours or from history? This chapter considers the processes underlying deforestation and degradation in a little more detail. If we can understand why degradation has occurred, we might be in a better position to prevent it and to overcome it where it has already occurred.

It is important to realize that forests ecosystems are always being affected by naturally occurring disturbances such as windstorms, fires or landslips. Most ecosystems become adapted to these disturbance regimes and are able to recover. However, recovery is far more difficult when the nature of the historical disturbance regime changes. That is, if there is a change in the intensity, frequency, or

D. Lamb, *Regreening the Bare Hills*, World Forests 8,
DOI 10.1007/978-90-481-9870-2_2, © Springer Science+Business Media B.V. 2011

extent of the disturbance to which the ecosystem is adapted. In such cases, the ecosystem must adapt and there are likely to be modifications to its composition and structure. Over time, the system may eventually recover its original condition. But severe disturbances can push the system over a threshold from which recovery is impossible. The new steady-state condition with new species and structural attributes may be more economically attractive, as in the case of an agricultural landscape, or it may be economically unattractive and represent a form of 'degradation'.

This chapter begins by reviewing different types of natural and human-mediated disturbances and the effects these can have on tropical forests. Deforestation has occurred as human populations have increased and have needed more land to produce food crops. In some cases, the transition from forest to cropland has been successful, while in others it has not. The chapter reviews the long history of this process in China and Japan, which have had contrasting responses to deforestation and degradation. It then examines a series of more recent case studies from across the Asia-Pacific region to explore why degradation is occurring now in order to understand what might be needed to overcome it in the future.

Natural Disturbances

Although tropical forests often appear ancient and relatively homogenous, most are made up of a mosaic of different successional stages and are recovering after a variety of naturally occurring disturbances (Chazdon 2003). Some of these disturbances occur quite frequently but are usually small in scale. Single tree falls are a common example. Other disturbances occur less frequently but cover much larger areas. Tropical storms are an example of such larger-scale disturbances. These storms regularly damage forests in many Pacific Islands, northern Australia, Philippines and coastal areas surrounding the South China Sea (Mueller-Dombois and Fosberg 1998). Their frequency varies; in the Southwest Pacific they can occur around 15 times per year but up to 25 times per year in the northwest Pacific (Scatena et al. 2005). Storm frequency in the north-west Pacific is greater in La Niña years (Kelly et al. 2001). In discussing frequency it is also necessary to specify storm intensity. Ash (1992) suggests cyclones with gusts up to 150 km/h occur across Fiji once every 5–10 years while those with gusts of up to 200 km/h occur every 30–50 years. These frequencies are clearly much shorter than the lifetimes of most forest trees.

Trees vary in their ability to tolerate strong winds. Most shallow-rooted species are easily blown over but some trees such as *Agathis* spp. are more resistant (Ash 1992; Mueller-Dombois and Fosberg 1998; Scatena et al. 2005). Keppel et al. (2010) suggest trees with higher wood densities and smaller leaves are most resistant. Many cyclones and other severe storms are accompanied by multiple landslides.

Nunn (1990) describes one storm in Fiji that triggered 620 separate landslips in a single catchment. The resulting soil loss from that storm exceeded the total soil loss from the site in the previous year. The effects of storms on forests can be long-lasting and the impact of one large storm that occurred in Sarawak in 1880 and which blew over forest covering hundreds of square kilometres has remained visible for more than 100 years (Whitmore 1984). In areas where storms are frequent so-called 'storm' forests develop in which vines dominate canopies and shroud tree regrowth (Webb 1958).

Fires are another disturbance that can affect even larger areas. This might seem surprising in wet tropical forests. However, droughts periodically occur (related to ENSO events in at least some parts of the region) and fires can burn during these periods (Knapen 1997). In more seasonal environments, fires can affect the boundaries between rainforest communities and more open forests and savannas as well as affecting the structure of the open forests (Bowman 2000; Rundel and Boonpragob 1995). In frequently burned areas, rainforests are restricted to wetter gullies or riverine areas. Historical evidence of the drought and fire history in the humid lowland rainforests of Papua New Guinea gathered by Johns (1989) demonstrates that a significant number of quite large fires have occurred over the last 100 years, including in places now occupied by dense rainforests. Fire records are often poor because the events are rarely documented, but dating of charcoal found in rainforest soils suggest they may have been much more common than is often suspected (Goldammer and Seibert 1989; Haberle et al. 2001; Hopkins et al. 1996; Whitmore 1984). Fire hazards increase in logged-over forests because of additional fuel in the form of logging debris (Siegert et al. 2001; Woods 1989). The hazard is also likely to increase as farmers move up logging roads and clear these areas for cropping.

Other natural disturbances that have an impact on forests include landslips, volcanic gas emissions and ash showers. These differ in frequency and scale but are especially common in seismically active areas such as Papua New Guinea and Indonesia. Finally, changing river alignments are common occurrences for many large, sediment-laden tropical rivers flowing across flatter lowland terrain. The significance of this in maintaining lowland forests in relatively early succesional stages can be appreciated by observing the extensive areas of braided streams, oxbows and shingle formations across the region. Similar changes have been observed in Amazonia (Salo et al. 1986). It goes without saying that some of these disturbances are very site specific (e.g. ash showers are only found around volcanoes). And, of course, places with overlapping hazards are more likely to have more frequently disturbed vegetation.

Any assessment about the rate at which forests are able to recover after such disturbances depends on how this is measured. Some ecosystem attributes such as tree heights may recover within decades, while others such as species richness may take centuries to return to their pre-disturbance condition (Chazdon 2003). Whether recovery does actually occur depends on the timing of subsequent disturbances. White (1975) suggested the frequency of naturally occurring disturbances in Papua

New Guinea is such that there are probably few old primary forests in the lowlands of that country. This view is supported by survey data showing that many lowland forests are dominated by long-lived secondary species and that truly 'primary' forest species able to regenerate and growth in shade are rare (White 1975; Johns 1989). The same may be true in other forests across the region (Brookfield 1997).

Human Uses of Forests

Hunting and Gathering

The hunting of wildlife for bushmeat and the gathering of plant foods is widely practiced in forests throughout the region (De Beer and McDermott 1989; Wollenberg and Ingles 1998). However, very few people living in rain forests are able to subsist by relying entirely on such activities despite the biological diversity present within these forests. The problem is that food resources are sparsely distributed and most plant species are present in low population densities rather than being clumped. In non-seasonal climates, these plants only flower and fruit irregularly, so they represent an unreliable food source. In seasonal climates, they are more likely to have a short-lived reproductive period, meaning that food supplies from these sources are episodic. Wildlife biomass may also be limited (Kikkawa and Dwyer 1992). These problems aside, the key difficulty is that while there might be fruit and nuts, there are relatively few naturally occurring energy-rich food resources such as carbohydrates present in most tropical forests.

This has led some to argue that it is simply not possible to live as hunter-gatherers in these forests and people who seem to do so are actually relying on trade with nearby shifting cultivators or sedentary farmers to obtain their carbohydrates (Bailey et al. 1989; Headland 1987). This suggestion prompted a rather vigorous debate among anthropologists, with Dwyer and Minnegal (1991) arguing there are some small groups who may be able to obtain sufficient carbohydrate from naturally occurring yams and from sago palms. Nonetheless, it seems few people living in rainforest areas rely entirely on hunting and gathering for their livelihoods. The situation is different in woodlands and savannas outside rainforests and Australian aborigines have successfully practiced hunting and gathering without resorting to agriculture. They do so by using fire to manipulate their environment and migrating to take advantage of seasonal changes in food resources (Bowman 2000).

But even if hunting and gathering largely provides a supplementary food source for many agriculturalists, it is, nonetheless, a very common activity. In addition to food, many people gather resources that are used domestically for medicines, fuel and building materials. Some of these are also traded and this trade in non-timber-forest-products (NTFPs) has a long history. Records describing trade between Laos, Cambodia, Vietnam and China go back over 2,000 years (Donovan 2003) while Dunn (1975) describes trade between Malaya, China and

Arabs as far back as the fifth century. The material traded includes animals and animal products as well as resins, nuts, rattans, specialty woods (e.g. camphor, sandalwood) and medicinal plants. Many of these still remain commercially attractive and trade in some has intensified as a consequence of new technologies (guns, motor boats) and improved road access (Dunn 1975). There is ample evidence that much of this hunting and gathering has been unsustainable and the populations of many species have shrunk or disappeared as a consequence (Donovan 2003; Sodhi and Brook 2006).

At a time when change is widespread, there is often a tendency to look back and regard traditional forest users as living in balance with their environments such that their rate of resource use matched the rate at which resources became available. Many traditional forest dwellers did, in fact, regulate hunting or gathering and often devised access or usage rules to decide who could use certain resources as well as when and how this could be done. These rules were not created for 'conservation' reasons. Instead, their real purpose was to benefit the group or clan rather than the wider community. Only rarely were traditional rules developed to conserve the *species* itself. If the ratio of land to people became unfavourable thereby increasing the risk of degradation, resources were redistributed by bending these rules or by forcing other people out of the area. Religious and magical rituals were also used to protect certain species or patches of forest. However, the beneficiaries of such practices were primarily the humans involved rather than the biota (because maintaining protection and the rituals leads to social or physical protection of the humans who established the protective regimes). This is not to argue that some traditional activities did not have environmental benefits. Rather, it points to the fact that any such benefits were secondary to the primary aims of these forest users. Bulmer (1982, p. 63) argues that:

> …traditional Papua New Guinea societies scored more points for adaption, innovation and development of new resources than they did for conservation. There is little evidence that Papua New Guineans were or are very different from the majority of humanity who have not been greatly concerned with the long-term conservation of their natural environment. What they were and still are concerned with, very directly and very profoundly, are the present and immediately foreseeable yields of their crops and catches and the amount of time, effort and care required to produce them. To these traditional primary concerns for yields, we must now add concerns for cash. Who can challenge the rationality of this view, given the current state of the world's economy, the rate of inflation, and the fluctuating prices of virtually every commodity.

Shifting Cultivation

Shifting agriculture or swidden agriculture has been commonly practiced across the region. It usually involves clearing a small patch of forest (often around 1 ha or less), burning the debris, planting a variety of food crops and then harvesting these over the next 1 or 2 years as they mature. In the tropical lowlands, up to 20 or 30 crop species can be used in a single garden; although one or two staple food crops usually predominate. Smaller numbers of plants might be used in upland regions.

The crops used by shifting cultivators vary across the region. Rice is the main crop in Asia, while root crops such as sweet potato (*Ipomoea batatas*), yams (*Dioscorea* spp.) or taro (*Colocasia esculenta*) are the main crops in the Pacific region. A large number of other plant species able to supply food, fiber and medicines are used in both areas. Rather than replanting these crops after harvesting, the site is abandoned and the gardeners move elsewhere to repeat the process. The abandoned site is quickly recolonised by forest trees which exclude weeds and restore soil fertility. Gardeners clearing the forest may deliberately leave fruit or nut trees or any larger trees that are difficult to fell. The presence of these, together with coppice from old stumps, helps accelerate the subsequent successional processes during the fallow period. Normally these sites were usually left under fallow for 15–20 years before being used again. By this time a secondary forest was well established. Most farmers preferred to re-use secondary forests on former garden sites since the smaller trees were easier to fell than undisturbed forests having larger trees. These basic patterns have been described by Freeman (1955), Clarke (1971), Geddes (1976), Kunstadter et al. (1978), Rappaport (1968) and Cairns (2007).

Not surprisingly, there are a large number of variations on this basic pattern. For example, burning may not be done in areas with heavy rainfall and Schiefflin (1975) describes a site in Papua New Guinea with a rainfall of over 5,000 mm where crops are sown prior to felling the trees. Some of the crop plants were subsequently damaged during clearing, but most survived. This form of shifting cultivation was not used because of a lack of dry weather to burn but rather to prevent the degradation and erosion of exposed bare soils and to prevent nutrient leaching losses. And farmers may not always prefer to use secondary regrowth forests; Freeman (1955) describes a form of shifting cultivation practiced by the Iban of Sarawak where new gardens were created in intact primary forest rather than secondary forest. These farmers might be thought of as 'forest pioneers' rather than 'rotational' shifting cultivators. This version appears to reflect the fact that the Iban were moving into essentially 'virgin land' and saw the forest as an expendable resource. Rather than settling in one place, they were moving rapidly to claim as much land as possible and could move 80–160 km in a single generation and not return.

The shifting agriculture system has several considerable advantages. Provided the sequence is maintained and fallow periods do not fall below 10 years, the system is sustainable and the crop plants do not need fertilizers, weedicides or pesticides. Nutrients are conserved on the site and weeds are excluded by shade. There is also an inbuilt form of insurance: if one crop species fails there are others still available. It is also highly efficient in terms of labour costs with the energetic value of the food output exceeding the energetic input cost (i.e. labour) by perhaps 15 or 20:1 (Rappaport 1971).

As population densities have increased and fallow periods have shortened, changes have occurred in the way shifting cultivation has been practiced (Cairns 2007; Clarke and Thaman 1993). Some of these involve the use of introduced shrubs or herbs to improve soil productivity, while others involve the introduction of additional trees that are able to provide various timber and non-timber products in the fallow stage. One of these systems, described as dispersed tree fallows, uses scattered plantings of woody legumes such as *Leucaena leucocephala*, *Sesbania* spp., *Falcataria moluccana* (known previously by a large number of synonyms including *Paraserianthese falcataria and Albizia falcata*), or *Erythrina* spp. Other versions of this system incorporate non-leguminous nitrogen-fixing species such as *Alnus nepalensis* and *Casurina oligdon*. Another type of tree-based fallow involves the well-known taungya system where crops are grown with trees for a few years until the tree canopies close. Crops are replanted when the trees are felled and are grown with the trees until canopy closure occurs again. A large variety of native and exotic tree species are also used in such systems. Finally, some fallows have been enriched to form essentially permanent agroforestry plantings. These latter systems will be discussed in more detail in Chapter 5.

Despite being potentially sustainable, shifting cultivation can lead to landscape degradation. This can happen in several ways. The most common is when populations increase and land becomes limiting (or where customary lands are taken over by the state leaving farmers with less lands available for the fallow cycle). Under these circumstances, the length of the fallow period shortens and the capacity of trees to re-establish and rejuvenate the site declines. Recolonisation by forest species is also more difficult as the overall area under cultivation enlarges and the dispersal distance for seeds from undisturbed natural forests increases. There is a risk that weeds, especially grasses, will persist once the length of fallow decreases to below 10 years. This, together with a lack of soil nutrient restoration, then makes the site difficult to use (Fig. 2.1).

A second trigger for degradation following shifting cultivation occurs when the duration of cropping lengthens. This was the case with the form of shifting

Fig. 2.1 Shifting agriculture progressing onto on steep slopes in northern Thailand

cultivation practised by the Hmong (or Miao) people in the hill areas of northern Thailand (Geddes 1976). They grew food crops and opium poppies in successive crops on the same piece of land for as long as 10 years depending on the rate at which productivity declined. Because of this, the sites become so degraded that a much longer fallow, perhaps up to 50 years, is probably needed before forests can be re-established (Geddes 1976). This system reflected the high cash value of opium at the time and the fact that people did not stay in the one location but moved significant distances once a particular site was abandoned. In that sense, they did not have to live with the adverse consequences of their system.

The permanent establishment of grasses usually marks the end of shifting cultivation and results in a complete turnover and replacement of species. The ecosystem can be said to have crossed a threshold and reached a new state-condition. This conversion is usually most likely in strongly seasonal climates that are more subject to wildfires that prevent natural forest regrowth. It may be still possible to continue farming in these grasslands and complex systems of tillage, composting and mounding are used in the highlands of Papua New Guinea to improve fertility and to deal with frosts. In these cases, grass fallows of 4–17 years are used (Vasey 1981; Waddell 1972). However, grasslands in lowlands across the region are usually much less productive and require draught animals for cultivation (Potter 1997). Conroy (1960) argues that the conversion of forest to grassland is much less common when annual rainfalls exceed 2,500 mm because, under these circumstances, the dry seasons are usually short and the fire frequency is then reduced.

Some shifting cultivators' plant trees within the shifting cultivation cycle but most of these are fruit trees or species providing something of particular subsistence value. The topic of such agroforests will be discussed further in Chapter 5. Few shifting cultivators have planted trees to simply restore the forest. When Papua New Guineans living in the large grassland valleys of the country's highlands were asked about other forest tree species they replied that these had been planted by ancestors and it was not their task to try to replace them (Meggitt 1960).

Sedentary Agriculture

Sedentary agriculture was first practised in Asia in the alluvial floodplains and only later spread into the hill areas where shifting cultivation was still practiced. The cause of the transition from shifting cultivation to sedentary agriculture has been the subject of considerable interest. Boserup (1993) argues that the change is primarily driven by increasing population densities in situations when people cannot expand their territories or migrate. Under these circumstances, fallows must shorten as populations increase. This means labour must be used to restore soil fertility. The key metric of efficiency shifts from the food produced per person or man-day of effort to the amount of production per hectare.

There is considerable empirical evidence supporting this view that population density is one of the key drivers of change though it clearly not the only one (Geertz 1963; Rasul and Thapa 2003; Stone 2001). In recent years, the decline in shifting cultivation has accelerated across the region, especially in lowland areas, and the role of some of these other factors may have strengthened. For example, road networks, market access and the provision of agricultural advice have all prompted the growing of new cash crops. Likewise, deliberate government policies aimed at changing land tenure arrangements and reducing the areas under fallow have also played a role in fostering more sedentary forms of agriculture (e.g. Midgely et al. 2007; Roder et al. 1995).

The conversion of forest to agricultural land is often quite wasteful with more forest being cleared than is actually needed at the time. This can be witnessed today in the clearings being carried out to establish oil palm. Fires associated with these clearings have sometimes become wildfires that burned through large areas of Borneo in the latter years of the twentieth century. But the same was true in the mid-nineteenth century, well before the tropical deforestation crisis was widely recognised. For example, the naturalist Henry Forbes lamented:

> As in Java the original forest is rapidly disappearing; each year sees immense tracts felled for rice fields, more than is actually neccessary, and also much wanton destruction by wilful fires.
>
> (Forbes 1885, p. 132)

Like their counterparts who practice shifting cultivation, sedentary farmers should not be seen as simply traditionalists who are stuck in a technological rut (though see Box 2.1). Instead, most are better described as people who are constantly experimenting, learning and modifying the production technologies to adapt to the unfolding circumstances in which they live (Kennedy and Clarke 2004). But, at the same time, there has been a trend towards simplification with more crops being grown in monocultures and a tendency for the number of varieties of each species to be lost and replaced by a smaller number of high-yielding cultivars (Clarke and Thaman 1993). Sometimes this combination of intensification with simplification has led to problems caused by diseases or declining soil fertility (see, for example, Henley 2005; Matson et al. 1997; Nibbering 1999). A common response in such cases is to change crops. In Papua New Guinea, sweet potato (*Ipomoea batatas*) appears to have been introduced some 400 years ago and was able to increase agricultural productivity when yields of traditional crops like yams and taro were declining (Allen et al. 1995). In Asia, cassava (*Manihot esculenta*) is often used at sites where fertility has declined following cropping with other species.

Sedentary farming sometimes fails entirely. Farmers growing cash crops are subject to market fluctuations and those who are entirely dependent on a single dominant crop can find themselves exposed when markets change. For example, Geertz (1963) describes how a collapse in international coffee prices in the 1930s caused the abandonment of coffee growing over large areas of southern Sumatra (needless to say, there have been many coffee 'booms' since then as well). Failures can occur and sites may be abandoned for a variety of other reasons as well,

Box 2.1 Errors in rice planting dates lead to food shortages

Inappropriate technologies can lead to sub-optimal or declining levels of productivity, especially when population numbers are increasing. Veldkamp (1979) describes how a government official once visited a village in Sumatra where the rice crop had failed several times. He noticed that a large post had been erected in the village with the Islamic Calendar on it. This calendar is lunar and not astral so that a few days are always 'lost' each year. Because the rice had been sown for many years on a certain Islamic date, the planting date had gradually shifted to an agronomically unfavourable time. The official's problem was how to persuade people to change without giving offence. He managed to get out of the predicament by finding a few old men and asking how they had known when to sow in the less enlightened days before this most excellent calendar had been introduced. After some hesitation, they started to chant a song which went something like 'at sundown point to the star and when the bracelet falls it is the time'. He asked them to point out this star and when the ivory bracelet fell to their elbows measured the angle. Subsequent information from the Bandung Observatory confirmed that the star had reached the correct position at the most appropriate date for rice sowing. And so, by cautiously suggesting that *adat* or traditional law should be followed again, he earned the name Tuan Padi (Mr. Rice). The incident shows how top-down prescriptions and conformity, in this case induced by religion, can have unexpected consequences.

including the use of inappropriate sites, warfare, short-term climate change or the arrival of new diseases. Some detrimental activities may initially produce an increase in farm income before increasing levels of degradation eventually leads to a productivity decline. Some landholders may consciously adopt this practice, secure in the knowledge they can move elsewhere when this becomes necessary. But others may have no fall-back position and use inappropriate practices because they have no other choice. In such cases degradation eventually leads to a decline in their standard of living. Evidence of abandoned former agricultural sites can be seen across the region. In the Pacific, this can sometimes be seen where old irrigation channels that once sustained extensive taro gardens are now enveloped in mature forest containing large trees (Hviding and Bayliss-Smith 2000). In other places, former intensively managed agricultural areas have reverted to grasslands.

A particular form of sedentary agriculture that forms a bridge between shifting agriculture and sedentary agriculture is agroforestry. A very large variety of agroforestry systems have been developed across the region. Some of these are simple home gardens involving small patches of trees intercropped with other species, but others are very extensive and species-rich agroforests that have been developed over long periods of time and which sometimes cover very large areas (Clarke and Thaman 1993; Kennedy and Clarke 2004; Michon 2005). These will be discussed further in Chapter 5.

Logging

Traditional forest dwellers made use of certain timber trees and there are records in Laos from as early as AD 200 of timber species such as eaglewood (*Aquilaria crassna*), ebony (*Diospyros* sp.), sappanwood (*Caesalpinia sappan*) and sandalwood (*Santalum* sp.) being traded with China, Cambodia and Vietnam (Donovan 2003). Much of this early log harvesting was opportunistic and unsustainable although it is likely that many of the people doing it had a good deal of knowledge about the basic biology of these species. Nonetheless it was probably only in the late 1800s and early twentieth century that sufficient was known about the silviculture of most tropical forest species to allow the development of the first management prescriptions.

If logging is to be commercially sustainable, managers must have a good understanding of the tree density and timber volumes of the preferred species, where these trees are located and how fast they are growing. In species-rich tropical forests, this information may take considerable time to assemble. Not all species are equally valuable in the timber market and the proportion of commercial species in a forest can vary with location. Statistically valid surveys are difficult to carry out in landscapes where roads may be sparse and the terrain is challenging. Measures of growth rates are especially difficult. Canopy trees exposed to light usually grow faster than trees in sub-canopy positions. But most trees grow faster after logging if the canopy is opened up around them to let in more light. Actual growth rates can only be assessed from successive measurements on the same tree. The usual way in which growth data is collected is to establish permanent plots and periodically remeasure these (Vanclay 1994). This is more easily said than done. E. Nir (personal communication 2003) described how Papua New Guinea foresters were able to ride motorbikes along former logging roads to establish permanent plots in recently logged forest. It is expensive to maintain these types of roads and, 5 years later when it was time to remeasure the plots, the roads were overgrown and they had to walk in. It took them a week to reach and then remeasure just one plot. Because of the difficulty of getting these basic volume and growth data, most logging across the region has been opportunistic and done without knowing the capacity of forests to provide a future timber yield.

Irrespective of future timber yields, logging should be done in such a way that regeneration is promoted. Several silvicultural systems have been developed that facilitate this objective although there a variety of variations and refinements to each of these (Baur 1964; Lamprecht 1993; Whitmore 1984). One involves harvesting all the trees in a stand at the same time to allow seedlings on the forest floor to grow in the improved light conditions present after logging. Such systems are referred to as Monocyclic or Uniform systems because there is a single logging cycle during the lifetime of the trees. A second harvest is possible after these seedlings grow up and reach a merchantable size, possibly after 80 years.

Another system, known as a Polycyclic system, depends on the fact that many forests contain a range of tree sizes including young saplings growing in sub-canopy positions and older trees that dominate the canopy. This system sets a cutting size limit and only takes commercially attractive trees with a diameter greater than this limit leaving behind the smaller (and mostly younger) trees to grow through and be harvested in the next cutting cycle. Growth of the remaining trees is usually enhanced after logging because harvesting opens up the canopy and reduces between-tree competition. A second harvest takes place once sufficient trees have grown in size and exceed the cutting threshold. This cutting cycle may take 30 years. This means there can be several logging episodes or cycles in the lifetime of a tree.

These two systems are not interchangeable but depend on the silvicultural characteristics of the trees in a forest. There are also a number of conditions that must be met if either of these systems is to work (Table 2.1). A variety of refinements have been developed to cater for situations where this is not possible (Baur 1964; Burgess 1991; Lamprecht 1993; Whitmore 1984).

Perhaps it is not surprising, given earlier comments about the difficulties in getting basic tree growth data, that there is often a good deal of difference between what should be done and what is actually practised. Logging is usually carried out by companies who hold a government concession to carry out logging in a particular area of land. The design of logging operations is rarely carried out by government or company foresters and is more commonly done at the whim of bulldozer operators who decide which path to take to reach a log. Damage to seedling pools and residual saplings and trees is common, while extensive soil disturbance and stream sedimentation is widespread. Further damage to vulnerable regeneration is caused if a second logging operation is carried out again before the forest has had a chance to recover. These are not problems that need further silvicultural research since there is ample evidence that such damage can be readily avoided using known prescriptions (Forshed et al. 2006; Sist et al. 2003). Instead, they require supervision to ensure these prescriptions are followed (Fig. 2.2).

The upshot is that the International Tropical Timbers Organisation found that only a very small proportion of the world's tropical forests were being managed sustainable (ITTO 2006). In the Asia-Pacific region, only 11.6% of the permanent forest estate in ITTO member countries was considered to be sustainably managed.

Table 2.1 Preconditions for using monocyclic or polycyclic silvicultural systems

System	Necessary pre-conditions	Failure if:
Monocyclic	An evenly distributed dormant seedling pool of commercially preferred species is present on the forest floor at the time of logging.	Seedling density is too low. Seedlings are destroyed by logging operation. Seedlings are swamped by weed growth.
Poly cyclic	Appropriate diameter limit set. An adequate number of trees of the commercially preferred species remain undamaged after logging. These residual trees are able to grow quickly.	Diameter limit is set too low. Too many residual trees are damaged by logging. Successive logging operations occur too frequent to allow regrowth. Tree growth is too slow leading to very long felling cycles.

Fig. 2.2 Poor forest regrowth in Sabah after intensive logging and fire (Photo: Robert Ong)

Table 2.2 Ways in which various forest uses can lead to degradation

Activity	Degradation and land abandonment likely if:
Hunting and gathering	Harvesting rates are too high driven by dense human populations and/or strong market demand for NTFPs.
Shifting cultivation	Fallow period shortens to less than 10 years and fertility declines, grasses encroach and recurrent fires occur.
Sedentary agriculture	Soil erosion leads to soil fertility declines, weeds encroach or if market prices decline abruptly.
Logging	If too many residual trees are damaged during felling (polycyclic system) or if seedlings of commercially preferred species fail to regenerate (monocyclic system). Degradation also occurs if successive logging occurs before recovery is complete (i.e. within <30 years).

Over most forests logging is of the cut-out-and-get-out variety, with the consequence being that any future harvests will now be far into the future. That is, the commercial value of many forests (not to mention their biodiversity values) has been unnecessarily reduced after the first logging cycle. Given time, many of these forests may eventually recover, but the roads left by logging can sometimes open up areas for agriculture and lead to the removal of forests on all but the steepest terrain.

In summary, all of these activities are potentially sustainable but most can also cause forest and land degradation (Table 2.2). Lands are degraded if shifting cultivation fallow period drops below 10 years or, in sedentary agriculture, when farming is extended on to marginal sites. Forests are unlikely to be significantly damaged by low-level harvesting, but many are now being over-exploited because of the unregulated use of intensive mechanised logging techniques which leaves them in a degraded condition.

Environmental Determinants of Deforestation

The likelihood that a particular site will be deforested or degraded depends on the types of land use activities being practiced. But this prompts several questions: are some sites more prone to deforestation than others? And, do certain environmental conditions predispose some sites to more deforestation than others? It is possible to explore these questions by examining the extent of deforestation occurring in islands scattered across the Pacific following their settlement by humans. Some islands were settled by migrants from the west more than 3,000 years ago, while others were colonised as recently as 1,000 years ago. It is a common observation that, across the Pacific, significant deforestation, erosion and biodiversity loss appears to have followed human settlement (Anderson 2002; Kennett et al. 2006). This means the colonisation of the Pacific can be seen as a gigantic natural experiment in which to explore the circumstances under which changes occur.

The scale of forest loss on many Pacific islands seems to have been disproportionately high compared with the agricultural demands of settlers. This may have been due to fires used to clear agricultural lands escaping and burning more forest than was intended, especially in some of the drier areas. Forests can recover after clearing but not if repeatedly burned and fires appear to have transformed many forests into degraded savannas and fern-grasslands (Clarke and Thaman 1993). At the time Europeans arrived, some islands still had large areas of forests (e.g. Samoa, Bismark Archipelago), but others had been completely deforested (e.g. Easter, Necker and Nihao islands). Forest loss has led to severe erosion on many islands and this has sometimes been followed by declines in human populations (Clarke and Thaman 1993). In extreme cases, islands appear to have been abandoned some years after settlement largely because of the degradation that occurred.

Some of these differences may have been caused by cultural differences between the various colonists. But might some have been driven by environmental factors? This was investigated by Rolett and Diamond (2004) who studied pre-European conditions at 81 sites on 69 Pacific islands. They explored the relationship between their estimate of the amount of deforestation that had occurred and a variety of environmental and geographic factors. In addition, they also examined the relationship between the amounts of reforestation (including replanting with exotic tree species) that had occurred and these same environmental variables. These variables included: (i) rainfall and temperature (as indicated by latitude), because these are primary determinants of plant growth; (ii) island age and volcanic ash fallout, because these are indicative of soil fertility (younger islands have less heavily weathered soils while regrowth is likely to be more rapid on fertile soils although these are also more favoured by farmers) and (iii) elevation, area and isolation, because these could have multiple effects.

Table 2.3 Environmental factors acting as significant predictors of deforestation or reforestation on Pacific islands (Rolett and Diamond 2004)

Factors increasing the likelihood of deforestation	Factors decreasing the likelihood of reforestation
Lower rainfall	
Higher latitude	Higher latitude
Older islands	
Distant from zone of aerial tephra	Distant from zone of aerial tephra
Low islands	Low islands
Small islands	Small islands
More isolated islands	More isolated islands

The results are shown in Table 2.3 and are much as might be expected. Deforestation was found to be more likely where growing conditions are less favourable and where soil fertility is likely to be poorer (i.e. more deforestation occurred on drier and cooler islands with less fertile soils). By contrast, islands with higher elevation were likely to have more forest because of orographic rains and because steeper terrain is less attractive to farmers. Deforestation was also more likely on smaller islands because these probably have a lower diversity of habitats and relatively fewer coastal resources. More isolated islands are likely to be more deforested because the human populations have fewer opportunities to obtain alternate resources by trading or to escape by emigrating. The extent of reforestation was affected by many of the same variables. Overall, the analysis suggested permanent deforestation was more likely where environmental stresses were greater.

While these findings make sense and are intuitively satisfying, they are not necessarily the most important causes of deforestation. Societies differ in their organisation, political institutions and forms of governance, as well as in their attitudes to environmental conservation. Over time, such difference can have profound effects on the rates of deforestation (and reforestation).

The Socio-Economic Context – a Short History of Deforestation in China and Japan

China and Japan are outside the main geographic scope of this book but the unique written records describing their long environmental histories provide some insights into how societal factors can influence deforestation and degradation.

China

When Europeans visited China in the nineteenth century, they found it bare and degraded with most of the natural forests destroyed. Because of this they described the Chinese as being 'destructive' or 'ruthless' and having an innate hatred of trees. In fact, the situation was far more complex than this and there have been periods of forest protection as well as periods of forest destruction (Menzies 1996). China has a long written history concerning forests and the effects of deforestation. One of the first texts on silviculture appeared in the Han period around AD 200 entitled 'On planting trees, storing fruit and caring for silkworms' (*Chung Shu Tsang Kuo Hsiang Tshan*) (Menzies 1996). Others followed. In AD 530, the agricultural text *Chhi Min Yao Shu* appeared with descriptions of techniques for breaking seed dormancy, striking cuttings and transplanting established trees from an early age. There was also knowledge about the effects of tree density on tree size and growth (Menzies 1996). Subsequent texts from the medieval period describe the silviculture of particular tree species and complex silvicultural systems where trees were inter-planted with food or perennials (Menzies 1996). These various books clearly reflect a significant and widespread knowledge about forests and tree growing that pre-date all comparable European texts.

There are few records from these early periods discussing the extent of deforestation or land degradation, although official concerns about diminishing natural resources appear from as early as the fifth century BC (Elvin 2004). Forests were nominally under the control of the Emperor and commoners were officially only allowed access under prescribed conditions. However, centralised control was difficult. In AD 500:

> The prefect of Yangzhou … reported to the Emperor 'Though the prohibitions regarding the mountains and lakes have been established since times past, the common people have become accustomed to ignoring them, each one of them following in this the example of others. They completely burn off the vegetation on the mountains, build dams across rivers, and act so as to keep all the advantages for their families… It should be reaffirmed that the old laws that defined what was beneficial and what harmful are still in Force.'
>
> (Elvin 2004, p. 55)

The Emperor demurred, saying that the prohibitions were rigorous but severe and so they should be eased so as to be in keeping with the spirit of the times. Besides, if the lands were taken back it would provoke anger and resentment. He went on to prescribe maximum land holdings for people. But attitudes and policies changed over time and other emperors had different views. Thus the Emperor Hsuan Tsung AD 800 was more 'conservation minded' and sought to maintain temple and mausoleum gardens and protect forests for watershed protection reasons (Schafer 1962). In an edict issued to protect the slopes of a mountain near his capital city from fuelwood cutters he declared:

> …from now and thereafter let the gathering of fuel be taboo there!
> Consider that a sealed precinct!
> Consider our will in this!
>
> (Schafer 1962, p. 295)

There are a number of early records describing attempts to reforest cleared lands. One of the earliest accounts describes the officially sponsored plantations established along Great Wall in 221 BC to hinder military invasions (Menzies 1996). Other early records describe how degraded shrublands were reforested using commercially attractive species.

> …the old country town was in the mountains. In the twentieth year of Khai Yuan (732AD) it was moved out of the mountains. Dense thickets and brush grew where the former town had been, where animals and poisonous snakes had their lairs. This had been troubling the

> townspeople for a long time so in the eight year of Yuan Ho (813AD) the District Magistrate Han Chen ordered that the grass and trees be burned over and the area planted to pine and cunninghamia
>
> (Menzies 1996, p. 577).

In other locations, reforestation was achieved through natural regeneration and this was done after the Ming capital moved to Peking in the early fifteenth century. In this case, a ban was imposed on recutting local forests and this was maintained long enough for reforestation to take place and overcome damage caused by the previous dynasty (Menzies 1996).

For most Chinese the medieval forests must have seemed inexhaustible. On the other hand, there were references as early as in the eleventh century to deforestation and local wood shortages caused by land clearing for farmland and timber harvesting for fuel and for building materials (Elvin 2004). Some reforestation was carried out on a small scale to supply local needs and there were state incentives to plant fruit trees or mulberry trees for silkworms but there were none for large-scale reforestation, such as, for example, of watersheds.

Forest losses accelerated in the seventeenth century largely as a result of population pressure. The population in AD 1000 was around 100 million but reached 200 million by the eighteenth century and 400 million by 1850. By the late nineteenth century most of the temperate forests had disappeared. Murphey (1983, p. 116) quotes:

> All boys in the village big enough to walk and carry a basket are sent out over the hillsides to gather grass, twigs and any kind of herbage that can be used as fodder or fuel. Each boy carries an iron grubbing hook, and thus equipped he clambers up the slopes working away at his task with cheerful energy. Through the industry of this army of human locusts the mountains are denuded of herbage and even roots are often grubbed up.

Despite activities such as these small pockets of forest persisted. Some of these were protected by geomancy while others were in hillier or more remote regions. These remnants were not enough to protect larger animals such as elephants which had disappeared from most of China by the early 1800s (Elvin 2004), but many wildlife species did persist. For example, Wilson (1986) describes an 'extra-ordinary wealth of species exists notwithstanding the fact that every available bit of land is under cultivation' in Hupeh and Szechuan provinces of western China in 1899. He also commented on the good 'sport' (i.e. hunting) still to be had in this area. Forests remained in the tropical south for a little longer than the remainder of the country. This may have been because the area was viewed as being a terrible place and a 'benighted land of exile where pestilential vapours and malaria constantly threatened the health of immigrants' (Menzies 1996). It was only after the beginning of the nineteenth century that substantial deforestation appears to have occurred in the south but, once started, it was soon completed.

Throughout China, the deforestation that occurred was primarily carried out to create food-producing agricultural land. Cash cropping was of minor importance (except for tea) and, by the nineteenth century, most of the timber taken from the forests then remaining was used for fuel or local building purposes rather than for sale or export. Once forests had been cleared peasants often burned the hillsides since the lands could not be used and the resulting wood ash might wash down the slopes onto fields and act as fertilizers. The fires may have also discouraged bandits and predatory wild animals (Murphey 1983). Although this seems wasteful, Murphey (1983) argues that the average farmer could not survive on the low yields likely from farming steeper slopes and that tree crops or grazing animals could only be marginal side-lines to which most farmers could give little time. Most non-agricultural land was eventually privatised, commonly by the rich and powerful (Elvin 2001). Some communities did have communally-owned woodlots, but most were small and these were far from universal. Many such woodlots were destroyed by marauding armies in the final years of the nineteenth century and in the early years of the twentieth century.

The picture that emerges from this 2000-year history is of the seemingly inevitable deforestation of China. It occurred despite a rich silvicultural knowledge base, an understanding of the functional and protective role of forests and a strong political apparatus that had built a unified state. There was also a philosophical tradition that emphasised a reverential attitude to nature and which stressed man's lesser status (Murphey 1983). This tradition required men to adjust to and respect nature and to not despoil it; nature was seen as a source of virtue, wisdom and internal peace and to be contemplated rather than dominated.

Given all these circumstances, why was deforestation so complete? Apart from the relentless demographic pressures, there appear to have been at least two reasons. One concerns the limited ability of the successive governments to use existing knowledge and impose their policies.

> All of the psychological conditions to produce sound policy for the protection of nature were present in T'ang times. But though enlightened monarchs issued edicts, conformable to the best morality of their times these were ignored by their successors. In short, there was no embodiment of these advanced ideas in constitutional form. And so they were ultimately ineffective.
>
> (Schafer 1962, p. 30)

The second was that the philosophical views about nature were limited to a very small proportion of the population who had either the means or status enabling them to be free of manual labour. Most peasants had neither and had to use nature and its resources for their own purposes in a continuous struggle to feed themselves and survive. The ever-increasing population of peasant farmers needed new lands, fuel and building materials and forests were the natural source of these. Over several thousand years, they were able to continue to find new forest lands to expand into and clear until, suddenly, there were no more.

> If we had to create a swift characterization of this Chinese style in the last centuries of the empire it would be in terms of a dynamic but relatively poor society that was constantly driven by population expansion to attempt to master nature in new environments, and which often achieved this in a skilful manner, marked by a patient tenacity but which in the long run more often than not damaged or even destroyed these environments. And yet, overall, a larger and larger population was supported. This can be seen, according to ones perspective, as either a disaster or a triumph.
>
> (Elvin 2001, p. 29)

There are some obvious parallels here between this history and the present-day situation in parts of the Asia-Pacific region; not only is there still an inability to use existing knowledge to manage forest resources but there is also a disconnect between the views of small farmers attempting to wrest a living from their own small patch of land and those whose circumstances mean they can afford to be concerned about global conservation issues.

Japan

The environmental history of Japan is quite different. Like China, Japan went through an extended period of forest exploitation with little concern about the extent of harvesting, or of the capacity of the residual forests to regenerate. But, unlike China, feedback from the problems induced by deforestation were eventually recognised and acted upon (Totman 1989). The geography of Japan imposes substantial limits on agriculture. There is a comparatively small area of alluvial soils and a large area of earthquake-prone mountain areas. The boundaries between the two are more clearly defined than in China and when agriculture was pushed beyond its limits, the result was expressed as severe ecological damage and human hardship.

In the twelfth century, Japan entered a period of unstable decentralised rule after a period of stability under an imperial court. This period of civil war lasted for several hundred years. But, during this time, the population grew substantially (nearly doubling from 7 million in 1200 to 12 million in 1600) and increased use was made of the forests outside the agricultural areas. They were used to collect fuel, fodder and as green manure to fertilise paddy fields. They were also exploited by the local elite for buildings such as temples and fortresses.

In late sixteenth century, a military dictatorship re-imposed order on the country. Overall control was in the hands of the Shogun, but there was a hierarchical but decentralized form of administration with local power being in the hands of daimyo or feudal lords (Totman 1983). Forests were controlled by the Shogun as well as the daimyo, but much of the forest land around villages was assigned communally to villagers. Peace resulted in a massive surge in the rate of forest exploitation by villagers and the noble elite. Towns, temples and fortress buildings were all constructed of timber. Many of these buildings were periodically burned, giving rise to a cycle of construction, loss and replacement, which accelerated the demand for timber even further. As forests were degraded and cleared there was widespread erosion, flooding and damage to downstream cropland. Disputes over rights to particular forests increased.

Eventually there was a turnabout or what Totman (1989) refers to as a 'negative regimen'. This took place in the period between 1630 and 1720. Constraints were imposed on access to forests and the tools that could be used to fell trees. There were also limits on wheeled vehicles that could be used to transport logs and timber. The intent was to take the pressure off forests and allow natural regeneration. But even these changes were not enough and, eventually, a period of tree planting began. Reforestation became increasingly widespread from the late eighteen century, especially in market-oriented areas near towns. This involved sophisticated and sometimes novel silvicultural techniques. Reforestation had both a protective and a production objective, although the type of planting carried out depended on who was doing it. The daimyo or lords tended to favour conifers able to produce large logs suitable for building purposes such as *Cryptomeria* and *Pinus*. Villagers tended to plant hardwoods such as *Quercus* for fuelwood or species such as chestnuts (*Castanea*) for emergency food. The changes were not cost-free and they tended to discriminate against lower status people by denying them access to lands they would have otherwise used.

The interesting question, of course, is why did these changes occur? Totman (1989) argues that a number of factors appear to have been involved. Firstly, the changes were a practical response to a problem from which they could not escape. Since Japan is an archipelago of islands, people could not move elsewhere, which is the usual response by people who have destroyed their resources. In addition, Japan adopted the policy of minimizing external contacts and so precluded the receipt of ideas from abroad. People took it for granted that they would have to solve their own problems.

Secondly, the changes were encouraged by a number of institutional arrangements that defined rights and regulated forest usage by both villagers and lords. These included the fact that households were recognised as the building blocks of society and the social status and location of households was hereditary. It was assumed that one's heirs would inherit the household estate and that the results of one's labours would benefit these heirs. In addition, the control of forests was placed in the hands of those with a vested interest and with the resources to pursue long-term forest regeneration. Parts of forests around villages were allocated to households to manage with strict conditions on their use. Forests in more distant locations were managed by imposing systems to prevent over-cutting and facilitate regeneration. There were also various forms of forest leasing and joint ventures between households and daimyo to encourage sustained management. All of these arrangements gave the community a reason to exercise responsible stewardship and were crucial to the natural and artificial reforestation that subsequently occurred (Totman 1983, 1989). As Williams (2006, p. 310) observed about the Japanese experience:

> …it was a unique social and environmental situation in which a disciplined and literate society sorted out its priorities, people of all classes knew that resources were limited and they simply had to make do with what they had

These necessarily short accounts show the Japanese experience stands in stark contrast to that of China. Over the long history of China, there was much inconsistency between successive dynasties and constraints on forest usage in one period could be replaced by policies offering incentives for land clearance and resettlement in another. In Japan, forest degradation was accompanied by recognition of the relationship between environmental problems and the uncontrolled use of forests. This recognition was accompanied by a rising consciousness of the need for conservation. A stable system of government and supportive institutional settings enabled forest regeneration to take place. This difference between the two countries in institutional settings, legal consistency and in the opportunities for those living in degraded lands to move elsewhere, appear to have contributed to their different environmental histories.

Deforestation and Degradation in the Asia-Pacific Region

Unlike China or Japan, most of the rest of Southeast Asia was comparatively untroubled by increasing population pressures until the late 1800s, with the exception of a few places like Java, the Khmer kingdom and some of the more fertile river deltas where substantial deforestation also occurred. As noted earlier, NTFPs were the primary resource harvested from forests and timber harvesting was generally restricted to a few key species (e.g. teak). This began to change during the nineteenth century as logging increased and states asserted ownership over forest resources. By 1880, many states had begun to establish Forestry Departments to regulate logging operations in order to obtain a greater share of the timber revenues. These forestry agencies also sought to define logging concession areas and the rates at which logging would be permitted.

But the need for more agricultural land also increased. Between 1880 and 1980, the population in Southeast Asia rose from 57 million to 356 million, a sixfold increase. The pattern of decline in overall forest cover and the matching rise in cultivated lands are shown in Table 2.4 which is based on a comprehensive study by Richards and Flint (1994). Statistics on forest cover are notoriously difficult to assemble and some of their data are imperfect or incomplete (a fact acknowledged by these authors). However, their study probably represents the best broad overview of changes in forest cover during an important 100 year period in Southeast Asian history.

Table 2.4 shows that total forest cover declined from 365.7 million hectares in 1880 to 274.2 million hectares in 1980, a loss of 91.5 million hectares of forest or 25% of the total forest cover (including forests/woodlands as well forested wetlands and 'interrupted woodlands' which they defined as forested lands with less than 40% canopy cover). The loss of largely intact forests (i.e. excluding interrupted woodlands and wetlands) was 83 million hectares or 32% of the area of this type present in 1880. The annual rate of loss increased sharply after 1950 rising from 0.81 million hectares each year to 1.3 million hectares per year. Much of this change was due to an increasing area of agricultural land. This rose from 16 million hectares to 78 million hectares over the same period. Most of this was used for annual crops such as rice but 17.9 million of the total (29%) was in permanent crops such as rubber plantations.

Table 2.4 Changes in cover (million hectares) of forest, cultivated lands and degraded lands[a] in Southeast Asia between 1880 and 1980 (Richards and Flint 1994)

	1880	1920	1950	1980
Forest				
Total[b]	365.7	337.5	313.3	274.2
Intact[c]	255.3	236.0	210.1	171.9
Annual loss		0.705	0.806	1.303
Non-forest				
Cultivated	16.1	34.0	45.7	78.0
Grasses	57.1	67.5	79.4	84.0
Population density[d]	0.13	0.23	0.39	0.79

[a] Includes grasslands, shrublands and barrens
[b] Includes forest/woodlands, interrupted woodlands and forested wetlands
[c] Forest/woodlands only
[d] (Persons/ha)

The Rise in Abandoned Former Agricultural Lands

An important finding from Richards and Flint's analysis was that only 68% of the cleared forest land was used for agriculture and, over time, the remaining 32% was presumably added to the categories they classify as grasslands, shrublands and barren grounds. Some of these lands may have been used for purposes such as grazing, but some many constitute 'degraded lands' and have suffered topsoil erosion, weed invasion and be subject to recurrent fires. These areas increased by nearly 27 million hectares over the period, to reach 84 million hectares by 1980. This degraded land represented 19% of the total land area and was equivalent to nearly half the area of remaining intact forest.

This pattern has continued and Houghton (2001) suggested that between 1980 and 1985, 59% of all cleared land is subsequently abandoned and becomes degraded land (although this estimate was for tropical Asia as a whole and not just Southeast Asia). Other estimates of the areas of shrub, brush, pasture, waste and 'other' land categories (some of which may be secondary vegetation) now present in Southeast Asia were described in Chapter 1. Fox and Vogler (2001) estimate these degraded areas range from 26% to 49% of all cleared lands. Such assessments are sensitive to differences in the definitions of forest cover and measurement methodologies used in various countries Nonetheless, they point to the creation of very large areas of under-utilised and 'degraded' land in a relatively short period. Many of these areas continue to be used for grazing or gathering thatching materials, or other purposes (Potter 1996). Nonetheless, they represent a change to a category of agricultural land with undeniably lower levels of productivity.

Populations and Deforestation

These changes occurred as population grew and more land was needed to raise food. However, the nature of the relationship between population and deforestation has been controversial (Carr et al. 2005; Geist and Lambin 2002; Kummer 1992; Kummer and Turner 1994; Mather and Needle 2000; Uusivuori et al. 2002). On the one hand, many have seen population growth in Malthusian terms so that it is self evident that larger numbers of people need more land on which to grow food. And over the long sweep of human history, it is clear that rising populations have been accompanied by decreases in forest cover. This pattern is exemplified by the history of land use in China. But not all deforestation is carried out for subsistence agriculture and the simple statement that population and forest cover are linked is not particularly useful in furthering an understanding of the processes involved when deforestation occurs, particularly at a regional or local scale. Sometimes, the present forest cover reflects changes that took place many years earlier and are not at all indicative of the present relationship.

Changes can also occur over time in the nature of the relationship so that forest cover increases at the same time as populations rise. This happened in Japan and Gilmour and Fisher (1991) and Lindblade et al. (1998) give some more recent examples from Nepal and Kenya respectively. Kummer (1992) also points out that there is considerable ambiguity in the terms 'population' and 'deforestation'. What measure of population is most relevant – the total population, the population density, density per unit of arable land, percent increase in population, absolute increase in population or amount of in-migration? Likewise, is it the present forest cover, the deforestation rate or the increased area of arable land that is most relevant?

In the majority of cases, the most useful terms are probably population density and deforestation (i.e. the rate of net forest loss). A high population density is likely to cause more deforestation than a low density but an area with a low population density (and low overall population) could still have rapid deforestation if a steady stream of immigrants kept arriving to match the forward movement of the deforestation zone. Although many early studies were based on pooled data sets involving many countries, most now recognise that it is usually more profitable to explore these relationships as changes over time within a particular national or local context.

Richards and Flint (1994) did such a study using the data set referred to above and this is shown in Fig. 2.3 Their data has been supplemented using recent updates of 1980 cover estimates for Philippines, Thailand and Vietnam and post-1980 forest cover estimates for all countries (FAO 2007). The results show that, across the region, the area of forest land has decreased over time with increasing population density.

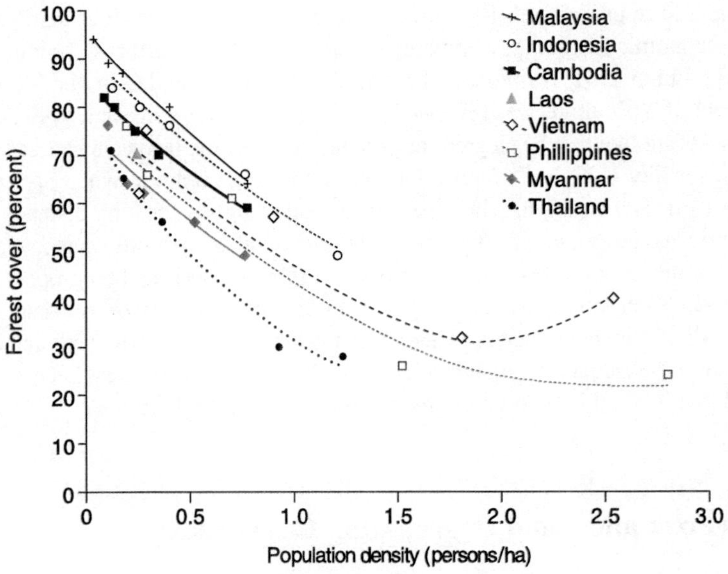

Fig. 2.3 Population density (persons/ha) and estimates of forest cover (including plantations) based on Richards and Flint (1994) and supplemented by 2005 data from FAO (2007). Updated information has been used to change the original 1980 estimates for Philippines (Kummer and Turner 1994), Thailand (Hurst 1990, Poffenberger 1990, Feeny 1988) and Vietnam (de Jong et al. 2006)

The general trends are similar but there are clear between-country differences in the relationship. For example, at a population density of around one person per hectare the forest cover in Thailand is around 30% but nearly 60% in Indonesia. These differences suggest the rate of deforestation is affected by factors other than just population density. Such factors might include the productivity of the farming systems (a function of soils, climate, technology etc.), the proportion of the national landscape suitable for agriculture, land tenure systems (landless people might be more inclined to migrate to find new land) or government policies affecting the extent to which cash crops or plantation estates are established. The Vietnam data also shows that at some point forest cover can begin to increase once more, despite the fact that population density continues to increase (Vietnam probably reached its lowest forest cover of perhaps 26% in the late 1980s when the population density was around two persons per hectare although the evidence is anecdotal). The process by which net deforestation changes to an increasing overall forest cover will be discussed further below.

Causes of Deforestation

A number of studies have explored the cause of deforestation and concluded there are usually a variety of factors responsible. Most studies have distinguished between (i) direct or proximate drivers such as agricultural expansion, logging, infrastructure development, fire or war and (ii) indirect or underlying drivers such as demographic changes, economic factors, government policies and cultural factors (Chomitz 2007; Geist and Lambin 2002; Kummer and Turner 1994; Nguyen and Gilmour 2000). The critical role of agriculture was reviewed by Kaimowitz and Angelsen (1998). They found forests are likely to be cleared for agriculture when the lands are accessible and when soil fertility is high. Deforestation was also more common when agricultural and timber prices were high, when there were opportunities for long distance trade, when there was a shortage of off-farm employment and when rural wages were low. They also found many forests are cleared when they are seen as being open-access resources. In such situations, clearing is a means by which a person can obtain property. All of this makes perfect sense. What is less obvious is why so much of this cleared land is subsequently abandoned? And why has so much forest been so badly degraded that it is unlikely to be productive for the foreseeable future?

Seven Forest and Land Degradation Case Studies

Seven case studies have been gathered to explore how forest and land degradation has occurred in different parts of the Asia-Pacific region and to tease apart the factors responsible. Three of these case studies describe how forests supposedly managed to sustain a local timber industry and protect watersheds and biodiversity have been degraded or deforested (Sarawak, Philippines, Thailand). Two other case studies involve situations where governments were deliberately seeking to clear forests to promote agriculture (North Queensland, Indonesia) or national development (Papua New Guinea) and where the programs ultimately failed. The final case study describes the degradation of forests supposedly managed as a common property resource (Samoa).

Case Study 1 – Intensive logging, Sarawak, Malaysia

Sarawak, together with Sabah, joined with the states of Peninsular Malaya to form Malaysia in 1963. As part of the negotiations, it was agreed that the states would retain control of land and forest resources while the new federal government would have responsibility for oil and gas reserves. Since then, Sarawak's forestry policies have differed sharply from those of the other states of Peninsular Malaysian. It did not join the National Forestry Council (made up of relevant federal ministers and chief ministers of each state) which was established in 1971 to coordinate forest policies across the country. Instead, the state chose to be represented by observers. Neither did it adopt the Council's 1978 National Forest Policy that sought to establish a permanent forest estate across Malaysia, promote sound forestry management and foster a sustainable industry. Instead, it has pursued a pattern of intensive logging that has had little to do with sound management or sustainable practices (Dauvergne 2001).

The state was well-endowed with forests, although these were not as commercially attractive as those in Peninsular Malaysia or Sabah (Ross 2001). On the other hand, there was no strong pressure for agricultural clearing because the state's population density was comparatively low (in 1980 the population density in Sarawak was 10.6 persons per km^2 while Peninsular Malaysia had 86 persons per km^2). In any case, only 28% of the land area is suitable for agriculture (with half of this being regarded as 'marginal'). This suggests the state could have developed a permanent and sustainable forest industry without the agricultural pressures experienced by many other countries in the region. This was not to be. Large scale logging lagged behind that in Peninsular Malaysia but by the late 1970s it had become a major component of the state economy. In 1972, a team from FAO estimated sustainable production could be achieved with an annual allowable harvest of 4.4 million cubic meters per year. But, by the late 1970s, the logging rate was nearly double this level and in 1982 FAO advised the government to take 'urgent action' to avoid the possibility that sustained yield management and the development of forest industries might be compromised (Ross 2001).

As the scale and intensity of logging increased so too did protests from some indigenous forest-dwelling people and a number of NGOs. In response to this criticism, a mission from the International Tropical Timbers Organization (ITTO) was invited to assess logging operations in 1989. This mission concluded that the then annual logging rate of 18 million cubic metres was well beyond their estimate of the sustainable yield of around 4 million cubic metres per year. Note that this ITTO estimate was very similar to the previous FAO estimate provided 17 years earlier (Ross 2001). ITTO suggested the logging rate should be decreased and the government agreed to do so. However, 9 years later, it was still 13 million cubic metres.

The puzzle is why such destructive logging should be allowed in a state with a low population density and only limited agricultural opportunities? The answer is that the forest industry had been captured by a corrupt political elite who were unlikely to slow the pace of logging to simply avoid deforestation or degradation (Dauvergne 2001; Hurst 1990; Jomo et al. 2004; Ross 2001). Sadly, it was also clear that the state was not benefiting as much from the industry as it should have and the royalties or resource rents being charged to loggers were very low compared with other Malaysian states such as Sabah (Gillis 1988; Vincent 1997).

The long-term impact of these logging practices on Sarawak's forests is not known. If properly managed, forests like these are probably capable of generating a sustained yield of timber in perpetuity. And again, if properly managed, much of the biodiversity contained in these forests could be conserved. But there appears to be no information on the state of Sarawak's forests after logging or about their capacity to sustain a second cutting cycle. The condition of similar forests in the nearby state of Sabah gives some cause for concern. In that case a World Bank study found 20% were virtually deforested by poorly supervised logging and a further 50% were poorly or very poorly stocked with residual trees of commercial value leaving only 30% in reasonable condition (Jomo et al. 2004). It seems entirely likely that a similar degree of forest degradation has occurred in Sarawak as well. Some have claimed that shifting cultivation has been the primary cause of deforestation in Sarawak (Lau 1979) but others strongly contest this and argue that shifting cultivation this has not been nearly as damaging as claimed (Cramb 1993; Hurst 1990; Jomo et al. 2004). The weight of evidence suggests that Sarawak is simply another example of a boom-then-bust logging cycle where the forests were treated as a free-good.

The next stage is unclear. Since the 1990s there has been an acceleration of large scale oil palm plantations and agricultural development. Most of the early developments have occurred on land logged in the 1960s and 1970s. These have been excised from the Permanent Forest Estate and re-gazetted as agricultural land (Hansen 2005). Given that only about a quarter of the land in Sarawak is thought to be suitable for agriculture, there is presumably a limit to the extent to which such agricultural developments can be extended onto other heavily logged land. Provided migrants do not move along old logging roads into these areas, the forests may eventually recover although the composition and nature of these future secondary forests is difficult to predict.

Conclusion: forest degradation has occurred because of the failure to enforce regulations and because of corruption. The forests have been regarded as a free public good and, consequently, the silvicultural and environmental costs of poor logging practices have been ignored.

Case Study 2 – Unregulated logging, Philippines

By the end of the twentieth century, most of the primary forests in the Philippines had been destroyed and the Philippines had gone from being the world's largest producer of tropical timbers in 1975 to being a net timber importer in 1994. This was despite the fact it was one of the earliest countries to begin developing silvicultural management systems aimed at preventing this. Both the scale of deforestation and the rate at which it occurred exemplifies just how easily technical knowledge can be over-ridden by poor governance.

Most of the significant early deforestation in the Philippines was carried out to enable the development of commercial agricultural activities such as sugar (Roth 1983). Substantial fuelwood logging was also carried out in some locations to power the new sugar mills. Much of the newly cleared land remained permanently under agriculture except when clearing was carried out on steeper land. In 1863, a Bureau of Forestry was established to help manage forests and prevent inappropriate logging and clearing on steeper lands. An active group of professional foresters were recruited, especially after the turn of the century. Significantly increased logging began at this time and large amounts of timber began to be exported. Forests then covered around 70% of the country. Most of the early logs were species of Dipterocarpaceae and studies were commenced to develop ways of ensuring these regenerated so logging could be sustainable (Roth 1983). By the 1920s, the Philippines had become Asia's largest timber supplier, a position it would hold for the next 50 years.

The pace of logging increased after the 1950s. As Ross (2001) notes, the country was in an enviable position. Forests still covered over 50% of the country and managers had access to some of the most advanced harvesting practices in the world. The Bureau of Forestry also had a corps of well-trained staff to supervise the industry and manage the forests and was regarded as one of the least-corrupt of the government's institutions. But the situation changed as the windfall profits being gained by loggers soon attracted unscrupulous politicians and businessmen. Politicians were able to ensure logging concessions were awarded to friends or companies controlled by themselves. In return, loggers contributed to their political campaigns. The profitability of logging was so great that no attention was paid to the previously assessed annual allowable cut or to the notion of a sustainable harvest; it became simply a cut-out then get-out operation. In 1954 loggers removed 3.6 million cubic meters of timber (which was believed to be the maximum sustainable harvest) but by 1964 this had increased to 11.4 million cubic meters. Concerns were raised by Philippines foresters as well as World Bank and United Nations agencies but these were ignored. By 1987 forest cover had declined from 50% in the immediate post-war years to around 22% (Fig. 2.4) which must rate as one of the fastest episodes of deforestation in the tropics. By then there was no unlogged forest remaining except for small patches in inaccessible areas. In 1985 the Ministry of Natural Resources estimated that one third of the country was 'severely' eroded (Hurst 1990).

The financial benefits to the national economy of this whole episode were small. Kummer (1992) quotes one estimate that the state probably received only 12% of the revenues which had probably amounted to around US$1 billion. This means abnormal profits of around US$820 million was diverted to private interests. Most of these were friends of President Marcos who was in power between 1965 and 1986 (Dauvergne 2001).

A striking feature of Philippines society is in the way land is concentrated in the hands of a wealthy minority. In the late 1980s about 6% of landowners owned 50% of the land. This elite gradually acquired even more power and the proportion of

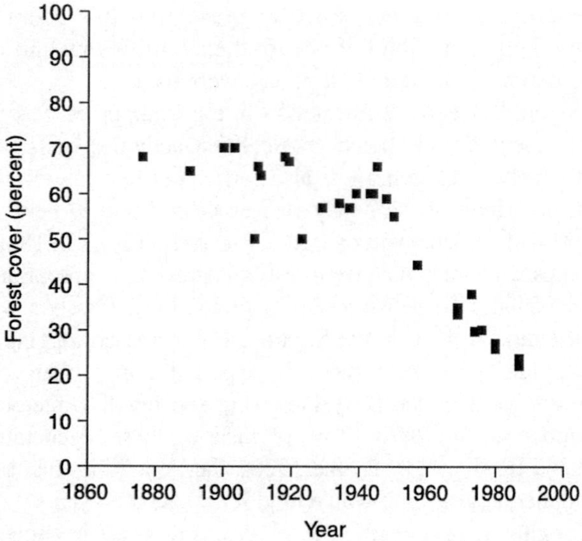

Fig. 2.4 Estimates of changes in forest cover over time in the Philippines based on data from Kummer (1992)

farmers having to rent from landlords rose from 37% in 1948 to 72% in the late 1980s (Leonen 1993). This situation led to large numbers of people moving out of lowland areas and into the uplands in search of farmland. Throughout this period the Philippines GNP grew but most indicators showed that people's income and standard of living actually declined. Large numbers of children remained malnourished and poverty worsened; by 1975 around 60% were living in poverty and this increased during the 1980s (Kummer 1992).

As noted earlier, logging does not necessarily lead to deforestation if it is managed correctly but this population movement meant few of the logged-over forests were able to regenerate. Kummer and Turner (1994) provide strong evidence that logging and agricultural deforestation went hand in hand in the Philippines. By the mid 1980s about 18 million people lived in the uplands and 77% of these were on land classed as public forest lands. This number had probably grown to 24 million by 1995. Large numbers of poor landless people used former logging roads to gain access to forest areas and quickly cleared the remaining vegetation to establish farms (Dawning et al. 1993). The highest rates of population growth were in areas with logging concessions. By the mid 1980s only 10% of land logged before 1955 was still occupied by forest (Kummer 1992; Kummer and Turner 1994).

At a superficial level it might seem that deforestation in the Philippines was simply caused by population pressure. However, Kummer (1992) argues that the primary cause of deforestation was not population pressure nor even misguided government policies. Instead it was caused by both wealth and poverty. A corrupt and 'predatory' economic and political elite deliberately manipulated government policies and regulations. These policies guaranteed a large land rent and those mismanaging the resource were not replaced or punished. Neither were illegal loggers stopped or penalised. On the contrary, there was large-scale collusion between government officials, military and loggers and a deliberate manipulation of statistical reporting system by agencies responsible for management to mislead the public and prevent the records being critically examined by outsiders. Farmers left landless by the inequitable distribution of land had no alternative but to use logged-over forest lands irrespective of its suitability for farming.

By 1987, a large area of degraded land described in forest survey statistics as 'other' had accumulated. This covered 10.9 million hectares (c.f. the 11.3 million hectares converted to farmland and the 6.7 million hectares remaining as forest). In other words, it represented 37% of all land. Most of this land was probably grasslands (Kummer and Turner 1994).

Conclusion: Forest and land degradation were caused by a failure to enforce regulations, low stumpage fees, corruption and a large landless rural population who saw the residual forest as being a free public good.

Case Study 3 – Spontaneous settlement in Uthai Thani Province, Thailand

The forests of Thailand, especially those in the north, were attractive to loggers from an early period because they contained commercially attractive species such as teak which has always fetched a high market price. During the nineteenth century loggers spilled over from what is now Myanmar to harvest teak for the booming local and international trade. By the late 1800s, Bock (1985) reported seeing 600 elephants employed at just one location in central plains of Thailand (Raheng, now called named Tak) bringing logs to the river where they were floated down the Chao Phraya river to Bangkok. The river banks were 'lined for a considerable distance with enormous piles of timber, awaiting a sufficient volume of water to carry them downstream'.

The first attempt to regulate this logging occurred in 1896 when a Forestry Department was established. Despite this, the rate of teak logging increased and may have reached a peak in 1907 before stabilising, although the rates which were mostly still higher than those achieved in the latter years of the nineteenth century (Feeny 1988). In the meantime, population increases created an increasing demand for agricultural cropland. It is difficult to get precise statistics on just how rapidly clearing took place in these early years. Feeny (1988) estimates the area under cultivation increased at an annual rate of 3% between 1906 and 1955 with most of the new cropland outside the Chao Phraya river delta being created by deforestation of state-owned forest land. The lowlands areas were mainly used, at least initially, for irrigated rice but shifting cultivation continued to be practiced in the northern uplands.

Faced with this rate of clearing, the Royal Forestry Department made a number of attempts to create a permanent forest estate. In the early 1950s the target was at 50% of the area of Thailand. However, continued agricultural clearing and illegal logging meant it had to be revised downwards and by 1971 it had dropped to 37% (Feeny 1988). Figure 2.5 shows the pattern of decline in forest cover over time.

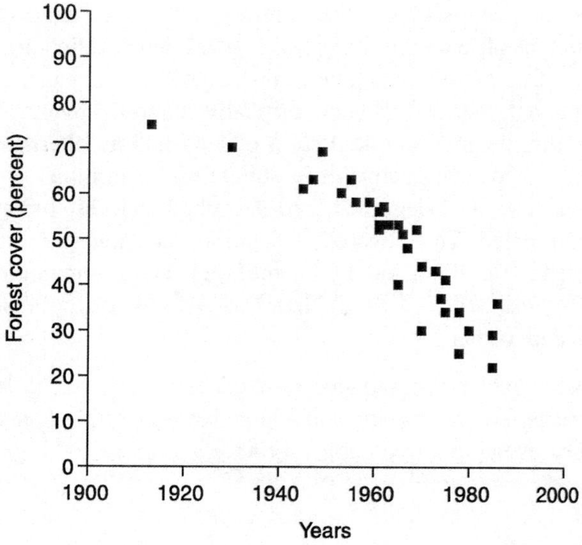

Fig. 2.5 Estimates of changes in forest cover in Thailand over time based on Feeny (1988) and Hurst (1990)

These statistics come from a variety of sources including some where 'forest' includes land that may have been heavily degraded. Nonetheless, the data show that the rate of forest loss increased sharply after 1950. Not all land cleared for agriculture continued to be used for this purpose and Feeny (1988) estimates that only 85% of the forest cleared between 1913 and 1975 was permanent converted to cropland. Most of the remaining 15% presumably reverted to shrublands and grasslands although it is possible some may have regenerated as forest.

Much of the new agricultural lands were cleared by farmers moving in along former logging roads. This meant that 'logged-over' but regenerating natural forest was quickly destroyed and the rate of deforestation then became equal to the rate of logging in the corresponding period (FAO 1981). But this also meant that few of these farmers had any title to the land they were farming. According to Hurst (1990), only 3.7 million hectares of the 24 million hectares under cultivation in 1981 were being farmed by owners with formal legal tenure that gave them secure access and usage rights.

These 'spontaneous' agricultural settlements were not simply the result of demographic pressure. Instead, the process was a rather more complex phenomenon reflecting a number of inter-related influences with the spontaneity modified by the timing of various policy settings. Perhaps the two most important factors were the awarding of timber concessions by the Royal Forestry Department to logging companies and the accelerated national road building program in the 1970s. Both created access into previously inaccessible forest areas.

Hirsch (1988) describes the process of spontaneous settlement that took place in the Lan Sak district of the Uthai Thani Province in central Thailand. This is probably representative of many other such deforestation episodes during this period. The dry dipterocarp and mixed deciduous forests in the area were originally used by shifting cultivators (mainly Karen people). However, in the 1950s, lowland Thai people moved in to tap resin from *Dipterocarpus* trees and to carry out hunting. In the early 1960s some small areas of rice were also being grown in lower lying areas. Access at this stage was difficult and the population remained small. Authority was maintained by local leaders who established rules relating to access to forest resources. In 1969 a logging concession was granted to the government-owned Thai Plywood Company and in 1973 a logging road was built. Although guidelines were in place to regulate logging practices, illegal loggers also moved in and caused severe damage to the forests remaining after the 'official' logging. New settlers also used the logging roads to move into the area encouraged by the then high prices for agricultural produce and by the relative fertility of the newly cleared lands. By 1980, large areas had been cleared and the population was growing rapidly. In 1975 it was 13,500 and the annual rate of increase was 30% (the population reaching 47,000 by 1987).

As more new settlers moved in, the number of land conflicts increased. Thirty people died in a single month in 1970. No legal title was available and land boundaries were not formally defined. Those asserting 'ownership' over a particular area had to enforce their claim against newcomers by personal authority. The area was still regarded as a forest reserve by the Royal Forestry Department but the Lands Department took a different view and demanded payment of a land tax from the settlers. This was often paid because settlers thought this might help them acquire legal tenure.

Over time, ecological changes caused changes in the forms of agriculture being practised. The species initially used by most settlers were low-input but high-yield crops like maize.

But the soil fertility gradually declined and, after the fourth year of cropping, fertilisers and ploughing were needed to maintain productivity. Few settlers had cattle to plough their fields and most were unable to afford the hire of a tractor. Because of this, maize was replaced by crops like mung beans and cotton but the former was too sensitive to occasional dry periods and cassava gradually became more prominent. Weeds such as *Imperata* grass and *Thysanolaena maxima* also began to be more troublesome.

Social changes occurred as well. Declining yields together with stagnant or falling crop prices meant increasing numbers of people acquired debts. Most of these were owed to influential villagers who simultaneously acted as money lenders, suppliers, tractor owners and middlemen. More of the smaller and poorer farmers were bought out by wealthier farmers.

In summary, spontaneous settlement at Lan Sak went through three stages. The first was when the forest and its NTFPs was more important than the timber it contained or the land on which it grew. At this stage the pressure on resources was low. The second stage occurred after logging commenced when the district was colonized by sedentary farmers and rapid population growth occurred. At this stage there was a breakdown of what had then become a managed common property resource. Competition for forest and land became intense. The third and final stage was a period of consolidation and degradation. By then, sedentary agriculture had completely replaced shifting cultivation but there was an increasing need for external inputs to sustain agricultural productivity.

Conclusion: deforestation in this region was caused by corrupt local officials unwilling to control illegal logging. In the scramble for land, poor people cleared areas that should have remained under forest cover. Some of this newly cleared land was quickly degraded because it was unsuitable for agriculture.

Case Study 4 – Planned agricultural settlements on the Atherton Tableland, Australia

When the first European settlers arrived in the wet tropics of north Queensland, the area was dominated by tropical rainforest. These first settlers were tin miners and timber cutters who moved into the Atherton tableland area, near Cairns in 1875. The stature of the forests was impressive suggesting the soils were productive so some settlers established farms in the area in the 1880s. More followed after 1907 when the government made land formally available to new farmers. Government policy at the time was to support agricultural development across the state. This was often done with little regard for economics or environmental realities. The official vision was of a rural community of yeoman farmers and this drove government policy for perhaps the next 60 years. The Atherton tablelands are 50 km from Cairns, the nearest port on the coast, and several thousand kilometres from the state capital, Brisbane. These long distances mean considerable care has to be taken to ensure farming is economically viable. Nonetheless, a large number of enthusiastic settlers moved in and cleared forest for farms. Large scale forest destruction was common because transport was difficult and much forest was simply burned where it fell (Frawley 1987). Maize growing was the dominant crop in the early years but dairying became more prominent as farmers and government researchers gradually learned – over a period of several decades – how to establish the best pasture species (Lamb et al. 2001). Land continued to be opened up for farmers by the government but, by the 1930s, government policies required roads be completed first so that any valuable timbers be extracted before land clearing could proceed. By the time of the Second World War most of the tablelands had been cleared though small patches of forest remnants scattered throughout the area. Forests on the hills and mountains remained as state forests or, later, as national parks.

A final allocation of land was made available for settlement after the war in the early 1950s. This was in an area known as Maalan and lay to the south of the dairy town of Milla Milla. These lands were around 1,000 m in elevation and covered by dense rainforest. The Forestry Department opposed the proposal because the area was occupied with some particular valuable timber stands and they were worried too much land was being cleared for agriculture. They complained:

> The logging and milling industry cannot be made permanent if it has to depend for existence on the desperate salvaging of logs (prior to clearing for agriculture) from the unique jungle forests of the North before their conversion to smoke and ashes.
>
> (Frawley 1987, p. 35)

However, the government believed the proposal was popular with the community and over-ruled the department. The Maalan block was made available in 1953.

Successful applicants received surveyed blocks of around 80 ha and were obliged to clear 10 ha each year and establish pasture or a crop in each of the first 3 years to fulfil purchase conditions. All clearing was done by hand using axes and crosscut saws. The Maalan area was often steep and had a high rainfall (annual rainfall was >2,500 mm with more than 150 rain days per year). Most settlers found it difficult to clear and even more difficult to establish pastures or crops. Conditions were too wet for the good burn that was needed to remove the felled trees and many of the soils were relatively infertile (compared with the rather better soils occupied by earlier settlers on the tablelands). Most farmers initially used the pasture grasses *Melinus minutiflora* (molasses grass) and *Panicum maximum* (Guinea grass). But frosts could kill *Melinas*. When, after a process of trial and error lasting several years, suitable crops or pasture species were eventually identified most farmers found the soil fertility had declined because of the heavy rainfall. Some found it difficult to continue and by 1960 five of the original 24 farms had changed hands. By 1970, a total of 13 had been sold. Frawley (1987, p. 1) quotes Otto Benecke, one of the original Maalan farmers as saying:

> To go down the Maalan, you need a fistful of dollars, a big heart, no brains, and a bloody big umbrella.

In the late 1960s, Gilmour and Reilly (1970) concluded there were up to 50,000 ha of degraded farmland in the southern Atherton tablelands (including the Maalan area) available for reforestation because agricultural productivity was so low. Most of the failed farms were in the higher rainfall areas with steep slopes and poorer soils. In the meantime, selective logging of the remaining forests continued without leading to deforestation. Indeed, Poore (1989); Vanclay (1993); Vanclay et al. (1991) regarded these operations, based as they were on a set of silvicultural prescriptions and a network of growth plots, as being an example of good and sustainable management. Logging ceased in 1988 when all the remaining Australian tropical rainforests were placed on the UNESCO World Heritage List and became part of the national protected area estate (Lamb et al. 2001).

Conclusion: political pressure led to land that should have remained under forest cover being made available to farmers. Agricultural systems able to use this land had not been developed and many farms failed.

Case Study 5 – The mega rice project, Central Kalimantan, Indonesia

During the Suharto era in Indonesia a number of very large agricultural resettlement schemes were carried out to help migrants from densely populated Java move and resettle in some of the outer Provinces of Indonesia. The so-called Mega-Rice project (or, more formally, the Peat Area Project) in Central Kalimantan was one of these. This scheme began in 1996 and aimed to convert 1 million hectares of wetlands and peat swamp forest into irrigated rice fields. It was seen as a way of helping Indonesia to become self-sufficient in rice.

The area chosen was covered by peat with the depth reaching up to 14 m in some locations (Page et al. 2009). The peat is represented by large 'domes' up to 50 km wide and rising several meters above river levels. Each dome is demarcated by rivers and forms its own hydrological unit. Deep peats are not good agricultural soils and are subject to subsidence and accelerated decomposition. Some can become acidified when they dry if the peat contains enough sulphidic compounds (Rieley and Page 2008).

Some selective logging was carried out in the area prior to the project commencing. The forests growing in these areas contained typical wetland communities including the valuable timber species ramin (*Gonystylus bancanus*) and meranti (*Shorea teysmanniana*, *S. platycarpa*, *S. uliginosa*). It is noteworthy that ramin has subsequently been listed as an endangered species under the CITES convention.

When the Mega-Rice project began a more intensive 'salvage logging' operation was carried out and then the remaining vegetation was cleared in order to cultivate rice. A series of drainage canals were constructed through the area. The total length of these was 4,600 km, with the largest being 25–30 m wide. The drainage canals were to remove water during the wet season and allow rice to be irrigated during the dry season. Following the start of the project around 50,000 people moved into the project area to take up land. Few if any of these people had experience in farming peatland areas and it was assumed they could transfer their knowledge of growing irrigated rice in Java to the peatlands.

Very little rice was ever produced because the drainage system led to over-drainage, which allowed the peat to dry out but failed to impound water for irrigation. Draining areas at edge of the peat dome also caused areas in the centre of the dome to dry out and become more susceptible to fire. The summer of 1997/1998 was an El Nino year and Borneo suffered from a series of large wildfires. Smoke from these fires blanketed Southeast Asia and resulted in a massive loss of greenhouse gases. One estimate suggests these represented 13–40% of global emissions from burning fossil fuels (Aldhous 2004). Around 80% of the Mega Rice Project area was damaged by these fires. The legacy was a treeless landscape with a network of malfunctioning canals and a huge peat deposit that was rapidly oxidizing. The project was closed in 1999, not long after President Suharto was forced to step down after the 1997 Asian financial crisis. The area has been since affected by further wildfires.

The failure would not have been surprising to the various specialists who had previously argued against the project going ahead. It appears no preliminary agricultural testing was carried out. The project also violated the government's own regulations which forbade clearing of lands with peat depths greater than 3 m and also required that an environmental impact assessment should be made before projects of this magnitude could proceed. An impact assessment was eventually carried out but not until 6 months after the project began (Boehm and Siegert 1999).

The former settlers were placed in a desperate situation and, not surprisingly, some fell back on logging to try to make a living. This was carried out in forests surrounding the area with funds being supplied by town businessmen. Others are now growing swidden rice and some have established rubber plantings. But the extent of degradation means that, over much of the area, forest restoration will be the best way of dealing with what has happened. Attempts are now being made to restore the forests by blocking the drainage canals at frequent intervals and exploring various ways of carrying out reforestation including using natural regrowth as well as planting (Page et al. 2009).

Conclusion: large-scale agricultural developments need careful planning. In this case an area totally unsuited for agriculture was cleared on the assumption that techniques developed elsewhere could be transferred and used. Even the government's own safeguards were ignored and there was a wilful disregard for the need to undertake preliminary field trials and experimentation.

Case Study 6 – Pulpwood logging in the Gogol Valley, Papua New Guinea

A large pulpwood logging operation commenced in the lowland rainforests of the Gogol Valley on the northern coast of Papua New Guinea in the early 1970s (Lamb 1990). This was established in the final years before independence when the country was being administered by Australia as a Trust Territory of the United Nations. The Administration was concerned that the soon-to-be independent country should be financially secure and the forests were seen as one national asset that might help this be achieved. But, at this time, the country's forests were not as attractive to loggers as those of nearby Indonesia, Philippines or Malaysia where fewer conditions were imposed. Eventually, an offer by a Japanese paper company to carry out pulpwood logging was reluctantly accepted. This proposal involved a clear-felling operation that removed all trees (except *Ficus* which could not be pulped) irrespective of size. The Administration purchased the rights to harvest timber from the traditional landowners and these were then passed on to the company.

The project took place within a very complex social setting. The proposed concession area covered 73,000 ha of forest in the Gogol Valley and an additional block in the valley of a nearby tributary, the Naru River. There were 3,000 people belonging to 330 land owning clans living in the area and, between them, these people spoke eight languages. All of these people practiced shifting cultivation and their isolation meant few were able to participate in the market economy or had access to government services such as schools or health clinics. At the time the project first began to be discussed, many of the able bodied men had already left their villages to seek work in towns and it appeared to the Administration that the social fabric in some communities was breaking down.

A considerable effort was needed to contact the land-owning clans and obtain their agreement to go ahead with the proposal. The social complexities meant lengthy negotiations were needed to define clan land boundaries, settle disputes over these and arrange a schedule of payments to individual clans as the land was progressively logged. Most village people had only an imperfect knowledge of events taking place beyond the village and few were aware of the coming political changes that independence would bring. This meant it was difficult to explain the advantages and disadvantages of logging or to discuss future land use options. Nobody, including the administration, had experienced logging of this intensity. Nonetheless, most people were willing to accept the proposal because of the perceived benefits it would bring. Of these, the most valued was road access to the outside world.

Both the colonial administration and its critics at the time were concerned about the post-logging land use. Since the administration had not purchased the land but only the rights to harvest timber, it had no control over what would be done after logging was completed. A number of alternative scenarios were proposed that allowed for different combinations of plantation reforestation (to permit a continuation of the timber industry) as well as rice growing, pastures for cattle and other land uses. After considerable debate, a plan was eventually adopted that included most of these land uses. The administration favoured reforestation with fast-growing pulpwood species but for this to be viable it had to arrange a land lease for 20,000 ha from the various landowners. This proved impossible to achieve and only 10,000 ha could be acquired for planting. Nor did the rice or the cattle industries develop. In each case, the reason was largely because of the reluctance of landowners to commit to new land uses about which they had little experience. It is a large step to change overnight from a simple shifting cultivator to a sedentary farmer growing cash crops. And, compared with coffee, trees were not a crop in which anyone had any confidence. Many also thought the logging operation was more damaging than they had expected and felt they had been 'tricked' by the administration into allowing it. This being the case, they were reluctant to take additional advice from the administration on what they should do next.

These issues meant key government agencies were unable to follow through with advice on how to make better use of the newly cleared lands. The options were limited. The nearest town, Madang, is small and was already well-supplied with fruit and vegetable. Some coffee could be grown for export but this would have to be the lower quality robusta variety because of the lowland climate. But, in any case, the ecological, agricultural and social changes underway were simply too great for people to make quick decisions. As a result, most landowners continued to practice shifting cultivation and most of the clear-felled forest land was simply abandoned. Fortunately, natural regeneration was rapid. After 10 years, regrowth had formed a closed canopy and trees were 20 m tall while the average tree species richness in 0.12 ha plots was 50–60 tree species (Saulei and Lamb 1991). This diversity was comparable with that in unlogged forests although, unsurprisingly, there were rather more secondary forest species represented in the regrowth. This suggested recovery was well underway but was still incomplete. The logging plan ensured a number of small forest reserves were left unlogged near villages to protect drinking water supplies and maintain the supply of various NTFPs. However, the advent of shotguns and the new road network appear to have led to the disappearance of many larger wildlife species.

The road network left behind after logging prompted many of the original households to shift so they could be closer to transport and this is likely to increase the risk of shorter fallows and more grasslands along these zones. Unlike parts of Asia, no immigrants from outside the area have used these roads to colonise the site since it was well known that all land was formally owned. The new landscape is now a patchwork of regrowth forest, post-farming fallow and some grasslands. Meanwhile, the commencement of several other large national development projects elsewhere in the country and the commencement of a system of provincial government meant the project began to receive much less government attention. This meant opportunities for modifying management practices and learning from the accumulating experiences declined.

Conclusion: despite an attempt to devise future land uses after logging, the government was unable to implement this because of the complex social circumstances prevailing at the time. Over time, it may have been possible to devise better use of land cleared by logging but political circumstances led to extension services and government support being withdrawn before this could happen.

Case Study 7 – Changed land systems, Western Samoa

Substantial deforestation has occurred in Western Samoa in the last 50 years and forest cover has declined from 74% in the 1950s to 40% by 1990. Population growth, logging and commercial agriculture have all played a role in this decline but were not the primary causes. Nor was corruption by political elites. Rather, the change appears to have been more a consequence of changes in customary land tenure (Paulson 1994; Ward 2002). Prior to 1960, most rural people originally lived in coastal villages because water was easier to get from coastal springs and because there were few interior roads. This meant it was easier to carry agricultural products (copra, cocoa) to market by boat than by overland transport.

A strong social system controlled how land was used. People belonged to an extended family or *aiga* (which could contain several households) with these being under the authority of a chief or *matai*. The chief controlled land allocation and use. They also mobilised the workforce and managed the distribution of produce from gardens.

The old systems began to change when the country achieved independence in 1962. Only *matai* were given the right to vote in the new legislature and, for largely political reasons, this prompted an increase in the number of *matai* titles. At about the same time, a new road network began to be developed across the inland areas and chainsaws became increasingly available. Piped water also followed the roads. Together, these changes prompted people to move inland and led to an increase in forest clearing. Under traditional practice, an *aiga* clearing land and using it in some way allowed the family to attain rights to this land although these rights were formally vested in the *matai*. Adjoining *matai* acknowledged the claim as long as the land continued to be used. The new roads enabled changes since these provided access to lands not already claimed by the tradition *matai*. Many new *matai* and some individual families began to clear this new land for themselves. They saw themselves achieving economic and social security through having personal control over their own land rather than through a system of traditional reciprocity and service. The changes were also facilitated by opportunities for new, non-traditional cash crops and by the fact that larger numbers of people had begun working away from the village for wages or a salary. This meant a ‘new’ tenure system rapidly developed in parallel with the existing ‘old’ system as people sought to establish landholdings they could pass on to their children under ‘new’ tenure rules. As time has gone by, large areas of land with tree crops (e.g. coconuts, cocoa) grown under the original tenurial arrangements are being abandoned because nobody is interested in investing time or effort in these communal lands.

The government is seemingly powerless to prevent these changes and conserve forests. This is because all land is supposedly under the control of the *matai.* Even though individual land holdings are widely evident they are not publicly acknowledged because most people still prefer to espouse the merits of traditional practice. Until this impasse between the two tenure systems is resolved, deforestation looks set to continue.

Conclusion: degradation has been caused by changing patterns of land tenure, by inappropriate policy settings and by a reluctance to acknowledge changing land use practices.

Lessons Emerging from These Case Studies About the Causes of Forest and Land Degradation

Logging does not necessarily lead to either forest degradation or deforestation. And neither does deforestation necessarily lead to land degradation. But these several case studies show how easily these transitions can occur, especially on the agricultural frontier. There are also some striking parallels with Chinese experiences.

Blaikie (1989) suggested the reasons why degradation occurs can be found at one or more points along a 'chain of explanation'. At the beginning of the chain are factors associated with site conditions such as soil fertility or climate. As previously noted, Rolett and Diamond (2004) concluded some Pacific islands are more prone to degradation than others simply because of their particular biophysical attributes. Among the case studies described above, the environmental conditions at some of the sites being cleared on the Atherton Tablelands in Australia (Case Study 4) and at the Mega-rice project in Indonesia (Case Study 5) were also such as to make them difficult to convert to productive agricultural lands. Under these conditions, some form of degradation was almost inevitable.

The next step in Blaikie's chain concerns the specific land use practices used and the resources available to managers at particular sites. These might be inappropriate practices caused by a lack of knowledge, or by restrictions imposed by poverty, or they might be the result of a wilful failure to use existing knowledge. The former might occur when a migrant farmer begins at a new site where they have little knowledge about the best species to grow or the local constraints on crop production. Such farmers might have a limited capacity to experiment or purchase resources such as fertilisers. The case of the immigrant farmers in the Thailand (Case Study 3) is an example. The poorly managed logging operations in Sarawak and the Philippines (Case Studies 1 and 2) are examples of where existing and well-established silvicultural knowledge and codes of practice were disregarded causing severe damage to the forest resources and biodiversity values.

The third stage in the chain concerns the nature of agrarian society. By this, Blaikie (1989) is referring to issues such as land tenure and the distribution of land between wealthy and poor landholders as well as population densities and local terms of trade (difference in prices received when selling goods and those paid when buying goods). Land tenure is especially important because those without some degree of certainty that they will benefit from investing for the long-term are less likely to engage in sustainable land use practices. The role of land tenure is illustrated by the dramatic changes in farming in Samoa when tenurial arrangements changed (Case Study 7).

Finally, there are several factors distant from the field that, nonetheless, have a major impact on how land users behave and on the likelihood that degradation will occur. One of these is the role of the state. The role of the state is substantial because it establishes the policies and institutions that determine how forest utilisation and land use planning shall be carried out. It also maintains the bodies responsible for administering these policies. If this framework is poorly conceived or is undermined by politics or corruption, such that laws are not enforced, then forest and land degradation is more likely. Most of the case studies outlined above point to some form of policy or administrative failure. In some cases planning or policies were simply inappropriate (Case Studies 3, 4 and 5) or faulty (Case Study 6). In others, government agencies did not respond quickly enough to feedback showing the original policies were not working or they simply ignored the feedback completely. The final factor in Blaikie's causal chain concerns international economic events. He was largely concerned with issues such as foreign debt, but other international factors such as a strong international market for timber will inevitably tempt some to ignore laws or local codes of practice. This seems to have been one of the factors underpinning the events taking place in both Sarawak and the Philippines (Case Studies 1 and 2). Similarly, strong agricultural export markets may lead to marginal lands being cleared that should have remained under forest. Subsequent fluctuations in these prices can have dramatic effects on rural economies and the abilities of land users to avoid degradation.

The chain of explanation is a useful conceptual approach, but forest and land degradation are usually not the result of single causal factors. More commonly there is often a mixture of causes from various points along the chain and often interactions between these various points. For example, inappropriate government policies can cause a cascade of changes that impact on rural societies, overturn local land use practices and ultimately affect the capacity of farmers to be able to survive on their farms. Nonetheless, several key drivers of degradation emerge from these case studies and from other reports (e.g. Barbier 1997; Blaikie and Brookfield 1987; Colombijn 1997). These have been separated into those largely concerned with forest degradation and those mostly concerned with land degradation though there is some overlap between both groups.

Underlying Drivers of Forest Degradation

1. *External costs are transferred*: if the beneficiaries of poor logging are able to transfer the external costs (e.g. erosion, river sedimentation, hydrological changes, biodiversity losses) to others, there is less reason for them to adopt more prudent and conservational management practices. Poor logging practices are less likely where this is prevented and there is a market for ecological services.

2. *Regulations are not enforced*: enough is already known to enable logging to be carried out in most forests with only limited environmental damage. But logging prescriptions, Codes of Practice and even the idea of a Permanent Forest Estate are all ineffective if governments are unable or unwilling to enforce them. Confusion allows considerable scope for people to 'interpret' regulations or laws to suit themselves.
3. *Corruption*: corrupt politicians or public officials seeking personal wealth from forest logging can disrupt or dismantle institutions meant to regulate forest management and maintain forest values. Corruption also means the financial resources of states and their capacity to deal with any subsequent degradation is reduced. Corruption leads to a lack of trust in public systems of governance and a reluctance to invest for the longer-term.
4. *Low stumpage and taxes*: if the fees charged on timber from forest concessions are too low, there may be little incentive to carry out careful logging. In addition, low financial returns may encourage governments to convert forests to other land uses that yield higher returns.

Underlying Drivers of Land Degradation

5. *Land users lack tenure*: Farmers without land are likely to move into and clear recently logged forest including marginal land more prone to degradation. Many of those lacking tenure will subsequently move to a new site once productivity declines rather than invest in improving productivity. Those practicing shifting cultivations are usually forced to use much shorter fallow periods when they lose access to their customary lands thereby increasing the risk of degradation.
6. *Inappropriate technologies and lack of knowledge*: farming often needs specialised knowledge; techniques developed for some situations may not work in others, especially where unfamiliar crops are being used or where soils are less fertile. Degradation may occur because a farmer uses inappropriate species or management systems. It can take time and expertise to develop appropriate methods and most migrant smallholders lack both of these.
7. *Poverty*: many people living in or about forested landscapes are poor. They may have a limited capacity to sustain the productivity of marginal soils which they find themselves using (e.g. by using fertilisers). As was the case with land tenure, poor farmers may find it easier (and cheaper) to abandon a site once productivity begins to decline and begin again elsewhere.
8. *Forests are often viewed as being 'endless'*: an apparently inexhaustible forest area can encourage the view that it is easy to move and clear more forest if degradation occurs. This attitude can be common in so-called frontier situations at the interface between forests and the expanding agricultural sector. It is less likely if some form of monitoring is carried out so that the actual state of forest and land resources are known. The perception is one that may be held by both individual farmers as well as by governments.

9. *Uncertain markets*: changes in commodity prices or labour costs can make previously successful farms uneconomic. This may prompt changes in ownership or cause sites to be over-exploited and degraded before being finally abandoned. The problem is likely to be most acute where lands are only marginally suitable for agriculture.
10. *Inappropriate land use policies*: not all lands are suitable for agriculture and agricultural settlements are sometimes promoted or permitted to develop on lands that should be left under forest cover (e.g. poor soils, steep hill slopes, difficult climate, deep peatlands or sites distant from markets). This often occurs when landless farmers use logging roads to move into recently logged forests and during the land rushes that often develop in these situations.

In short, degradation can arise from attempts to maximise the short-term benefits from using forests or land, irrespective of the consequences. It can also occur when managers have neither the capacity nor the resources to invest in practices that will maintain productivity. Climatic changes and changes in market prices can both trigger degradation at marginal sites. But perceptions also matter and land can be abandoned as wasteland – irrespective of its condition – if farmers believe they will be better off moving and starting again in another location. Population density is sometimes seen as a cause of land degradation and it may be so where farmers have insecure tenure. But it was not a principle cause of degradation in any of these case studies.

Thresholds and Forest Transitions

Much degradation occurs quickly and before government bureaucracies are aware that it has happened, or of the scale at which it is occurring. But at what point do they become concerned enough about deforestation and accumulating areas of wastelands to reverse these trends and undergo what Rudel et al. (2005) refer to as the 'forest transition'? The historical evidence is that some societies such as China have been unable to prevent almost complete deforestation while others such as Japan have been able take action at a much earlier stage when significant amounts of natural forest still remain. The turnabout occurs when increases in the area of new forests, either natural regrowth or plantations, outweigh losses of natural forest such that the overall forest cover begins to increase. Several explanations have been offered to explain this change. One is that the reversal is largely driven by changes associated with economic development. Some deforestation is required to enable this development but, at a certain point, the wealth acquired allows reforestation to occur. This transition has been likened to an environmental Kuznet's curve with the relationship between forest cover and economic development resembling a U shape. Bhattaria and Hammig (2004) found empirical evidence supporting this view but concluded that the nature of the relationship was also influenced by governance and the quality of governmental institutions and not just wealth.

Rudel et al. (2005) approached the problem in a slightly different way and suggested there are two potential triggers. One they labelled the 'economic development pathway'. This occurs when the growth of employment in towns and cities is sufficient to draw people away from the countryside, leading to farmland being abandoned. Under these circumstances, any reforestation is unplanned and largely takes place because of natural regrowth rather than because of any deliberate government reforestation policy (in fact it reflects the absence of any such policy). Perhaps the best tropical example of this actually occurring is in Puerto Rico where regrowth has flourished as large numbers of people have left the land (Aide et al. 1995). Something similar may be beginning to happen in parts of Peninsular Malaysia where Jomo et al. (2004) describe the development of a pool of 'idle land' now reaching 890,000 ha or 22% of the land cultivated by smallholders. Many of these farms have become increasingly small and it seems some owners or tenants have left them to seek more remunerative opportunities in urban areas.

The other trigger they termed the 'forest scarcity pathway'. In this case, an increasing shortage of natural forests eventually drives up the local prices of forest products and prompts reforestation. Under these circumstances, landowners find it profitable to plant trees for their own use or for commercial reasons. A striking example of this occurring was provided by Holmgren et al. (1994) who described how Kenyan farmers began planting trees on farms and created a timber resource that became greater than that in natural forests. Contrary to expectations, there was a positive relationship between population density and planted woody biomass. Similar findings have been reported from India (Foster and Rosenzweig 2003). An example from the Philippines is described in Box 2.2. In each case, forest resources were suddenly in short supply and farmers found it financially rewarding to adopt timber trees as a new cash crop.

There is little doubt that reforestation for commercial gain does take place when market conditions are attractive, but governments have a crucial role to play in establishing a supportive policy framework and providing the technical assistance needed to overcome degradation. Many governments have sought to foster reforestation to improve economic outcomes but also to overcome erosion, improve watershed conditions, prevent floods and conserve biodiversity resources. Some governments have simply promoted reforestation but allowed deforestation to continue while other have tried to prevent further deforestation and also foster reforestation. Examples of such government driven reforestation programs in Asia are those in Japan (Totman 1983) and more recently in China and India (Mather 2007) and Korea (Tak et al. 2007). The Korean example is dramatic with more than 2 million hectares, or around 30% of the landscape, being reforested since the 1950s, bringing the total forest cover to 63% of the land area (Tak et al. 2007). This suggests the two models identified by Rudel et al. (2005) are insufficient to describe all transitions.

The only country in Southeast Asia that has undergone the forest transition is Vietnam (see Fig. 2.3). There is some uncertainty about just how far deforestation progressed before change occurred. Some suggest forest cover may have fallen to as little as 15–17% (De Koninck 1999), although a cover of 24–26% in the early 1980s appears to be more widely accepted (De Jong et al. 2006; Meyfroidt and Lambin 2008; Nguyen and Gilmour 2000). Whatever the precise figure, forest

Box 2.2 The forest transition at a local level in Northern Luzon, Philippines

The dipterocarp forests in the mountain areas of northern Luzon have been heavily logged and most of the area is now without a commercially attractive timber resource. Workers attracted to the area during the height of the logging boom have either left the area or taken up farming on former logging concession areas. Some have also carry out illegal logging in the remaining forests but favour the local *Pterocarpus indicus* rather than the dipterocarp species taken by the original logging companies. This timber is used by local furniture makers and now commands a much higher price than the local dipterocarps (that were previously used for plywood). The new supplies of this timber have led to a doubling in the number of furniture factories. Not surprisingly, this is causing a gradual depletion of the *Pterocarpus* resource. In the meantime, a number of former loggers have begun planting trees on their own land in addition to crops like maize, rice and bananas. Seki (2003) describes this as 'spontaneous' reforestation because it has occurred without significant government assistance. The reason for this reforestation is the changing economic and ecological conditions surrounding the ex-logging workers. A survey found most of those interviewed planted trees (mostly *Gmelina arborea*) because they believed it would be profitable for them or because they saw the need for a new natural resource to replace that destroyed by logging. Tree planting has became more profitable as the natural forest shrunk and became confined to less accessible areas where the costs of transporting logs is increasingly high. Under these circumstances, local sawmills have found *Gmelina* an acceptable alternative. Growing *Gmelina* is also attractive to growers for other reasons other than there being a ready market; seed is easily obtainable, the species is resistant to fire and harvesting can be carried out after only 8 years. Seki (2003) considers the success of this reforestation phenomenon is unrelated to the government's community-based forest management strategy. The latter is more interested in large-scale tree plantation programs managed by cooperatives but this has sometimes led to land disputes. While cooperatives may be appropriate for managing residual secondary forests, they are much less attractive to former logging workers who prefer managing their own farms rather than being employed as plantation workers.

cover had increased to 40% by 2005 (FAO 2007). The early stages of this turnaround occurred when Vietnam was a centrally planned economy and reforestation was carried out under the direction of the state. But the process changed after 1989 when Vietnam began transforming itself into a market economy. Households were granted long-term access and rights to use land and assistance was given to households to protect existing forests and replant new production and protection forests (De Jong et al. 2006; Do and Le 2003; Nguyen and Gilmour 2000). This alone may not have been enough to encourage reforestation (Sikor 2001) but the government has also promoted the idea through policy and legislative changes and funded a series of national reforestation programs. The most recent of these programs aims to reforest an additional five million of land (MARD 2001). It should be noted that Vietnam's attempts to reduce deforestation and increase the national forest cover have been assisted by the importation of large volumes of illegally logged timber from neighbouring countries (Meyfroidt and Lambin 2009).

This transition does not immediately fit either the simple 'environmental Kuznet's curve' of Bhattaria and Hammig (2004), or the 'economic development pathway' of Rudel et al. (2005). On the other hand, there is some evidence that elements of the 'forest scarcity pathway' were at work. Based on a regression model of landscape changes, Meyfroidt and Lambin (2008) suggest changes in land tenure arrangements have allowed the intensification of agriculture in valleys and leaving land available for reforestation on upper hillslopes. In areas with road access it has been profitable for farmers to reforest these slopes because of a market for forest products. They refer to this as being a 'smallholder agricultural intensification pathway'. Markets and road access have been undeniably important but there is little doubt that the evolving national policy framework in Vietnam has also been instrumental in dealing with some of the problems described in the case studies and facilitating the turn-about (De Jong et al. 2006).

Conclusions

This chapter provides additional evidence concerning the extent of forest degradation and loss in the region. Traditional agricultural practices have damaged forests in the past but recovery has usually occurred because the scale of the damage was limited. More recently, the rate and scale of clearing has accelerated. Some of the cleared land has been successfully used for agriculture but other areas have been used only briefly and then abandoned in a degraded state. It is not easy to determine the proportion of land affected in this way because of the difficulty of defining and mapping degraded lands, but estimates range from 26% to 49% of all cleared land.

Some lands may be more prone to degradation than others because of their biophysical attributes but forest and land degradation has often been caused by a wilful disregard for regulations and of well-established practices designed to prevent degradation from occurring. It is often a sign of poor governance with the distribution of costs and benefits being grossly inequitable and with neither ecological nor economic information being used in decision-making. In some cases, poor farmers have been the agents as well as the victims of land degradation because they did not have the knowledge to manage their land or were unable to afford the investments needed to make their land productive. The economic costs of degradation are generally invisible in national accounts but the effects are usually profound, especially for the poorer members of society, since they are the ones most dependent on the natural resource base.

At a national scale, there appear to be at least three pathways by which reforestation might occur. One of these is the 'economic development pathway' that develops when agricultural land is abandoned as people move to urban areas. A second is the 'forest scarcity pathway' where reforestation is triggered by a shortage of forest goods or services. The third pathway is where governments initiate and facilitate reforestation. In each of these cases, reforestation will be assisted if the policy and institutional failures that led to degradation can be overcome. But new land use policies and silvicultural techniques will also be needed that match the economic and ecological circumstances now present in these degraded lands. These will need to address the needs of smallholders as well as large industrial plantation owners. They should also ensure the new forests are capable of producing a wide range of ecosystem services, including biodiversity conservation, as well as just timber.

The decline in forest cover and the increase in areas of degraded land raises two important questions. One is how to conserve the biodiversity still remaining in tropical landscapes? Is the present reliance on protected areas sufficient or are other approaches also needed? The second question is what can be done to hasten the forest transition in order to improve the livelihood of people now living in these deforested and degraded areas? Might some forms of reforestation be beneficial or must further deforestation occur if their living standards are to increase? Both these issues will be discussed in the next chapter.

References

Aide TM, Zimmermann JK, Herrera L, Rosario M, Serrano M (1995) Forest recovery in abandoned tropical pastures in Puerto Rico. For Ecol Manage 77:77–86

Aldhous P (2004) Borneo is burning. Nature 432:144–146

Allen BJ, Bourke M, Hide RL (1995) The sustainability of Papua New Guinea agricultural systems: the conceptual background. Global Environmental Change 5:297–312

Anderson A (2002) Faunal collapse, landscape change and settlement history in remote Oceania. World Archaeol 33:375–390

Ash J (1992) Vegetation ecology of Fiji: past, present, and future perspectives. Pac Sci 46:111–127

Bailey RC, Head G, Jenike M, Owen B, Rechtman R, Zechenter E (1989) Hunting and gathering in tropical rain forest: Is it possible? Am Anthropol 91:59–82

Barbier E (1997) The economic determinants of land degradation in developing countries. Philosphical Transactions of the Royal Society of London B 352:891–899

Baur GN (1964) The Ecological Basis of Rainforest Management. Forestry Commission of New South Wales, Sydney

Bhattarai M, Hammig M (2004) Governance, economic policy and the environmental Kuznets curve for natural tropical forests. Environment and Development Economics 9:367–382

Blaikie P (1989) Explantation and policy in land degradation and rehabilitation for developing countries. Land Degrad Rehabil 1:23–37

Blaikie P, Brookfield H (1987) Land Degradation and Society. Methuen, London

Bock C (1985) Temples and elephants: the narrative of a journey of exploration through Upper Siam and Lao. White Orchid Press, Bangkok (originally published by Sampson Low, Marston, Searle, and Rivington. London, 1884)

Boehm H, Siegert F (1999) Remote sensing verification by aerial surveys and ground truth campaigns 1997 and 1998 in Central Kalimantan, Indonesia – peat swamp forest, Mega-Rice Project and fires. Workshop on Tropical Forest and Remote Sensing The Center for Remote Sensing, Imaging and Processing, National University of Singapore, 5 March 1999

Boserup E (1993) The Conditions of Agricultural Growth: the Economics of Agrarian Change under Population Pressure. Earthscan, London

Bowman DMJ (2000) Australian rainforests: Islands of Green in a Land of Fire. Cambridge University Press, Cambridge
Brookfield H (1997) Landscape history: land degradation in the Indonesian region. In: Boomgaard P, Colombijn F, Henley D (eds) Paper Landscapes: Explorations in the Environmental History of Indonesia. KITLV Press, Leiden, pp 27–59
Bulmer RNH (1982) Traditional conservation practices in Papua New Guinea. In: Morauta L, Pernetta J, Heaney W (eds) Traditional conservation in Papua New Guinea: implications for today. Institute of Applied Social and Economic Research; Monograph 16, Port Moresby, pp 59–777
Burgess P (1991) Natural rainforest management. In: Collins M, Sayer JA, Whitmore TC (eds) The Conservation Atlas of Tropical Forests: Asia and the Pacific. MacMillan Press, London, pp 43–50
Cairns M (ed) (2007) Voices from the Forest. Resources for the Future, Washington
Carr DL, Suter L, Barbieri A (2005) Population dynamics and tropical deforestation: State of the debate and conceptual challenges. Popul Environ 27:89–113
Chazdon RL (2003) Tropical forest recovery: legacies of human impact and natural disturbances. Perspectives in Plant Ecology, Evolution and Systematics 6:51–57
Chomitz K (2007) At Loggerheads? Agricultural Expansion, Poverty Reduction and Environment in the Tropical Forest. World Bank, Washington
Clarke WC (1971) Place and People: an Ecology of a New Guinean Community. Australian National University Press, Canberra
Clarke WC, Thaman R (1993) Agroforestry in the Pacific Islands: Systems for Sustainability. United Nations University Press, Tokyo, New York, Paris
Colombijn F (1997) The ecological sustainability of frontier societies in eastern Sumatra. In: Boomgaard P, Colombijn F, Henley D (eds) Paper Landscapes: Explorations in the Environmental History of Indonesia. KITLV Press, Leiden, pp 309–340
Conroy W (1960) The evolution of the agricultural environment in Papua and New Guinea. Symposium on the impact of man on humid tropics vegetation; Goroka Territory of Papua and New Guinea. Administration of the Territory of Papua and New Guinea and UNESCO Science Cooperation Office for South East Asia, pp 94–97
Cramb R (1993) Shifting cultivation and sustainable agriculture in East Malaysia: a longitudinal case study. Agric Syst 42:209–226
Dauvergne P (2001) Loggers and Degradation in the Asia-Pacific: Corporations and Environmental Management. Cambridge University Press, Cambridge
Dawning SL, Iverson LR, Brown S (1993) Rates and patterns of deforestation in the Philippines: application of geographic information system analysis. For Ecol Manage 57:1–16
De Beer J, McDermott M (1989) The Economic Value of Non-timber Forest Products in Southeast Asia. Netherlands Committee for IUCN, Amsterdam
De Jong W, Sam DD, Hung TV (2006) Forest Rehabilitation in Vietnam: history, realities and future. Center for International Forestry Research, Bogor
De Koninck R (1999) Deforestation in Vietnam. International Development Research Center, Ottawa
Do DS, Le QT (2003) Forest policy trends in Vietnam. In: Inoue M, Isozaki H (eds) People and Forest - Policy and Local reality in Southeast Asia, the Russian Far East and Japan. Kluwer Academic Publishers, Dordrecht
Donovan D (2003) Trading in the forest: Lessons from Lao history. In: Lye T-P, De Jong W, Abe K (eds) The Political Ecology of Tropical Forests in Southeast Asia: Historical Perspectives. Kyoto University Press, Kyoto, pp 72–106
Dunn FL (1975) Rainforest collectors and traders: a study of resource utilisation in modern and ancient Malaysia. Monograph No. 5. Malaysian Branch of the Royal Asiatic Society, Kuala Lumpur
Dwyer P, Minnegal M (1991) Hunting in lowland tropical rainforest; towards a model of non-agricultural subsistence. Hum Ecol 19:187–212
Elvin M (2001) Three thousand years of unsustainable growth: China's environment from archiac times to the present. East Asian History 6:7–46
Elvin M (2004) The Retreat of the Elephants: An Environmental History of China. Yale University Press, New Haven and London

FAO (1981) Tropical Resources Asessment Project: Forest Resources of Tropical Asia. Food and Agriculture Organization of the United Nations and the United Nations Environment Programme, Rome
FAO (2007) State of the World's Forests 2007. Food and Agriculture Organisation of the United Nations, Rome
Feeny D (1988) Agricultural expansion and forest depletion in Thailand 1900 – 1975. In: Richards JF, Tucker RP (eds) World deforestation in the twentieth century. Duke University Press, Durham, pp 112–143
Forbes HO (1885) A Naturalist's Wandering in the Eastern Archipelago: a Narrative of Travel and Exploration from 1878 to 1883. Sampson Low, London
Forshed O, Udarbe T, Karlsson A, Falck J (2006) Initial impact of supervised logging and pre-logging climber cutting compared with conventional logging in a dipterocarp rainforest in Sabah, Malaysia. For Ecol Manage 221:233–240
Foster AD, Rosenzweig MR (2003) Economic growth and the rise of forests. Quarterly Journal of Economics 118:601–637
Fox J, Vogler J (2001) Land-use and land-cover change in montane mainland southeast Asia. Environ Manage 36:394–403
Frawley K (1987) The Maalan Group Settlement North Queensland: A Historical Geography. Department of Geography and Oceanography, University College, University of New South Wales, Australian Defence Force Academy, Canberra
Freeman JD (1955) Iban Agriculture: A report on the Shifting Cultivation of Hill Rice by the Iban of Sarawak. H.M. Stationery Office, London
Geddes WR (1976) Migrants of the Mountains: the cultural ecology of the Blue Miao (Hmong Njua) of Thailand. Clarendon Press, Oxford
Geertz C (1963) Agricultural Involution:the Process of Ecological Change in Indonesia. University of California Press, Berkley
Geist HJ, Lambin EF (2002) Proximate causes and underlying driving forces of tropical deforestation. Biotropica 52:143–150
Gillis M (1988) Malaysia: public policies and the tropical forest. In: Repetto R, Gillis M (eds) Public Policies and the Misuse of Forest Resources. Cambridge University Press, New York, pp 115–164
Gilmour DA, Fisher RJ (1991) Villagers, Forests and Foresters. Sahayogi Press, Kathmandu
Gilmour DA, Reilley JJ (1970) Productivity survey of the Atherton Tableland and suggested land use changes. Journal of the Australian Institute of Agricultural Science 36:259–272
Goldammer JG, Seibert B (1989) Natural rainforest fires in eastern Borneo during the Pleistocene and Holocene. Naturewissenschaften 76:518–520
Haberle SG, Hope GS, van der Kaars S (2001) Biomass burning in Indonesia and Papua New Guinea: natural and human induced fire events in the fossil record. Palaeogeogr Palaeoclimatol Palaeoecol 171:259–268
Hansen TS (2005) Spatio-temporal aspects of land use and land cover changes in the Niah catchment, Sarawak, Malaysia. Singapore Journal of Tropical Geography 26:170–190
Headland TN (1987) The wild yam question: How well could independent hunter-gatherers live in a tropical rain forest ecosystem? Hum Ecol 15:463–491
Henley D (2005) Agrarian change and diversity in the light of Brookfield, Boserup and Malthus: historical illustrations from Sulawesi, Indonesia. Asia Pacific Viewpoint 46:153–172
Hirsch P (1988) Spontaneous land settlement and deforestation in Thailand. In: Dargavel J, Dixon K, Semple N (eds) Changing Tropical Forests: Historical Perspectives on Today's Challenges in Asia, Australasia and Oceania. Center for Resource and Environmental Studies, Australian National University, Canberra, pp 359–376
Holmgren P, Masakha EJ, Sjoholm H (1994) Not all African land is being degraded - a recent survey of trees on farms in Kenya reveals increasing forest resources. Ambio 23:390–395
Hopkins MS, Head J, Ash JE, Hewett RK, Graham AW (1996) Evidence of a Holocene and continuing recent expansion of lowland rain forest in humid, tropical North Queensland. J Biogeogr 23:737–745

Houghton R (2001) Agriculture. In: Woodwell GM (ed) Forests in a Full World. Yale University Press, New Haven and London, pp 35–50
Hurst P (1990) Rainforest Politics: Ecological Destruction in South-East Asia. Zed Books, London
Hviding E, Bayliss-Smith T (2000) Islands of Rainforest: Agroforestry, Logging and Eco-tourism in Solomon Islands. Ashgate, Aldershot
ITTO (2006) Status of Tropical Forest Management 2005. International Tropical Timbers Organisation; Technical Series no 24, Yokohama
Johns RJ (1989) The influence of drought on tropical rainforest vegetation in Papua New Guinea. Mt Res Dev 9:248–251
Jomo KS, Chang YT, Khoo KJ (2004) Deforesting Malaysia: The Political Economy and Social Ecology of Agricultural Expansion and Commercial Logging. Zed Books in Association with United Nations Research Institute for Social Development, London and New York
Kaimowitz D, Angelsen A (1998) Economic Models of Tropical Deforestation: A Review. Center for International Forestry Research, Bogor
Kelly PM, Hoang MM, Tran VL (2001) Responding to El Niño and La Niña: averting tropical cyclone impacts. In: Adgers WN, Kelly PM, Ninh NH (eds) Living with environmental change: social vulnerability, adaptions and resilience in vietnam. Routledge, London, pp 155–181
Kennedy J, Clarke WC (2004) Cultivated landscapes of the Southwest Pacific. Resource Management in Asia-Pacific Working Paper No 50, Resource Management in Asia-Pacific Program; Research School of Pacific and Asian Studies; The Australian National University, Canberra
Kennett D, Anderson A, Winterhalder B (2006) The ideal free distribution, food production and the colonization of Oceania. In: Kennett D, Winterhalder B (eds) Behavioural Ecology and the Transition to Agriculture. University of California Press, Berkeley, Los Angeles, London, pp 265–288
Keppel G, Buckley Y, Possingham H (2010) Drivers of lowland rainforest community assembly, species diversity and forest structure in islands in the tropical South Pacific. J Ecol 98:87–95
Kikkawa J, Dwyer P (1992) Use of scattered resources in rainforests of the humid tropical lowlands. Biotropica 24:293–308
Knapen H (1997) Epidemics, droughts and other uncertainties in Southeast Borneo during the eighteenth and nineteeth century. In: Boomgaard P, Colombijn F, Henley D (eds) Paper Landscapes: Explorations in the Environmental History of Indonesia. KITLV Press, Leiden, pp 121–154
Kummer DM (1992) Deforestation in the Postwar Philippines. University of Chicago Press, Chicago
Kummer DM, Turner BL (1994) The human causes of deforestation in Southeast Asia. Bioscience 44:323–328
Kunstader P, Chapman EC, Sabhasri S (1978) Farmers in the Forest: Economic Development and Marginal Agriculture in northern Thailand. East-West Center and University Press of Hawaii, Honolulu
Lamb D (1990) Exploiting the Tropical Rainforest: An Account of Pulpwood Logging in Papua New Guinea. UNESCO and The Parthenon Publishing Group, Paris and Carnforth
Lamb D, Keenan R, Gould K (2001) Historical background to plantation development in the tropics: a North Queensland case study. In: Harrison SR, Herbohn J (eds) Sustainable farm Forestry in the Tropics: Social and Economic Analysis and Policy. Edward Elgar, Cheltenham, pp 21–34
Lamprecht H (1993) Silviculture in the tropical natural forest. In: Pancel LR (ed) Tropical Forest Handbook. Springer-Verlag, Berlin, pp 728–810
Lau BT (1979) The effect of shifting cultivation on sustained yield management for Sarawak national forests. Malaysian Forester 42:418–422
Leonen M (1993) Philippines: dwindling frontiers and agrarian reform. In: Colchester M, Lohman L (eds) The Struggle for Land and the Fate of the Forests. Zed Books, London, pp 264–290
Lindblade K, Carswell G, Tumuhairwe JK (1998) Mitigating the relationship between population and land degradation: landuse change and farm management in southwest. Uganda Ambio 27:565–571
MARD (2001) Five Million Hectare Reforestation Program Partnership: Synthesis Report. International Cooperation Department, Ministry of Agriculture and Rural Development, Hanoi

Mather AS (2007) Recent Asian forest transitions in relation to forest-transition theory. International Forestry Review 9:491–502

Mather AS, Needle CL (2000) The relationship of population and forest trends. Geogr J 166:2–13

Matson P, Parton WJ, Power AG, Swift MJ (1997) Agricultural intensification and ecosystem properties. Science 277:504–509

Meggitt MJ (1960). In discussion, p.137. Symposium on the impact of man on humid tropics vegetation. Goroka, Territory of Papua and New Guinea. Administration of the Territory of Papua and New Guinea and UNESCO Science Cooperation Office for South East Asia, Papua and New Guinea

Menzies NK (1996) Forestry. In: Needham J (ed) Science and Civilisation in China. Cambridge University Press, Cambridge

Meyfroid P, Lambin EF (2008) The causes of reforestation in Vietnam. Land Use Policy 25:182–197

Meyfroid P, Lambin EF (2009) Forest transition in Vietnam and displacement of deforestation abroad. Proc Nat Acad Sci 106:16139–16144

Michon G (2005) Domesticating Forests: How Farmers Manage Forest Resources. Institut de Recherchepour le Developpement, Center for International Forestry Research, World Agroforestry Center, Paris, Bogor

Midgely S, Blyth M, Mountlamai K, Midgely D, Brown A (2007) Towards Improving Profitability of Teak in Integrated Smallholder Farming Systems in Northern Laos. Australian Center for International Agricultural Research, Canberra

Mueller-Dombois D, Fosberg FR (1998) Vegetation of the Pacific. Springer-Verlag, New York

Murphey R (1983) Deforestation in modern China. In: Tucker RF, Richards JF (eds) Global Deforestation and the Nineteenth-Century World Economy. Duke Press Policy Studies, Durham, pp 111–129

Nguyen VS, Gilmour DG (2000) Forest rehabilitation policy and practice in Vietnam. In: Anon (ed) Forest rehabilitation policy and practice in Vietnam: Proceedings of a National Workshop, Hoa Binh. International Union for Conservation of Nature, Hanoi, pp 4–34

Nibbering JW (1999) Tree planting on deforested farmlands, Sewu Hills, Java, Indonesia: impact of economic and institutional changes. Agroforest Syst 46:65–82

Nunn PD (1990) Recent environmental changes on Pacific Islands. Geogr J 156:125–140

Page S, Hoscilo A, Wösten H, Jauhiainen J, Silvius M, Rieley J, Ritzema H, Tansey K, Graham L, Vasander H, Limin S (2009) Restoration ecology of lowland tropical peatlands in Southeast Asia: current knowledge and future research directions. Ecosystems 12:888–905

Paulson DD (1994) Understanding tropical deforestation: the case of western Samoa. Environ Conserv 21:326–332

Poffenberger M (1990) Keepers of the Forest: Land Management Alternatives in Southeast Asia. Kumarian Press, Connecticut

Poore D (1989) Queensland, Australia: an approach to successful sustainable management. In: Poore D, Burgess P, Palmer J, Rietbergen S, Synnott T (eds) No Timber Without Trees: Sustainability in the Tropical Forest. Earthscan, London, pp 28–39

Potter LM (1996) The dynamics of Imperata: historical overview and current farmer perspectives, with special reference to South Kalimantan, Indonesia. Agroforest Syst 36:31–51

Potter LM (1997) The dynamics of Imperata: historical overview and current farmer perspectives, with special reference to South Kalimantan, Indonesia. Agroforest Syst 36:31–51

Rappaport RA (1968) Pigs for the Ancestors: Ritual in the Ecology of a New Guinea People. Yale University Press, New Haven

Rappaport RA (1971) The flow of energy in an agricultural society. Sci Am 225:117–132

Rasul G, Thapa GB (2003) Shifting cultivation in the mountains of South and Southeast Asia: regional patterns and factors influencing change. Land Degrad Dev 14:495–508

Richards JF, Flint J (1994) A century of land-use change in south and southeast Asia. In: Dale VH (ed) Effects of Land Use Changes on Atmospheric CO2 Concentrations: South and Southeast Asia as a Case Study. Springer Verlag, New York, Berlin, pp 15–66

Rieley J, Page S (2008) The science of Tropical Peatlands and the Central Kalimantan Peatland development Area. Master Plan for the Rehabilitation and Rehabilitation of the Ex-Mega Rice Project Area in Central Kalimantan. Euroconsult, Mott MacDonald and Deltares; Delft Hydraulics

Roder W, Keoboualapha B, Manivanh V (1995) Teak (*Tectona grandis*), fruit trees and other perennials used by hill farmers of northern Laos. Agroforest Syst 29:47–60

Rolett B, Diamond J (2004) Environmental predictors of pre-European deforestation on Pacific Islands. Nature 431:443–446

Ross ML (2001) Timber booms and institutional breakdown in Southeast Asia. Cambridge University Press, Cambridge

Roth DM (1983) Philippines forests and forestry: 1565-1920. In: Tucker RP, Richards JF (eds) Global deforestation and the Nineteenth-Century World Economy. Duke Press Policy Studies, Durham, pp 30–49

Rudel TK, Coomes OT, Moran E, Achard F, Angelsen A, Xu JC, Lambin E (2005) Forest transitions: towards a global understanding of land use change. Global Environmental Change-Human and Policy Dimensions 15:23–31

Rundel PW, Boonpragob K (1995) Dry forest ecosystems of Thailand. In: Bullock SK, Mooney HA, Medina E (eds) Seasonally Dry Forests. Cambridge University Press, Cambridge

Salo J, Kalliola R, Häkkinen I, Makinen Y, Niemala P, Puhakka M, Coley K (1986) River dynamics and the diversity of the Amazon forest. Nature 322:254–258

Saulei S, Lamb D (1991) Regeneration following pulpwood loggingin lowland rainforest in Papua New Guinea. In: Gomez-Pompa A, Whitmore T, Hadley M (eds) Rain Forest Regeneration and Management. UNESCO and The Parthenon Publishing Group, Paris, pp 313–322

Scatena FN, Planos-Gutierrez EO, Schellekens J (2005) Natural disturbances and the hydrology of humid tropical forests. In: Bonell M, Bruijnzeel LA (eds) Forests, Water and People in the Humid Tropics. Cambridge University Press and UNESCO, Cambridge, pp 489–512

Schafer E (1962) The conservation of nature under the T'ang dynasty. Journal of the Economic and Social History of the Orient 5:279–308

Schiefflin EL (1975) Felling the trees on top of the crop: European contact and the subsistence ecology of the Great Papuan Plateau. Oceania 46:25–39

Seki Y (2003) Adaptability of community based forest management for ex-logging workers in the Philippines. In: Inoue M, Isozaki H (eds) People and Forest - Policy and Local Reality in Southeath Asia, the Russian Far East, and Japan. Kluwer Academic Publishers, Dordrecht, pp 287–298

Siegert F, Ruecker G, Hinrichs A, Hoffmann AA (2001) Increased damage from fires in logged forests during droughts caused by El Nino. Nature 414:437–440

Sikor T (2001) The allocation of forestry land in Vietnam: did it cause the expansion of forests in the northwest? Forest Policy and Economics 2:1–11

Sist P, Fimbel R, Sheil D, Nasi R, Chevallier MH (2003) Towards sustainable management of mixed dipterocarp forests of South-east Asia: moving beyond minimum diameter cutting limits. Environ Conserv 30:364–374

Sodhi NS, Brook BW (2006) Southeast Asian birds in peril. The Auk 123:275–277

Stone GD (2001) Theory of the square chicken: advances in agricultural intensification theory. Asia Pacific Viewpoint 42:163–180

Tak K, Chun Y, Wood PM (2007) The South Korean forest dilemma. International Forestry Review 9:548–557

Totman C (1983) The forests of Tokugawa Japan: a catastrophe that was avoided. Transactions of the Asiatic Society of Japan 18:1–15

Totman C (1989) The Green Archiplelago: Forestry In Pre-industrial Japan. University of California Press, Berkley

Uusivuori J, Lehto E, Palo M (2002) Population, income and ecological conditions as determinants of forest area variation in the tropics. Global Environmental Change 12:313–323

Vanclay JK (1993) Tropical rainforest logging in North Queensland: Was it sustainable? Annals of Forestry 1:54–60

Vanclay JK (1994) Modelling Forest Growth and Yield: Applications to Mixed Tropical Forest. CAB International, Wallingford

Vanclay JK, Rudder EJ, Dale G, Blake GA (1991) Sustainable harvesting of tropical rainforests: reply to Keto, Scott, Olsen. J Environ Manage 33:379–394

Vasey D (1981) Agricultural systems in Papua New Guines: adapting to the humid tropics. In: Denoon D, Snowden C (eds) A History of Agriculture in Papua New Guinea: a Time to Plant and a Time to Uproot. Institute of Papua New Guinea Studies, Port Moresby

Veldkamp JF (1979) Obituary of Charles Monod de Froideville. Flora Malesiana Bulletin 197 32:3180–3183

Vincent J (1997) Resource depletion and economic sustainability in Malaysia. Environmental Research and Economics 2:1937

Waddell E (1972) The Mound Builders: Agricultural Practices, Environment, and Society in the Central Highlands of New Guinea. University of Washington Press, Seattle

Ward RG (2002) Deforestation in Western Samoa. Pacific Viewpoint 36:73–93

Webb LJ (1958) Cyclones as an ecological factor in lowland tropical rain-forests, North Queensland. Aust J Bot 6:220–228

White KJ (1975) The Effect of Natural Phenomenon on the Forest Environment. Presidential Address to the Papua New Guinea Scientific Society, Port Moresby

Whitmore TC (1984) Tropical Rain Forests of the Far East. Clarendon Press, Oxford

Williams M (2006) Deforesting the Earth: From Prehistory to Global Crisis. University of Chicago Press, Chicago, IL

Wilson EH (1986) A naturalist in Western China. Cadogan Books, London (first published 1913 as A Naturalist in Western China with Vasculum, Camera and Gun: being an account of eleven years travel, exploration, and observation in the more remote parts of the flowery kingdom). Methuen & Co., London

Wollenberg E, Ingles A (1998) Incomes from the forest. Center for International Forestry Research and the International Union for the Conservation of Nature, Bogor

Woods P (1989) Effects of logging, drought, and fire on structure and composition of tropical forests in Sabah, Malaysia. Biotropica 21:290–298

Chapter 3
Reforestation, Conservation and Livelihoods

> Trees owned and grown by the poor are not a panacea, but the evidence assembled indicates that they have more potential for reducing deprivation than has been recognized, and their potential is increasing. Seen from the point of view of the poor themselves, they are like savings bank accounts with low initial deposits and high rates of appreciation. Where ownership and rights to harvest and sell are secure, poor people plant more and harvest less than expected.
>
> Chambers and Leach (1989, p. 341)

Introduction

Conservationists tend to view deforestation largely in terms of its impact on biodiversity. But, because of deforestation, there is an equally compelling human tragedy unfolding as well. Many people used these forests in the past and are now trying to make a living in the degraded forests and lands that have replaced them. Their numbers are large and some estimates suggest there are 300 million people across the tropical world dependent on degraded or secondary forest for their livelihoods (ITTO 2002). Within Asia, Poffenberger (2006) has estimated there are 140 million 'forest-dependent' people (or 30% of the population) in Cambodia, Indonesia, the Philippines, Thailand and Vietnam alone. Some of these people live within the residual forests as shifting cultivators and hunter-gatherers. Others, such as farmers and artisans, live outside the forests but draw on them for various resources. When forests are degraded or lost it is the poorer people in these rural communities who are usually the most adversely affected. It is true that they themselves have sometimes contributed to the degradation process. But, as seen in Chapter 2, more often than not, degradation has been caused by the activities of the rich and more powerful members of society or by a lack of concern by governments about how forests and lands are managed.

The question is, what to do about this unfolding tragedy? The decline in forest cover has led some conservationists to press for more of the remaining forests being placed in protected areas (Terborgh 1999). Others have argued

D. Lamb, *Regreening the Bare Hills*, World Forests 8,
DOI 10.1007/978-90-481-9870-2_3, © Springer Science+Business Media B.V. 2011

that it is equally important to find ways of dealing with the large numbers of poor people now living in rural areas (Adams et al. 2004; Colchester 2000; Fisher et al. 2008). But might both problems be addressed by reforesting some of the degraded lands that have now accumulated? New forests around parks could act as buffers and protect the parks from agricultural clearings and fire. And new forests could create an economic resource that could help improve rural livelihoods. In the context of the Millennium Development Goals (World Bank 2004), reforestation might be a way of reconciling Goal 1 (to eradicate extreme poverty and hunger) and Goal 7 (ensuring environmental sustainability).

However, who could do this? Government forest departments have traditionally undertaken some plantation establishment but few have either the funds or the capacity to mount reforestation operations on the scale now needed. In fact, in recent years the rate of planting by government agencies appears to be declining. On the other hand, private timber companies are becoming more involved in plantation establishment and many of their plantations are large and exceed 100,000 ha. But the value of most of these plantations is often limited; their economic benefits are not widely distributed across the community and they usually provide only modest conservation benefits because they rely on exotic species grown on very short rotations.

A third group potentially able to undertake reforestation are smallholders living in these deforested lands. In the past many have planted trees in gardens or woodlots. Most of these plantings have been carried out for subsistence purposes but their planting designs have usually generated rather more conservation benefits than the simple monocultures used by government agencies and private companies. Might they also undertake commercial reforestation in a way that improved their livelihoods but also generated some conservation outcomes?

One obvious constraint on them doing so is that many of these people are poor; food is sometimes scarce, their access to health services is limited and they commonly live in a harsh economic environment. Most have no access to financial institutions and high inflation means the cost of saving money is high. If they are to contribute to the reforestation of degraded lands and benefit from doing so they will need assistance.

This chapter examines the relationships between rural smallholders, biodiversity conservation and reforestation. It begins by examining the ways in which many communities have traditionally used the resources supplied by natural forests for subsistence and trade. It then goes on to consider how, following deforestation, tree planting might be carried out to re-supply certain forest resources in a way that improves livelihoods and, at the same time, generates some conservation benefits as well. It concludes by examining the way reforestation might be carried out by communities and that done by individual households. But first, since many smallholders are considered to be living in 'poverty', it is useful to start the discussion by clarifying just what this means.

Defining and Assessing Rural Poverty

There is a good deal of debate over just how to define and measure poverty (Angelsen and Wunder 2003). One definition describes poverty as being subject to a pronounced deprivation of well-being. This is related to a lack of income and is accompanied by low levels of education and health, high levels of exposure to risk and a high degree of powerlessness (World Bank 2001). A commonly used and practical expression of this is to say that people living on less than US$1 per day are living in poverty because they are unable to satisfy their basic nutritional needs. Other measures define 'poverty lines' for particular places on the basis of the money needed to purchase a certain amount of food (e.g. 2,100 calories per person per day – the 'food poverty line') or a 'general poverty line' which includes the money needed to buy food plus a certain amount of non-food items. Some poverty analysts assess poverty using surveys of education, occupation or asset ownership (e.g. the number of cattle the household owns, whether a household has a motorbike or TV, etc.). The incidence of poverty in a region is then the frequency of people in a community who fall below these various thresholds whether they are externally or locally defined.

But poverty is more than just having a low income and poor people themselves often have a more multi-dimensional view of their problem. Thus some Laotian villages regarded 'poverty' as not having enough rice or having too few cattle. But they also recognized that a number of other factors including land tenure arrangements and that natural disasters contributed to declines or deficiencies in both these indicators (Whiteman 2004). Likewise, in the colonial era, many Papua New Guineans living in some of the more isolated parts of the country had adequate land and food but would have seen themselves as being 'poor' because of their lack of road access. They knew that without roads it was difficult for them to earn money by selling garden produce in a market, get access to health clinics or provide their children with an education.

In seeking a description of poverty that encompassed more than just per capita income or food intake Fisher et al. (2008) built on the earlier Sustainable Livelihoods Framework of the British Department of International Development (DFID 1999) and the analyses of the World Bank (2001) and suggested three broad descriptors – a lack of assets or capital, a sense of powerlessness and increased vulnerability to natural or economic crises. Fisher et al. (2008) provided a number of dimensions to each of these descriptors and ways in which these problems might be addressed (Table 3.1).

Hobley (1996) also made a distinction between the extremely poor, the coping poor and the improving poor. These represent a range of vulnerabilities and capacities for taking risks. As well, Rigg (2006) drew attention to the distinction between what he called 'old poverty' and 'new poverty'. The former represents the more traditional view of poverty and is the consequence of a lack of access to markets and government facilities such as education and health service. The latter is caused by the development

Table 3.1 Factors contributing to poverty and some potential methods of overcoming these (Based on Fisher et al. 2008)

	Dimensions of poverty	Potential methods of overcoming poverty
Lack of Assets	Natural capital	Expanding assets of poor
	Human capital	Encouraging private investments
	Financial capital	Increasing market access
	Physical capital	Improving technical knowledge
	Social capital	Debt relief
		Restructuring aid
Powerlessness	Social differences	Addressing social inequities
	Inequitable access to resources or the benefits of using these	Enhance ability to participate in decision making
	Inequitable access to legal resources	Pro-poor decentralization
	Unresponsive public administrations	Public administration reform
	Corruption	Legal reform
Vulnerability	Economic crises	Diversify asset base
	Natural disasters	Develop forms of risk management
	Social disasters	Provide safety nets

process (e.g. loss of customary land, deforestation by loggers) and people's engagement with the market and the cash economy on terms that have often been unfavourable to them. In some cases traditional communities relying on subsistence agriculture were well nourished and led normal lives by their own standards but suffered once the old social networks and systems of land tenure were changed.

Natural Forests and Livelihoods

The role of forests in improving livelihoods or in reducing poverty is complex. On the one hand, natural forests often contain commercially valuable resources. But, on the other, local people have rarely benefited from the more valuable of these resources. The most obvious example is the case of valuable timber trees where Governments have usually asserted ownership of and ignored the claims of traditional land owners. Some forest-dwelling people have even been statutorily excluded from access to timber resources to ensure privileged access by large timber companies (Sunderlin et al. 2005).

This is not the case in the Pacific where, in contrast to Asia, governments of most countries recognize traditional land and forest ownership claims. However, even under these circumstances, it has been hard for traditional communities to obtain a commercial benefit from the resources they own. Harvesting timber trees requires access to heavy machinery, finance and market knowledge and most traditional communities lack these. Communities have usually ended up selling their timber to logging companies who have these resources. Governments have sometimes taken it upon themselves to act as intermediaries between landowners and these logging companies arguing that this is necessary to protect people from unscrupulous operators. However, in most cases, this has still meant that only a small proportion of the revenue generated by logging finds it way back to the owners with most being taken by the foreign logging companies, well-placed politicians and unprincipled village leaders. In theory these landowners are economically rich but, unfortunately, they are also politically weak.

The situation is different with non-timber-forest-products (NTFPs) such as fruits, nuts, resins, bushmeat, building materials and medicinal plants (Fig. 3.1). These have been collected for sustenance and as well as for sale and people have often benefited from doing so (Wollenberg and Ingles 1998). Mayers (2006) quotes a study by Vedeld et al. (2004) who carried out a meta analysis of 54 studies undertaken across the tropical world and found 'forest environmental income' (largely fuelwood, wild foods and fodder for animals) made an average contribution to rural household incomes of 22%. Similar findings have been reported by Wollenberg and Ingles (1998) and Lopez and Shanley (2004). NTFPs can be particularly important in making up food shortfalls at certain times of the year or during periods of emergency such as floods, droughts or in times of war.

Despite this, there has been some debate concerning the overall importance of NTFPs for people's livelihoods. Dove (1993) has argued that most forest dwellers

Fig. 3.1 Gathering fuel and thatching material from forests and plantation areas in Vietnam (Photo: Sharon Brown)

are usually only able to get a very small proportion of the final market value of the more valuable products they harvest because middlemen and long market chains siphon off most of the profit. And others have argued that while NTFPs might be regarded as safety nets when other alternatives are limited they can also be poverty traps because people cannot significantly increase their income using these resources alone (Sunderlin et al. 2005). Morris et al. (2004) disagree and argue that, on the contrary, NTFPs can be 'ladders out of poverty'. They describe examples from Laos where simple changes in marketing arrangements for NTFPs (e.g. selling goods by weight) combined with other small-scale interventions led to substantial improvements in livelihoods over a relatively short time and a reduction in the proportion of 'poor' households in the community.

The importance of NTFPs to livelihoods can change over time; some are relatively inferior goods and are abandoned once per capita incomes rise while others are specialty products or luxury goods and the consumption of these may rise as incomes increase. In such cases the collectors often shift from working part-time to working full-time and specialise in a smaller number of products. The collection of medicinal plants is an obvious example.

There are four ways in which natural forest resources, including NTFPs, might improve the livelihoods of rural communities (Table 3.2). The first is if communities can be given legal access to the resources that forests contain and they can be assisted to manage, harvest and market these resources. This could be done by

Table 3.2 Ways by which natural forests might be used to help alleviate rural poverty (After Sunderlin et al. 2005)

Action	Mechanism or policy
Provide legal and fair access to the forest and its resources	Protect the forest (from illegal loggers and squatters) and allow people to make use of and benefit from its resources. Provide assistance in developing appropriate forms of silviculture and management that ensure production is sustainable and that benefits are equitably shared
Increase the value of forest products (timber and NTFPs)	Use technology to increase productivity (machinery rather than hand tools), enhance prices (via improved market access and marketing arrangements) and increase local value-adding activities (e.g. small rural sawmills and furniture factories. This may require access to new sources of financial assistance
Pay for the ecosystem services provided by forests	Develop mechanism for transfer payments to people for services such as watershed protection and clean water, carbon sequestration, recreational opportunities or the biodiversity provided by their forests
Clear forest and develop alternative land uses	Use land for more profitable uses such as agriculture

providing training and loans to the community so they can carry out harvesting themselves. Alternatively, it might be done by forming partnerships between communities (who supply the resource and protect it against illegal loggers or harvesters) and, say, a timber company (who provide finance and technical skills).

A second way is if the value of forest products being harvested from the forest can be increased. For example communities would obtain a better return if they produced sawn timber rather than simply logs (Fig. 3.2). Sawn timber might be produced using small, portable sawmills (Filer and Sekhran 1998; McGrath 1998). These have had a chequered history in the Pacific because the machinery has to be maintained in often isolated situations and the timber has to be carried out of the forest and sold to a buyer willing to accept small volumes of sawn timber arriving at irregular intervals. In theory sawmilling should generate a greater cashflow but, to date, there appear to be few situations where small portable sawmills they have done so over an extended period. In the case of NTFPs, marketing cooperatives and small scale factories or processing plants could increase the return to communities (Morris et al. 2004). Some NTFPs may even be sufficiently valuable to be worth domesticating and growing in plots on farms.

A third way in which forests might used to improve livelihoods is through arranging for payments to be made for the provision of the ecosystem services supplied by forests. These services might include watershed protection, clean water, carbon sequestration and biodiversity protection. There are relatively few markets available to rural people for any of these services at present although they are likely to become more common in future. This topic will be discussed in more detail later in Chapter 9.

Fig. 3.2 Sawn timber being produced using a chain saw and a portable frame in Solomon Islands. This allows traditional forest owners to produce boards and flitches from their own logs and so increase their income

The fourth and final way in which natural forests might be used to improve livelihoods is by clearing them and converting the land to another use such as growing cash crops. This choice has been widely adopted but is not available to everyone because most cash crops need roads to reach a market. Logging and the roads created by loggers remove this limitation which is the reason why there is often such a flurry of land clearing after logging. And cropping is not feasible if the land is unsuitable because it is too steep or relatively infertile. Needless to say, large areas of such marginal lands are still being cleared and much of the abandoned and degraded land now present in the world's tropical regions has originated in this way. By clearing these marginal lands people are increasing their risk of failure and, at the same time, losing the safety net the forests once provided.

In the early 1960s Westoby (reprinted in Westoby 1987) tried to counter such agricultural clearing by arguing that a permanently conserved and properly managed forest estate was an ideal vehicle by which rural development could be achieved. He noted that forests were a renewable resource and the use of these would sponsor industrialization (through the establishment of sawmills). This, in turn would generate rural employment and facilitate light industries associated with this timber processing (e.g. vehicle maintenance, sawmill machinery maintenance, etc.). Benefits would then flow through the rural community and improve rural livelihoods. The argument was seen as a powerful rejoinder to those who saw forests as impediments to rural development and a largely residual land use. However, some years later at the 1978 World Forestry Congress in Jakarta, he sadly acknowledged that he had been wrong (Westoby 1987). Forests had not lifted the poor out of poverty but instead the benefits had often been captured by a small expatriate and indigenous elite. Industrialisation had not developed to the degree he had expected and forests were not being managed as renewable resources. Instead, most had continued to shrink in area or become degraded, including those located within a supposed permanent forest estate.

Since that time there has been considerable debate about the role forests could have in improving rural livelihoods and reducing poverty. But the debate has become further complicated by rising concerns about the global biodiversity losses arising from deforestation.

Biodiversity Conservation or Livelihood Improvements?

There is no question that biodiversity is under threat throughout the tropical world and the gravity of the situation in the Asia-Pacific region has already been discussed. Two solutions have been proposed. One is to substantially increase the size of the protected area network. In tropical forests this currently stands at about 10% (although Table 1.9 showed much higher values were present in many Asia-Pacific Countries). An increased protected area network would preferably involve new large contiguous protected areas including strict nature preserves within a matrix of 'soft' protected areas (e.g. IUCN category I–IV areas within a matrix of category VI reserves). The proponents of an enlarged protected area network say it is only widespread ignorance of the magnitude of the threat to biodiversity and a lack of political will that prevents this being achieved.

However, enlarging the protected area network will be difficult. Many forest areas still contain significant human populations and most of these are likely to resist being forced from their customary lands. This response is already clear in the Pacific where traditional land ownership is legally recognized and no community has yet agreed to the creation of large protected areas that would constrain their future land-use options. Where governments have asserted state ownership over forest lands the establishment of large National Parks has frequently caused gross violations of human rights and the economic and the political marginalization of large numbers of people (Fisher et al. 2008). Some of these have been forcibly relocated causing a collapse of traditional management systems and a loss of access to livelihood resources. In many cases there have been increases in rural conflict and famine.

A second suggested way of improving biodiversity conservation is to enforce the protection of existing National Parks and nature reserves more strictly than has occurred in the past and prevent rural people from continuing to use forest resources within these reserves. According to this view, governments even have a duty to limit individual freedoms and move people to protect the 'common good' that undisturbed nature represents. This means removing customary landowners living within National Parks as well as illegal squatters and loggers moving in from outside (Terborgh 1999; van Schaik et al. 1997; Wilshusen et al. 2002). Terborgh (1999) has proposed the establishment of an internationally financed, elite policing group which could take over park protection since many national governments have shown they are unable to provide it to the extent needed.

> If local park guards are too weak or too subject to corruption and political influence to carry out their duties effectively, internationally sponsored guards could be called in to help. As foreigners, they would be independent of local pressures and thus better able to exercise authority
>
> (Terborgh 1999, p. 201)

Opponents have responded by saying these authoritarian views ignore the social and political realities present in most tropical forest areas (Brockington et al. 2006; Wilshusen et al. 2002). It is unlikely many governments would be in a position to adopt the stronger policing model even if they wanted to. Moreover, the numbers of people living in and around many parks are simply too large to be easily removed and re-settled. In Thailand, for example, up to one third of rural villages are close to or within protected areas and depend on them for forest or marine resources (ICEM. 2003a). The same is true in other places across the region. Others sympathetic to the problem of parks being gradually degraded over time say that that carrots are needed as well as sticks and that alternatives need to be found to draw people away from parks rather than using police to force them out.

But perhaps the key problem with both suggestions is the moral issue: is it ethically acceptable that so many millions of people continue to remain living in poverty in forested landscapes of the tropics when the standard of living in the developed world is so high? The issue is highlighted by a recent case in Vietnam where a large

World Bank project was established targeting 540 of the poorest communes in the country for various infrastructure projects (clinics, schools, roads, bridges). However, 86 selected communes were excluded because they are partially or totally within protected areas and there was concern that the new infrastructure might have adverse effects on these reserves (ICEM 2003b). It is not clear just what park managers have to offer these traditional land owners apart from sharing in the management of eco-tourism. Across the region few forest people derived any benefit from the logging that has led to the deforestation crisis and it is unfair that they should now be expected to carry the burden of conserving much the world's remaining biodiversity. Nor should they be expected to live in biological museums (Colchester 2000; Schwartzman et al. 2000; Wilshusen et al. 2002). As Kaimowitz and Sheil (2007, p. 572) argue:

> For hundreds of millions of people, biodiversity is about eating, staying healthy and finding shelter. Such needs, in addition to those of the tiger and other endangered species must also be considered a conservation priority. Clearly it must not a question of either/or but rather of finding a better balance.

Just what sort of balance might be possible? Adams et al. (2004) outlined four alternative perspectives:

- *Biodiversity conservation in forests and poverty reduction are two separate realms.* Both are important but each must be pursued independent of the other. Following this viewpoint leads to a program of protecting forests by creating strictly protected nature conservation areas with some kind of matching rural development work in the agricultural matrix away from the protected areas. This carries the implication that the areas immediately surrounding protected areas are not very relevant for biodiversity conservation.
- *Biodiversity conservation is important but attempts to achieve this should not compromise efforts to eradicate poverty.* This acknowledges that conservation activities can sometimes have adverse effects on people (e.g. by forcing them off their ancestral homelands) and takes the view that ways must be found to avoid or at least compensate people who suffer in such ways. An example of this might be the development of eco-tourism activities within the new reserve that are managed by former land owners or the payment to former landowners for the ecosystem services arising from forest conservation.
- *Poverty is a critical constraint on forest conservation.* This acknowledges that conservation and poverty eradication are not separate realms. Rather, it assumes it will be impossible to achieve biodiversity conservation unless poverty is also overcome because people will have no alternative other than to continue to use the resources in protected areas. A consequence of this viewpoint might be to seek ways to improve livelihoods in buffer zones around protected areas so people have alternatives to using the protected areas.
- *Any reduction in poverty is dependent on conserving living resources and thus, on protecting biodiversity.* This view argues that there is an intimate connection between improvements in human livelihoods and the conservation of biodiversity including that found in forest areas outside the formal protected area network. Poor ecosystem health will undermine social and economic stability

and the livelihoods of rural people. It points to the need for not only protecting this biodiversity but also developing methods of achieving sustainable use of these forests to generate goods and services and improve human livelihoods.

The last two are closely linked and suggest forest conservation and poverty reduction could be tackled together through reforestation. There might even be scope for a 'win-win' outcome or, perhaps more realistically, what Fisher et al. (2008) referred to as a 'win-more-lose-less' outcome since some trade-offs will be required.

Considerable effort has been made to develop forms of interventions that explore these trade-offs. These have been termed 'integrated conservation and development projects' (ICDPs) and are essentially people-oriented approaches to conserving biodiversity. Most have been linked in some way with an existing protected area and have been based on the assumption that landholders would switch to practices not causing a loss of biodiversity if offered alternative opportunities to make a living. There has been considerable debate about the efficacy of ICPDs with many arguing that the evidence largely shows they have not achieved their purpose (i.e. the compromises struck have outweighed the benefits). Even proponents admit this; despite considerable effort there are few unambiguously successful projects that have achieved improved biodiversity conservation and also improved livelihoods (McShane and Wells 2004; Wells and McShane 2004; Wilshusen et al. 2002).

There are several possible reasons for this apparent failure. One is that too much may have been expected in too short a time. Attempts to change the economic circumstances and social relationships among large numbers of people dispersed over big areas takes time and most development projects are inherently unsuited to this. A second potential explanation is that the success of any ICDPs invariably depends on having in place the right policies, institutional frameworks and laws. But this means that failures may have been caused by policy settings and institutions operating at a national rather than local scale and few ICDPs have tackled these larger issues. Berkes (2007) suggests more success has come from projects involving networks with multiple partners and especially when these interactions involve four or more levels of organization (e.g. local communities, regional government agencies, regional NGOs, national bodies, international groups, etc.). He also argues (p. 15191) that much 'of the so-called community-based conservation of the last 2 decades or so has been half-hearted, misdirected and theory-ignorant'. In short, some of the past problems with ICPDs have had less to do with the fundamental concept and more to do with the ways these projects have been carried out (Wilshusen et al. 2002; McShane and Wells 2004).

Reforestation to Enhance Livelihoods and to Foster Biodiversity Conservation

Reforestation of degraded or under-utilized land outside the context of a single externally funded project such as an ICDP may offer more opportunities. Provided the type of reforestation used suits the particular circumstances of farmers there is good reason to expect tree-growing can often improve farmer's livelihoods. It could do so by improving household assets, diversifying income sources and providing a cushion against economic shocks or other unexpected contingencies. In this sense, trees on farms are like bank deposits in places where there are no banks (Chambers and Leach 1989). Likewise, reforestation could provide significant conservation benefits. Most protected areas are too small to sustain all the species they contain and are not big enough to withstand the changes that global warming may bring. Reforestation of areas around and between protected areas could provide buffer zones at park margins, improving the connectivity between forest patches and allow population movement and genetic inter-change between separated populations. Reforestation could also help conserve biodiversity in areas outside the park network by sustaining those species still present in small forest remnants or enabling species to recolonize other remnants. And, perhaps most important of all, reforestation has the potential to eventually become a self-sustaining enterprise unrelated to the constraints of project cycles but driven by the self-interest of landholders.

Reforestation does have an opportunity cost and a shift away from growing food to growing trees might make some households more vulnerable to unexpected events. But not all farmland is equally productive and trees might be grown on less fertile or steeper areas of land less suited for growing food (Fig. 3.3). And, unlike many crops, trees require little labour once they are established. In short, the benefits of tree-growing may be significant while the spatial and temporal costs may be modest.

Fig. 3.3 Plantations in Vietnam are often established on less fertile soils on hills while rice is grown in the more fertile valleys

Tree plantations currently account for 7% of global forest cover and are increasing at the rate of five million hectares each year (FAO 2010). Mayers (2006) argues that the overall environment for reforestation is currently improving because:

- Natural forests able to supply species favoured by the market are shrinking; many of the remaining forests have been over-cut in the first cutting cycle and will be unable to support another harvest in the immediate future. Countries that were once ‘forest-rich’ are becoming ‘forest-poor’. The looming timber scarcity should also lead to higher financial returns from plantations.
- There are growing demands for forest products as populations and standards of living increase.
- There are technical and market developments that permit the use of smaller logs derived from plantations so that shorter rotations are possible. This should make tree-growing more financially attractive to smallholders. Note, however, that these developments might also mean small trees remaining in cut-over natural forests can also be marketed and this increases the risk that these forests will be prematurely logged a second time.
- There are increasing demands for the ecosystem services provided by forests such as carbon sequestration and watershed protection.
- There are opportunities for niche markets in a globalised world for higher-value, and hence more profitable, timbers.
- There is increased attention to and, possibly a reduction in, corruption and illegal logging in natural forests meaning forest growers can benefit more from tree growing.

It is usually assumed that the majority of reforestation is carried out by larger industrial groups or state forestry agencies. In fact smallholders play a much greater role than is generally appreciated. One ‘rough estimate’ suggested there were 500–1,000 million smallholders who grow trees on farms or manage remnant forests for subsistence and income (Scherr et al. 2004). A more quantitative estimate comes from the results from an international survey and reported by del Lungo et al. (2006) and Carle (2007). This showed smallholders are currently responsible for significantly more plantations that corporate or industrial groups. This was true at a global level (where smallholders had established 49.9 million hectares of plantations while corporate groups had around planted 27.2 million hectares of plantations) but was even more the case in Southeast Asia where smallholders have established 2.3 million hectares while corporate groups have established only 0.6 million hectares of plantations (Table 3.3). These data have to be qualified since neither Indonesia nor Myanmar recognised any corporate or smallholder plantations. Likewise, Thailand claims state ownership of all plantations except for rubber. And national data like this is almost certainly likely to under-estimate a resource made up of small and scattered plantations. Nonetheless, the data point to a significant role being played by smallholders.

The conventional view is that large reforestation projects such as those carried out by government agencies and corporate groups are more successful than small-scale plantings. This is because of economies of scale and because governments and industrial plantation owners can afford more technically sophisticated management. This is not always true and many government sponsored planting across the region

Table 3.3 The areas (×1,000 ha) of productive plantation established by different owners in 2005 (Del Lungo et al. 2006)

	Public	Corporate	Smallholder	Other
Global	77,352	27,176	49,980	492
Southeast Asia	6,758	637	2,302	65
Indonesia	3,399	0	0	0
Malaysia	263	227	1,084	0
Myanmar	696	0	0	0
Philippines	186	43	75	0
Thailand	1,723	274	0	0
Vietnam	491	93	1,143	65

have failed (Chokkalingam et al. 2006; Nawir et al. 2007). And while well-managed government and corporate plantations can be very productive, the same can also be true of well-managed smallholder plantations. In addition, most of these are less likely to suffer from wildfires than are industrial plantations because owners can afford to look after their plantations on an almost tree-by-tree basis.

Of course these types of comparisons also depend on how 'success' is defined. In many cases timber productivity is only one of the measures that might be used. Mayers (2006) argues that with technical advice and with appropriate policy settings in place, small and medium forestry enterprises offer some considerable advantages to rural communities and households including spreading wealth more widely, empowering local communities and making greater commitments to operating in specific areas. The large overall area of smallholder planting activity also means that even small improvements in silviculture and productivity in farm plantations can have a very significant collective impact.

Types of Reforestation

Reforestation is usually defined as the re-establishment of forests on sites deforested by human activities or natural disturbances within the last 50 years (Carle and Holmgren 2003). Many people equate reforestation with large industrial plantations of eucalypts or pines. But the options are much greater than this even if they are relatively unknown amongst many rural communities. As well as industrial plantations, there can be farm plantations, agroforestry plantings and various kinds of environmental or conservation plantings. In the present context reforestation is also taken to include forests that have regenerated naturally. Some of the features of these main types of reforestation are as follows:

Plantations of Pulpwood and Commodity Grade Timbers

Fast-growing pulpwood plantations are being established in many parts of the Asia-Pacific region. These are often grown for pulpwood using rotations of less than 10 years. By 2000 there were 14 companies in Indonesia alone with pulpwood plantations exceeding 100,000 ha (Effendy and Hardono 2000). Across Asia as a whole these industrial plantations may now cover 4.5 million hectares (Mayers 2000). Most of these use only a single species drawn from exotic genera such as *Eucalyptus* or *Acacia* (Cossalter and Pye-Smith 2003). These species are chosen because their seed are easy to get, their timbers are known to be suitable for the pulpwood market and the silvicultural knowledge needed to grow these species in plantations has already been developed. As a result of tree-breeding some eucalypt plantations can now produce yields over of 40 m^3 ha^{-1}year. Many more farmers would carry out reforestation if they were more familiar with the technology. Managers of National Parks would use more suitable forms of reforestation to rehabilitate degraded areas within their parks if they knew how to do so. How can existing knowledge be shared? How can it be communicated in a way that makes sense to the people who might use it?

In practice the average plantation usually achieves rather more modest yields than these and the productivity of some widely planted species found in a regional survey by ITTO are shown in Table 3.4. Though these yields are less than the levels reached in experimental plantings or in well-managed sites, they are still much greater than those produced by most natural forests.

While these types of plantation are clearly profitable for industrial groups their value to smaller growers depends on the plantation location and on the relationship these growers have with a market. Small farm plantations producing cheap timber and located some distance from markets are unlikely to be financially rewarding to a grower because the volumes they can produce are small, harvesting is infrequent and transport costs are high. To be successful these types of growers need to be part of a marketing cooperative or they need to develop a long-term arrangement with an industrial partner as outgrowers or in some form of joint venture. There are a number of examples where such arrangements have been made. Some have been very beneficial to growers while others have been exploitative relationships. Accounts of some recent experiences are provided by Mayers (2000); Mayers and Vermuelen (2002); Angelsen and Wunder (2003); Nawir et al. (2003) and Scherr et al. (2004).

Commodity grade timbers can be grown in monocultures but on longer rotations (20–30 years) to produce sawlogs. Species such as *Gmelina arborea* or *Pinus* spp. are often used for this purpose. Timbers of these species usually attract only modest prices which means that, like pulpwood plantations, these plantations are unlikely to be profitable unless growing close to a market. The profitability of different types of plantations will be discussed further in Chapter 9.

Table 3.4 Average productivity in the Asia-Pacific region of some commonly used plantation genera based on survey commissioned by ITTO (STCP 2009)

Species	Productivity ($m^3ha^{-1}year^{-1}$)
Eucalypts	15–20
Acacia	6–8
Pine	10–12
Teak	4–6
Rubber	10–15

The conservation value of both of these types of plantation is modest. Some critics have labelled them as 'green deserts'. This is a little ungenerous since it depends on the plantation design and they can provide habitats for some wildlife and native plant species under appropriate circumstances. The Grand Perfect plantation estate in Sarawak, for example, plans to use less than half of its total area for plantation trees with the remainder being used for nature reserves, riverine strips, corridors and other purposes (Cyranoski 2007). A similar partitioning between plantations and natural forest is being developed at the Sabah Forest Industries plantations in north western Sabah where 62% of the 290,000 ha concession will remain as natural forest protecting steep slopes and riverine strips (Wooff 2009).

Any assessment of the relative merit of a particular plantation also depends upon what it replaces. Even a pulpwood plantation may offer some conservation benefits if it replaces a degraded grassland or impoverished shrubland. On the other hand, a plantation established by clearing undisturbed natural forest is another matter, especially when there is so much degraded land and forest across the region. One estimate suggests 15% of Southeast Asian pulpwood plantations have been established by replacing natural forest (Cossalter and Pye-Smith 2003) while Tyynela et al. (2003) suggest the comparable statistic for Indonesia is 22%. The situation becomes less clear when already degraded forests are cleared. How important are these for biodiversity conservation? Would they have been allowed to recover if plantations had not been established or would they have been converted to another land use such as oil palm? Under these circumstances the Grand Perfect and Sabah Forest Industries plantation designs seem a useful compromise because they guarantee large areas of secondary forest remain protected.

Plantations of High-Value Timbers Grown on Longer Rotations

Plantations of 'higher-value' sawlog or plywood species grown on longer rotations of 30 years or more are less common because of the longer time periods needed to achieve a cash return and their generally lower productivities. But they offer the prospect of much higher prices than pulpwood timbers and have attracted attention from researchers and some state forestry agencies, particularly since supplies from natural forests have begun to diminish (Fig. 3.4). The term 'high-value' is a loaded term because all trees can be 'high-value' in certain circumstances. It is used here to refer to species able to produce specialty timbers or what are sometimes called 'cabinet timbers' and which have a high commercial price. Some of the more popular species include teak (*Tectona grandis*), rosewood (*Pterocarpus indicus*) and mahogany (*Swietenia macrophylla*) but a very large range of native species have been tested and used in state plantations across the Asia-Pacific region (Appanah and Weinland 1993; Do and Nguyen 2003; Erskine et al. 2005; Evans and Turnbull 2004; Krishnapillay 2002).

Some of these species have also been grown by smallholders and they can be especially attractive in more isolated locations where high transport costs make lower value timbers uneconomic. Thus in the Solomon Islands where transport is

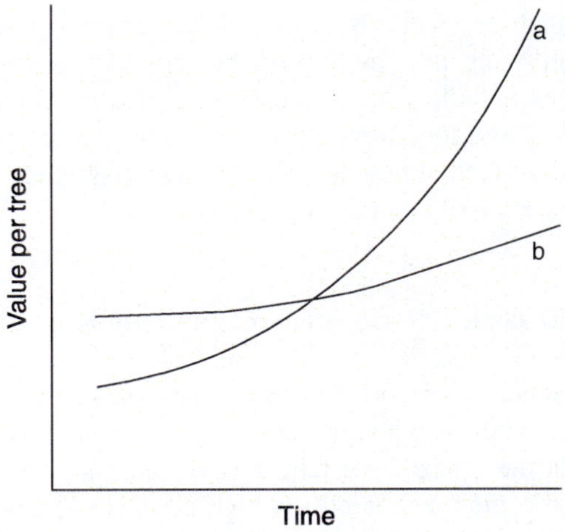

Fig. 3.4 High-value timber species (**a**) are less profitable than pulpwood species (**b**) in the short term but the value of each tree is likely to be considerably greater over time

expensive it is still profitable to market even a single shipping container of plantation timber provided it contains a high value timber such as teak (Raymond and Wooff 2006). The quality of timber in logs coming from smallholder plantations is likely to be less than that of logs from natural forests but the price advantage is considerable and there are silvicultural opportunities (e.g. via pruning and thinning) to improve log and timber quality over time.

These types of plantations offer some conservation advantages over short rotation pulpwood plantations because their longer rotations enable more structurally complex understories to develop. These advantages increase when native species are used in preference to exotic species. They are also likely to provide better watershed protection because the longer rotations mean disturbance occur less frequently and protective understories and litter layers are able to accumulate on the forest floor.

Multi-Species Plantations of High-Value Species

Multi-species plantations or polycultures have not been greatly favoured by silviculturalists in the past because they are more difficult to establish and manage than monocultural plantations (Wormald 1992). On the other hand, these more complex designs have been widely used by farmers in the tropics in traditional agroforestry plantings because they increase the variety of goods produced and, at the same time, reduce the risk of total crop failure. A plantation mixture might include high value timber trees as well as multi-purpose tree species able to produce NTFPs. Although multi-species plantations are essentially versions of agroforestry plantings they have the potential to be more financially rewarding than most traditional agroforestry plantings. They may also have some financial advantages over simple plantation monocultures. For example, they might include species or crops able to mature and be sold at an early age leaving slower growing and more valuable species to continue growing until they reach a marketable size. Mixed-species planting are likely to offer rather more biodiversity and functional benefits than the two types of monoculture plantations already described although the nature and magnitude of these benefits will depend on the species and plantation designs used.

Ecological Restoration or Environmental Plantings

Some species-rich plantings are established to restore biodiversity and protect watersheds without there being any intention of a harvesting operation some time in the future. Tree felling is not normally allowed in such forests. This, together with their comparatively high establishment cost suggests such plantings are more likely to be used by government agencies and NGOs to reforest small strategic locations than by poor farmers. But some communities undertake these types of plantings for religious and subsistence reasons and Box 3.1 describes such a case in northern Thailand. The conservation advantages of these types of plantings can be very high most especially when they attract seed-dispersing birds which bring in other species to supplement those that were planted.

Natural Regeneration

Sometimes forests are able to regenerate at degraded sites such as former farmlands if they are protected from further disturbances. These forests can vary enormously in composition and value (both for subsistence and for commercial purposes). Some may be rich in species and have large numbers of rapidly growing timber trees or high densities of NTFP species. Others may have relatively few of these species but can be enriched with them. There are many examples across the Asia-Pacific region of such agroforests enriched with species able to produce NTFP (Clarke and Thaman 1993; Michon 2005). There are also many examples where heavily logged forests have been enriched with timber trees (Dawkins and Philip 1998). Properly managed, these forests can have benefits for both livelihoods and biodiversity.

Each of these reforestation alternatives differs in their capacity to improve human livelihoods and conserve biodiversity and these differences are summarized in Table 3.5. The most promising approaches that contribute to both objectives are those involving natural regeneration or native species grown in plantation mixtures while pulpwood monocultures and restoration plantings are the least attractive (although they may be highly suited for just timber production or just biodiversity conservation). All of these options are open to smallholders or communities although they are most likely to be attractive in regulatory environments with low costs of entry and operation, limited numbers of regulations, few subsidies to large industrial groups (including loggers) and secure forest rights (Scherr et al. 2004).

Box 3.1 Forest Restoration in Northern Thailand by Villagers Lacking Formal Land Tenure

Many communities living in deforested areas are interested in establishing multi-species forests near their villages. An example is the forest established by villagers of Jabusee village near the town of Mae Salong in the Province of Chiang Rai in northern Thailand. The 30 households are all Akha people who probably settled there in the 1970s and now grow corn and raise livestock. When they first arrived the area was highly degraded and covered by *Imperata* grasslands even though it is supposedly part of the national forest estate administered by the Royal Forestry Department. This deforestation had probably resulted from shifting cultivation carried out by earlier inhabitants. With the help of an NGO, the Hill Areas Development Foundation, the community reforested about 10 ha of this land using a variety of tree species including many native species. Some of this was done by planting seedlings and some by natural regeneration. The outcome has been the creation of a species-rich forest with a closed canopy growing close to the village. Most of this forest is now more than 10 years old and is able to supply a variety of NTFPs. These are utilized by households according to a set of specified conditions. The village has also developed an eco-tourism business based around the forest. Being on a ridgetop the site was probably not very good for cropping but there appear to be several other reasons for undertaking this particular form of reforestation. One is that such forests are part of the Akha religious tradition and are used to protect sacred areas. But another may be that villagers knew that the government classes the area as an important watershed and hoped that the reforestation they have carried out will eventually assist them in their search for formal land tenure and Thai citizenship. An account of the biological diversity in another newly created village forest not far from Jabusee village is given in Box 5.2.

Some Qualifications

There are several qualifications that need to be made about these four alternatives.

Environmental Conditions Will Constrain Silvicultural Choices

The more degraded a site, the fewer the choices. Perhaps only a handful of exotic species will be able to grow at highly degraded sites. In these cases it may be possible to grow monocultures of timber trees but the ability to restore much of the original biodiversity may be limited (although it may be possible to improve ecological processes and functioning). In such situations a multi-step process may be needed with more native species introduced at a later stage once environmental

Table 3.5 Forms of reforestation that overcome degradation and improve conservation values

		Overcome degradation and improved conservation values	
		More beneficial	Less beneficial
Capacity to improve livelihoods	Higher	Natural regeneration and enriched secondary forests Mixed species plantations of mostly high-value species grown on long rotations Monoculture plantations of high-value native species (timber or multi-purpose) grown on long rotations	Monoculture plantations of fast growing exotic species grown on short rotations (especially if close to markets and/or private growers have a long-term and mutually beneficial contracts with industry)
	Lower	Restoration plantings for strictly biodiversity purposes	Company owned plantations established at the expense of existing land owners Farmer owned plantation established but failing because of inappropriate species choices, management methods, pests, disease or wildfire

conditions have been ameliorated by the exotics. Some silvicultural options for dealing with this issue are discussed further in Chapter 7.

Not All Smallholders May Be Willing or Able to Engage in Reforestation

Not everyone may be interested in timber tree planting. Those without land tenure are unlikely to do so because of the risk they will not be present to benefit when the trees mature. And of those with tenure, the poorest farmers may not engage in reforestation because they need all the land they have to produce food or they do not have enough money to pay the initial planting costs. For them the opportunity costs of any kind of reforestation may simply be too high (although agroforestry may be attractive – see further below). Poverty alleviation for such people may require land redistribution, off-farm employment or some kind of social protection. Other people with land may not be able to participate in reforestation because they don't have, or are unlikely to get, road access to enable them to harvest and market their logs. This is because even small-sized logs need tracks that small trucks or tractors can pass over. Some of the factors influencing landowner's decisions about tree planting are discussed in Chapter 10.

The Most Appropriate Silvicultural Systems Are Complex and Many Are Still Being Developed

Plantation monocultures such as those used to produce pulpwood are relatively simple to establish and manage. However, the types of reforestation able to deliver both livelihood and conservation benefits must be rather more complex. It is because of this that they are rather less attractive to most large industrial growers. On the other hand, this complexity is not necessarily an impediment to smaller growers who are interested in 'quality' as well as 'quantity' and in creating forests generating a variety of goods. The wealth of agroforestry practices already being used provide a starting point for those interested in developing these multi-purpose plantation designs (Michon 2005). The issue is discussed further in Chapter 4.

These Options May Vary with Time

The difference between the four cells in Table 3.5 may not be as definite as they seem. Circumstances may change during the life of a plantation and make it less or more profitable than expected. For example, markets for cheap pulpwood timbers may decline because of oversupply while to price of higher value timbers may rise as the supplies from natural forests decrease. Similarly, a restoration planting established largely for biodiversity restoration (a potential win-lose situation) may also generate livelihood benefits because of the NTFPs they produce or the development of eco-tourism thereby becoming a win-win situation.

There May Be Other Silvicultural Options

The several alternatives described here do not adequately represent the full range of silvicultural options that could be available. For example, plantations differ in whether native or exotic species are planted or in the lengths of the rotations used, but they also differ in the types of trees planted (timber trees, wildlife attracting species or multi-purpose species able to supply various NTFPs) and whether they are also planted with non-tree species such as food crops or medicinal plants in a system perhaps better described as agroforestry. Different species and different planting designs will have different consequences for the way these ecosystem function and the ecological processes they are able to sustain. All these variables affect both the conservation benefits arising out of reforestation as well as the contribution reforestation can make to improving livelihoods. These options will be discussed in more detail in later chapters.

Of course not all options will be available in every ecological or social circumstance and the variety of choices is likely to be greater in, say, a humid location with a low population density than in a seasonally dry area with a dense population.

There Must Be Trade-Offs Between Production and Biodiversity Conservation

Finally, it is important to acknowledge that plantations designed to conserve biodiversity as well as generate cash incomes run the risk of doing neither particularly well. This means trade-offs will be needed. These may easier for smallholders to make than large industrial plantation growers since many smallholders will be growing trees for reasons other than simply maximising timber production. For example, they may be concerned with reducing financial risks by using several species and prefer to use trees that produce fruit as well as timber.

There is also a scale issue involved. Not all trade-offs must be made at every site since landscapes are not uniform and biodiversity 'hotspots' and poverty 'hotspots' may not necessarily overlap. This means trade-offs may be easier to make at a landscape scale and the type of intervention made at particular locations may depend on the context such as population densities, market access, opportunity costs of alternative land uses or the amount of remnant forest still present. These landscape issues will be discussed further in Chapter 11. But trade-offs cannot always be made and there may be some situations where it is not possible to use market mechanisms alone to achieved the required conservation outcomes. In such cases special payments will be needed to compensate growers for doing so ('if biodiversity cannot pay then biodiversity must be paid for').

In short, Westoby may have been right in his original 1962 hypothesis that under the right conditions, forestry – but in this case plantation forestry – should have a beneficial role to play in rural development even if it cannot solve all livelihood problems (Westoby 1987). And, at the same time, at least some forms of reforestation should also be able to improve conservation outcomes on degraded lands even if these outcomes may take some years to appear. However, as already noted, the fact that one or other forms of reforestation may have some potentially useful contributions to make to rural development does not necessarily mean any of these they will be taken up by landholders. As past experience with ICDPs has shown, there are a host of factors other than technological factors that influence such choices. One of the most important of these factors is the system of land tenure.

The Role of Land Tenure

Tenure refers to a right of access to land and the use of the resources found on this land. There are many forms of tenure but these usually determine the individuals or communities who can use various resources, the conditions under which they can do so and the lengths of time that they can have control (FAO 2002). Thus rights might involve the rights to use land, the right to derive an income from the land and the right to lease or sell that land. Some farmers have legal title to the land they are using while others live in a more ambiguous situation where they only have *de facto* tenure based on the fact that their claim is recognized by the community in which they live but not by the government.

Patterns of land tenure vary across the region and are currently undergoing significant changes in many countries. In reviewing the relationship between land tenure and reforestation it is useful to begin by examining traditional forms of tenure. In traditional societies quite complex systems usually developed linking individual tenurial rights and communal obligations (Cleary and Eaton 1996; Crocombe 1982; Eaton 2005). These various systems often shared certain characteristics. One was that a person's rights were usually acquired through membership of a particular clan, lineage or group. Within this group, rights could be allocated and transferred according to traditional practices. Decisions could be made by the group as a whole or by chiefs or other community leaders. Rules usually regulated who had access to the group's forest or land and, in some cases, how these resources might be managed. Some people had usufruct rights (i.e. to enter a site and use the products growing there) or control rights (i.e. to plant crops or trees or to control who else uses the site). Sometimes systems differentiated between the rights of a landowner and the rights of someone who has planted a tree on that land. In some cases the act of clearing land or planting a tree was seen as conferring 'ownership' of that land on the planter.

These rights could last for years or might lapse if not maintained by regular visits. For example, Freeman (1955) noted the rights of Iban farmers in Borneo often lapsed about 7 years after a garden has been abandoned. In Papua New Guinea in the 1970s it was noticeable that most male university students living in the capital, Port Moresby still took care to periodically visit their often distant home villages to maintain their usufruct rights. Byron (2001) quotes an extreme example of the sometimes complex nature of customary tenure. This involved a single mango tree in Bangladesh over which seven families had tenurial claims. These included the family who had planted it, the current and previous owners, the family engaged in marketing the fruit and the local sawmiller who had a 'lien' over the tree once it was ready for felling. Despite this complexity the system worked although, as Byron wryly notes, any breakdown would almost certainly lead to the demise of the tree.

A second characteristic of many customary tenurial systems is that household land holdings are flexible and can vary over time. Few customarily owned lands have formally recorded clan boundaries or individual holdings and most boundaries were defined by natural features such as rivers or ridge lines. This often led to disputes over ownership which sometimes led to warfare and changes in land ownership. However, most customary systems were also flexible and allowed some transferring of use or ownership between members of different groups. These mechanisms included gifts, exchanges or the adoption of children. Land ownership patterns within groups were normally adjusted by re-distribution after death and the factors determining how this was done included gender (of both the transferee and heir), age, seniority of birth and personal standing (Crocombe 1982).

These basic arrangements could be overlain by more complicated local arrangements. For example, kings, sultans and other local rulers in Asia often claimed sovereignty over certain lands (mostly coastal lowlands and river floodplains rather than the hills). In practice, villagers usually claimed rights to actually use the lands and worked out amongst themselves how these resources might best be distributed in order to serve the interests of individuals, families and the state. Such arrangements were established in customary law and practice and were strongly defended against external challenges (Chandler et al. (1987). In the Pacific, and especially Polynesia, royal families and noblemen also claimed certain special privileges over land ownership and use.

Many of these traditional systems have either changed or are changing and the present patterns of rural land ownership across the Asia-Pacific region now include traditional practices derived from customary and feudal societies as well as those developed during the colonial period and from more recent political systems. In most Asian countries the state has usually asserted its rights to own or manage the ownership of much of the land within its boundaries. As a consequence, land has often been compulsorily taken from customary owners when it suited the State's purposes. Some has subsequently re-allocated to others, including foreigners, through a system of freehold purchases, grants or leases. The process has been described by Jacoby (1961); Chandler et al. (1987); Cleary and Eaton (1996); Sato (2000) and Eaton (2005).

This assertion of state ownership of land in Asia has meant that customary owners have often become squatters on their own land. The areas involved and numbers of people affected by this transfer of ownership are very large. For example, Li (2002) quotes estimates of 60% of the national territory of the Philippines and 75% of Indonesia were taken over in this way. In the case of the Philippines, the land in question was home to about 24 million people or one third of the country's population. Being officially landless these people became subject to eviction or displacement if confronted by new large-scale agricultural projects or other state-sponsored enterprises. Unsurprisingly, this has sometimes led to conflict. In Sarawak, logging operations in the newly claimed state forests and forced resettlements have prompted many protests. These have been widely publicized (Colchester 1993; Dauvergne 2001). The Sarawak state government argued it was bringing economic growth to the province and that all citizens would benefit. The customary owners responded by erecting barricades to prevent the takeover of their ancestral lands by what they saw as an urban elite.

Sometimes government sponsored changes have led to conflict between different ethnic groups. This occurred in Central Kalimantan where a national government transmigration program had brought in large numbers of people from the island of

Madura (near Java). They were granted legal rights to land traditionally owned by Dayaks. The migrants were seen by local Dayak people to not only represent the state's monopoly over land but also to show insufficient regard for local laws and prerogatives. The uneasy relationships between the two groups eventually led to a series of riots with an especially serious disturbance occurring in 2001 when 500 Madurese were killed and 150,000 were forced to leave the area (Bouvier and Smith 2006). The Dayaks believed the migrants were taking over 'their' province with state support while the Madurese believed the Dayaks wanted their farms and jobs.

These conflicts over land use and tenure are often made more difficult to resolve because the responsibility for land is sometimes spread across several government departments or agencies (agriculture, mining, forestry, the military). This leads to overlapping functions and a plethora of often conflicting legislation (Sato 2000). Not all of these agencies have the will, let alone the capacity, to enforce the various laws. Further, many of these agencies are in competition with one another. For example, forestry departments are usually supposed to be the agencies administering national forest estates and protecting watersheds. But once the great logging boom of the 1960–1990 period began to wind down, many agricultural departments began looking at these logged-over forests as places to settle landless people, establish oil palm plantations or implement other agricultural schemes, thereby fostering national 'development'. This potential loss of control was resisted by forestry agencies who had expected to continue controlling and managing the forests recovering after logging. Unfortunately for them, agricultural crops are usually more profitable than timber production, especially in heavily logged forest and large areas of former forest land have been converted into plantation crops such as oil palm, especially in Malaysia and Indonesia.

These inter-agency rivalries have mostly left customary owners in limbo (although a number are probably unaware of their changed legal circumstance). Not that this troubled many government agencies. In fact it probably suited most of them. As Li (2002) observed, if a forestry department acknowledges that there are millions of people living in 'state forests' they must then also acknowledge that they are not really in control and cannot implement their own laws. Over time, the response of customary owners and more recent migrants has been to seek opportunities to realign their relationship with the state system using whichever agency seemed appropriate. Their primary purpose in doing this has been to consolidate or secure their hold over the lands they are using.

Land Tenure and Reforestation

The importance of the role of land tenure in fostering reforestation is widely recognized (Byron 2001; Chokkalingam et al. 2006; Enters et al. 2003; Mercer and Soussan 1992; Sanchez 1995). In the absence of some form of tenure most farmers will be hesitant about making an investment from which they may not receive any return. Plantations with long rotations are especially risky. This means that national reforestation schemes or joint ventures between smallholders and private timber companies are unlikely to succeed unless households have some kind of formal and legally enforceable form of land ownership or tenure including the rights to harvest forest products such as timber. Unless this is the case the only rational course of action for a farmer would be to make short-term investments and engage in opportunistic and possibly exploitative behaviour. Vietnam has recently embarked on a large-scale process of land allocation including allocating land specifically for reforestation (Box 3.2). The program appears to have helped increase forest cover although it has been clear from this experience that tenure alone is not sufficient to foster tree-growing.

The importance of tenure for plantation forestry might seem self-evident so it is interesting to find examples of reforestation that have been carried out by smallholders without formal legal tenure. Many of these cases involve some form of enrichment planting of state-owned forests by local communities who regard themselves as the traditional owners. Examples include the damar (or resin) forests of Sumatra (Michon 2005), the jungle rubber systems of Sumatra and Borneo (Schroth et al. 2004), the mixed fruit and timber forests of West Kalimantan (de Jong 2002) and the 'tea' forests of northern Thailand (Sasaki et al. 2007). But other cases involve farmers who have reforested degraded lands over which they have nor formal title such as those in the uplands of northern Luzon in the Philippines (Schuren and Snelder 2008).

At first these plantings seem paradoxical. Why should someone invest time and resources in planting trees without any assurance they will be able to benefit by harvesting the products it produces? One of the answers is that these farmers perceive the risk of dispossession as being low even though they do not have legal tenure. From their point of view the most significant thing is that their ownership is widely recognized within their community. Recent events in the damar forests of southern Sumatra are illustrative of the situation. These complex and species-rich forests have been created in south west Sumatra over the last 100 years. Their purpose is to produce damar from the dipterocarp *Shorea javanica* and they have attracted the attention of researchers because of their biological diversity and importance to local communities (Kusters et al. 2007; Michon 2005). Their location and the resources they contain meant they also began to attract the attention of loggers and people who wanted to convert the area into oil palm plantation. In 1998, as a result of lobbying by NGOs and scientists, the Government of Indonesia agreed to prevent these outside interventions and enable the customary owners to register to acquire concession rights over the forests. This would not grant formal land ownership but would guarantee the villagers rights to management and benefit from the forests they had created. The agreement was regarded by outside observers as something of a breakthrough in reconciling the disjunction between de facto and de jure land ownership in Indonesia.

However, the farmers thought otherwise. By 2005 not one application for registration had been lodged. From the farmer's point of view they would be acknowledging the legitimacy of the government's claim of ownership of their land if they signed. They might have security but it would be at the expense of their traditional rights. Besides, the offer itself had recognized the primacy of the traditional owner's usage and people believed this alone had been sufficient to prevent outsiders moving in. In short, farmers felt empowered that their rights

Box 3.2 Land Tenure and Reforestation by Smallholders in Vietnam

After a period of collectivized agriculture Vietnam has begun allocating land to households for farming and reforestation. The amounts of land are relatively small and average less than 5 ha per household. Certificates, known as 'Red Books' are issued giving legal rights to land for up to 50 years. The reforms have led to significant improvements in agricultural productivity and a major reduction in rural poverty. The changes also have a reforestation component and degraded land unsuited to cropping is also being allocated to households on long leases for tree planting. Areas of up to 30 ha per household are being allocated depending on location. In addition, farmers living near natural forests are being offered contracts to earn money by protecting and managing these.

Significant reforestation (via natural regeneration and plantations) has occurred since these changes with the national cover increasing from around 25% in the 1980s to 35% by 2002 (Meyfroidt and Lambin 2008). However the results have been uneven and have varied with location. Sunderlin and Huynh (2005) describe positive as well as negative assessments of the impact of the land allocation process on reforestation. Perhaps the key issue is that a farmer's decision on whether or not to plant trees depends on factors other than just land tenure. These other factors include the need by many farmers to first establish food security. As well there are the opportunity costs of tree-planting, the need for technical knowledge about species and silvicultural methods, the availability of capital, the perception of markets for forest products and the availability of transport to bring forest products to markets. In some places none of these problems were especially significant and the granting of tenure soon prompted tree-planting. But in others it did not or farmers were more cautious about what, for many, was a new land use activity (Castella et al. 2002; Sikor 2001).

The process of forest land allocation has been difficult to implement and is still on-going. One problem has been the difficulty in developing rapid and robust means of carrying out participatory land use planning prior to land allocation. Another has been the challenge in changing the culture of forestry authorities from one that emphasized control over forested lands to one emphasizing facilitation and partnership with local communities or individual households (Nguyen and Gilmour 2000). Sunderlin and Huynh (2005) also point to a number of other problems including the frequent incompatibility of the changes to local livelihood practices, the fact that many sites were badly degraded (and so are hard to reforest) and that policies are often altered. In short, land tenure is an important precursor to reforestation but must be accompanied by other policies relevant to the people whom are to carry it out if it is to trigger reforestation.

had been recognized and believed these were now relatively secure enough even though a process that would have granted them legal rights was never completed (Kusters et al. 2007).

The perception of tenurial security by these farmers is necessarily subjective. Kusters et al. (2007) suggest it is affected by the existence of external threats, the extent to which people are aware of their actual legal status, the degree of external support received by the community and the role of local officials who may be sympathetic or hostile to their cause. People facing few threats or those with sympathetic and powerful allies are more likely to be optimistic about their opportunities and be prepared to take a longer-term view than those without these advantages. Of course the reverse is also true and Byron (2001) describes an interesting case in Vietnam before the Doi Moi reforms in 1989 where farmers had been given tenure documents but behaved as if they had no tenure because they feared the government would one day renounce these documents and re-appropriate any new forests that had been established. Finally, the case from northern Thailand described in Box 3.1 shows how reforestation by recent migrants might also be carried out as a means of demonstrating good citizenship. By planting trees farmers believed they would be able to eventually acquire land tenure and citizenship.

These examples do not imply that tenure is unimportant; legal certainty is always preferable to tacit or implied agreements and reforestation is always likely to be more attractive for farmers owning land than those without some form of tenure. However, the lack of formal ownership need not imply that any kind of reforestation is impossible.

Community Forestry

There are several ways in which rural communities might undertake reforestation of areas outside protected areas. One is by the community as a whole doing it and the other way is through the action of individual households. In discussing the possible role of communities it is useful to first consider how they have begun to play an increased role in the management of natural forests.

Community Forestry Within Existing Natural Forests

The historical record outlined in earlier chapters suggests that, with some exceptions, most government agencies charged with managing forests have not been particularly successful. They have neither protected them nor addressed the needs of communities living in or around them. This has prompted some governments to devolve responsibility for protecting and managing the remaining forests to local communities. In most cases these have been secondary forests regenerating after logging but some cases have included primary forests. In return, these communities have been allowed a greater share of the benefits of harvesting resources from the forests. This change has several potential advantages. Firstly, those living in or near the forests are likely to be more successful in protecting them for the national benefit than government agents commuting from a district headquarters. Secondly, livelihoods are likely to be improved because the benefits of harvesting are shared more equitably than in the past when most benefits were captured by an urban elite. And, thirdly, the process can sometimes take advantage of traditional methods of natural resource management to create robust management systems which are in everyone's interest to maintain.

The approach has become known as community forestry. Fisher (2003) defines this as 'some element of community participation in forest management and some commitment to improved or secure provision of at least some forest products to rural people living in and near forests'. The effect of these changes has been to turn people previously regarded as squatters or illegal collectors of NTFPs into legitimate forest managers. Community forestry is now part of the forestry dialogue throughout the region and most (though not all) countries have some kind of policy framework and have explored the idea through various approaches (Hobley 1996; Lynch and Talbot 1995; Poffenberger 1990; RECOFTC and FAO 2003; Shackelton et al. 2002; Victor et al. 1997).

In practice the process of devolution has taken several forms. One approach involves the government forestry agency offering contracts to communities or households to protect a defined area of existing state forest for a certain time. As well as receiving a cash payment, participants may also be offered the rights to harvest timber or NTFPs under prescribed conditions. This may not be sufficient for those households with no other land (Fig. 3.5). Another version allocates these

Fig. 3.5 Natural forest on a hillside in southern Vietnam is gradually being converted to orchards by poor farmers even though this is supposedly a Protection Forest and they have contracts to protect it. Each household in this particular area has an average of 0.8 ha of forest to protect. Their dilemma is that they have little other land from which to make a living

forest lands to the community or household on a long-term lease (e.g. 20–30 years with the possibility of renewal). Again, the community is responsible for protecting the forest but is granted the right to harvest timbers and NTFPs under agreed conditions. In both cases the government retains land ownership and prescribes the conditions that the community or household must fulfill not least of which is that they must maintain the area as forest and not clear the land and use it for agriculture. The full transfer of ownership is also being explored in some countries although the state usually retains ownership of the higher quality forests (Castella et al. 2002; Cornista and Escueta 1990; Lynch and Talbot 1995; Penafiel 1996; Scherr et al. 2004; Sunderlin et al. 2005).

The methods communities use to manage these forests also vary but usually involve some kind of supervisory committee. This body decides who shall have access to the forest's resources and the rates at which these resources shall be harvested. The committee usually establishes penalties for those who break these rules.

The situation is different in the Pacific because people have always had full legal ownership of their forests. In the past governments usually took it upon themselves to manage these forests on behalf of the owners which usually meant simply allocating logging concessions to overseas companies. There have been some recent examples, however, of local communities managing their own forests for commercial purposes. Most of these involve using small portable sawmills which enable communities to sell timber and not just logs (Filer and Sekhran 1998; McGrath 1998). These community management systems have had some modest successes but have not reached the stage where they have supplanted overseas logging companies. In fact they are often seen as being complementary with rather than an alternative to large-scale logging. Filer and Sekhran (1998) describe some of the complex history of forest exploitation and community forestry in Papua New Guinea and the relationships between landowners, government, NGOs and donors. They suggest Papua New Guinean landowners are gradually learning that they cannot rely on other stakeholders for solutions and, through a process of learning-by-doing, are gradually moving towards a stage of being more self-reliant.

While some of these community forestry programs are already yielding promising outcomes others have failed. A number of problems frequently occur.

1. *Decision-making and governance*: many government forestry agencies have been reluctant to pass on authority for decision-making and management to communities believing that they cannot have the technical capacity to succeed. This is especially the case in countries where all forest land is regarded as being State Forest. This has meant that much community forestry is often still a top-down process with key decisions still being made by government staff.
2. *Agreements*: there is a need to finalize agreements between the government and other stakeholders that define the rights and responsibilities of each party. It is often difficult to do this since there can be a substantial difference in the expectations of governments and people over just what community forestry is about and what it can deliver. Sometimes communities can be trapped in one-sided

collaborative agreements that force them to take on a large share of the responsibility while getting little benefit.
3. *Equity*: communities are not necessarily homogenous entities and sometimes wealthy or politically powerful individuals can control decision-making or commandeer a disproportionate share of the benefits arising from the devolution process (e.g. they acquire all the better quality forest or land located nearest the road network). If some equity is to be achieved there must be some kind of supervisory committee whose role and membership is accepted by the community.
4. *Legal back-up*: local institutions and agreements often need to be supported by some form of legal authority so that the parties are held accountable and there may be a need to control outsiders such as local entrepreneurs or loggers who may ignore rules established by the community and continue to act outside any agreement.
5. *Stability*: most countries have had difficulty in implementing community forestry because government policies keep changing. Hence many people – including forestry department field staff – are often unaware of current policies. This means national laws or policies may not be implemented at a district level.
6. *Technical and market knowledge*: many communities have a limited understanding of the species and silvicultural options available to them. Similarly they may have a incomplete knowledge of markets and marketing so that they can not receive the full financial benefits potentially available to them.
7. *The forest area is too small or too distant from the community to be worth managing*: small patches of forest may not be large enough for people to make a living from NTFPs alone. Unless the forest can be enriched in some way there is a risk it will be gradually converted to agriculture or abandoned (Fig. 3.5). People allocated more distant patches of forest may find it is simply not worth the effort involved in travelling to them.

Fisher (2003) is of the opinion that there has been a systematic tendency to exaggerate the contribution that community forestry has made, so far, to livelihoods although there may have been some improvements in forest condition. Community forestry will of necessity be an evolving process where participants, including government agencies learn, adapt and develop local institutions to suit their circumstances.

Community Forestry on Cleared or Degraded Lands

To date most community forestry has involved managing natural forests regenerating after disturbances. The process is relatively passive and involves protecting the regenerating trees from disturbances and establishing rules concerning access and harvesting rights. There are cases where the community has reforested cleared lands for subsistence or religious reasons (e.g. Box 3.1) but it appears rather less common for communities to undertake the reforestation of cleared or degraded lands for largely commercial purposes. Where this has happened most early approaches were simply contractual arrangements in which a landowner (usually the government) initiated the process, paid the community to reforest a prescribed area and provided them with the seedlings to do so. Many of these plantings subsequently failed because the community was not concerned whether the plants survived or not and had no interest in follow-up maintenance.

A more promising approach has been to develop joint ventures between the community and government or a private company using lands granted to or leased by the community. In these cases both parties receive a share of the financial benefits so there is more of an incentive for the community to ensure the planting is a success. An alternative version of this model operates in Papua New Guinea where land owned by the villagers is leased to the reforestation company and both parties share in the profits (Hunt 2002).

The simplest approach to involving communities in reforestation involves granting communities either full ownership or a long lease over land and assisting them to reforest these areas. The communities then retain all the eventual financial benefits generated by the new forest (Peluso et al. 1990; Penafiel 1996; Pragtong and Thomas 1990). A significant community-based program has developed in the Philippines using this approach to reforest degraded lands and Poffenberger (2006) estimates that agreement agreements of this kind now cover around 37% of forest land in the country. The program is not without problems but appears to be more successful than much of the reforestation carried out directly by government agencies (Box 3.3).

Another apparently successful community reforestation scheme was that carried out in north eastern Thailand and described by Hafner (1995). The area is one with

Box 3.3 Community Forestry in the Philippines

There has been a long history of attempting reforestation in the Philippines. Most of the early attempts were carried out by government agencies or by logging concession holders (as part of their concession obligations). Many of the early reforestation projects necessitated evicting upland farmers from their lands or simply using them as labourers (Chokkalingam et al. 2006). The planning was top-down and success rates were modest, especially when payments for contract planting were delayed. Some of the plantations were subject to continual degrading pressures such as fires and illegal harvesting.

Since the late 1980s there has been a strong shift away from government reforestation towards community-based reforestation. The number of programs has been large (Harrison et al. 2004) and the plethora of programs and policies, the frequency of changes and the inconsistencies between various policies have led to a good deal of uncertainty amongst government field staff as well as communities over just what the policy conditions are. The two most recent national programs have been the Community-Based Forest Management Program (1995) and the Community-Based Resource Management Program (1998). Both have involved

(continued)

Box 3.3 (continued)

a number of distinct sub-programs all of which were aimed at communities and it is likely that most of the tree-planting carried out in the Philippines between 1996 and 2002 was accomplished through these two programs. The programs have sought to ensure the sustainable development of the national forest resources, reduce rural poverty and overcome environmental degradation. They have mostly followed a deliberate and participatory planning process in which communities were involved. Communities have been given leasehold land (initially 25 years but renewable for a further 25 years) although the government has retained effective control over the timing and manner of timber harvesting (Harrison et al. 2004; Chokkalingam et al. 2006). Such tenure rights are conditional and the government cancelled a number of agreements in 2006 following reports of some logging violations (Chokkalingam et al. 2006).

These two programs were heavily dependent on outside funding and community enthusiasm appears to have declined, at least in some areas, as these external funds have dried up. There may be several reasons for this. One may be that tree growing is perceived as being financially unattractive without such subsidies and continued technical and marketing support is needed until profitability can be demonstrated. Another is that there are often significant constraints placed on the rights of communities to sell the trees they have planted even though they may have secure tenure over their land. Felling permits are required that often involve a lengthy bureaucratic process and these can be affected by unclear harvesting policies for watersheds and the frequent suspensions of harvesting rights in response to environmental and political crises. Finally, there is evidence that some farmers, especially migrants in non-traditional communities, prefer to grow trees on land of their own rather than as part of a more cumbersome community effort provided they can acquire appropriate land tenure (S. Harrison: pers. comm.).

Chokkalingam et al. (2006) carried out an assessment of recent forest rehabilitation in the Philippines and concluded these community-oriented programs had the potential to produce rather better outcomes than the plantings carried out by state forestry agencies, NGOs or private sector initiatives. Many of the latter were destroyed or failed (for largely social, institutional or financial rather than technical reasons). The community-based plantings also appeared to have achieved some rather better, though still modest, biodiversity conservation benefits.

high levels of rural poverty, significant population pressures and a forest cover of less than 5%. The NGO-sponsored project operated over 10 years and created 165 ha of plantations and involved 89 villages. It had four phases. The first was an organizational and implementation phase where the focus was on gaining community support and commitment to the idea of establishing community woodlots on common land. Interested volunteers were sought to form a Community Forestry Committee and this group chose the initial species to plant and arranged the land to be planted together with the necessary labour to do so. In this early stage community involvement was encouraged by developing a festival-like atmosphere in which food and entertainment were used to reward participants. The second phase was concerned with sustaining interest in the project once the trees had been established. This was done by encouraging intercropping with food plants to generate short-term income and by providing information on agroforestry techniques. The third phase dealt with managing the first harvest (after about 5 years) and sharing the funds generated to individual farmers and for village projects. This was carried out with the use of a revolving fund. The final phase was one in the sites are being replanted using a more diversified group of tree species and attempts are being made to increase the levels of tree planting throughout the community without the need for external support. The apparent success of the project has been due

Box 3.4 Community Reforestation on New Georgia Island, Solomon Islands

Most of the logged-over land in the Solomon Islands has been left to recover through natural regeneration and only small areas of plantation have been established by the government or private companies. This is largely because the difficulty in acquiring land from traditional owners. Most of the existing plantations were established during the colonial era and the Solomon Islands government subsequently sold these to overseas corporate owners after independence in 1978. However an unusual community-owned plantation has been established by the Ngrassi, Dukerna and Lunga people living on the northern coast of New Georgia Island in the Western province. The project was initiated by the influential and charismatic leader of the local Christian Fellowship Church who persuaded the community to pool the funds received from logging in the community's forests instead of distributing them among individual clan members (Hviding and Bayliss-Smith 2000). The project commenced in October 1999 with the building of a nursery. Since then over 1,000 ha of plantations have been established using mainly *Eucalyptus deglupta, E. tereticornis, Gmelina arborea* and *Acacia mangium*. Some technical support came from an NGO (the Rural Development Trust Board) who helped develop the nursery and get seed. The trees have grown well and the main problem now lies in marketing the timbers. The island of New Georgia lies off the main shipping routes and it may be difficult to find buyers for comparatively small volumes of logs. On the other hand, the community might be able to form a relationship with the company that now owns some the former government and now company-owned plantations on nearby (ca. 10 km) Kolombangara Island. Similar species are being grown there and are being sold to Japan, Indonesia, Korea and Vietnam. The success of this community project undoubtedly lies in the strong community structures created by the church and its leaders. It remains to be seen whether similar community tree-planting programs can be developed elsewhere in the Solomon Islands. There is evidence that many people in other villages favour individually owned plantations.

to the level of local participation, a flexible operational strategy and an emphasis on building local capacities so future plantings can be managed without external support. In this case it is assumed that the initial group-centred approach will eventually be replaced by individual household tree-planting projects.

Finally, an intriguing form of community-based reforestation is currently underway in the Solomon Islands on land fully-owned by the community (Box 3.4). This is a close-knit traditional community with strong leadership and a clear idea of what it hopes to achieve through reforestation. The importance of the latter point is illustrated by the more disappointing experience of a community living near Port Moresby in Papua New Guinea. This community acquired a well-established teak plantation covering 1,500 ha that had been previously owned by the government's Forestry Department. In the late 1970s the Department was forced to hand over the land and trees after a court ruled it had mistakenly purchased the land from people who were not the legal owners. This is rare example of alienated land being re-customized. But, rather than manage the asset, the new owners gradually let it be (prematurely) felled and within a comparatively short time the plantation had disappeared. In this case it seems the community had neither the leadership nor the internal management structures necessary to maintain what could have been a significant and valuable community asset. And, perhaps because the newly independent government was still establishing itself, it was not able to develop some form of joint venture that could have enabled the plantations to continue.

Community or Private Reforestation?

There may be a limit to the extent to which degraded lands can be reforested by communities. One of the reasons why various forms of shared land tenure probably evolved was because it enabled the provision of mutual protection in the face of endemic warfare and feuding. In addition, people often needed help beyond that which could be provided by the nuclear family when larger farm or construction work had to be done. Under these circumstances the benefits that arose from being a member of the community outweighed the advantages of being a free agent. The institutions and moral norms in villages and clans might be seen as a way of penalizing opportunistic behavior that threatened this natural insurance system. Agriculture could be carried out as a private household activity but it was done so on communal land.

However, in more recent times, changed economic incentives are prompting people to reassess these institutional relationships. One important factor has been the advent of perennial cash crops such as coffee, cocoa, rubber and oil palm. Most traditional systems recognized that special rights were acquired by someone who felled a patch of forest and planted a crop. Similarly the special rights of a farmer who established a rice paddy field were recognized even while the nearby natural forest remained a common property resource. This was because the community realized that a person had invested effort to create the new resource and that this deserved recognition. But the intensification of agriculture, the spread of cash crops and the rise of land values has prompted a consolidation of permanent individual property rights. More families now prefer to pass their agricultural assets on to their biological children rather than to the community as a whole. In some cases membership of the community may still be useful such as when the new assets are distant from the village and need special protection which the community can provide more easily than an individual. But even this advantage shrinks when the lands are near the household's own dwelling.

Tree-planting certainly triggers these types of problem and most households are likely to prefer to invest effort in reforestation on lands that they alone control. Part of the reason is because not all members of a community may be equally interested in tree plantations and it is difficult to develop sharing mechanisms that reflect the contribution individuals make to establish new plantings. A second is because community control restricts rapid access to these resources in times when a household suddenly needs cash for medical emergencies or because of unexpected crises such as funerals. Community reforestation may continue to succeed in strong traditional communities but is less likely to be successful among communities of recent migrants lacking leadership and strong social cohesion.

The privatisation of traditional communal land for tree planting poses a number of problems. Not the least of these is that while some people may benefit others may lose out; elites can gain control of some of the best of the land once owned by the community. This means that, whatever the economic advantages, privatization can lead to a variety of social problems within the community involving social cohesion, stability and identity. In these fluid situations it is very difficult for bureaucracies to keep track of ownership claims and land boundaries. One middle pathway appears to be developing in some parts of the Pacific. In this case, forms of individualized, secure and transferable property rights based on customary ownership are being developed rather than idividualised freehold land (Fingleton 2005). For some communities this approach may offer the best of both worlds.

Conclusions

The last 100 years of land use practices in the world's tropical forest landscapes have left an impoverished and increasingly threatened biota as well as many poor people living in and around these areas. The circumstances of many of these people have been worsened by 'development' because they have they lost the forest's resources as well as legal access to these lands. There has been considerable debate between conservationists and development specialists over how these twin problems of biodiversity loss and poverty should be solved. Some argue for simple unilateral approaches (i.e. tackle either conservation or poverty but do not try to achieve both at the same location) while other have sought to find methods of solving both problems at once. These Integrated Conservation and Development Projects have had limited success to date although this is not necessarily because the concept is flawed. On the contrary, it may be because it has simply been approached in the wrong way.

There are reasons for thinking there may be opportunities to tackle both problems by reforesting some of the large areas of degraded lands that have now accumulated across the region. Reforestation would add habitats and heterogeneity to these landscapes and help complement the existing network of protected areas. In addition, reforestation could improve the asset base of rural households, diversify their incomes and help reduce their vulnerability to economic and other unexpected contingencies.

But it is important to keep these potential advantages in perspective. Some forms of reforestation will make only modest contributions to biodiversity conservation (and may take time to do so). Likewise not all poor people will be interested in reforestation, especially those with only small areas of land to use or with limited resources with which to carry it out. Nonetheless, carefully devised forms of reforestation could help with both tasks. Further, there is empirical evidence that many rural households are interested in being involved once they have what they believe is an appropriate degree of tenurial security over the land they are using. Tenure can be a problem because many people now live in an ambiguous political and institutional landscape where land tenure and usufruct rights can be limited or entirely absent.

The type of reforestation employed and its capacity to generate conservation or livelihood benefits will depend on the type of land available and its landscape context. Much of the land likely to be available for reforestation will be marginal agricultural land. This will determine the types of species and planting systems that can be used and in some cases only exotic species may be able to tolerate these sites. The landscape context is important because it will determine the extent to which new plantations will facilitate the movement of native biota (wildlife as well as seeds) across the landscape. The landscape context is also important because it influences the economic value of any plantation; isolated plantations distant from roads or transport will be less attractive than those closer to markets. Plantation owners must take these matters into account when setting their objectives. The next chapter discusses the different forms of reforestation in a little more detail as well as their advantages and disadvantages.

References

Adams WM, Aveling R, Brockington D, Dickson B, Elliott J, Hutton J, Roe D, Vira B, Wolmer W (2004) Biodiversity conservation and the eradication of poverty. Science 306:1146–1149

Angelsen A, Wunder S (2003) Exploring the forest-poverty link: key concepts, issues and research implications. Center for International Forestry Research, Bogor

Appanah S, Weinland G (1993) Planting quality timber trees in Pensinsular Malaysia: a review. Forest reserch Institute, Malaysia, Kepong

Berkes F (2007) Community-based conservation in a globalised world. Proc Natl Acad Sci 104:15188–15193

Bouvier H, Smith G (2006) Of spontaneity and conspiracy theories; explaning violence in Central Kalimantan. Asian J Soc Sci 34:475–491

Brockington D, Igoe J, Schmidt-Soltau K (2006) Conservation, human rights, and poverty reduction. Conserv Biol 20:250–252

Byron RN (2001) Keys to smallholder forestry in developing countries in the tropics. In: Harrison SR, Herbohn JL (eds) Sustainable farm forestry in the tropics: social and economic analysis and policy. Edward Edgar, Cheltenham, pp 211–226
Carle J (2007) Vunerabilities of smallholder plantings. Unasylva 58:59
Carle J, Holmgren P (2003) Definitions related to planted forests. Working Paper 79. FAO, Forestry Department Rome
Castella JC, Boissau S, Nguyen HT, Novosad P (2002) Impact of forest land allocation on agriculture and natural resources management in Bac Kan Province, Vietnam. In: Castell JC, Dang DQ (eds) Doi Moi in the mountains land use changes and farmers livelihood strategies in Bac Kan Province, Vietnam. The Agricultural Publishing House, Hanoi, pp 121–148
Chambers R, Leach M (1989) Trees as savings and security for the rural poor. World Develop 17:329–342
Chandler D, Roff WR, Small JRW, Steinberg DJ, Taylor RH, Woodside A, Wyatt DK (1987) In search of Southeast Asia: a modern history. University of Hawaii Press, Honolulu
Chokkalingam U, Carandang AP, Pulhin JM, Lasco RD, Peras RJJ, Toma T (2006) One century of forest rehabilitation in the Philippines: approaches, outcomes and lessons. Center for International Forestry Research, Bogor
Clarke WC, Thaman R (1993) Agroforestry in the Pacific Islands: systems for sustainability. United Nations University Press, Tokyo, New York, Paris
Cleary M, Eaton P (1996) Tradition and reform: land tenure and rural development in Southeast Asia. Oxford University Press, Kuala Lumpur, New York
Colchester M (1993) Pirates, squatters and poachers – the political ecology of dispossession of the native peoples of Sarawak. Global Ecol Biogeogr Lett 3:158–179
Colchester M (2000) Self-determination or environmental determinism for indigenous peoples in tropical forest conservation. Conserv Biol 14:1365–1367
Cornista LB, Escueta EF (1990) Communal forest leases as a tenurial option in the Philippines uplands. In: Poffenberger M (ed) Keepers of the forest: land management alternatives in Southeast Asia. Kumarian Press, Connecticut, pp 134–144
Cossalter C, Pye-Smith C (2003) Fast wood. Center for International Forestry Research, Bogor
Crocombe RG (1982) Land tenure and agricultural development in the Pacific Islands. In: Bay-Peterson J (ed) Land tenure and the small farmer in Asia. Food and Fertilizer Technology Center for the Asian and Pacific Region, The Council for Agricultural Planning and Development, The Land Reform Training Institute, Taipai, pp 28–48
Cyranoski D (2007) Biodiversity: logging: the new conservation. Nature 446:608–610
Dauvergne P (2001) Loggers and degradation in the Asia-Pacific: corporations and environmental management. Cambridge University Press, Cambridge
Dawkins HC, Philip MS (1998) Tropical moist forest silviculture and management: a history of success and failure. CAB International, Wallingford
De Jong W (2002) Forest products and local forest management in West Kalimantan, Indonesia: implications for conservation and development. Tropenbos International, Wageningen
Del Lungo A, Ball J, Carle J (2006) Global planted forests thematic study. Results and analysis. FAO, Rome
DFID (1999) Key sheets for sustainable livelihoods – overview. In: Carney D (ed) Department for International Development, London. www.odi.org.uk/resources/default.asp; accessed 22 September 2010
Do DS, Nguyen HN (2003) Use of indigeneous tree species in reforestation in Vietnam. Agricultural Publishing House, Hanoi
Dove MR (1993) A revisionist view of tropical deforestation and development. Environ Conserv 20:17–24
Eaton P (2005) Land tenure, conservation and development in Southeast Asia. Routledge Curzon, London/New York
Effendy A, Hardono D (2000) The large scale private investment of timber plantation development in Indonesia. Proceedings of the international conference on Timber plantation development. Forest Management Bureau of the Philippines Department of Environment and Natural

Resources, International Tropical Timbers Organisation, Food and Agriculture Organisation of the United Nations, Manila
Enters T, Durst P, Brown C (2003) What does it take? The role of incentives in forest plantation development in the Asia-Pacific region. Unasylva 54:11–18
Erskine P, Lamb D, Bristow M (2005) Reforestation in the tropics and subtropics of Australia using rainforest tree species. Rural Industries Research and Development Corporation, Canberra, https://rirdc.infoservices.com.au/items/05-087; accessed 19 September 2010
Evans J, Turnbull JW (2004) Plantation forestry in the tropics; the role, silviculture and use of planted forests for industrial, social, environmental and agroforestry purposes. Oxford University press, Oxford
FAO (2002) Land tenure and rural development. Food and Agriculture Organization of the United Nations, Rome
FAO (2010) The global forest resource assessment 2010. United Nations Food and Agriculture Organization, Rome
Filer C, Sekhran N (1998) Loggers, donors and resource owners: Papua New Guinea country study. International Institute for Environment and Development, London
Fingleton J (2005) Privatising land in the Pacific: a defence of customary tenures. Discussion Paper No 80. The Australia Institute., Canberra
Fisher RJ (2003) Innovations, persistence and change: reflections on the state of community forestry. Community Forestry: Current Innovations and Experiences Regional Community Forestry Training Center and Food and Agriculture Organisation of the United Nations, Bangkok, Thailand
Fisher RJ, Maginnis S, Jackson W, Barrow E, Jeanrenaud S (2008) Linking conservation and poverty reduction: landscapes, people and power. Earthscan, London
Freeman JD (1955) Iban agriculture: a report on the shifting cultivation of hill rice by the Iban of Sarawak. H.M. Stationery Office, London
Hafner J (1995) Beyond basic needs: participation and village reforestation in Thailand. Commun Develop J 30:72–82
Harrison S, Emtage N, Nasayao E (2004) Past and present forestry support programs in the Philippines, and lessons for the future. Small-scale Forest Econom Manage Policy 3:303–317
Hobley M (1996) Participatory forestry: the process of change in India and Nepal. Overseas Development Institute, London.
Hunt C (2002) Production, privatisation and preservation in Papua New Guinea forestry. International Institute for Environment and Development, London
Hviding E, Bayliss-Smith T (2000). Islands of rainforest: agroforestry, logging and eco-tourism in Solomon Islands. Ashgate, Aldershot
ICEM (2003a) Thailand national report on protected areas and development. Review of protected areas and development in the lower Mekong River region. International Center for Environmental Management, Indooroopilly, Australia
ICEM (2003b) Vietnam national report on protected areas and development. Review of protected areas and development in the lower Mekong River region. International Center for Environmental Management., Indooroopilly, Australia
ITTO (2002) ITTO guidelines for the restoration, management and rehabilitation of degraded and secondary tropical forests, vol No 13, ITTO Policy Development Series. International Tropical Timbers Organization, Yokohama
Jacoby EH (1961) Agrarian unrest in Southeast Asia. Asia Publishing House, London
Kaimowitz D, Sheil D (2007) Conserving what and for whom? Why conservation should help meet basic humid needs in the tropics. Biotropica 39:567–574
Krishnapillay B (2002) A manual for forest plantation establishment in Malaysia. Forest Research Institute, Malaysia, Kepong
Kusters K, de Foresta H, Ekadinata A, van Noordwijk M (2007) Towards solutions for states vs. local community conflicts over forestland: the impact of formal recognition of user rights in Krui, Sumatra, Indonesia. Human Ecol 35:427–438
Li TM (2002) Engaging simplifications: community-based resource management, market processes and state agendas in upland Southeast Asia. World Develop 30:265–283

Lopez C, Shanley P (2004) Riches of the forest: foods, spices, crafts and resins of asia. Centre for International Forestry Research, Bogor

Lynch OJ, Talbot K (1995) Balancing acts: community-based forest managment and national law in Asia and the Pacific. World Resources Institute, Washington

Mayers J (2000) Company-community forestry partnerships: a growing phenomenon. Unasylva 51:33–41

Mayers J (2006) Poverty reduction through commercial forestry: what evidence? What prospects? The forest dialogue. Yale University, New Haven

Mayers J, Vermeulen S (2002) Company-community forestry partnerships; from raw deals to mutual gains? International Institute for Environment and Development, London

McGrath M (1998) Community-based forest conservation and management in the Pacific Islands. In: Victor M, Lang C, Bornemeier JE (eds) Community forestry at a crossroads: reflections and future development of community forestry. Regional Community Forestry Training Center, Bangkok, pp 59–70

McShane TO, Wells MP (2004) Getting biodiversity projects to work. towards more effective conservation and development. Columbia University press, New

Mercer DE, Soussan J (1992) Fuelwood problems and solutions: policy options. In: Sharma N, Rowe R (eds) Managing the world's forests: looking for balance between conservation and development. Kendall/Hunt Publishing, Falls Church, VA, pp 177–214

Meyfroidt P, Lambin EF (2008) The causes of reforestation in Vietnam. Land Use Policy 25:182–197

Michon G (2005) Domesticating forests: how farmers manage forest resources. Institut de Recherchepour le Developpement, Center for International Forestry Research, World Agroforestry Center, Paris, Bogor

Morris JD, Hicks E, Ingles A, Ketphanh S (2004) Linking poverty reduction with forest conservation. Case studies from Lao PDR. International Union for the Conservation of Nature, Bangkok

Nawir AA, Anyonge C, Race D, Vermeulen S (2003) Towards equitable partnerships between corporate and smallholder partners: relating partnerships to social, economic and environmental indicators: workshop synthesis. Food and Agriculture Organisation of the United Nations, Rome

Nawir AA, Murniat, Rumboko L (2007) Forest rehabilitation in Indonesia: where to after three decades? Center for International Forestry Research, Bogor

Nguyen VS, Gilmour DG (2000) Forest rehabilitation policy and practice in Vietnam. In: Anon (ed) Forest rehabilitation policy and practice in Vietnam: proceedings of a national workshop, Hoa Binh. International Union for Conservation of Nature, Hanoi, pp 4–34

Peluso NL, Poffenberger M, Seymour F (1990) Reorienting forest management on Java. In: Poffenberger M (ed) Keepers of the forest: land management alternatives in Southeast Asia. Kumarian, Connecticut, pp 220–236

Penafiel SR (1996) Opportunities for income generation in the Baggio community forestry project. In: Victor M (ed) Income generation through community forestry. Regional Community Forestry Training Center, Bangkok, pp 150–156

Poffenberger M (1990) Keepers of the forest: land management alternatives in Southeast Asia. Kumarian, Connecticut

Poffenberger M (2006) People in the forest: community forestry experiences from Southeast Asia. Int J Environ Sustain Develop 5:57–69

Pragtong K, Thomas D (1990) Evolving management systems in Thailand. In: Poffenberger M (ed) Keepers of the forest: land management alternatives in Southeast Asia. Kumarian, Connecticut, pp 167–186

Raymond DH, Wooff WG (2006) Small-scale forest plantations are the key to the future of the Solomon Islands forest industry. Int Forest Rev 8:222–228

RECOFTC and FAO (2003) Community forestry: current innovations and experiences. Regional Community Forestry Training Centre and Food and agricultural Organisation of the United Nations, Bangkok

Rigg J (2006) Land, farming, livelihoods, and poverty: rethinking the links in the rural south. World Develop 34:180–202
Sanchez PA (1995) Science in agroforestry. Agroforest Syst 30:5–55
Sasaki A, Takeda S, Kanzaki M, Ohta S, Preechapanya P (2007) Population dynamics and land use changes in a Miang (chewing tea) village in northern Thailand. Tropics 16:75–85
Sato J (2000) People in between: conversion and conservation of forest lands in Thailand. Develop Change 31:155–177
Scherr SJ, White A, Kaimowitz D (2004) A new agenda for forest conservation and poverty reduction: making markets work for low-income producers. Forest Trends, Washington
Schroth G, Harvey C, Vincent G (2004) Complex agroforests: their structure, diversity and potential role in landscape conservation. In: Schroth G, da Fonseca G, Harvey C, Gascon C, Vasconcelas H, Isac A-M (eds) Agroforestry and biodiversity conservation in tropical landscapes. Island Press, Washington, pp 227–260
Schuren SHG, Snelder DJ (2008) Tree-growing on farms in northeast Luzon (The Philippines): smallholders' motivations and other determinants for adopting agroforestry systems. In: Snelder DJ, Lasco RD (eds) Smallholder tree growing for rural development and environmental services. Springer, Netherlands, pp 75–97
Schwartzman S, Moreira A, Nepstad D (2000) Rethinking tropical forest conservation: perils in parks. Conserv Biol 14:1351–1357
Shackelton S, Campbell B, Wollenberg D, Edmunds D (2002) Devolution and community-based natural reource management: creating space for local people to participate and benefit? ODI Natural Resource Perspectives No 76. Overseas Development Institute, London
Sikor T (2001) The allocation of forestry land in Vietnam: did it cause the expansion of forests in the northwest? Forest Policy Econ 2:1–11
STCP (2009) Encouraging industrial forest plantations in the tropics: report of a global study, vol No 33, ITTO Technical Series. International Tropical Timbers Organization, Yokohama
Sunderlin WD, Angelsen A, Belcher B, Burgers P, Nasi R, Santoso L, Wunder S (2005) Livelihoods, forests, and conservation in developing countries: an overview. World Develop 33:1383–1402
Sunderlin WD, Huynh TB (2005) Poverty alleviation and forests in Vietnam. Center for International Forestry Research, Bogor
Terborgh J (1999) Reqiem for nature. Island Press, Washington
Tyynela T, Otsama R, Otsama A (2003) Indigenous livelihood systems in industrial tree plantation areas in West Kalimantan, Indonesia: economics and plant species richness. Agroforest Syst 57:87–100
van Schaik C, Terborgh J, Dugelby B (1997) The silent crisis: the state of rain forest nature preserves. In: Kramer R, van Schaik C, Johnson J (eds) Last stand: protected areas and the defense of tropical biodiversity. Oxford University Press, Oxford, pp 64–89
Victor M, Lang C, Bornemeier J (1997) Community forestry at a crossroads: reflections and future directions in the development of community forestry. Proceedings of an international seminar. Regional Community Forestry Training Centre.
Wells M, McShane TO (2004) Integrating protected area management with local needs and aspirations. Ambio 33:513–519
Westoby J (1987) The purpose of forests: follies of development. Blackwells, Oxford
Whiteman A (2004) Review of plans and policies concerning forestry and poverty alleviation in Laos, Vietnam and Cambodia. RETA No 6115-REG. Asian Development Bank
Wilshusen PR, Brechin SR, Fortwangler CL, West PC (2002) Reinventing a square wheel: critique of a resurgent 'protection paradigm' in international biodiversity conservation. Soc Nat Res 15:17–40
Wollenberg E, Ingles A (1998) Incomes from the forest. Center for International Forestry Research and the International Union for the Conservation of Nature, Bogor

Wooff W (2009) Sabah forest industries experiences in plantation forestry. Conference on the current state of plantation forestry in Malaysia: a special focus on Sabah, Forestry Department Headquarters, Sandakan, 18–20 November 2009

World Bank (2001) World development report 2000/2001: attacking poverty. Oxford University Press, New York

World Bank (2004) Global monitoring report: policies and actions for achieving the millenium development goals and related outcomes. World Bank, New York

Wormald TJ (1992) Mixed and pure forest plantations in the tropics and subtropics. Food and Agriculture Organisation of the United Nations, Rome

Chapter 4
Different Types of Reforestation

The art of forestry is different from that of paddy or dry field. Though one may be spared flood, drought, frost or snow, he still must give general care to the area for about ten years before withdrawing human effort. If this is done, the forest will be as though filled with a treasure whose virtue is so immense it will reach to one's children and grandchildren. Truly, one's prosperity will be eternal.
Mikami Gennosuke, forester from Tsugaru during Japanese Edo period.

(Quoted by Totman 1989, p. 124)

Introduction

Previous chapters have argued there are a number of potential advantages in reforesting degraded lands and that such reforestation has the potential to improve human well-being and help conserve biological diversity. But there are different ways of achieving this. In the recent past most large-scale industrial reforestation schemes have relied on even-aged plantations involving a single species. Many of these species were fast-growing exotics used for pulpwood and the rotation lengths used were often less than 10 years. Such plantations can produce large amounts of a homogenous timber product very efficiently and are ideally suited for industrial enterprises. However, they are as useful in situations where landholders have other objectives. For example, some growers might wish to produce higher value timbers that take longer to grow while others, including many smallholders, might wish to produce goods other than timber. Likewise, some government agencies and NGOs may be more interested in forms of reforestation that protect watersheds or provide habitats for threatened wildlife and have no intention of harvesting timber or NTFPs from their plantings. These quite contrasting objectives mean the standard industrial model should not be seen as the only way in which reforestation can be done. Rather, it is simply one of a variety of silvicultural options that might be used depending upon the land owner's objectives.

D. Lamb, *Regreening the Bare Hills*, World Forests 8,
DOI 10.1007/978-90-481-9870-2_4, © Springer Science+Business Media B.V. 2011

The situation is similar in agriculture. In discussing the reasons why large state-sponsored agricultural schemes often fail. Scott (1998, p. 262) wrote:

> The simple 'production and profit' model of agricultural extension and agricultural research has failed in important ways to represent the complex, subtle, negotiated objectives of real farmers and their communities. That model has also failed to represent the space in which farmer's plant crops – its microclimates, its moisture and water movement, its microrelief, and its local biotic history. Unable to effectively represent the profusion and complexity of real farms and real fields, high-modernist agriculture has often succeeded in radically simplifying those farms and fields so they can be more directly apprehended, controlled, and managed.

The objective of this chapter is to review the main methods that can be used to reforest cleared or degraded land. It emphasizes that there are a number of silvicultural approaches that might suit the 'profusion and complexity of real farms and real fields'. This chapter identifies three broad forms of reforestation. It also explores how new forests might be buffered against ecological and economic changes that could occur in the future and the implications this has for silvicultural practices.

A Conceptual Model of Degradation and Forest Restoration

We are primarily concerned here with the reforestation of 'degraded' land. As discussed earlier, 'degradation' is a term that is fraught with definitional problems and, depending on their condition, 'degraded' lands will differ in their ecological attributes and in their capacity to recover unaided. In Chapter 1 degradation was described as occurring when human activities had caused a reduction in the productivity, economic value or amenity of a site. This is shown conceptually in Fig. 4.1. At point A the undisturbed ecosystem has a certain level of biodiversity and structure or biomass. Changes caused by deforestation reduce both biodiversity and structure leaving the site in a degraded state (B). Further disturbances such as wildfire or overgrazing may lead to even more degradation (C). At this point few of the original species remain and the site is occupied by a variety of grasses and broad-leaved weeds. Logging (rather than agricultural clearing) may also cause changes although these are usually much less transformative. So, carefully managed Reduced Impact Logging might move the system to D while unregulated and poorly managed logging might move the system to E. Compared with the situation at D, some species may have been lost and there would be substantial changes to forest structure. In some cases a number of new, so-called secondary species may colonize the site. Some of these may be exotic weed species. Many would regard E as also being degraded like B and C.

Natural recovery can occur at some of these new states but not from others. So, regrowth from D may be sufficiently rapid to allow a subsequent logging operation after, say, 30 years and most of the original species will have remained present at the site. Indeed this is what happens in a well managed logging operation. Recovery after the poorly managed logging (E) may take much longer (and there may be some change in the final species composition because new species, possibly including some exotics, may have permanently colonized the site). Recovery from the degraded conditions at B and C are likely to be more problematic. Natural recovery may occur relatively

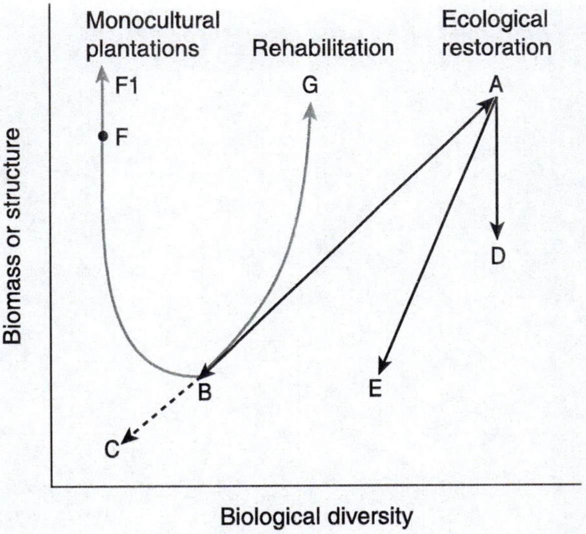

Fig. 4.1 A conceptual diagram showing the relationship between ecological restoration, rehabilitation and monocultural plantations. At point A the original forest has a certain biomass/structure and biodiversity. Various types of disturbance can change its condition. It is considered degraded when it loses both biomass/structure and biodiversity and arrives at point B. See text for further explanation

rapidly where the degraded site is not large, soils remain intact and where species remain on the site (as seed stored in the soil, as seedlings or as old root material) or can disperse into the site from nearby intact forest. Such might be the conditions after a site has been briefly used for, say, shifting cultivation. But this recovery may not occur where the site has been cropped for a number of years or has been occupied by grasses. In this case changes to key processes (e.g. nutrient cycling) or natural feedback mechanisms (e.g. seed dispersal) may have caused the system to move to an alternative state from which recovery is difficult or, at best, very slow.

Under these circumstances there are three ways in which reforestation might be undertaken. One is to restore the original forest and re-establish the former composition and structure. This means promoting the transition from state B or C to state A. This can be done by facilitating natural regrowth or by planting seedlings of the original species. This approach will be described here as *Ecological Restoration.* The second is to forgo trying to regain state A but to plant a monoculture timber plantation (or agricultural crop) using a species that is commercially attractive and able to tolerate the conditions now present (e.g. the site might now have less fertile soils). In this case, a new state (F) is established. If various forms of site amelioration including fertilizers are used the biomass may increase beyond that of the undisturbed forest (F1). There is no particularly appropriate term to describe this and so it will be simply referred to here as a *Monoculture Planting* (cf. Lamb 2001; Lamb and Gilmour 2003). The third approach lies between these two. It involves fostering the establishment of some, but not all, of the original species such that the biomass and most of the structure are re-established though not the original biodiversity. The new state (G) may eventually have a similar biomass or structure

Fig. 4.2 A plantation monoculture of *Eucalyptus urophylla* in Vietnam. Over time a thick groundcover of grasses and herbs develops and provides good protection against erosion

to that of the original forest but a lower level of biodiversity. This approach will be referred to as *Rehabilitation*. Examples of the three approaches being used in the field are shown in Figs. 4.2, 4.3, and 4.4.

The three approaches are necessarily a simplification of the much wider variety of ways in which reforestation of a degraded site might be undertaken and each will be described in rather more detail below. All are similar in that they attempt to develop productive forests. However, they differ in the extent to which biodiversity or structural complexity is regained and in their capacity to supply various goods and ecosystem services. They also differ in the rate at which their objectives are likely to be achieved. Many Monoculture Plantations achieve their objective and are felled after less than 10 years while some Ecological Restoration projects may take more than 100 years to be completed. Some of these terminological issues are discussed further in Box 4.1.

Choosing Between Ecological Restoration, Plantation Monocultures and Rehabilitation

The choice between these three reforestation alternatives depends on the land owner's objectives and whether they are interested in forests producing goods, ecosystem services or a mixture of both. The advantages of each reforestation approach are

Fig. 4.3 Ecological restoration of rainforest in central Thailand. The site was restored using seedlings and seed and is now about 15 years old and contains a large number of the original tree species

reasonably clear but any choice must also pay attention to some of the disadvantages each has. Some of these advantages and disadvantages are outlined below.

Advantages and Disadvantages of Ecological Restoration

Restoring forests on degraded lands to recreate the former forest is surely a worthy goal since it will restore biodiversity and generate a variety of ecological services although not necessarily commercial goods. It might be achieved using natural regeneration or by planting seedlings (Table 4.1). But restoration, as defined in this way, can present a host of difficulties. The first of these is that the target may be unclear, especially when deforestation took place many years earlier and no remnants of the original forest now remain. This is an obvious problem for those in highly modified and long-settled landscapes such as those in Europe but it also applies to many locations in the Asia-Pacific region where all that may be known is perhaps the names of a handful of the more dominant former canopy tree species.

A second difficulty concerns changes to the physical environment. Degradation can change soil chemical and physical properties, hydrological conditions and fire regimes. Such changes may make it impossible for the original species to re-establish at the site,

Fig. 4.4 Forest rehabilitation at a former open-cut bauxite mine in northern Australia. A new forest with a variety of trees and understorey plants has been established. After 15 years it resembles the original open monsoonal forest although the species composition is different because of the changed environmental conditions (Photo: Peter Erskine)

at least in the short term, because these can no longer tolerate the present site conditions. Again, a species-rich forest may develop but it will not match the original forest.

A third problem is that, following deforestation, some original species may have been lost through extinction or exotic species may have invaded and become naturalized and be impossible to eradicate. Such changes can affect ecological processes within the new ecosystem which, in turn, affect some of the remaining native species. Obvious examples are where a missing species was an important seed disperser or a new species is an aggressive weed. These may not prevent a species-rich new forest being established but it will be qualitatively different from the original forest. And fourthly, environmental conditions at the site may be changing as part of a longer-term climatic change perhaps induced by global warming. Thus there might be might be changes in temperatures, rainfall seasonality or fire regimes. These may alter the capacity of some of the original biota to regenerate or reproduce at particular sites and

Box 4.1 Some Definitions

There is considerable variation in the terminology used to describe the ways forests can be established and this debate continues (Carle and Holmgren 2003). The terms below will be used as follows:

Reforestation is used here as an all-embracing term covering the development of forests by both natural regeneration and planting irrespective of the species or planting designs that are used. Thus it includes forests created using Monoculture Plantations, Rehabilitation and Ecological Restoration. Although the term *restoration* is also often widely used as a general descriptor of reforestation it will be avoided here to avoid confusion with the more specific term Ecological Restoration (see below). In most of the cases discussed it is assumed that forests occupied the sites being reforested within the previous 50 years. It contrasts with the term *afforestation* which is generally used to describe reforestation at sites that have never been occupied by trees or have not had trees for >50 years.

Natural regeneration is the re-establishment of native trees and other plants by self-sown seed or by vegetative regrowth.

Monoculture plantations are plantings of single species carried out at the same time. The species may be indigenous or exotic species and are commonly established at densities of around 1,100 trees per hectare. Most are grown for a fixed period or rotation after which time the plantation is harvested and re-established. Only some of the natural processes and functions are recovered. The productivity of the plantation may exceed that of the natural forest because of the species used, site preparation or fertilizer applications.

Rehabilitation describes the development of new forests made up of some, but not necessarily all of the original species at a site. Rehabilitated forests may also include some exotic species. Most are developed by planting or seeding but some natural regeneration may also be allowed to develop. There can be considerable variation in the number of species used and in the management methods applied. The former productivity and many of the original ecological processes are usually recovered.

Ecological Restoration is the process of assisting the recovery of an ecosystem that has been degraded, damaged, or destroyed (SER 2004). Recovery takes times and an ecosystem can be said to be restored when it contains sufficient biotic and abiotic resources to continue its development without further assistance or external subsidy. In contrast to those establishing monocultural plantations or rehabilitating degraded lands, those seeking to ecologically restore forests often aspire to re-establish the pre-existing biotic integrity in terms of species composition, community structure and ecological functioning. Ecological Restoration might be achieved through natural regeneration or by a combination of planting, seed sowing and natural regeneration.

Table 4.1 Reforestation methods to suit different objectives

	Reforestation objective		
Reforestation method	Monoculture plantings	Rehabilitation	Ecological restoration
Natural regeneration (discussed further in Chapter 5)		Is the outcome when complete natural regeneration is not possible; may also be achieved through enrichment planting using native or exotic species	Likely to be achieved where undisturbed natural forests are nearby
Single-species plantings (discussed further in Chapter 6)	Achieved with native or exotic species grown using short or long rotations		
Mixed-species plantings (discussed further in Chapter 7)		Achieved when multiple species of trees and shrubs grown in temporary or permanent mixtures at the same site	
Restoration plantings (discussed further in Chapter 8)			Likely to be achieved when a high proportion of native plant species are planted or sown and colonists from nearby intact forests are able to reach the site

so change their spatial and altitudinal distribution. On top of these difficulties are the practical problems inherent in trying to re-assemble an ecosystem about which the restorationist has incomplete knowledge and, moreover, having to do this at a scale that will allow some of the key ecological processes to operate. In the case of plants, how might one regenerate the many hundreds of species and life forms once present?

All this means that the task of restoring former forests is indeed a formidable one. But it is not necessarily an impossible undertaking and some very promising attempts have been made despite the difficulties listed here. These have involved using natural regeneration, plantings and direct seeding and will be discussed further in Chapters 5 and 8 respectively. Interestingly, they do not necessarily involve tracking directly back up the B-to-A pathway as Fig. 4.1 implies. Because of this it is useful to have a series of benchmarks with which to monitor the system's development and show whether the new system is on an appropriate successional trajectory. Some possible benchmarks are shown in Box 4.2.

The extent of changes induced by humans and the difficulty of restoring degraded lands led Oosthoek (2008) to argue (under a sub-heading 'Nature is finished;

Box 4.2 Attributes of Restored Ecosystems (Society for Ecological Restoration International 2004)

1. The restored ecosystem contains a characteristic assemblage of species that occur in the reference ecosystem and that provide appropriate community structure.
2. The restored ecosystem consists of indigenous species to the greatest practical extent.
3. All functional groups necessary for the continued development and/or stability of the restored ecosystem are present or have the potential to colonize by natural means.
4. The physical environment of the restored ecosystem is capable of sustaining reproducing populations of the species necessary for its continued stability or development along the desired trajectory.
5. The restored ecosystem apparently functions normally for its ecological stage of development and signs of dysfunction are absent.
6. The restored ecosystem is suitably integrated into the landscape with which it interacts through abiotic and biotic flows and exchanges.
7. Potential threats to the health and integrity of the restored ecosystem from the surrounding landscape have been eliminated or reduced as much as possible.
8. The restored ecosystem is sufficiently resilient to endure the normal periodic stresses in the local environment that serve to maintain the integrity of the ecosystem.
9. The restored ecosystem is self-sustaining to the same degree as its reference ecosystem and has the potential to persist indefinitely under existing environmental conditions although the composition and other attributes may evolve as environmental conditions change.

conservationists admit defeat') that future restoration would largely be concerned with re-assembling new ecosystems using non-native species rather than trying to return to the historic state. Hobbs et al. (2009) have referred to these as 'novel' ecosystems. As a generalization this may be an excessively gloomy prognosis but it is likely to be correct in at least some degraded landscapes. In these cases the best options may be to develop multi-species, self-sustaining and resilient ecosystems that contain as many as possible of the original biota but which also make use some non-indigenous species. Although these will not be identical with the original ecosystems they may be able to restore most of the original functionality and provide a good starting point for adapting to future changes such as those induced by global warming. In the present terminology these types of plantings might be described as 'rehabilitation' and will be discussed further below.

The problems involved in restoring wildlife populations deserve particular comment. Deforestation and fragmentation will have made some species locally extinct but allowed the population of some others to increase. In most cases restorations can only seek to restore habitats and food supplies and hope that sites will be naturally recolonized from residual populations of these species still present elsewhere in the region. Such recovery may or may not occur. When it does occur it will usually take time because some habitat features only develop slowly (e.g. hollow-bearing trees, logs on ground). Wildlife translocation programs are rarely possible even though these species may influence pollination, seed dispersal, seed predation and regulate trophic structures. And some wildlife such as large herbivores (e.g. elephants) or large top-level carnivores (e.g. tigers) are unlikely to be welcomed by nearby human communities. The functional consequences arising from the absence of species such as top-order predators in newly restored forests are mostly unknown although Soule and Terborgh (1999) argue they may be profound. Large areas of fully restored forests are needed for the conservation of these species but may be hard to re-establish. On the other hand, some species may be able to use the so-called novel ecosystems referred to above and survive in a mixture of fully restored forest and rehabilitated forest.

Advantages and Disadvantages of Plantation Monocultures

Large plantations involving a single tree species have been established in many parts of the world and they are common throughout the Asia-Pacific region. These can use native or exotic species and may be grown on short or long rotations (Table 4.1). Some have been profitable and regarded by their owners as highly successful. Others have failed because the species chosen were unsuited to the site, seedlings were of poor quality, site preparation was insufficient or weed control, fire control, pests, diseases or a host of other issues were not dealt with. Few have involved the species forming natural mono-dominant forests (Box 1.1) presumably because any ecological advantages these species have is outweighed by their economic disadvantages.

Intensively managed plantation monocultures can suffer productivity declines over time. Pulpwood plantations that use fast-growing exotic species are very prone to nutrient losses because the logs being harvested contain a high proportion of nutrient-rich sapwood and because many nutrients can be lost by leaching each time the site is cleared and replanted. This nutrient drain can lead to productivity declines in later rotations unless the loss is remedied. Of course such problems face all those seeking to reforest degraded sites but the problems can be more acute if only a single species is being relied upon.

But other kinds of failure have occurred as well. In some cases market conditions have changed after the plantations were established and the species prove to be unsuited to the new timber markets. In other cases the expectations of society change as living standards rise. People want cheap timber but they also want recreational opportunities, wildlife conservation and aesthetically pleasing landscapes. Monoculture plantations are efficient at producing particular goods but may be much less able to generate these various ecosystem services.

Advantages and Disadvantages of Rehabilitation Plantings

Rehabilitation plantings form much of the continuum between restoration plantings and monoculture plantations. These plantings are not attempts to restore a forest to some bygone condition nor do they necessarily seek to maximize the production of a single product. Instead they can be seen as a way of accommodating the objectives of a variety of stakeholders and as a means of adapting to the new environmental and economic conditions now present (or likely to develop in future). Plantings such as these were referred to earlier as 'novel ecosystems'. Rehabilitation can involve planted seedlings, natural regeneration or a combination of the two (Table 4.1). If they are well-designed rehabilitation plantings can improve both human well-being and ecosystem integrity (Lamb and Gilmour 2003). The former occurs when there are direct financial benefits generated by reforestation. The latter is improved by increased functional effectiveness and ecological naturalness. The dilemma for those interested in using this approach lies in designing types of reforestation that achieve both elements. What form should these take? Just how many species are needed? Which particular species should be used and in what proportions should these be planted? The answers to these questions depend on the circumstances at particular sites meaning that the label 'rehabilitation' covers a variety of silvicultural approaches and techniques.

How do people make choices between Ecological Restoration, Monoculture Plantations or Rehabilitation? Some people will have an over-riding preference for one particular approach because of their wish to generate a financial return or to improve ecosystem functioning at a particular location. Others will choose after considering what each alternative offers and what it might cost to implement. But attitudes and preferences can change over time as changes occur in the economic or ecological environment or as landowners personal circumstances change. For example, extensive natural regeneration in the understorey of a plantation might lead to the decision that a production forest has more value for conservation that for timber production. Likewise, a landowner may choose to delay felling a mature plantation forest because of it's aesthetic appeal. Other ways in which the balance may change over time as new forms of management are adopted are shown in Fig. 4.5. One consequence is that while it might not be possible to achieve the preferred balance immediately, it may be possible to work towards this over several decades as economic and social circumstances allow (Lamb et al. 2005).

Degradation and Resilience

There is an additional element that can help inform this design process and that concerns the desirability of making the new forests more resilient to future disturbances. Ecologists use the term resilience to refer to the capacity of any system to absorb disturbances and remain in the same state with essentially the

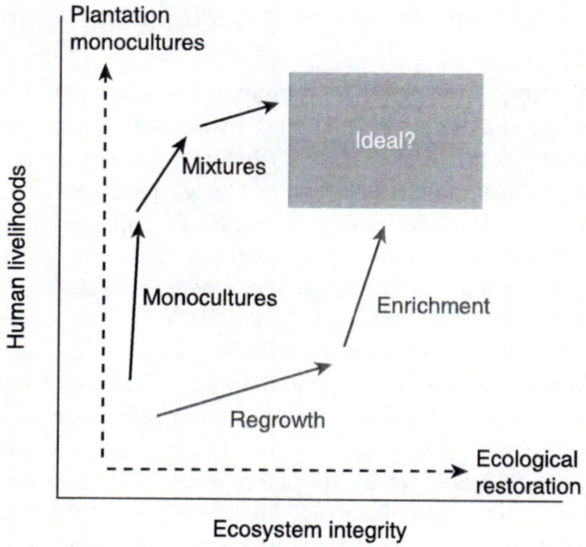

Fig. 4.5 Those undertaking reforestation must make a trade-off between improving human livelihoods and improving ecosystem integrity. Plantation monocultures can help improve livelihoods while Ecological Restoration is best able to improve ecosystem integrity. But, over time, it may be possible to modify the forms of silviculture being used to achieve elements of both goals

same structure, functioning and feedback mechanisms (Berkes et al. 2003; Gunderson 2002). The more resilient the system, the greater the amount of change it can undergo and still maintain these same controls. A little background about what is known as resilience is necessary before exploring the implications it has for reforestation choices.

The current 'best practice' in agriculture and plantation forestry seeks to increase the productivity of certain species. This is done by breeding productive varieties of preferred species, fertilizing these and taking the system to some optimum state. Managers then seek to hold that level of productivity and make that state sustainable (Walker and Salt 2006). Productivity is maximized by controlling each aspect of the production cycle. Managers usually assume that incremental improvements can be made and that change is linear (thus better site preparation and improved fertilizer technology will lead to more production) but rarely take account of what might be happening elsewhere in the landscape away from the farm or in processes operating at smaller scales such as among microbial populations in topsoils. In addition, they often ignore the changing environmental and economic conditions in which their production system is enveloped. This approach contrasts with the various forms of agriculture practiced by most traditional farmers where a variety of species were grown, often on the same piece of land to generate a diversity of foods and other goods and build a degree of insurance into the system.

Over the years, the 'best practice' model has worked reasonably well in agriculture as well as in plantation forestry and productivity levels in both have increased. Or, at least the model has worked until recently. There are now signs in many parts of the world that it has some critical weaknesses and that it is working less well now than it did in the past. Sometimes the model has even failed after a comparatively short period of farming. Examples of these failures are the increasing levels of chemicals (fertilisers and pesticides) needed to sustain productivity and the increasing areas of degraded lands that are beginning to appear in many agricultural landscapes.

Walker and Salt (2006) point to a paradox. Optimizing agricultural or forestry production is supposed to be about promoting efficiency. This might be expressed as greater food production or timber volumes per hectare. But optimization is also about reducing redundancies by eliminating all those species that are not immediately valuable. The problem with this is that ecological systems are usually configured by interactions between a number of species and these relationships are mostly defined by extreme events and not average conditions. Many species may appear redundant but, in fact, play an important role in maintaining the system when environmental circumstances change (Folke et al. 2004; Walker et al. 1999). In other words, systems with many such species are more resilient. As a result, the more a manager seeks to reduce diversity and optimize components of a production system in isolation from the remainder of the ecosystem as a whole, then the more vulnerable such a system becomes to changes and disturbances. That is, the optimized systems lack insurance. These simplified systems may be temporarily 'efficient' but they are also fragile. This sounds counter-intuitive but it appears to be the conclusion emerging from a number of studies. As Walker and Salt (2006, p. 7) note:

> The paradox is that while optimization is supposedly about efficiency, because it is applied to a narrow range of values and a particular set of interests, the result is major inefficiencies in the way we generate values for societies

In the present context one of the aims of reforestation is to improve the livelihoods of smallholders by reducing their vulnerability to future shocks. If Walker and Salt (2006) are correct the 'maximum sustained yield' model may be a flawed and risky way forward.

Resilience in Social-Ecological Systems

This issue forms part of a broader question concerning the way ecosystems function in the face of change or disturbances. It is well-known amongst ecologists that ecological systems are non-linear in their trajectories of change and have the capacity to exist in a number of alternative, stable states or regimes in which their structure, function and feedback mechanisms are different (as are the goods and services they are able to provide).

Systems are thought to move through four stages of what is known as an adaptive cycle (Gunderson 2002). The commencement of the cycle is a colonisation or exploitation stage when un-utilised resources are acquired. This is followed by a conservation stage as the system matures and inter-connections between components of the system develop. But, the more inter-connections and the stronger these are, then the less flexible the system becomes and the more susceptible it is to external shocks. Eventually a disturbance will cause the system to break up and pass into a re-organization phase and the cycle begins again. The release and re-organization phases are both chaotic and rapid. It is during these stages that innovations and adjustments can be made.

The operation of this cycle can be seen in natural ecosystems that acquire biomass and diversity as they mature. Over time, a greater proportion of the system's nutrients are immobilised in biomass and more of the species become long-lived habitat specialists to the exclusion of shorter-lived, generalist species. Many form highly specialised mutualistic relationships. Eventually the system loses resilience and becomes less able to tolerate disturbances or shocks and the system collapses when the inevitable fire, storm or insect outbreak eventually occurs. A similar pattern can be found in socio-economic systems. In the early stages of a cycle the participants are innovative and non-hierarchical. Over time there is an increase in social and economic capital. However, the society gradually evolves into a more staid and socially conservative system with strong conventions and less flexibility. There are connections across a network of relationships in this system but information in these tends to flow from a centralised decision-making body. Innovation and experimentation decline. Over time the system becomes increasingly brittle until, finally, it is confronted by political or economic challenges it has not faced before and is unable to respond (Homer-Dixon 2008).

If the disturbance forms part of the historical disturbance regime the system will probably recover and the cycle will begin once more. If, on the other hand, the disturbance or shock is unusually severe the system may be pushed over a threshold into a new state from which recovery is slow or impossible. The conversion of forests to grasslands seen in some tropical areas and caused by the unusual combination of agricultural clearing and fire is an example of such a transition.

Rather than thinking of just ecological or social or economic systems it is more useful to think of a combined entity or what Gunderson (2002) and Walker et al. (2006) have referred to as a social-ecological system. Following a disturbance a social-ecological may recover and re-establish the same adaptive cycle with essentially the same biota and controlling economic variables. But if the system has been forced across a threshold then an entirely new set of biological communities, socio-economic structures and controlling variables will develop. Such changes occur when the adaptive capacity of the system has been exceeded and it is 'degraded'. Crossing one threshold in a social-ecological system can trigger changes in other components of the system. This means that ecological degradation may be caused by socio-economic events but this, in turn, is likely to generate other economic changes and force the system to cross additional economic and social thresholds as well.

In assembling new social-ecological systems or re-organising degraded ones it is important to find ways by which resilience is enhanced to avoid the development of fragile conditions that pre-dispose the system to collapse. Diversity is at the core of resilience. In ecological systems there are three types of diversity that are important. One is the diversity of *functional types* or species having a similar impact on ecosystem processes. For example, whether the ecosystem has shade-tolerant as well as light-demanding species, nitrogen-fixers, decomposers, herbivores, carnivores, pollinators and seed dispersers. Representatives of all these groups are needed if the system is to function effectively. A second type is the diversity of species able to generate a particular *functional response*. Resilience is increased if there are several species able to perform each of these various functions with some being most effective under some environmental conditions (e.g. dry weather) and different species able to do so in other conditions (e.g. in wet weather). A seemingly redundant species may, under changed environmental conditions, become very important to the way a system functions (Diaz and Cabido 2001; Elmqvist et al. 2003). A third type of diversity is that occurring at a *landscape* level rather than just at a site level. A species-rich landscape means that local extinctions can be overcome by recolonisation from populations elsewhere in the landscape. A small amount of diversity can often restore a significant proportion of ecosystem functioning but, in the longer term, and over larger areas, a much greater degree of functional diversity is needed to ensure ecosystems are able to function consistently.

Within the economic and social components of a social-ecological system a diversity of markets, institutions and sources of knowledge is also important. Thus a system where income is derived from a variety of goods and services that are sold into a number of separate markets is preferable to a system that depends on a single product and a single buyer. Likewise, management systems that use knowledge gained from a diversity of sources, including external sources and traditional ecological knowledge, and that use inputs from a variety of stakeholders to make decisions about natural resources are usually more resilient than top-down forms of management informed from a single perspective. Diversity in social-ecological systems increases the systems capacity for self-organization following a disturbance or shock.

Resilience has a cost. In the short term it is likely to be far more profitable to maximize production and not worry about building resilience. But the longer a system is managed in this way the more likely it is there will be an unexpected ecological or economic shock that will push the system across a threshold (Anderies et al. 2006). Somehow managers must strike a balance between the cost of the short-term benefits foregone by building resilience and the longer-term likelihood of the system collapsing and moving to a new state when resilience is ignored.

Building Resilience During Reforestation

Overcoming degradation usually involves transforming the system to a new state which can generate a larger amount of natural, financial and human capital. As capital increases so does flexibility. There are several implications arising from resilience theory for the ways in which reforestation should be carried out.

Ecological: The first is that patches of remnant forests or areas of secondary regrowth should be protected, however small these are. Such forests can help protect the genetic diversity of plant species needed in reforestation programs. They may also provide habitats for wildlife such as birds or bats able to carry seeds across the landscape. This will be discussed further in Chapter 5. The second implication is that any plantings should involve a variety of species and functional types. Ideally, this diversity should be sought at every site but this might not always be realistic. When it is not possible then diversity should be sought at a landscape scale (i.e. if not alpha diversity then gamma diversity). This will be discussed further in Chapter 7.

Economic: The third implication is that plantings should take account of economic circumstances and, where-ever possible, those designing plantations should seek to provide goods and services for a variety of markets. A plantation producing a single product sold to a single buyer places a grower in a highly vulnerable position and sensitive to economic as well as biological misfortunes. Agricultural and forestry history is littered with examples of problems arising from over-reliance on a single species (Boxes 4.2 and 4.3).

Social: Finally, resilience requires that any reforestation program should ensure that people and institutions are in place to absorb feedback and to innovate, research and develop new knowledge rather than being largely dependent on an external source of technical advice. Ways must also be found to spread this newly developed knowledge amongst those carrying out reforestation. This is often most easily done by developing learning networks which bring together researchers and practitioners. This will be discussed further in Chapter 10.

Reforesting degraded lands is an uncertain business. In many cases new silvicultural techniques must be developed and it is inevitable that mistakes will be made. Adaptive management treats the management process as a series of experiments which are carefully monitored such that adjustments to management inputs can be made if this is necessary (Anderies et al. 2006). The process involves learning-by-doing. Resilient systems are those that use this approach and have the stakeholder networks and monitoring systems in place to respond to ecological, social and economic feedback. They also enable the institution and policy settings to be adjusted where this is found to be necessary. The point of all these interventions is to generate flexibility so that the system can adjust to change and not be degraded again in future.

Some Problems for Those Seeking to Design Resilient Forms of Reforestation

The task of building resilience raises several interesting questions for those undertaking reforestation.

What Sort of Resilience – Specific or More General?

When building resilience should one seek to build resilience towards a particular form of disturbance that is perceived to be more likely such as a wildfire, a windstorm or a change in the market for certain timber products? Or should a more general form of resilience be sought that buffers the system against a variety of disturbances? Choosing to guard against specific stresses such as wildfires may reduce the overall resilience of the system to a wider variety of disturbances or changes such as climate change.

Box 4.3 The Hazards of Single Markets

Fluctuations in prices of agricultural products such as coffee, cocoa or sugar cane are well-known but similar price fluctuations can occur in forest products. Large numbers of people have sometimes been affected when these occur. In the nineteenth century NTFPs rather than timber were the major products harvested from tropical forests and a number of these went through boom and then bust cycles in Southeast Asia. These include gutta percha from *Palaquium* (Knapen 1997; Potter 1997), jelatong or rubber from *Dyera* (Potter 1997) and gambier from *Uncaria gambier* (Colombijn 1997). In all cases attempts were made to domesticate the crop but these attempts eventually failed. The failures were caused by alternative products entering the market (e.g. *Hevea brasiliensis* from Brazil replacing gutta percha and jelatong) or site degradation (e.g. gambier).

A more recent example of a fluctuating market is that of rattan in Kalimantan (De Jong et al. 2003; Michon 2005). In some cases this has been due to drought but in others it was caused by misguided government policies that attempted to regulate export markets. These eventually led to a market collapse. Indonesia has tended to dominate the international rattan trade so that these factors, as well as changes in the Indonesian exchange rate, have had dramatic effects on the profitability of growing rattan elsewhere in the region.

Smallholder production of *Gmelina arborea* in parts of the southern Philippines is an example of a heavily promoted timber species becoming unprofitable. In this case large numbers of farmers successfully grew the trees but were unable to obtain a worthwhile price when timber from these plantations flooded the market at the same time. The experience has driven many farmers in this region out of tree-growing (Pasicolan and Macandog 2007). Something similar appears to have occurred in parts of Vietnam where farmers were encouraged to grow *Eucalyptus* spp. The fast growth of eucalypts and their ability to tolerate degraded sites made them attractive to many farmers. They remain so for farmers near pulpwood markets but are now regarded much less favourably by growers distant from these markets because the market value of small eucalypt logs is low (Hawkes 2000; McElwee 2009; Raintree et al. 2002; Rambo and Le 1996). Both *Gmelina* and *Eucalyptus* remain popular and important plantation species in other places but the examples show that an over-reliance on even widely-used species can sometimes have unexpected consequences, especially where transport cost preclude long distant transport.

How Much Diversity is Needed in Plantations to Generate (Sufficient) Resilience?

Should growers focus on just one single plantation species that is productive in the sites they have available and for which there is presently a good market? Or should they include additional species that are not necessarily as productive or valuable in order to hedge their ecological and economic bets in case present circumstances change? The sheer number of sparsely distributed species in tropical forests suggests many are probably truly functionally redundant. So what is the risk of not using many of these? Risk involves two elements; one is the chance that an event will occur and the other is the magnitude of the adverse consequences if it does. Different plantation owners are likely to have quite contrasting perspectives on both elements with large industrial plantation owners taking a different view than, say, a small landholder with a limited income. The former, having assessed their circumstances, may decide they are able to continue growing trees for pulpwood using monocultures, especially if they are able to use short rotations and have various financial instruments to shelter them from risk. Some of the latter might well take a different view especially if they are living some distance from industrial markets and the identity of their future market is still unclear. But, even so, these farmers are unlikely to seek to duplicate the diversity present in natural forests.

How to Encourage the Development of Resilient Forms of Reforestation?

Many farmers from across the Asia-Pacific region traditionally practiced forms of agriculture such as shifting cultivation that were highly resilient. These systems evolved over time through experience of change and crisis. But many of these practices are being swept away by deliberate government policies prompting a switch to more sedentary forms of agriculture as well as by exposure to new cash crops, new markets and changes to land tenure systems. There is now a tendency to simplify and intensify cropping systems. In the light of these trends, what decision should a smallholder proposing to reforest part of their land make about resilience? Should they simply focus on maximizing productivity and generating an early cashflow or should they try mimic their former agroforestry systems and reduce vulnerability by establishing mixed-species plantations? And who can advise them? Government agencies advocating simplification and intensification are unlikely to be supportive of a system that gives weight to resilience. These questions are discussed further in Chapter 10 in a discussion about farmers and the partnerships they may form.

Might It Be Easier to Enhance Resilience at a Landscape Scale Rather Than at a Particular Site?

It may be difficult – or unnecessary - to establish highly resilient plantations at every site and perhaps the diversity of functional types may be more easily achieved by aiming to develop a variety of types of plantations in different parts of the landscape? Thus the landscape may become a mosaic of vegetation types including undisturbed natural forests, regrowth forests, plantation monocultures and perhaps rehabilitation plantings. Designing such a mosaic to balance financial and ecological needs is likely to be difficult when only a single landowner is involved but will be even more difficult when there are a range of landowners and other stakeholders. The topic is discussed further in Chapter 11.

Conclusion

Notwithstanding the simple monocultures of exotic species that are commonly used there are, in fact, a variety of ways in which degraded lands might reforested. These differ in the numbers of species planted and in the extent to which they restore ecosystem integrity and improve human well-being. They also differ in their functional effectiveness and in their resilience. Some forms of reforestation are very suitable for producing large quantities of industrial timbers but are less suited for producing the variety of forest goods that are desired by many smallholders. Some forms of reforestation are able to generate ecosystem services such as protecting watersheds but will be much less able to create the habitats needed by certain wildlife.

The circumstances and objectives of landowners or manager will determine which type of reforestation is ultimately carried out. In the past the dominant factor determining this choice for most industrial growers was the expected financial return. But smaller private growers may take a different view. Planted forests differ from most other land uses because of the length of time between when an investment is made and there is a benefit to growers. This means risks are greater and more resilient types of reforestation that can minimize these risks deserve greater consideration.

The following chapters provide a more detailed examination of different forms of reforestation. At its simplest there are two ways in which reforestation can be achieved and these are by natural regeneration or by some form of planting. Natural regeneration is the least costly form of reforestation where it is able to occur although its capacity to produce particular goods and services varies a good deal. Natural regeneration and so-called secondary forests will be discussed in the next chapter. Subsequent chapters will address some of the ways reforestation might be carried out using planted seedlings.

References

Anderies JM, Walker BH, Kinzig AP (2006) Fifteen weddings and a funeral: Case studies and resilience-based management. Ecol Soc 11(1):21, http://www.ecologyandsociety.org/vol11/iss1/art21/

Berkes F, Colding J, Folke C (2003) Navigating social-ecological systems: Building resilience for complexity and change. Cambridge University Press, Cambridge

Carle J, Holmgren P (2003) Definitions related to planted forests. Working Paper 79, FAO, Forestry Department, Rome

Colombijn F (1997) The ecological sustainability of frontier societies in eastern Sumatra. In: Boomgaard P, Colombijn F, Henley D (eds) Paper landscapes: Explorations in the environmental history of Indonesia. KITLV Press, Leiden, pp 309–340

De Jong W, Belcher B, Rohadi D, Mustikasari R, Levang P (2003) The political ecology of forest products in Indonesia: a history of changing adversaries. In: Tuck-Po L, De Jong W, Ken-ichi A (eds) The political ecology of tropical forests in Southeast Asia: Historical perspectives. Kyoto University Press, Kyoto, pp 107–132

Diaz S, Cabido M (2001) Vive la difference: Plant functional diversity mattters to ecosystem processes. Trends Ecol Evol 16:646–655

Elmqvist T, Folke C, Nyström M, Peterson G, Bengtsson J, Walker B, Norberg J (2003) Response diversity, ecosystem change and resilience. Frontiers Ecol 1:488–494

Folke C, Carpenter SR, Walker BH, Scheffer M, Elmqvist T, Gunderson L, Holling CS (2004) Regime shifts, resilience and biodiversity in ecosystem management. Ann Rev Ecol Systemat 35:557–581

Gunderson LH (ed) (2002) Panarchy: Understanding transformations in human and natural systems. Island Press, Washington, DC

Hawkes M (2000) Conservation and sustainable use of medicinal plants in Yen Do and Yen Ninh communes, Thai Nguyen province, northern Vietnam. Thesis, School of Natural and Rural Systems Management, University of Queensland, Brisbane

Hobbs RJ, Higgs E, Harris JA (2009) Novel ecosystems: Implications for conservation and restoration. Trends Ecol Evol 24:599–605

Homer-Dixon T (2008) The Upside of Down. The Text Publishing Company, Melbourne

Knapen H (1997) Epidemics, droughts and other uncertainties in Southeast Borneo during the eighteenth and nineteeth century. In: Boomgaard P, Colombijn F, Henley D (eds) Paper landscapes: Explorations in the environmental history of Indonesia. KITLV Press, Leiden, pp 121–154

Lamb D (2001) Reforestation. In: Levin SA (ed) Encyclopedia of biodiversity. Academic Press, San Diego, CA, pp 97–108

Lamb D, Erskine P, Parrotta J (2005) Restoration of degraded tropical forest landscapes. Science 310:1628–1632

Lamb D, Gilmour DG (2003) Rehabilitation and restoration of degraded forests. International Union for the Conservation of Nature and World Wide Fund for Nature, Cambridge

McElwee P (2009) Regforesting 'bare hills' in Vietnam: social and environmental consequences of the 5 million hectare reforestation program. Ambio 38:325–333

Michon G (2005) Domesticating forests: How farmers manage forest resources. Institut de Recherchepour le Developpement, Center for International Forestry Research, World Agroforestry Center, Paris, Bogor

Oosthoek S (2008) Nature 2.0. New Scientist 199:32–35

Pasicolan PN, Macandog DM (2007) Gmelina boom, farmers' doom: Tree growers' risks, coping strategies and options. In: Harrison SR, Bosch A, Herbohn J (eds) Improving the triple botton line from small-scale forestry: Proceedings of IUFRO 308 Conference. Ormoc City, Leyte, the Philippines, University of Queensland, Brisbane, pp 313–318

Potter LM (1997) A forest product out of control: Gutta percha in Indonesia and the wider Malay world, 1845–1915. In: Boomgaard P, Colombijn F, Henley D (eds) Paper landscapes: Explorations in the environmental history of Indonesia. KITLV Press, Leiden, pp 281–308

Raintree J, Le TP, Nguyen VD (2002) Marketing research for conservation and development: Case studies from Vietnam. Forest Science Institute of Vietnam, Hanoi

Rambo AT, Le TC (1996) Rural development issues in the upland agro ecosystems of Vinh Phu Province. In: Le TC, Rambo AT, Fahrney K, Tran DV, Romm J, Dan TS (eds) Red books, green hills: The impact of economic reform on restoration ecology in the Midlands of North Vietnam. Center for Natural resources and Environmental Studies, Hanoi University; East-West Center, Program on Environment; Southeast Asian Universities Agroforestry Network, University of California Hanoi, pp 117–127

Scott JC (1998) Seeing like a state: How certain schemes to improve the human condition have failed. Yale University Press, New Haven and London

Society for Ecological restoration International (2004) The SER International Primer on Ecological Restoration. Society for Ecological Restoration International, Tucson

Soule ME, Terborgh J (1999) The policy and science of regional conservation. In: Soule ME, Terborgh J (eds) Continental conservation: Scientific foundations of regional reserve networks. Island Press, Washington, pp 1–17

Totman C (1989) The green archiplelago: Forestry in pre-industrial Japan. University of California Press, Berkley, CA

Walker B, Kinzing A, Langridge J (1999) Plant attribute diversity, resilience and ecosystem function; the nature and significance of dominant and minor species. Ecosystems 2:95–113

Walker B, Salt D (2006) Resilience thinking: Sustaining ecosystems and people in a changing world. Island Press, Washington, DC

Walker BH, Anderies JM, Kinzig AP, Ryan P (2006) Exploring resilience in social-ecological systems: Comparative studies and theory development. CSIRO Publishing, Collingwood

Chapter 5
Natural Regeneration and Secondary Forests

In a country where the forests are formed of few species, and the trees are gregarious, the destruction of the primeval vegetation does not produce a great alteration in the flora; but in a tropical country covered with virgin forests, where hundreds of species of trees, lianes, and epiphytes can be found crowded together in an area of a few square miles, a clearance of the forest produces a complete change in the character of the flora, and should such destruction be extended and continued, there can be no doubt that not a few species would be rendered totally extinct. It is very probably, indeed almost certain, that in the long run the truly forestal species would gain possession of the secondary forest, once more forming a forest of the primitive type.

(Beccari 1904, p. 382)

Introduction

The easiest way of reforesting degraded lands is to take advantage of the capacity of many disturbed areas to recover unaided. The forests originating in this way are often referred to as secondary forests and these now cover large areas across the Asia-Pacific region. Many seemingly pristine forests are actually mosaics of undisturbed primary forest with patches of secondary forest of various ages arising from past shifting cultivation or natural disturbances. Indeed, some of these should be referred to as tertiary or quaternary forests because they have been disturbed on many occasions. It is difficult to make precise measures of the amount of secondary forest because of definitional problems (see below) and difficulties in mapping these forests using remote sensing technologies. However, de Jong et al. (2001) quote global estimates ranging from 340 to 600 million hectares and ITTO (2002) suggest secondary forests represent roughly 60% of the area now defined as tropical forests. In some countries the area of secondary forest far exceeds the area of undisturbed primary forest (Brown and Lugo 1990). In the Philippines, for example, secondary forests and brush-lands are thought to cover more than five million hectares while primary forests cover only 2.9 million hectares (Lasco et al. 2001).

D. Lamb, *Regreening the Bare Hills*, World Forests 8,

DOI 10.1007/978-90-481-9870-2_5,

Secondary forests are continuing to increase in area and Mayaux et al. (2005) estimate that around one million hectares of secondary forests are being produced around the globe each year with nearly half developing in Southeast Asia. These regrowth forests are equivalent to 17% of the area deforested around the world each year and 21% of that annually deforested in Southeast Asia.

The variety of ecological histories means that secondary forests vary considerably in both structure and composition. Surprisingly little is known about the areas of secondary forests in different age classes. For example, is there a preponderance of younger secondary forests or are the age classes evenly distributed across a range from young to older secondary forests? Nor is much known about the dynamics of the secondary forests now present across the Asia-Pacific region including the longevity of the species present and their rates of turnover. This chapter discusses the circumstances under which secondary forests develop and ways in which these new forests might be managed to generate economic benefits and improve conservation outcomes.

Defining Secondary Forests

There is some confusion over just how to recognize and define secondary forests. Secondary forests tend to have a more simple structure, a more even upper canopy layer and be shorter in stature than primary forest growing in the same areas. However, further generalisations are difficult because there are many types of secondary forest arising from differences in site histories and landscape contexts.

In defining secondary forests some researchers have distinguished between forests regenerating after natural disturbances and those regenerating after man-made disturbances (Brown and Lugo 1990). Some have also distinguished between forests regenerating after logging in primary forest and that regenerating on abandoned farmland (Chokkalingam et al. 2001). Corlett (1994) and ITTO (2002) both suggest it is useful to distinguish between forests regenerating at a site where there has been a break in canopy cover for some years (e.g. after farming) and those without a break in the continuity of canopy cover (e.g. after logging). They suggest the latter might be referred to as degraded primary forest. This is a useful distinction to make but there are always situations in which some uncertainty remains. For example, some poorly managed logging operations can effectively destroy the upper canopy cover but still leave scattered trees of many species at the site. In this they differ from better managed logging operations that retain a high proportion of the original forest canopy. But some shifting cultivation areas may also retain a few trees within an otherwise cleared garden area. The notion of canopy continuity is unclear in both cases.

In the discussion that follows the term ‘secondary forests’ will be used in its broadest sense to include forests regenerating after all severe disturbances including poorly managed logging and agricultural clearings. This is similar to the definition used by Whitmore (1984) which is that secondary forest is simply forest developing in big gaps and consisting of light demanding or pioneer species. However, the importance of site history in determining the recovery processes in different types of secondary forests will be discussed further below.

One of more distinctive features of young secondary forest is the higher proportion of so-called pioneer trees such as species of *Macaranga, Homalanthus, Mallotus*, and *Trema* that dominate the early stages of secondary successions. These species differ from those normally found in undisturbed primary forest by being shade-intolerant, fast-growing and mostly short-lived (Table 5.1). These pioneers are able to regenerate and grow more quickly than primary forest species and so flourish at sites where canopy gaps allow more light to reach the forest floor. This simple dichotomy masks the fact that some species have attributes falling mid-way between these two extremes and some ecologists recognize a class of long-lived (ca. 80 years) but still shade intolerant species that are often found in secondary forests (Whitmore 1984). Nonetheless, the simple classification is a useful device. This early dominance by a handful of species has led some people to regard secondary forests as being biologically depauperate. In fact, many older secondary forests are quite diverse and there is some evidence these can be more diverse than old-growth or primary forests that have not been disturbed for a long period (Connell 1978; Sheil and Burslem 2003). But our knowledge about these ecosystems is still surprisingly modest and much of what knowledge we do have comes from studies done in the earliest stages of secondary successions (Chazdon and Coe 1999). Notwithstanding this modest scientific understanding, rural communities usually value secondary forests and use them as sources of foods, medicines and building material (Chazdon and Coe 1999; ITTO 2002; Wollenberg and Ingles 1998). So too have foresters. Indeed, some of the tropical world's most useful plantation species originate in secondary forest with Asian examples including *Eucalyptus deglupta, Falcataria moloccana, Gmelina arborea* and *Pterocarpus indicus.* These are examples of the shade intolerant but longer-lived pioneer species referred to earlier. Well-known timber trees from secondary forests in other tropical regions include *Swietenia macrophylla, Cedrela odorata* and *Khaya spp.*

Table 5.1 Generalised attributes of pioneer and primary forest tree species

Attribute	Pioneer species	Primary forest species
Soil seed stores	Abundant, long-lived	Less common; short-lived
Understorey seedling pools	Rare	Common
Coppice and root suckers	Common?	Common?
Seed production	Regularly	Less regularly
Seed dispersal ability	Good	Poorer
Seedling shade tolerance	Intolerant	Tolerant
Plant longevity	Short-lived[a]	Long-lived

[a] Most <15 years although longer lived (<100 years) secondary species also occur

Natural Forest Regeneration at Disturbed Sites

Tropical forests are often able to regenerate at disturbed sites as soon as a major disturbance ceases, provided sufficient soil is present. This can be seen at shifting cultivation areas across the region and sometimes at old archeological sites. For example, when first seeing the ruins of Angkor Wat in Cambodia in 1860, Mouhot (1966, p. 99) observed:

> Within this vast enclosure, now covered by almost impenetrable forest, are a vast number of buildings, which testify to the ancient splendour of the town

Similar comments on dense forests enveloping the remains of old irrigation canals and village fortification in various parts of the Pacific testify to the regenerative capacity of tropical vegetation (e.g. Hviding and Bayliss-Smith 2000).

Once established at a new open site, seedlings often grow rapidly seeking to capture light and other resources (Fig. 5.1). The density of trees and the leaf area indices of new regrowth forests are commonly high from a relatively early stage and biomass can reach 100 t ha within 15 years if soil conditions are appropriate (Brown and Lugo 1990; Richards 1952; Whitmore 1984). Structural complexity and the canopy height of the new forest also increase such that by around 80 years it is often difficult to distinguish a secondary forest from an undisturbed primary

Fig. 5.1 Early regrowth of various ages on old agricultural fields in northern Laos PDR. Most of the regeneration is probably from the soil seed bank and perhaps old roots. Some species may be dispersed from the natural forest on hills in the distance

forest (giving rise to the definitional problem mentioned earlier – at what age does a secondary forest become a primary forest?). Changes also occur in the composition of the forest although these changes are rather less predictable than the structural changes.

Sources of Plant Colonists

The species able to initially regenerate at disturbed sites originate from a variety of sources with the relative proportion of pioneers and primary forest species depending on how many of the latter were able to persist at the site during the disturbing event. The ways plant species might be able to take part in the recovery process are from:

Seed Stored in Topsoil

Some new seedlings originate from seed already stored in the topsoil. Rainforest soils often contain a significant number of dormant seed and densities of 400–500 seeds per m^2 are not uncommon (Garwood 1989). These can make a major contribution to the new forest community provided topsoils are not too eroded when forests are disturbed. Most of these seed are of pioneer species rather than primary forest species (but see Jankowska-Błaszczuk and Grubb 2006) and just one or two tree species can sometimes dominate the seedbank although there may be as many as 30–80 species present. These seed remain dormant until germination is triggered by a disturbance such as the forest canopy being opened, by soil disturbances or by fires. These environmental changes trigger germination and allow the pioneers to take advantage of the light and soil resources made available by the disturbance. Pioneers grow quickly and usually over-top seedlings of any primary forest species that are also present in the seedling pool remaining after the disturbance.

Many pioneer species can flower and fruit at a relatively young age. This means the seedbanks of young secondary forests are quickly replenished and become much larger than those in undisturbed primary forest. Recurrent disturbances favour those species able to reproduce quickly and the seed of pioneers (and exotic weed species) can come to dominate soil seed banks if disturbances are common. Seeds of pioneer species can be relatively long-lived and Erskine et al. (2007) quote a study by Abdulhadi who found above-average densities of viable seed of early pioneer species still being present in soils supporting secondary forest in Queensland some 60 years after it had regenerated and long after the parent plants had died.

Seedlings Remaining on Forest Floor

A second source of species for the new forest is the large number already present at the site as seedlings (<50 cm tall). Undisturbed forests commonly contain a large seedling pool or bank in the understorey. The seedling density is usually larger in humid forests than in drier, more seasonal forests since survival is constrained in dry, open areas. Most of these seedlings are necessarily shade-tolerant species representative of more mature successional stages and (shade intolerant) pioneer species are less common. The diversity of species and their population sizes vary over time. This is a function of the rate of recruitment following seed rain and the longevity of different species. For example, some dipterocarp species are able to maintain a more or less continuous population of seedlings while others fluctuated between having large seedling densities immediately after mast fruiting events to having virtually no seedlings in the periods between successive fruiting periods (Fox 1976; Whitmore 1984). This is the result of a trade-off between traits that enhance persistence in shade and those that allow rapid growth in gaps (Brown et al. 2000). Thus, one species may be able to grow quickly in newly created gaps but be able to persist in the seedling bank for only a short period while another might grow more slowly but persist in shade for a much longer period.

The longevity of seedlings on the forest floor can vary from months to years. One of the longer survival records is that reported by Connell and Green (2000). These workers monitored seedling populations of an Australian rainforest canopy tree species *Chrysophyllum* spp. over a 32 year period and found 6% of those recruited in 1969 were still present as seedlings 27 years later. The half-life of any particular cohort of this species was 66 months.

But the capacity of the seedling pool to contribute to the recovery process in secondary forests depends on the numbers that survive the initial disturbance. This is a function of the intensity of the disturbance and its timing in relation to the last period of seed rain. A disturbance that created canopy gaps but left the seedling pool intact would enable these to quickly grow and initiate a new succession dominated by these species then present. On the other hand, one that destroyed the seedling pool would obviously mean colonists would have to come from other sources.

Stumps, Rhizomes and Roots

Some types of disturbance allow stumps or old root systems to persist and many species are able to reproduce vegetatively from these (Stocker 1981; Vest and Westoby 2004). Whether they do so or not depends on the type of disturbance; some species may be able to produce suckers from stumps or roots after being damaged by storms but not if burned by fires. The proportion of species able to reproduce in this way is unclear although there is some evidence that it might be more prevalent in species from drier climates than in those from wetter areas (Murphy and Lugo 1986; Viera and Scariot 2006). The size of the stump may also be important with suckers from smaller stumps being more likely to persist and grow into a sapling than those of the same species growing from a larger stump. Pioneer species and primary forest species both seem capable of these forms of vegetative reproduction. The new shoots or suckers are usually able to grow quickly because they can take advantage of an established root system so that remnant stumps are, potentially, a powerful means by which secondary forests can recover from disturbances. Disturbances that leave many old stumps and roots intact enable these to make a major contribution to the recovery process but more intense or long-lasting disturbance such as cropping tend to diminish the pool of vegetative material.

Seed Dispersed into the Site from Outside

Some species produce seed every year while others do so only episodically. Some species produce large amounts of seed while others do not. Both these differences affect the likelihood that the seed of a particular species will form part of the seed rain reaching a newly exposed site. But perhaps the more important factor is the dispersal mechanism each species uses to distribute its seed.

Many forest species are able to disperse seeds over large distances using either animal vectors or wind (although some species have no particular dispersal mechanism and are simply shed around the base of the parent tree). Animals such as birds and bats are common seed dispersers in many wetter areas (Corlett 1998, 2002; Muscarella and Fleming 2007; Shilton et al. 1999; Wunderle 1997). This process allows species to recolonise disturbed sites at which they have become locally extinct. The capacity of wildlife to disperse seed and enrich secondary successions depends to a very large extent on the landscape context. What are the relative proportions of primary and other secondary forests in the landscape and what is the spatial distribution of these forests amongst other agricultural land uses? Answers to such questions will determine the types of seed available for dispersal to a particular site and the types of wildlife able to carry these seed. This will be discussed in more detail below.

Wind dispersal is a less common dispersal mechanism in most forests in the region. However, it is still important for some species, most notably members of the Dipterocarpaceae family which dominate forests of Malaysia, Philippines and parts of Indonesia. Seed of dipterocarp species have long wings (elongated sepals) which slows their decent and may assist their dispersal but the actual distance over which seed are dispersed depends on the tree height and the degree to which the crown is exposed to wind. Dispersal distances are short in intact closed-canopy forests and most seed are carried less than the equivalent of the canopy height but greater distances are possible from isolated trees with crowns exposed to stronger winds. But, even in these situations, most dipterocarp seed are probably dispersed over relatively short distances. Whitmore (1984) quotes an example where it took 23 years to colonize a patch of regrowth forest only 180 m from a parent tree. This means dipterocarps are invariably slow to recolonize sites from which they have been removed. Other moist forests also contain some wind dispersed species and around 20% of the tree flora at one site in the humid lowlands of Papua New Guinea (the Gogol Valley near Madang) are wind dispersed (Lamb 1990). Wind dispersal appears to be more common in drier forests (Viera and Scariot 2006; Whitmore 1984).

Many seed are lost or damaged during the dispersal process and some wildlife are better seen as seed predators rather than as seed dispersers (although many species are both). In Southeast Asia the main mammalian seed predators are columbine monkeys, rodents, pigs and deer while parrots and some pigeons are also important (Corlett 1998). Of course, these species are not necessarily present in all degraded habitats and in sites such as grasslands, the most important seed predators are rats, other small rodents and insects. The proportion of seed removed by predators (both before and after dispersal) can be high although there is considerable variation depending on the type of seed and the habitat into which the seed is shed. Factors thought to influence predation rates include seed or fruit size and endocarp hardness but contrasting findings about both attributes have been reported by different authors. For example, in Hong Kong, Hau (1997) found seeds with thick or hard endocarps were less likely to be eaten than fleshy fruited species but Doust (2004) working in tropical Australia found examples where seed with hard endocarps were readily eaten. Similarly, Osunkoya (1994) in tropical Australia and Dirzo et al. (2007) in Mexico found small seed were more readily removed than larger seed while Brewer (2001), working in Belize, found the reverse and Holl and Lulow (1997) working in Costa Rica found no relationship. The reason for these inconsistent patterns may be that predation is also affected by other factors such as chemical defences and the nutritional reward presented by the seed of different species. Seasonal differences are also likely to be important since these will affect the availability of other foods resources and the population sizes of potential seed predators. In the face of these inconsistent patterns perhaps the main conclusion to be drawn at present is simply that predators can destroy many of the seed dispersed into a site and so reduce the rate at which natural regeneration occurs.

The relative importance of these four sources of plant colonists in the development of secondary forests depends on the type of disturbance that has occurred. Recovery at some heavily degraded sites may depend entirely on the seed of species able to reach the site from some external source while recovery at less severely degraded sites may involve all four mechanisms.

The Landscape Context and Its Influence on Seed Dispersal

The types and amounts of tree seed being dispersed into new secondary forests depend on the landscape context. The two most important factors are the types of forests present in the particular landscape and the spatial distribution of these. Gaps formed in primary forests are likely to receive seed of mainly primary forest species. On the other hand, the seed rain in a long-established agricultural landscape with patches of secondary forest will be dominated by seed of species from these forests and will contain few primary forest species. Since many pioneer species produce large volumes of seed, the overall seed rain at these sites may be larger. More recently cleared agricultural landscapes with patches of residual forest and new secondary forests will receive a mixture of both types of seed (Fig. 5.2).

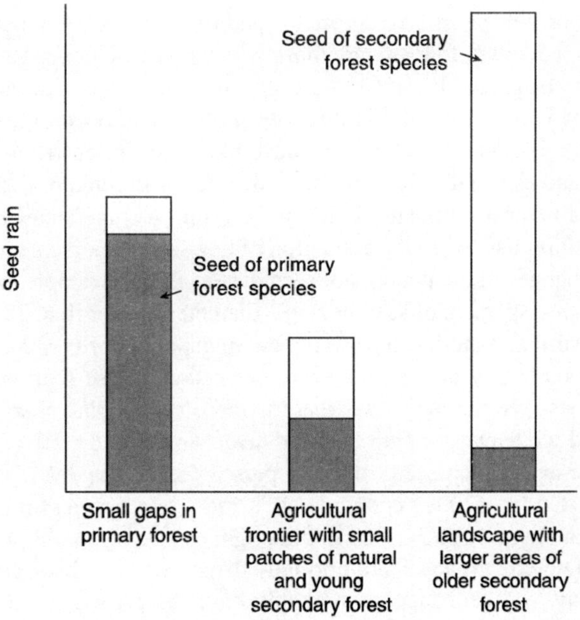

Fig. 5.2 Hypothesized seed rain in three contrasting landscapes. Seed rain in a primary forest will be dominated by primary forest species while that in a highly disturbed agricultural landscape containing older secondary forest will be dominated by secondary forest species. An agricultural landscape with patches of natural forest and young secondary forest will probably contain balanced mixture of both seed types. Pioneers and species from young secondary forests tend to be more prolific producers of seed than species from primary forests and this will be reflected in the relative abundance of the seed rain at the three sites

The extent to which any of this seed is dispersed to new regrowth areas depends on the spatial distribution of these existing source forests with the seed of many species being only dispersed over relatively short distances and the seed of far fewer species being distributed over longer distances. Much depends on how wildlife species are affected by the spatial distribution of the forest still present in this landscape.

Most of the wildlife species in secondary forests are likely to be habitat generalists. These can often cross agricultural areas between forest patches and so disperse seed of secondary forest trees over large areas. Wildlife species usually found in the interiors of undisturbed forests are less likely to be found in secondary forest or be capable of crossing open agricultural areas. This means the species most widely dispersed across agricultural landscapes are likely to be only a sub-set of the total forest flora. The issue is made more complex because of the importance of patch size. Many potential seed-dispersers require forest patches to exceed a certain minimum area and will not use very small fragments. The interplay between these various factors (i.e. types of forests able to produce seed and types of dispersers able to distribute it) means there is great scope for quite different patterns of seed dispersal to develop depending on the landscape context.

The most poorly dispersed tree species – and thus the species most at risk of not regenerating in secondary forests – are those with larger fruit or seed as well as many of those that are dispersed by wind. Some species with larger fruit are usually dispersed by larger wildlife (e.g. hornbills, fruit pigeons and doves, cassowaries, gibbons) but these species are the ones most likely to suffer from deforestation, fragmentation and especially from hunting. Very few will remain in landscapes that retain only small forest remnants. This can be seen on some small Pacific islands where deforestation has led to the extinction of key seed dispersers and has resulted in significant changes in the patterns of seed dispersal. For example, on Tonga there are now no avian dispersers of seed having a diameter greater than 28 mm (Meehan et al. 2002). A similar pattern can be found on Singapore (Corlett 2002). This means that the more commonly dispersed species are usually those with smaller fruit or seed because these are the only sizes that the more vagile generalists such as birds like bulbuls and white eyes or fruit bats and civets are able to carry (Corlett 2002).

The distance over which seed can be dispersed depends on the dispersal mechanism or agent. Corlett (2009) concludes that most tree species in the region are dispersed over a range of 100–1,000 m although, as already noted, wind-dispersed species of the Dipterocarpaceae are normally dispersed over distances of less than 100 m. Fruit bats and fruit pigeons probably have the greatest potential for long-distance dispersal (Corlett 2009). Birds often take fruit or seed they have collected to a nearby perch tree to consume. Much of this seed is then dropped below these perch trees making them focal points for seedling regeneration. In fact, some have suggested seed dispersal in degraded landscapes is most often limited by the availability of perch trees rather than dispersers (Corlett 2002; Driscoll 1984). There is some evidence that trees in degraded lands become more attractive as perch trees once they exceed around 5 m height (Toh et al. 1999). Once new tree seedlings become established and grow in stature the process is likely to accelerate because the site becomes more attractive to birds and because grasses are excluded.

Wind can be an effective dispersal agent for some species in recently cleared lands although the distances over which seed are dispersed by wind are usually small and the direction depends on the prevailing wind direction. The effectiveness of this as a dispersal mechanism also declines once regrowth develops and wind speeds within the canopy slow. Nonetheless, 41% of the early colonists on the new islands formed after Indonesia's Krakatau erupted were wind-dispersed and only 25% were dispersed by animals (Richards 1952). It is interesting to note that a different pattern is found on the island of Jarak which is 64 km off the Malaysian coast in the Malacca Straits and is thought to have been blanketed by volcanic ash 34,000 years ago. In this case most tree species now present are those with animal dispersed seed (Whitmore 1984).

The Fate of New Seedlings Colonizing After a Disturbance

Not all new seedlings regenerating after a disturbance are able to survive. Some are eaten by herbivores, some die of drought or fungal diseases and many succumb to

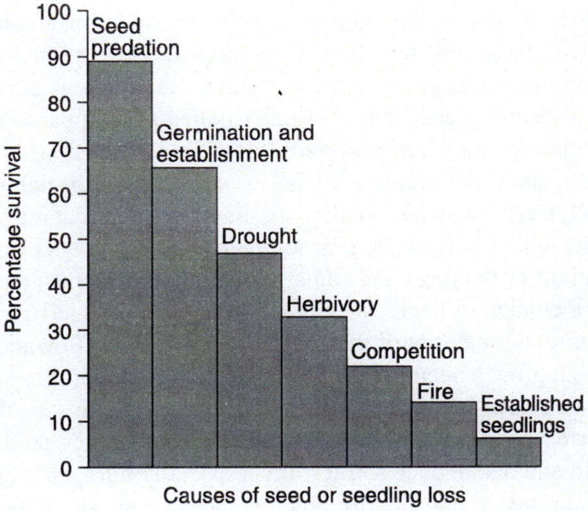

Fig. 5.3 A variety of factors reduce the number of seeds that can germinate and become established as seedlings. The relative importance of these differs at different sites. This generalized description shows the cumulative impact can be large

root competition from their neighbours. The number eventually able to establish may be a very small proportion of the seed dispersed (Fig. 5.3). Cattle and other farm livestock can be specially damaging where open grazing is practiced. These may eat new seedlings or simply trample them. But native herbivores such as deer can also be damaging. Seedlings of some large-seeded species can recover from single episodes of herbivory if their meristem tissue is not damaged (Nepstad et al. 1991) but otherwise these seedlings will be lost. Vegetative shoots from old roots systems will usually be more resilient. Herbivore damage to seedlings usually declines as seedlings grow taller.

Drought and root competition can limit the establishment of all seedlings but are especially damaging to seedlings regenerating in grasslands. Topsoils in these areas can change from being fully saturated with water to being very dry in relatively short periods. Seedlings exposed to full sunlight experience a high radiation loads and leaf temperatures. The evapo-transpirational stress is also high and those without deep roots may not be able to withstand this. Under these circumstances shade from nearby patches of grass can protect the seedling (Aide and Cavelier 1994; Hardwick et al. 1997). But grasses are usually seen as providing serious competition for seedlings because most have very dense root systems in topsoils. This means the above-ground advantage of temporary shelter can be outweighed in the longer term by the disadvantages arising from below-ground competition. At the very least, higher initial survival rates may be matched by much slower growth rates once the dry period passes. This means that seedling establishment may be initially favoured by some shading but, once established, seedlings usually benefit from more open conditions (Viera and Scariot 2006). Of course these generalizations depend on the height of the grasses and very few woody plants can establish under tall grasses because of the light levels although some shade tolerant primary forest species with large seed may be able to succeed. Once a few trees are established their canopies can begin to shade out grasses and there is evidence that the survival rates of seedlings are higher beneath isolated trees in pastures is probably due to the reduction in competitive pressure (Holl 2002; Toh et al. 1999).

Fire is also a major constraint on seedling establishment, especially in areas with seasonal rainfall. Fires can occur regularly in grassland areas but may also occur in disturbed forests where grasses are less common. Few seedlings of tropical forest species are tolerant of wildfires and some years without fire must pass if trees are to become large enough to begin shading out grass (i.e. the fuel) and become tall enough or acquire sufficient bark thickness to escape fires (Bowman 2000). The rate at which such growth occurs will clearly depend on soil fertility so that heavily degraded sites that have lost much of their topsoil may require much longer fire-free periods than sites with more fertile soils. Repeated fires tend to eliminate woody regrowth and also reduce soil fertility, especially nitrogen which is volatilized in fire (in contrast to phosphorus which remains on the site in wood ash).

Types of Secondary Forest Successions

The relative importance of seedbanks, seedling pools, vegetative growth and dispersal in successional development depends on the intensity and the scale of the disturbance. The intensity affects the numbers and diversity of species left in seedbanks, seedling pools and as old roots while the scale affects the distances over which dispersal from intact forest must occur. This is illustrated in Fig. 5.4 which shows how an increasingly intense disturbance progressively reduces the contribution of residual seedlings, roots and soil seedbanks while increasing distance from intact forest reduces the contribution from seed dispersal. Regeneration is usually rapid and most complete at lightly disturbed sites near intact forest but is limited at more distant sites that have been heavily degraded. The effect of these factors on recovery rates is shown conceptually in Fig. 5.5.

Recovery is also determined by the extent to which the physical and chemical properties of soils have been affected by the disturbance regime. Disturbances that leave the topsoil essentially intact are more likely to allow rapid recovery. On the other hand, those that lead to erosion, compacted topsoils and nutrient loss via leaching or volatilization after fire may change the site to the point where the original species cannot recolonise it again and productivity is reduced.

Some of the different types of secondary forest successions are listed in Table 5.2. A well-managed selective logging operation represents an example of a low-intensity small-area disturbance (Type 1). In such cases most of the original canopy remains intact and virtually all of the original canopy tree species remain and are represented by a range of size classes (and often by seedlings on the forest floor). Distances over which seed must be dispersed to reach gaps are short (Cannon et al. 1998; Pinard and Putz 1996). Most topsoils are retained although there may be

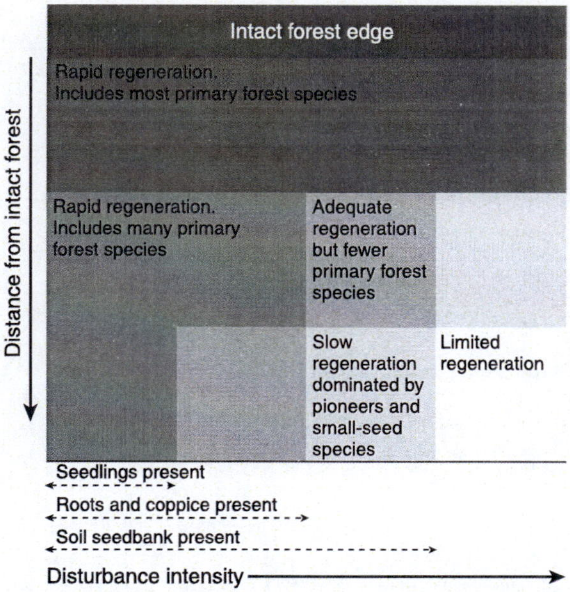

Fig. 5.4 Effect of intensity of disturbance and the distance a site is from intact forest on the ways regeneration will occur. As the disturbance intensity increases residual seedlings and then roots and coppice play a progressively smaller role. This means fewer primary forest species are likely to be present in any regeneration. Soil seedbanks are often persist at highly disturbed sites but have only pioneers and secondary species. Many species from intact forest can reach nearby sites but fewer will be dispersed to more distant sites

localized disruptions. In such cases most species remain on the site and structural recovery can be rapid.

Likewise, recovery can be relatively rapid after a severe storm even though trees will have been blown over and much of the canopy along the cyclone path is destroyed (Type 2). Despite this damage most of the original species are likely to be still present at the site as scattered saplings, seedlings or old stumps able to produce coppice (Burslem et al. 2000; Chazdon 2003). Again the soils remain essentially unchanged. In many cases the early recovery stages are dominated by a small number of pioneer species (e.g. *Macaranga*) but these are soon overtopped by a variety of primary forest species from seedlings or coppice. In other cases the new canopy might be occupied by a smaller number of longer-lived species that happened to be fruiting at the time or that were especially common in the seedling pool on the forest floor. In such cases these more common species may dominate the community for some years (Whitmore 1984).

Regrowth after shifting cultivation can also be relatively rapid (Type 3). These sites may be able to recover most of their original complement of species provided the area cleared is small and recolonisation from the undisturbed forest surrounding the area can take place. The recovery process may take more time in other situations where there have been many cycles and the fallow period has shortened to only a few years (Chazdon 2003; Clarke 1971; Geddes 1976). This may be because

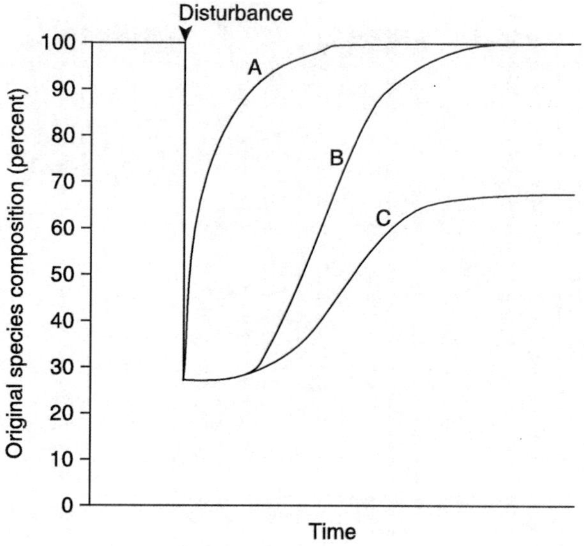

Fig. 5.5 Affect of the landscape matrix on forest regeneration. Curve A: site is a small gap entirely surrounded by forest and recovery can be rapid. Curve B: site subjected to a moderately intense disturbance but many forest patches are nearby. Recovery is slower but is eventually complete. Curve C: site subject to an intense and large-scale disturbance. There are few nearby forest patches and some seed dispersers are absent. In this case recovery is slow and incomplete

soil fertility is reduced, there are more weeds, fewer residual stumps or coppice and a greater distance to seed sources in intact forest. In each of these three cases secondary successions can develop with trajectories that are likely to lead to a convergence in floristic composition with that of the original primary forests (Norden et al. 2009; Webb et al. 1972).

In contrast to these cases, structural recovery is much slower and floristic recovery may be incomplete at more heavily or frequently disturbed sites. Sites that have been subject to badly managed or frequent logged are examples of this situation (Type 4). In these cases the canopy gaps are larger, the intensity of soil disturbance is greater and the distances to undisturbed forest are longer. Not only are many canopy species removed but a high proportion of sub-canopy species are damaged as well. Soil erosion or compaction may be widespread and spatial heterogeneity is often large. These differences usually lead to much greater changes in the structure and floristic composition of the regrowth and a reduction in the diversity of life forms present. Under these circumstances the secondary forest that develops typically has fewer primary forest species and a much greater representation of shade-intolerant pioneer species such as short-lived species of *Macaranga, Mallotus* and *Trema* or longer-lived pioneer species (perhaps <80 years) such as *Gmelina arborea, Octomele sumatra, Falcataria moluccana* or species of *Endospermum* or *Campnospermum.* Depending on seed or seedling availability, one or two of these species may dominate the new canopy. Bamboos, giant herbs such as bananas or gingers as well as tangles of vines and large woody light demanding climbers are

Table 5.2 The extent to which different types of secondary forest may develop after disturbances of various kinds

Type of community	Original canopy cover remaining	Forest attributes	Ecosystem can recover unaided?
1. After well-managed logging	Mostly intact	Most species and life forms remain on-site represented by seed, seedlings or saplings. Residual trees cover a range of size classes. Most wildlife still present	Yes, provided cutting cycle >30 years
2. After storm or cyclone	Little of original canopy remains	Most species remain on site as scattered residual trees, seed, seedlings or old stumps	Yes, although the new canopy may be dominated by long-lived pioneers
3. Fallow periods forming part of swidden cycle	Localized canopy gap created (<1 ha)	In short term species present include pioneers and any residual trees or stumps able to produce coppice. In longer term colonized by seeds from nearby natural forest	Yes, provided fallow period >10–15 years
4. After unmanaged or frequent logging	Many gaps in original cover	A reduction in the density of residual primary forest trees and of the diversity of species represented; a major change in tree size class distribution. Seedling pool badly damaged. Canopy gaps filled by pioneers and vines	Maybe, but only over a long period
5. Post-fire secondary forest	Nil	Most trees usually killed by fire. Only seed likely to be pioneer species stored in topsoil. Site likely to be dominated by fire-prone grasses	No
6. Former agricultural cropland	Nil	No residual forest species remain on site. Site likely to be dominated by fire-prone grasses	Slow
7. Former agricultural land with scattered trees	Low	No residual forest species remain on site except for scattered trees. Site otherwise dominated by fire-prone grasses	Slow
8. Former mine site	Nil	No forest species present.	No

also common. The rates at which such sites recover depend on how many of the original canopy species remain and are able to regenerate. If these were largely removed or damaged by logging it may take some considerable time for them to recolonise the site and rebuild their populations. Structural or floristic recovery can also be slowed if bamboos, vines, climbers such as *Merremia* or other exotic weeds become dominant. In some cases the main canopy species can recover but there will be significant changes in the composition and relative abundance of other species.

Successive logging events or other disturbances at frequent intervals will, of course, slow recovery even more. In such cases, any new regeneration is destroyed before it has a chance to reproduce. But these types of disturbances can advantage life forms other than trees. Lianas are often common in early stages of successions resulting from openings in the canopy. Recurrent disturbances or storms may damage tree regeneration but have less effect on these lianas. This can be seen in some coastal forests in northern Queensland in Australia that are subject to frequent storms. Many of these areas now have extensive vine tangles and climber towers (Webb 1958). Over time trees can grow through these vine covers if sites are not continually disturbed (Letcher and Chazdon 2009).

The capacity for self-recovery is most limited when the intensity of the disturbance is higher. Fires are especially damaging form of disturbance (Type 5) and secondary forests develop only slowly at sites that have been burned, especially when the areas affected are large. The logged-over areas in East Kalimantan that were burned in 1982–83 and again in 1997–98 are good examples (Box 5.1). Repeated fires within a short time period will eventually transform a forested landscape into grassland. This can be seen when fallow period in shifting cultivation cycles shorten and fires are used to clear sites and initiate a new cropping period. Recovery is also likely to be slow at former croplands (Type 6) since these are unlikely to have topsoils containing tree seeds or old root stocks let alone seedlings of native tree species. The fertility of soils at many of these sites may have also declined (Fig. 5.6). Some will be occupied by grasses such as *Imperata cyclindrica* making it difficult for woody species to recolonize even when seed sources are nearby and dispersers are available. Reforestation in these situations may be facilitated if some scattered trees remain (Type 7). Such tree act as perches for birds able to disperse seeds and can become important as nodes for tree recolonization (Elliott et al. 2006; Scott et al. 2000; Toh et al. 1999). Even so, the process of successional development is still very slow. Finally, recovery may simply not occur at severely degraded sites where erosion has been extensive or at former minesites simply because of the hostile site conditions (Type 8). These sites often remain as grasslands or shrublands.

There have been relatively few comparative studies of the secondary forests developing after different types of disturbances in the same community. Nepstad et al. (1991) observed Amazonian forests recovered relatively quickly at sites cleared for grazing but where pasture had failed to become established. In this case many woody species were able to regenerate vegetatively. On the other hand, they observed that recovery at similar sites was much slower where grazing had persisted for a number of years or where the sites had experienced grazing, weedicides and fire. In this case the pool of seedlings and old stumps and roots had been eradicated.

Box 5.1 Secondary Forests Arising After Successive Wildfires in Borneo

A severe drought in 1982–1983 was followed by fires that covered large areas of the lowland rainforests of Borneo (Dennis et al. 2001; Woods 1989). Some of the forests burned had been disturbed beforehand by logging but other had not. Fire killed most primary forest trees but especially those in smaller diameter classes and where logging had occurred. However, not all forests were equally affected and more primary forest trees were able to survive in lower topographic positions, presumably because of moister conditions and lower fire intensities at these locations. In the first few years after the fire species such as ferns, gingers and grasses dominated the burned sites. These declined in prominence as woody plants, especially species of *Macaranga,* became established. Depending on the location, other common woody plants found included *Omalanthus populneus, Duabanga moluccana, Callicapa* spp. and *Euodia* spp.

Fifteen years later, in 1998, a second low intensity surface fire burned through part of the area (Slik et al. 2008; Toma et al. 2005; van Nieuwstadt et al. 2001). Surprisingly, there was little difference in the composition of woody species regeneration between sites burned only once and those burned twice. Once again, species of *Macaranga* and other pioneers dominated the early succession. But there was a difference in tree density. Following the second fire there were fewer ferns and gingers and much denser stands of trees than after the first fire. This difference was probably due to the much greater soil seed bank created by the pioneer tree species that became established and reproducing after the first fire. Although forest cover was quickly regained there were few primary forest species, except in lower topographic positions, and no real recovery towards the pre-fire species composition. It seems this will take many years to occur.

A study carried out in Nicaragua by Boucher et al. (2000) assessed the regeneration developing on an old agricultural field and compared it with that in a forest following a major cyclone. At age five years the two secondary forests emerging from these contrasting disturbances were quite different. The old-field succession was dominated by a small number of early pioneer species while the post-cyclone forest was largely primary forest species that regenerated from resprouting and from seedlings and sapling already present; there were more species present and no single species dominated the site. A similar pattern was observed in the Amazon by Mesquita et al. (2001) who investigated successional development after clear-cutting and abandonment with that occurring after a former crop site was abandoned. Again, primary species recovered most quickly at the site that had not been cropped. Both studies clearly showed the nature of the successional community was a function of site history. An overview of possible successional pathways following different types of disturbances in forests in southwestern Sri Lanka is given by Ashton et al. (2001).

Fig. 5.6 An extreme example of a loss of soil fertility at Baku, Sarawak. In this case all topsoil has been eroded from a site cleared for agriculture over 100 years earlier. Only bedrock remains and regrowth has been severely limited

Ecosystem Services Provided by Secondary Forests

Some view secondary forests as being ‘degraded’ and valueless, especially governments and so-called developers who are seeking to acquire land cheaply. These forests are more likely to be protected and allowed to recover if they can be shown to be providing goods or some kind of environmental service. Young secondary forests have a limited capacity to provide commercially important goods but can provide certain ecosystem services from an early age. These include the habitats for wildlife, watershed protection and a capacity to sequester carbon.

Secondary Forests as Habitats for Old-Growth Forest Species

As noted earlier, secondary forests change in biomass, structure and tree composition as they mature. Wildlife species often track these changes and recolonize the regenerating forest as suitable habitats develop and food resources become available. In the case of insectivores, the process will be mediated by the insect faunas supported by secondary forests and in the patterns of flower and fruit production as the forests age. In the case of carnivores, herbivore numbers must increase first to allow carnivore populations to develop. Many generalist wildlife species begin using secondary forests when they are relatively young but the age at which they begin to provide habitats and breeding areas for species with more specialized habitat requirements such as those found in old-growth or primary forests is less certain. Part of the problem lies in identifying just which wildlife species are indeed old-growth specialists (Chazdon et al. 2009).

A study of the birds present in the extensive areas of young (mostly around five years) regrowth developing after pulpwood logging in Papua New Guinea found that even by this early age, there was a 75% overlap in the canopy species found in regrowth and adjacent undisturbed forest although few of these species were forest interior specialists and there were large difference in the relative proportions of species in each community and in their feeding niches (Driscoll and Kikkawa 1989). Vertical partitioning of foraging heights by birds contributes to the high species diversity found in rainforest. This partitioning was also found in these secondary forests even though they were much shorter than the intact forests. Most of the birds in the regrowth areas were generalists and few could be classed as regrowth specialists. This contrasted with observation in other regions and Driscoll and Kikkawa (1989) suggests this is because extensive areas of regrowth forests have not been present before in Papua New Guinea.

One of the most intensive investigations of the composition of secondary forests was that carried out in Brazil by Barlow et al. (2007). The forests studied were still relatively young and varied in age from 14 to 19 years old. They found that, even at this age, the proportion of nominal primary forest species present in these secondary forests was quite high in some groups such as orchid bees, large mammals and scavenger flies but was much lower in groups such as birds, moths and small mammals. Chazdon et al. (2009) reviewed a number of other studies in secondary forests (mostly less than 30 years old) and concluded that many volant fauna (e.g. species able to fly such as butterflies, birds and bats) could become established from a relatively early age probably because many are adept at crossing gaps but the recovery of non-flying fauna and plant species was rather slower. There was considerable variation in the data reflecting differences in spatial patterns of regrowth and intact forest and in the nature of the landscapes between these forests.

There have been very few studies of the composition of older secondary forests (e.g. > 40 years) which means we still have a poor understanding of the capacity of these forests to support and conserve old-growth species. Because of this it is premature to judge whether secondary forests might act as safety nets in which biodiversity is ultimately conserved as suggested by Wright and Muller-Landau (2006) or whether this an unrealistic expectation as argued by Laurance (2007), Brook et al. (2006) and Gardner et al. (2007). It will ultimately depend on the range of age classes and types of secondary forests present as well as the landscape context in which these forests are able to persist.

Of course complete recovery is not always assured. As noted earlier, the capacity of forests to recover from disturbances depends on the types of disturbances. This also means that not all secondary forest successions will necessarily converge on the original forest. Some original species may never recover while some exotic species might colonize and become part of the new ecosystems. These forests then become the 'novel ecosystems' of Hobbs et al. (2009) and Lugo (2009). Some of these novel ecosystems are a result of chance events but others such as agroforests arise from deliberate management decisions. This latter group will be discussed further below.

Watershed Protection and Hydrological Flows

Disturbances increase soil erosion and stream sedimentation but the increases usually decline within a few years as revegetation and successional development takes place on the disturbed sites. This is because plant growth is rapid and the structurally complex vegetation close to the soil surface limits the overland flow of water. An example of this is the rate at which stream turbidity declined after a clear-fell logging operation in Papua New Guinea (Lamb 1990). The sedimentation increased immediately after logging but declined rapidly after one or two years. Thus, only 27% of the weekly stream samples in a logged catchment had turbidity levels below ten units during the first year after logging but 75% of the weekly samples had these low levels in the second year. On the other hand, erosion may persist in secondary forests growing on steep slopes as a consequence of slumping and mass wasting following heavy rain or earth tremors because deep-rooted tree species are absent.

Changes in water flows are much less clearly understood because very few studies have actually been carried out in secondary forest and because differences in site histories affect both the type of forest present and the hydrological characteristics of the soils. One study in 3–4 year old regrowth in Sabah found streamflow was higher in the regrowth than in primary forest largely because of differences in the interception of rainfall by tree canopies. On the other hand, measurement in the Amazon found evaporation in secondary forest was similar to that in mature primary forest implying that run-off would be similar (Holscher et al. 2005). These and other results examined by Holscher et al. (2005) suggests the patterns of water use by young tropical forest regrowth might differ from that in temperate regrowth forests where increased evapotranspiration often reduces stream run-off. In these situations the reduced levels of streamflow persist for some decades until the regenerating forest matures. Note that in these temperate forests the species in the regrowth are the same as those in the old-growth stages which is not the case in the tropical forests.

Patterns of water movement are critically dependent on the hydraulic conductivity of topsoils. Many former agricultural lands have compacted soils and much rainfall is lost as surface flow rather than infiltrating into the soil. Different prior land uses can generate large differences in hydraulic conductivity and these differences may last for many years (Zimmermann et al. 2006). Giambelluca (2002) quotes work in Vietnam illustrating the rate at which changes can occur. In this case saturated hydraulic conductivity rates in secondary forests in Vietnam rose from 20 to around 60 mm h^{-1} over an eight year period of regrowth. Over the next 30 years conductivity improved even further and eventually matched the rate of 90 mm h^{-1} found in relatively undisturbed forest. This meant that, with age, a greater proportion of rainfall was infiltrating into these soils rather than being lost from the site as overland flow.

Such changes can affect the seasonality of flow since more water reaches the ground water store. But, as discussed earlier in Chapter 1, these effects on water yield are dependent on scale and may only be observed in small areas. At larger scales they are usually masked by differences in rainfall and other land use patterns.

Carbon Sequestration

The term carbon sequestration is used here to refer to the process of taking up and immobilizing atmospheric carbon. The annual growth or productivity of young secondary forests is high but this productivity declines as they mature and the rate of biomass accretion slows. Eventually a point is reach where biomass ceases to increase and new growth can only occur when tree deaths or disturbances create canopy gaps. That is, biomass increases must then be matched by losses. The role of the forest, then, is not so much to take up additional carbon but, rather, to simply store it.

Secondary forests have the capacity to sequester large amounts of atmospheric carbon far in excess of those in crops or pastures. This carbon is mostly immobilized in woody biomass and soils. The amount of carbon contained in plant biomass is affected by wood density and many short-lived pioneer species have a low wood density compared with primary forest species. This means they are less effective at storing carbon than trees of species from more mature successional stages. On the other hand, the annual rate of growth of these species is higher than primary forest species so that large amounts of carbon are temporarily absorbed before being recycled to the soil. In a review of productivity of secondary forests Silver et al. (2000) found biomass accumulated at the rate of 6.2 Mg ha^{-1} y^{-1} in the first 20 years. This subsequently slowed giving an overall rate of 2.3 Mg ha^{-1} y^{-1} over the first 80–100 years of regrowth. The carbon content of this biomass is roughly 50%. They noted these amounts are less than the amounts sequestered in some fast-growing plantation forests which can absorb between 0.8 and 15 Mg ha^{-1} y^{-1} within the first 26 years of establishment (although these rates depend on the species used and also decline as the plantations age).

Using an enlarged database, Marin-Spiotta et al. (2008) studied the influence of climate and prior land use history on the rates of biomass accumulation in secondary forests. They found that biomass accumulates in dry forests until a plateau is reached after about 50 years but that it continues to increase until about 100 years in wetter forests. The rate of accumulation depends on the productivity of the forest with those regenerating on infertile soils being less productive, and hence less effective, in sequestering carbon than those growing on more fertile soils. Site fertility in secondary forests is often associated with time since the original forests were cleared with more recently cleared sites being more fertile than those that may have been cropped or grazed for some time.

Carbon is also sequestered in soils supporting secondary forests because of litterfall and root turnover (Lugo and Brown 1993). This carbon is found in a variety of forms. It is only briefly immobilized in microbial tissue and light fraction organic carbon (plant and animal material undergoing decomposition) but can be immobilized for much longer periods in the much larger fraction of organo-minerals developing after organic material has been decomposed (Post and Kwon 2000). Silver et al. (2000) estimated soil carbon accumulated at an average rate of 0.4 Mg C ha^{-1} y^{-1} over the first 100 years (to 25 cm depth). The rate reached 1.3 Mg C ha^{-1} y^{-1} during the first 20 years and 0.2 Mg C ha^{-1} y^{-1} for then next 80 year period. This is less than 20% of the rate at which carbon accumulates in above-ground biomass. Further changes may be occurring at greater soil depths but there is little information about these. Most soil carbon is found in upper horizons and that present at greater depths is not likely to be affected by short-term changes to above ground vegetation although the reforestation of old shallow-rooted grasslands by deeper- rooted trees may enrich carbon contents of these deeper horizons (Jobbagy and Jackson 2000).

The quality of soil carbon sequestered in secondary forests may be different to that in some plantations. Li et al. (2005) found soils under 20 year old secondary forest regenerating on old farmland sites in Puerto Rico had more 'heavy fraction' organic carbon than did soils beneath a pine plantation grown at the same site and of the same age although there was no difference in total organic carbon.

The reviews by Silver et al. (2000) and Marin-Spiotta et al. (2008) both found there was considerable variation in the extent of above- and below-ground sequestration of carbon by secondary forests and that the actual rates depended on species and communities as well as climates and prior land use history. In some cases these variables are more critical in determining carbon stocks than the age of the regrowth. Overall, however, it is clear that secondary forests are able to sequester large amounts of carbon, both above- and below-ground, at a cost likely to be much lower than if plantations were used.

Using Natural Succession to Overcome Degradation

The foregoing suggests that the best way of overcoming degradation is to take advantage, where ever possible, of the capacity of many sites to recover through natural means. Recovery is more likely if some residual trees remain at the sites or if natural regeneration is already occurring. And recovery will be faster if there are patches of natural forest nearby. A list of some of the pre-conditions necessary for natural recovery to take place is given in Table 5.3. Perhaps the single most important of these is that the site can be protected from further disturbances.

Protecting the Site from Further Disturbances

Many secondary forests are relatively young because they are repeatedly cleared or disturbed. Sometimes they are cleared because they are seen as unused wastelands that can be better used for some other purpose such as agriculture. In other cases they may not be cleared but are degraded further by unregulated harvesting of the timbers and the other NTFPs they still contain or by hunting.

Table 5.3 Pre-conditions necessary if natural regeneration is develop at degraded sites

Pre-condition	Reason	Consequences of not achieving pre-condition
Further disturbances can be prevented	No sucessional development unless disturbances are excluded	Reforestation likely to fail
Weeds can be eradicated	Weed competition limits regeneration of native species	Weeds prevent regeneration or limit growth
Animal pests can be controlled	Act as seed predators or herbivores	Regeneration limited
Soils not degraded (or adverse conditions can be ameliorated)	Changes in fertility or physical properties may limit ability of original species to recolonize	Only tolerant (exotic?) species can establish at site
Representatives of plant and wildlife remain on site	Residual trees, seedlings or coppice allow rapid regeneration	All species will have to be re-introduced to site
Natural forest nearby; a range of seed-dispersers able to move across intervening landscapes	Allows new species to recolonize degraded site populations of existing species to be supplemented	Limits opportunity for species enrichment or population enhancement

Ownership is critical and long term protection is more likely if land is owned, the ownership claim is widely recognized and the owner(s) believe its value as a forest exceeds its value under alternative land uses. Most forests owned by communities persist because the community has devised rules governing who can use the resources contained within the forests and the circumstances under which these might be used (Gibson et al. 2004). For example, all members of the community may be allowed to harvest medicinal plants or items such as mushrooms whenever they wish but must apply to the village management committee if they want to fell a large tree for building materials (e.g. Santasomsat 2003). Note that the protection of a community forest may lead to the increased use of other nearby secondary forests not managed in this way. But even traditional management systems can be overwhelmed if outsiders try to move in. Momberg et al. (2000) describe how people in East Kalimantan had great difficulty in protecting their community forests which contained large numbers of gaharu or eaglewood trees (*Aquilaria* spp.). As the price of eaglewood increased, more outside collectors arrived to search for it. In one case the outsiders even used helicopters to ferry in collectors and ship out the eaglewood. Once this happened the villagers felt obliged to accelerate their own harvest of eaglewood trees to prevent outsiders reaping the benefits.

An unusual approach to protection has developed in parts of northern Thailand where some villagers have taken to 'ordaining' some of their trees as Buddhist monks as a way of protecting them from logging. The idea is not to fully ordain the trees since this is a ritual only applicable to men but, rather, to sanctify an area and build a kind of symbolic fence about it. The movement has subsequently been caught up in wider and more complex political manoeuvres concerning land tenure and community forestry in Thailand in which forest dwelling villagers have sought to demonstrate they are not destroyers of forest but forest guardians (Isager and Ivarsson 2002).

Wildfires are an especially difficult problem and a single fire can destroy large areas of secondary forest (Dennis et al. 2001). Most tropical rainforests species are easily killed by fires but those found in dry seasonal forests are usually better adapted because fires are more common in these areas and sensitive species would have been excluded. The fire problem become greater if a secondary forest is adjacent to, or is surrounded by, grasslands. Some fires may be the consequence of an overly casual attitude to the use of fire by hunters or those clearing land to use for agriculture. People 'tidying up' grave sites in Hong Kong regularly cause wildfires that prevent reforestation occurring in many hill areas (Hau, pers. comm, 2009). Elsewhere fires are sometimes deliberately started because of a land disputes while Potter (2001) describes how certain oil palm companies in Indonesia have deliberately lit fires to burn village forests in order to take over their now degraded land.

The most common method of protecting regenerating forests from fires is to use annually renewed firebreaks and Box 5.2 describes how villagers in a seasonally dry area in northern Thailand used firebreaks to allow the development of their community forest. In this case a large number of tree species were able to regenerate through natural processes. Others have promoted the planting of buffer strips of fire tolerant trees that shade out grasses in order to form firebreaks (Friday et al. 1999). Wibowo et al. (1997) proposed a list of species (e.g. *Acacia auriculiformis, A. mangium, Calliandra calothyrsus, Gmelina arborea, Vitex pubescens, Schima wallichii, Macadamia hidebrandii)* that might be appropriate in various climatic zones of Indonesia (although one other species on their list – *Leucaena leucocephala* - can spread and become a weed in fire prone environments). Farmers in the Philippines have used 10 m wide belts of the fire-tolerant exotic species *Senna spectabilis* as firebreaks (Friday et al. 1999) and it would be relatively easy to draw up similar lists for other fire-prone areas.

Rural people sometimes have conflicting attitudes towards fire. One study in Thailand by Maneeratana and Hoare (2007) found that most wildfires in this particular area were caused by the overly casual attitude people had to forest fires (Table 5.4). Respondents took the view that fires cause little damage, that most established trees generally survive and that understories recover the following wet season. On the other hand, many of these same communities have developed systems of regulations and fines designed to limit fire damage to community assets. For example, there could be penalties for lighting fires without telling the village committee or for burning without firebreaks (Hoare 2004). Many villages also form fire-watching teams. Unfortunately community-based forest guards have sometimes been known to take action against fires threatening their own forest but ignore other fires burning nearby that don't (ignoring the possibility that the wind could change and it may come back the next day as a bigger and more difficult problem). These mixed attitudes towards the protection of public and private assets are probably common across the Asia-Pacific region.

A dramatic attempt to control wildfire in order to allow regrowth to develop was carried out in U Minh Thuong National Park in the Mekong Delta region of Vietnam. The forest in this park is dominated by *Melaleuca cajuputi*. This species is adapted to an annual inundation that occurs every wet season and lasts several months.

Box 5.2 Natural Regeneration When Fires are Excluded

Some communities have relied on natural regeneration to establish village forests. An example is the people of Pakhasukjai village in Chiang Rai Province of northern Thailand. These people belong to the Akha ethnic group and originally migrated from Myanmar and southern China and eventually settled in the mountains of Chiang Rai in 1976. The area they settled was mostly grassland dominated by *Imperata cyclindrica*. The ecological history of the area is unknown although the grasses presumably arose from the activities of earlier shifting cultivators. Forests are important to the Akha for religious and subsistence reasons and one of their first community activities was to begin establishing a village forest. This was done by defining an area and building fire breaks around it. Not all households were happy to give up land for this purpose but were pressured to do so by the majority. Each year every household was required to contribute labour to build and maintain the firebreak. Over time natural regeneration began to appear from old stumps and roots. In time a species-rich and structurally complex forest has appeared containing both trees and shrubs. Durno et al. (2007) undertook a survey of the forest when it was 18 years old and covered 580 ha. Some 260 species of trees, shrubs, lianes, herbs and grasses were recorded and they believed this was only a partial list of the species present. The rate of increase in the species-area curves for tree species exceeded that for natural forests at comparable elevations in Doi Suthep-Pui National Park because of the larger number of secondary forest species. Plots of 1,600 m^2 had 55 tree species present while those of the same area in Doi Suthep-Pui National Park had around 40 tree species. Around half of the species at the Pakhasukjai sites were classed as species representative of primary forest. The density of trees was higher compared with the undisturbed forest (1,700 tph compared with 700 tph) although, not surprisingly, the average diameter was smaller. At one point some attempts were made to extend the forest using exotic species such as *Acacia mangium, A. auriculiformis, Artocarpus heterophyllus, Cassia spectabilis, Cassia siamea, Diospyros spp., Eucalyptus camaldulensis, Eugenia sp., Leucaena leucocephala, Mangifera indica, Pinus kesiya, Plumeria acutifolia, Prunus cerasoides* and *Tamarindus indica.* When a 400 m^2 sample plot was surveyed five years later only six trees survived out of the 100 trees originally planted. But, during the same period 136 indigenous trees from the existing forest had regenerated naturally in the same plot (Durno et al. (2007)). Overall, the villagers had shown that within this landscape context, simply protecting this grassland site from fire was sufficient to allow a species- rich forest to regenerate.

Most of the park's forests have been badly damaged in the past by various disturbances but were recovering. However, successional development was being constrained by frequent wildfires. Park managers decided to exclude wildfires by blocking the network of canals within the area and keeping the forest permanently flooded.

Table 5.4 Causes of wildfire inside and outside forests in northern Thailand in 1995 (Hoare 2004)

Cause	Percent of fires
Gathering of NTFPs in forest. Fires lit to market travel and collection easier	24
Burning to prepare agricultural land	18
Incendiary or grudge fires deliberately lit because of conflicts or to clear forests for agricultural purposes	20
To drive game during hunting	15
Carelessness	14
Unidentified	9

The results were dramatic: within a few years many of the trees had died or had fallen over creating large gaps in the forest canopy (Fig. 5.7). Wildlife populations within the park began to plummet. Recognizing that permanent flooding was inappropriate, park managers opened the sluice gates and re-instated the former seasonal hydrological cycle. This has allowed natural regeneration to take place and recovery is now underway. New methods of fire control are now being developed based on the canal network and water pumps (Fig. 5.7).

Just how long must secondary forests be protected before they are 'safe' and unlikely to burn? Many secondary forests growing in humid areas eventually become relatively 'fire-proof' once the tree canopy closes. But, even then, the risk of damage by fire may not be entirely removed, particularly for secondary forests growing on slopes above grasslands. The need for protection against other disturbances such as illegal logging, pilfering or agricultural clearing may persist for much longer. Indeed, the risk may even increase as the forest matures and trees become more valuable. In these cases the need for protection depends more on the nature and stability of the community relationships in the areas surrounding the new forest and on the effectiveness of the police and judiciary.

Protection of state-owned secondary forests can be especially difficult because many local communities may believe they have a traditional ownership claim to the land and that the state has stolen it from them. The difficulty of protecting state-owned secondary forests is illustrated by experiences in the highlands of northern Vietnam (Alther et al. 2002). Substantial deforestation has already occurred in this area and the government's current policy has been to try to protect the remaining forests by paying nearby villagers to act as guards. In this way the villagers are supposed to derive some benefit from a national conservation program. Areas of forests have been allocated to individual households with the areas varying from around 10–30 ha. Villagers are not allowed to clear this land for shifting cultivation or to fell trees for sale although they can collect certain NTFPs.

There have been several, quite contrasting, responses by villagers to this policy. People in one community were simply not interested in participating. Most of them grew paddy rice and the payment being offered for protection was seen as being too low. Besides, the forests were distant from the village meaning that the opportunity cost for individuals (in particular, the time taken to travel to and from the forests)

Fig. 5.7 The result of an attempt to exclude wildfires from seasonally flooded *Melaleuca* forests in the Mekong Delta of Vietnam. Water was prevented from draining from the area and the forests were kept permanently flooded for six years. The result was widespread tree death. Recovery is underway now that the seasonal flooding cycle has been re-instated

was too high. Since individuals did not want to participate village leaders were forced to adopt a joint community-based protection scheme and share the payments amongst community members. Some people left the village as a consequence of the policy to seek alternative opportunities. People in a second community were even less interested and ignored the scheme entirely. Unlike the first case they were some distance from the district administrative centre and too far away for authorities to enforce the policy. A third community was more accessible, being located on the main road, and could not afford to ignore the scheme because farmers feared the repercussions of doing so. But they also had fewer agricultural options than the other two villages. In their case the population density was relatively low and the areas of protection forest allocated to each household were rather larger than average and so households were able to receive a correspondingly higher annual payment. In this case the system worked well and forests in the area have begun to recover. In short, attempts by governments to use local people to protect forest areas may, or may not, be successful depending on circumstances.

Ideally, regenerating forests should be protected from all kinds of disturbance but different categories of protection will yield different outcomes. For example preventing hunting but allowing wildfire will generate a different outcome than allowing hunting but preventing wildfires.

Removing Weeds and Pests

The next pre-requisite for secondary forest development is that weeds and pests are excluded where ever this is possible. Invasive plants but especially exotic weeds are widespread in secondary forests across the region and are especially serious problems on many of the smaller islands of the Pacific where they are leading to the simplification of secondary forests and a loss of biodiversity (Clarke and Thaman 1993). Weeds are particularly influential in early stages of successions and can arrest successions or redirect the successional trajectory along entirely new paths (Norton 2009). Figure 5.8 shows a site in northern Thailand where ferns, grasses and shrubs have entirely excluded woody plant growth and are likely to block successional development for many years. The grass *Imperata cylindrica* is a particularly troublesome species because its presence increases the risk of fires which terminate successional development. Sometimes one weed is replaced by another and there are widespread reports of the grass *Imperata cylindrica* being supplanted by the broad-leaved *Chromalaena odorata.* Many farmers regard this as a positive change since *Chromalaena* is less competitive than *Imperata* and may

Fig. 5.8 Dense weed growth in northern Thailand. Successional development can be arrested by heavy weed growth that out-competes any tree seedlings able to establish at such sites. Reforestation at such sites can only be carried out if these weeds are removed and prevented from re-growing long enough for planted seedlings to become established

even act as a soil improver (Potter 1997; Roder et al. 1995). In time grasses and herbaceous weeds such as these can be colonized and replaced by woody species.

Woody weeds can be especially problematic and exotic species such as *Leuceana leucocephala ssp. leucocephala* have become invasive, especially after fire which promotes seed germination (Hughes and Jones 1998). Bamboos can sometimes be a prominent component of secondary forests and in some cases it may make sense to accept these forests as the new vegetation type and devise appropriate management systems rather than seek to eradicate the bamboos (Mohamed and Othman 2003).

Exotic animal such as rats, rabbits and goats can be serious seed predators, and eat young regeneration while free ranging cattle can be especially damaging. But native wildlife such as deer and elephants can also damage some regrowth species (Bawa and Seidler 1998). Other animal pests are those acting as predators of local wildlife. Like weeds, animal pests have been especially difficult in some of the smaller Pacific islands. They are difficult to eradicate although experience in New Zealand shows control is sometimes possible (Saunders and Norton 2001).

Soil Constraints

The fourth pre-condition necessary before natural regeneration will be successful is that changes in soil physical or chemical properties are not so large as to prevent the regeneration of the original flora (Table 5.3). Soil conditions may be a major constraint on natural regeneration at sites where there has been substantial erosion or at sites disturbed by extreme events such as mining. In such situations natural regeneration may not be a reliable method of reforesting the site because too few species can tolerate the new site conditions. Where this is the case some form of planting involving exotic species able to tolerate the soil conditions will be necessary.

Source of Colonists Nearby

The final constraint on natural recovery is that there may be too few species remaining at the site or able to colonize it from patches of residual forest. The nature of the constraint was shown earlier in Fig. 5.4. In the absence of residual species or colonists the type of forest able to regenerate is likely to be one dominated by only a few species many of which may be more easily-dispersed exotics.

Accelerating Successional Development

The constraints outlined above represent a formidable series of impediments. Nonetheless, across the region, secondary forests do regenerate. But at what point can one say regeneration is well and truly underway? It may be easy to argue for the retention and protection of a well established and species-rich secondary forest but it is more difficult to do so at a grassland site with only a few scattered woody plants and where the likelihood of success is rather less certain. Should such a site be left to continue regenerating by natural means or should reforestation be carried out by replanting the whole area? One determinant might be the density of tree seedlings that are already present. In a discussion on 'assisted natural regeneration' Friday et al. (1999) suggest there should be at least 200–600 seedlings per hectare present if there is to be any chance of forest recovery taking place in a reasonable time. They argued that, at these densities, it is worth making a special effort to exclude fires and control weeds (especially around each seedling) until canopy closure begins to exclude grasses. In marginal cases it may also be useful to plant scattered trees or clumps of trees to act as bird perches (Toh et al. 1999).

In many cases successional development is rapid once tree canopy closure finally occurs and the changed environmental conditions allow a large number of new colonists to become established. But sometimes successional development stagnates. This may happen when a relative small number of species come to dominate a site and hinder further successional development. Some long-lived, exotic, woody weeds can do this but it may occur with some native species as well. The indigenous species *Acacia mangium* and *A. aulacocarpa* are both early pioneers in secondary successions in different parts of Queensland in northern Australia. Both are easily dispersed by birds and both can regenerate from soil seed stores, especially after fire. When old agricultural sites are abandoned these often regenerate and form forests 20–30 m tall with an almost mono-specific upper canopy layer. Over time a diverse community of other species develops in the understorey but is unable to prosper (Tracey 1982). These *Acacia* overstoreys can persist for perhaps 50 years until the trees senesce. Only then can the trees in the understorey grow into the upper canopy. Similar delays have been observed elsewhere. Taylor (1957) describes the development of a multi-layered forest developing after a volcanic eruption 80 years previously at Mt Victory in Papua New Guinea. The upper canopy remained dominated by a few species such as *Octomeles sumatrana* and *Falcataria moluccana* while a rather more diverse community of species was present in the understorey. Taylor commented that the understorey species would not be able to replace the *Octomeles* or *Falcataria* for some years because of the inhibiting effect the early colonists were having. Bruenig (1996) describes a similar forest developing on an abandoned pepper farm in East Kalimantan. In this case the forest was 35 years old and the upper canopy layer was composed of pioneer species while the understorey was almost entirely primary forest species. He estimated it might take more than a century before the original forest structure would be restored.

In most of these cases the simplest solution would be to do nothing but wait and allow the succession to proceed. However, the succession could be accelerated by removing patches of overstorey or thinning the upper canopy layer to increase light to species in the lower strata and promote their growth. Such a treatment forms the basis of some of the Tropical Shelterwood silvicultural systems described by Baur (1964) and Finegan (1992). From a timber production point of view the key dilemma with such treatments is the possible damage the treatment could cause to the trees in the understorey. This might be limited by simply girdling trees and leaving them to die *in situ* rather than harvesting the trees for sale. Alternatively, fellings might be carried out in scattered large patches (say 20–50 m wide) rather than uniformly over the forest as a whole which would be the case in Tropical Shelterwood systems. The location and size of these patches would be dictated by the presence of seedlings of the desired target species and care would need to be taken that these did not simply trigger regeneration and growth of more short-lived pioneers. This was the approach recommended by Kuusipalo et al. (1997) for use in dipterocarp forests in Indonesia. The way the resulting forest is then managed would depend on the objective. If an economic benefit was needed then some kind of selection system might be practiced once trees reached a commercially acceptable size. If the primary objective was to conserve biodiversity or protect watersheds then the forest would simply be protected from further disturbances. Some of these issues are discussed in more detail by Finegan (1992) and Kuusipalo et al. (1997).

Sometimes interventions of this type are needed at an earlier stage. Cohen et al. (1995) describe a situation in Sri Lanka where a degraded site has been occupied by the fern *Dicranopterus linearis.* After perhaps 20 years this single species dominates the site and interwoven fronds as tall as 2 m had arrested successional development even though there was a forest nearby able to supply new colonists. Clearing the ferns and disturbing soils allowed a variety of colonists, including tree species, to occupy the site.

The timing of interventions aimed at encouraging successional development varies. Some are done best at an early stage of the succession while others can only be carried out once the forest stand structure has developed. These differences are outlined in Table 5.5.

Table 5.5 Timing of interventions needed to encourage recovery

Action	In younger regrowth forest?	In older regrowth forest?
Protect from further disturbances	Protection essential	Protection less necessary (though see text)
Thin to encourage growth of target trees	Less appropriate since the identity of target trees may not be clear	Preferable since identity of target trees more obvious
Create canopy gaps to 'unblock' successional development	Should not be done because it can encourage competitive pioneers	May be necessary to prevent truncation of successions
Enrich with key species	Preferable since canopy gaps more frequent	Less optimal; only possible if naturally occurring gaps (e.g. old logging tracks) are still present

Managing Established Secondary Forests

In many situations the best way of ensuring secondary forests persist is to ensure they have a commercial value. If not they will be cleared and the land will be used for other purposes such as agricultural crops where the financial returns are more obvious. The decision is not always simple because many governments are quite aware of the importance of the ecosystem services provided by secondary forests and are seeking to balance the benefits from these with the need for economic development. There are several ways in which the economic or conservation value of secondary forests could be increased. One way is to improve the growth rates of the existing timber trees to increase overall productivity. The second way is to increase the proportion of species in the forest having commercial or conservation value.

Increasing Timber Productivity in Existing Forests

There is a large forestry literature describing various ways of increasing timber productivity in logged-over forests. Much of this is relevant to secondary forest (which, according to the definition used here, includes forests degraded by poorly managed logging). This treatment has been described as 'timber stand improvement', 'refinement' or 'liberation fellings'. Many of these techniques arose during the 1950s and 1960s when logging was beginning to create large areas of secondary forest (Baur 1964; Bruenig 1996; Dawkins and Philip 1998; Fox 1976; Leslie 1989). Most of these approaches sought to promote the regeneration and growth of favoured species in order to promote the overall commercial profitability of the forest. This was usually done at the expense of less-favoured species. Several approaches were developed. One was to remove vines that threatened to overwhelm or distort particular trees. A second was to reduce competition by cutting, girdling or poisoning competing trees of lower value. These treatments were applied across the whole forest and, while they were often effective in boosting commercial productivity, their costs eventually became prohibitive. This was because they needed skilled workers or intensive supervision and because the treatment costs were high relative to the production gains being realized. By the turn of the century very few of these extensive silvicultural prescriptions were still being maintained (Dawkins and Philip 1998).

One additional problem with this approach was that the identities of the commercially preferred and 'useless' trees have changed over time. An illustration of the problem is provided by the changes that occurred in the Australian state of Queensland. In the 1880s only one species was sought (*Toona australis*) but this rose to 36 species by 1940, then 100 species by 1945 and finally around 160 species by 1970s. The increase in the number of commercially attractive species made the old treatment rules increasing difficult to implement.

An alternative to the blanket treatment of the whole area would be to treat only those areas where an adequate stocking of commercially preferred seedlings or saplings was already present. Trials undertaken in Indonesia found good responses where small canopy gaps of <500 m^2 were created. These promoted both seedling survival and diameter growth. Larger gaps fostered the growth of pioneers which out-competed the preferred seedlings (Kuusipalo et al. 1997; Tuomela et al. 1996).

Secondary forests outside the national production forest estate represent a different situation. Local communities often claim ownership of these forests and these communities are likely to take a different view on management than a government forestry department. In such cases the importance of particular species might not be decided on their timber value alone but also on their capacity to produce NTFPs or because they are important as a food tree for wildlife. These additional qualities make the identification of preferred 'target' trees rather more difficult to judge. Apel and Sturm (2003) describe a study carried out in a community forest in northern Vietnam. In this case the prospective target trees were 12–15 m tall and their crowns occupied the canopy layer. Three tree classes were recognized: Class 1 were high quality timber trees, Class 2 produced industrially relevant NTFPs while Class 3 trees were only useful for subsistence purposes. Smallholders were asked to identify candidate target trees from a sample of 80 trees and, on the whole, they chose a similar number to that chosen by a team of technical advisers. Around 75% of smallholders agreed with the technical advisers about the identity of Class 1 trees but differed in their classifications of the other classes. There were also differences amongst observers and no tree was picked for the same class by all smallholders. The implication of the study is that the identity of trees to be favoured by treatment may depend on the person making the decision and that there are likely to be differences between those interested in sawlogs and those more interested in subsistence uses.

Opportunities to improve the commercial value of trees in secondary forests may be greater in more seasonal areas. Trees in these drier forests are more likely to have multiple leaders or trunks with poor form. The value of these forests can be improved by treatments that remove or prune poorer trees to improve stand log quality. If grazing animals are common, some form of pollarding may also be attractive as a way of producing fodder. Some of the forms of silviculture used to manage secondary forests are described further by Gilmour and Fisher (1991) and Hobley (1996).

Modifying the Composition of Secondary Forests

A second approach to improving the value of secondary forests is to improve the proportion of desired species. The site history and landscape context may mean that some secondary forests have low populations of certain plant species. These species may be those of particular commercial value or species with some ecological or conservation significance. The overall value of the forest can be improved by increasing the populations of these species (or introducing them if they are absent) using a practice known as enrichment planting. Not only does this increase the productive value of the forest to the owner but it may also demonstrate to outsiders that the forest is 'owned' thereby decreasing the likelihood it can be taken over. It may also increase the economic resilience of the forest.

Various forms of enrichment planting have been tested over the years in a number of countries (Baur 1964; Dawkins and Philip 1998; Lamb 1969). Most of these have been carried out by cutting lines at relatively wide spaced intervals through the secondary forest and planting seedlings of the preferred species along these lines at short intervals. These seedlings eventually grow up and form part of the canopy. In the case of commercially valuable species these can then be harvested when they reach a merchantable size. A list of recommendations on how to carry out enrichment planting is listed in Box 5.3.

Though the principle behind enrichment planting is simple a balance has to be struck between (a) removing sufficient of the competing shrubs and trees in the secondary forest to allow light to reach the newly planted seedlings, and (b) minimizing the extent of weed control to reduce the cost of labour. Too little clearing will mean the new seedlings will be swamped by existing plants. Too much clearing will

Box 5.3 Requirements for Enrichment Planting in Timber Production Forests

Trials testing methods of enriching logged over production forests with timber trees have been carried out over many years in a number of silvicultural situations (Appanah and Weinland 1993; Dawkins and Philip 1998; Lamb 1969). Some of the main conclusions arising from these trials are that:

- There must be rules in place that regulate access to the site and guarantee harvesting rights.
- The species to be planted must be commercially (or ecologically) valuable.
- The species used must be capable of fast growth (meaning that most will be light-demanding).
- Seedlings should have well-established roots meaning that container-grown seedlings are preferable to bare rooted seedlings or wildlings.
- Species should have a low crown ratio (ratio of crown diameter to stem diameter).
- These species should be self-pruning and have good form.
- Planting lines should be oriented in an east-west direction and be separated at a distance about the same as the crown diameter of the species when mature (e.g. around 10–15 m).
- Seedlings should be planted more closely along these lines (i.e. <10 m) to allow for deaths and perhaps thinning.
- All overstorey competition should be removed before planting (to avoid damaging young seedlings).
- Weeds along the planting line should be removed at least three times in the first year in a strip about 2 m wide.
- The technique will fail if seedlings are susceptible to grazing by wildlife.
- The regrowth between the planting lines should not be flammable.

be expensive and, in any case, may allow new light-demanding weeds to flourish. The extent of the canopy opening required depends very much on the height of the surrounding vegetation. If the surrounding tree are tall (e.g. in a selectively-logged forest) a large canopy gap might need to be created before much direct sunlight will reach the forest floor. On the other hand, if the forest is shorter, such as young regrowth regenerating on old farmlands, there will be fewer overstorey trees and the main competition may come from low shrubs and vines.

The light environments in these two situations are compared in Fig. 5.9. This model is based on direct sunlight reaching the forest floor passing through a simple cylindrical hole in an opaque canopy (Stocker 1988). A canopy gap in forest at the equator with a tree height-to-gap diameter ratio of 1 (i.e. 15 m tall forest with a gap 15 m wide) allows sunlight to directly illuminate most of the forest floor at the equinox but leaves a substantial proportion of the gap shaded in midsummer and mid winter. A canopy gap with a height-to-gap diameter ratio of 4 (e.g. 20 m tall forest and a 5 m wide gap) has some limited direct light at the equinox but is shaded at other times of the year. Even more shading occurs in gaps in forests at higher latitudes away from the equator. The patterns for a wider range of height-to-diameter ratios are shown in Fig. 5.10. These indicate that annual illumination declines rapidly once the ratio exceeds a value of 1. Although indirect light can be important as well, in practice these patterns mean enrichment is usually best done immediately

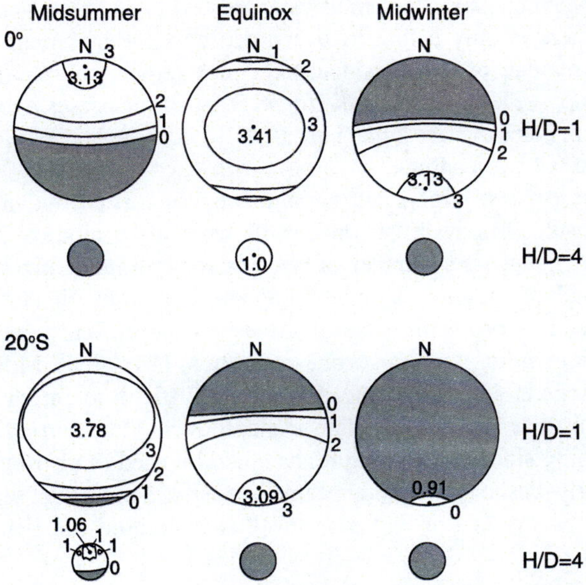

Fig. 5.9 Spatial patterns of sunlight on the forest floor in gaps at the equator (*top row*) and at 20° S latitude (*bottom row*), for two gap sizes and at three times of the year. At each latitude the larger circle represents gaps with tree height/gap ratios = 1; the smaller circle represents gaps of with tree height/gap ratios = 4. Isopleths join points with 0, 1, 2 and 3 h of direct sunlight (assuming cloud free conditions). Points marked with a dot represent the maximum values. Grey areas are those not receiving direct sunlight (After Stocker 1988)

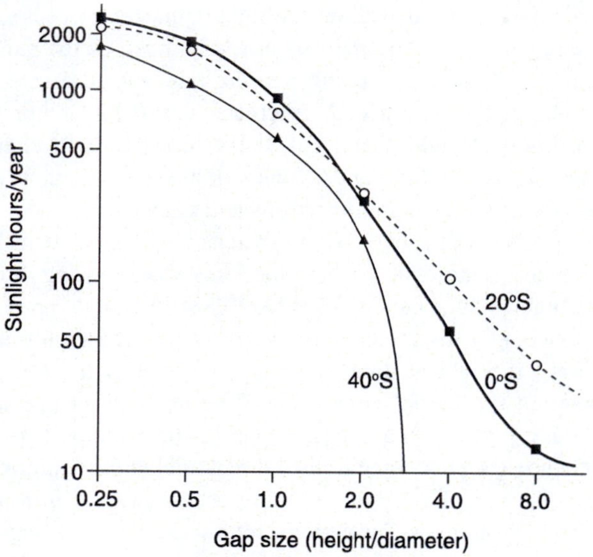

Fig. 5.10 The maximum potential number of sunlight hours received at the forest floor in gaps of various sizes plotted for 0°, 20° and 40° south latitude (After Stocker 1988)

after logging when canopy gaps are at their greatest or in short, young, regrowth forest where there is only a limited overstorey cover. Enrichment in secondary forests with a number of taller residual trees may require substantial girdling or poisoning to improve light levels on the forest floor. In such cases it might be better to only enrich in places where large gaps already occur such as old logging tracks and roads or on old log landings.

Some of the earliest large-scale enrichment was carried out in logged-over forest in Peninsular Malaysia but the results were often modest. Appanah and Weinland (1993) suggest a number of reasons for this including poor seedling quality and a lack of attention to maintaining adequate light. Ng (1996) describes some other practical problems. One of these was the difficulty workers had in carrying seedlings in to planting sites, especially in hilly areas. Most could only carry 20 seedlings in a backpack meaning it was difficult to enrich areas distant from roads. This was less of a problem if planting could be carried out immediately after logging since old logging roads could be used to distribute seedlings. But these quickly washed away and became impassable in the wet season. Access was easier in the dry season but planting then increased the risk of seedling drought deaths.

Some comparatively large areas of logged-over forests in the Solomon Islands were also enriched. In this case the early results were promising but tending costs rose when the logging intensity increased because vines such as *Merremia peltata* and *Operculina riedeliana* were able to flourish in the improved light conditions. Because of this enrichment planting was eventually dropped in favour of normal plantation establishment (Bennett 2000).

Enrichment planting has also been carried out in rather more degraded dipterocarp forests in Sabah where fires had followed an earlier logging. The site retained some trees which varied between 20 and 35 m in height. But, in addition, there was a substantial understorey varying in height between 2 and 20 m and dominated by *Macaranga*. Several treatments were tested including selectively felling or girdling of the overstorey trees and removing (brushing) all understorey plants. Felling just the pioneer trees in the understorey initially promoted faster growth in the planted seedlings but the effect gradually declined after a few years. Growth was enhanced for a longer period when the overstorey trees were girdled and the understorey was also removed. The trial lasted only 30 months and, given the complexity of the canopy structures, probably needed more time for a complete analysis (Romell et al. 2008). Details of another large post-logging enrichment planting operation that has been underway in Sabah for a number of years are given in Box 5.4.

More promising results with enrichment have been reported by Adjers et al. (1995) in South Kalimantan. Unlike many enrichment planting studies, this was carried out in much shorter (around 3 m tall) secondary forest regenerating after logging and shifting cultivation. These researchers found line orientation and line width had little effect on survival but that it did affect growth with the best result being found for lines that were 2 m wide and oriented in a SE-NW direction.

The advantages of enrichment are that it is cheaper than clearing remnant forest and establishing a plantation and it takes advantage of the existing seedling, saplings and residual trees of economically preferred species. It also conserves biodiversity and protects soils and soil carbon. It is most likely to be successful if implemented immediately after logging when canopy gaps are most extensive or in short secondary regrowth where it is possible to improve the light environment for newly planted seedlings. It is likely to be less successful in tall secondary forests especially those growing at higher latitudes. The chief disadvantage is that it can still be a risky operation to apply over a large scale unless the silvicultural requirements of the species being used are known and the cost of removing sufficient overstorey cover can be controlled. Although the focus of most enrichment planting has been on planting commercially attractive timber trees the techniques could be equally applied to enrich a secondary forest with wildlife food species.

A note of caution is needed here. The need to supplement the productive capacity of logged-over or degraded forests by enriching them with commercially important species has been recognized for more than 50 years. There have been many studies on how this might be done and these continue to this day (though mostly concerned with the early stages of enrichment planting than in the later stages). There are, however, few locations where enrichment planting has become a routine practice and persisted over time. This is not because the technique does not work. Rather, it is probably more to do with the costs involved, the productivity of the enriched forest and the delay in the return on this investment. An alternative approach to dealing with this dilemma is outlined in Box 5.5.

Box 5.4 Enrichment Planting in Secondary Dipterocarp Forest, Sabah

A large scale enrichment planting program was started in logged-over secondary forest in Sabah by the Innoprise Corporation (the forestry arm of the Sabah Foundation) and FACE (Forests Absorbing CO2 Emissions) Foundation of the Dutch Electricity Board (Bebber et al. 2002; Moura-Costa et al. 1996; van Ooschot et al. 1996). The aim of the project is to sequester carbon to offset emission by Dutch power stations. By 2007 the program had treated 10,000 ha and aimed to cover another 20,000 ha before it was completed. The secondary forests being enriched varied in age and structure. Some had dense canopies dominated by *Macaranga* while others had more grasses, weeds, shrubs, vines and bamboos. Planting was done along 2 m wide lines cut through the forest at 10 m spacings. Most vegetation along the lines was removed except for remnant trees, natural regeneration of dipterocarp species or fruit trees. The planting lines were oriented in an east-west direction and seedlings were planted at 3 m intervals in holes that were 10 cm diameter and 20 cm deep. Fertiliser in the form of rock phosphate was applied at a rate of 100 g per seedling.

Around 33 species have been used. Most were dipterocarps but a number of fruit trees have also been planted to provide food sources for wildlife. Timber species included *Shorea leprosula, S. parvifolia, S. ovalis, S. johorensis, Parashorea malaanonan, Drybalanops lanceolata.* Fruit trees included species of *Durio, Artocarpus, Mangifera, Diospyros* and *Dacyodes.* Planting material included wildlings (naturally occurring seedlings collected from the forest) as well as seedlings raised the nursery. These were grown in polybags (7 × 21 cm) and planted out when seedlings were between 30 and 60 cm in height. Assessments carried out after two months found that survival averaged 89% and low survival rates were generally associated with planting carried out during dry weather. Over a two year period most seedlings grew by 60 cm but there were considerable variations between species as well as within species. *Shorea leprosula* had the best growth and had relatively low mortality while species of *Dipterocarpus* had lower growth rates and higher mortality levels.

Subsequent investigation found that variations in the growth of seedlings were strongly related to differences in light. This is hardly surprising given the patterns shown in Figs. 5.9 and 5.10. In some cases too much light was found to lead to photo-inhibition and reductions in growth. Large gaps have hot, dry microclimates that can increase mortality. But, more commonly, excessive shade reduced growth and survival. Seedlings responded to variations in light along the planting lines but not to variations across the area between the lines where vegetation remained. There is now increasing interest in making more use of gaps (>10 m wide), both natural and artificial, to plant seedlings.

Box 5.5 Maintaining Secondary Forests in the Face of Land Clearing Threats

Much of the natural forests of Sabah have been logged and many of these logged-over areas have then been cleared and planted with oil palm. The current profitability of oil palm has meant that the Forestry Department has found itself in a difficult position in arguing that these forests should be retained as production forests. Intensive surveys in one 57,000 ha logged-over forest area (Sg. Pinangah Forest Reserve) found there was considerable variation in the density and size class distribution of the remaining commercially attractive timber trees. In some compartments there were large numbers of regenerating trees and advanced growth but in other places there were not. But, even in the best cases, a further harvest would not be possible for another 20 years (after which time a 40–50 year cutting cycle should be possible). In the current economic climate this meant there was a distinct risk the whole area could be cleared and replaced by oil palm. Enrichment planting would not shorten the time until another harvest was possible so, instead, the strategy adopted was to identify a small area representing less than a few percent of the total and convert this to a plantation of fast growing trees species. These are expected to generate a cash flow within ten years and so tip the balance in favour of retaining the overall area as forest.

The area to be converted to a plantation was chosen on the basis of the previous forest survey and included the most degraded forest with the poorest stocking of commercially valuable trees. Although this was the primary criterion used, the location of the plantation was also chosen to be near the boundary of the forest. This was to minimize damage to newly planted seedlings caused by browsing deer and elephants since many still remain in logged-over forest and can cause substantial damage.

Two concession holders were each given an initial 1,000 ha block to reforest. In both cases steep lands (slopes > 25°), riparian strips and roadside buffers were excluded from the planting area. These could account for 20% of the total area. Provided the concession holders fulfil their contractual obligations they are each assured of two further allocations of 1,000 ha so that the overall planting area could reach a maximum of 6,000 ha in a total forest area of 57,000 ha.

The species planted in the cleared areas include a variety of mostly fast-growing native species such as *Octomeles sumantrana, Anthocephalus cadamba, Duahanga moluccana* and *Paraserianthes falcataria* although some exotics such as *Khaya ivorensis* were also planted. All these produce commercially valuable timber and are capable of providing an early financial yield prior to the next cutting cycle being carried out in the natural forest. The different species have been planted in monocultures that collectively form a spatial mosaic across the plantation area.

(continued)

Box 5.5 (continued)

From the Forestry Department viewpoint this approach represents a pragmatic approach to a difficult management problem. It involves clearing parts of an existing secondary forest to establish simple monocultural plantations but it does so under carefully controlled conditions with constraints being established to protect watersheds and to limit the proportion of the area being cleared. More to the point, it offers a way by which forest cover can be maintained across the broader landscape. Without this approach the area would be replaced by oil palm. Quite apart form the implications such a change would have for species conservation, oil palm in this area is currently established with few environmental constraints such as riparian areas or buffer strips so that the overall environmental impact of converting forest to oil palm would be substantial.

Using Secondary Forests to Create Agroforests

Most enrichment planting with timber trees has been done by state forestry agencies. But many traditional forest communities have also enriched secondary forests using food and other NTFP species (e.g. de Jong 2002; Michon 2005). These provide households with a wider range of products and an earlier supply of these than unmodified secondary forests. Some of these enriched forests also provide significant cash incomes. This process can be seen as a silvicultural version of the process of agricultural intensification that Boserup (1993) argues usually follows increasing human population densities. Enrichment of this kind has mostly been practiced where the areas of natural forest have declined or where population densities have increased and shifting cultivators find long fallow periods are increasingly difficult to sustain.

These cultivated forests have been referred to as 'agroforests', 'improved fallows', 'polyculture plantations' or 'forest gardens' and a variety of different types are found throughout the Asia-Pacific region (Table 5.6). Similar forests have been developed in many other parts of the world. Michon (2005) has reviewed much of the literature describing enriched forests in the Asia region and argues they should not be seen simply as versions of the widely-used home gardens or as simple secondary forests but as a distinctly new type of man-made forest. Clarke and Thaman (1993) take a similar view based on their extensive analysis of agroforestry systems in the Pacific.

Types of Agroforests

Some agroforests focus on just one or two commercially important species and these dominate the tree canopy (e.g. *Shorea javonica* for damar gum, *Styrax* spp.

Table 5.6 Examples of various types of agroforests in the Asia-Pacific region

Location	Principle tree used	Reference
Indonesia	Rubber (*Heavea brasiliensis*)	Dove (1993); Gouyon et al. (1993)
	Damar resin (*Shorea javanica*)	Michon (2005)
	Benzoin resin (*Styrax* spp.)	Garcia-Fernandez and Casado (2005)
	Rattan (mostly *Calamus* spp.)	Weinstock (1983)
	Mixed fruit trees	Sunderland and Dransfield (2002) De Jong (2002); Michon (2005)
Thailand	Tea (*Camellia sinensis var. assamica)*	Sasaki et al. (2007)
Vietnam	Fruit, medicinal plants	Dao et al. (2001)
Laos PDR	Benzoin resin, Cardamon	Michon (2005)
Australia	Fruit	Hynes and Chase (1982)
Pacific – Melanesia	Fruit and nuts	Clarke and Thaman (1993); Hviding and Bayliss-Smith (2000)
Pacific – Polynesia and Micronesia	Fruit	Clarke and Thaman (1993)

for benzoin resin, *Hevea brasiliensis* for rubber or *Durio* spp. for durian fruit) or the understorey (e.g. the *miang* or 'chewing' tea of northern Thailand). Other agroforests are enriched with a much wider variety of trees or shrubs such as fruits and nuts, medicinal plants, building timbers or canoe trees and are mostly used for local consumption. Perhaps the simplest agroforests are the grassland fallows in the highlands of Papua New Guinea which are 'enriched' with just the putative nitrogen fixing tree *Casuarina oligodon* (Bourke 1997).

Agroforests have been established in several ways. One way is to plant seedlings of the preferred species into the early fallow stage of the shifting cultivation cycle. These introduced trees then form part of the new succession and are surrounded by naturally occurring regeneration of other forest species. This is the approached used by those growing fruit and nut species (de Jong 2002), canoe trees (White 1976), rubber (Dove 1993), damar (Michon 2005) and benzoin (Garcia-Fernandez and Casado 2005). An alternative approach is to establish plants under an existing secondary forest. This is the approach used for species such as tea, cardamon and the many medicinal plants grown in forest understories (Michon 2005; Santasomsat 2003; Sasaki et al. 2007). All of these agroforests have been developed by traditional farmers and are owned by individuals, households or clans. In most cases the person who carried out the planting has the pre-eminent harvesting claim although rules vary widely. Most farmers grow their agroforest on hill areas and use flatter lands for growing foodstuffs such as rice.

The size of areas converted into agroforests vary from relatively small and isolated areas of former shifting cultivation land to the much larger contiguous areas of agroforest found in parts of Indonesia. Michon (2005) estimates there are two to three million hectares of more or less contiguous areas of 'jungle rubber' (*Hevea brasiliensi*), 80,000 ha of damar (*Shorea javanica*) agroforests and 100,000s hectares of fruit forests in Indonesia alone. Most of these forests are rich in species because natural regeneration has grown up around the planted 'crop' species.

Agroforests have both material and ecological advantages. Some farmers are able to obtain a significant income from their agroforests (with some relying on this income to buy food) while all farmers with agroforests benefiting from the variety of foods and other resources they provide. Further, once established, little work is needed to maintain these forests and any such work rarely competes with the timing of other on- or off-farm activities. The forests also represent a significant asset to be passed on to heirs. The ecological benefits are obvious and come from the establishment and maintenance, at little cost, of biologically diverse and complex forests that increasingly restore key ecological processes as they mature. Agroforests maintain the productive capacity of the site and retain flexibility such that other species can be added if circumstances change. For example, some farmers in Indonesia began adding clove trees to their dammar forests when the clove price began rising in the 1980s (Michon 2005).

Surveys within these forests show they are rich in both plant and wildlife species even though the forest canopy may be dominated by a few tree species (de Jong 2002; Gouyon et al. 1993; Michon 2005). For example, Schroth et al. (2004) report studies in old damar forests and jungle rubber forests in Sumatra that found they contained more than 30 and 60 tree species respectively (with undisturbed primary forest containing around 150–216 species). De Jong (2002) also reported very high tree species numbers in agroforests in Kalimantan. This diversity, together with the associated structural complexity, provides habitats for many wildlife species. Thiollay (1995) reported damar and rubber agroforests had many more birds than monocultures of rubber or oil palm though still many less than undisturbed natural forest. Not all enrichment fosters enhanced biodiversity and some such as the tea gardens of northern Thailand may protect the upper canopy trees but result in many understorey species being removed during tending operations.

Conditions Favouring the Development of Agroforests

Agroforests appear to be an attractive method of overcoming degradation and offer both livelihood and conservation benefits. But there are certain conditions that must be satisfied before agroforests will be developed. These are summarized in Table 5.7. A key condition is that there must be a demand for the goods such as fruits, rubber, tea, damar, resins etc. that these types of forest can supply. Many agroforests were originally developed to supply local needs but commercial markets provide an added incentive. Agroforests then become attractive even when population densities are not high and there is sufficient land to grow rice or other primary food crops. Landowners deriving much of their income from off-farm work may

Table 5.7 Conditions under which agroforests may develop (Based on de Foresta and Michon 1993)

Requirement	Reason
Low population density	Below about 150 person km^2 otherwise the opportunity cost may be too high
Technical knowledge	Growers need sufficient technical knowledge about growing the crop and a willingness to try new options when agricultural or ecological circumstances change
Markets available	Agroforests become attractive when natural forests are unable to maintain the supply of NTFPs. In the past the goods produced by agroforests were locally consumed. Now cash markets are more important. The composition of agroforests is likely to change as markets change though many farmers will try to ride out fluctuations
Roads or other transport	The absence of transport will severely limit the commercial attractiveness of agroforests
Land tenure	Farmers must have confidence they will benefit from growing trees. Legal tenure is obviously preferable but de facto tenure may be sufficient. At the same time, many farmers believe the act of tree planting asserts land ownership in the face of attempts by migrants or governments to take over their land
Additional sources of foodstuffs	Most agroforests are complemented by nearby rice fields although some farmers also purchase food using income from their agroforests
Recognition of the insurance value of multi-species agroforests	Price fluctuations have exposed the financial vulnerability of some long-lived monocultures
Natural forest is nearby	These are needed to maintain the supply of seeds required to initiate and sustain the secondary succession

also find it convenient to establish a low-maintenance agroforest on their otherwise unused land to maintain a supplementary income stream. In some places farmers have been attracted to multi-species agroforests because they have seen how agricultural prices can fluctuate and see agroforests as providing a buffer against such variations.

Many agroforests, such as those in Sumatra, have been established for a number of years but new forms are continuing to develop. Fujisaka and Wollenberg (1991) give an example from the Philippines where agroforests were not particularly well developed in the past. In this case a forest once logged by commercial timber companies was logged a second time by a wave of small-scale illegal loggers. These effectively deforested most of the area. Immigrant farmers then moved in and began producing charcoal and growing rice on small farms. However yields were low because of the high rainfall (> 4,500 mm), weeds, insect pests and diseases. Many farmers switched to tomatoes but prices were too variable and yields were not sustainable. The tomatoes were replaced by root crops (cassava and sweet potato) mixed with bananas and fruit trees. The root crops were sustainable but not very productive. So, over time the farms developed into multi-storey, perennial tree systems also involving coffee, citrus and cocoa. In the meantime, some of the native forest tree species began to appear as regeneration and have been included into the farming system. The landscape as a whole has evolved from a highly disturbed forest left after logging to become one with patches of secondary forest regrowth and areas of perennial tree crops.

Agroforests can change over time. One of the most interesting examples of change and how systems can evolve are the *Styrax* forests of Sumatra used to produce the benzoin resin. A system of enriching swidden fallows with *Styrax* was developed in lowland areas of Sumatra perhaps 200 years ago. Benzoin was originally used for incense and perfumeries although it has now been replaced in these industries by synthetics. However it is still used in Indonesia's fragrant and ubiquitous *kretek* cigarettes (Michon 2005). The rubber boom in the 1920s led to the replacement of *Styrax* at these lowland sites because rubber was more profitable. However, the market for benzoin was still sufficient to encourage a number of farmers to move to alternative sites at higher elevations and create new agroforests (Katz et al. 2002; Michon 2005; Schroth et al. 2004). *S. benzoin* prefers altitudes below 600 m and most of the new sites exceeded this elevation so *S. benzoin* was replaced by *S. paralleloneurum.* This necessitated a change in the nature of the agroforest. In the lowlands the shade-intolerant *S. benzoin* was grown in swidden fallows. However, *S. paralleloneurum* is shade tolerant so it has been grown by planting it beneath an intact forest canopy and then girdling these trees once the *Styrax* are established. The change illustrates a high degree of adaptability on the part of the growers in devising a new agroforest system. But it probably also shows a determination to assert ownership of the mountain forests (by creating a new crop) in the face of government moves to take them over and award the land to a paper company (Katz et al. 2002, Michon 2005). Interestingly, some *Styrax* farmers in Sumatra have begun to take renewed pride in their farms once they became the subject of scientific study (Katz et al. 2002).

The Uncertain Future of Agroforests

Agroforests have an uncertain future. In some places they will probably persist but in others these seem destined to disappear. Within Indonesia Michon (2005) is of the view that established agroforests are likely to persist and that these are often remarkably tolerant of market fluctuations. She gives examples of cases where a downturn in crop prices has led to inactivity but not the replacement of an agroforest. Also surprisingly, many Indonesian farmers appear to prefer to keep crops such as their 'jungle rubber' in a multi-species agroforests even though they might get higher yields from a more intensively managed monoculture. According to Michon

(2005) the difference is more than made up for by the variety of other NTFPs produced by the agroforest (but see further below).

Changes will occur, however, when the price of a key species is thought to have declined permanently or when the market situation changes (e.g. because road access improves) making another species more attractive. This is the case in parts of Indonesia where oil palm is now replacing agroforests. Potter and Badcock (2006) studied communities in West Kalimantan seven years after oil palm was first introduced. Most communities still retained their agroforests but the areas covered and the species composition had both declined. The balance between adopting the oil palm monocultures or retaining the species-rich agroforests appeared to be a balance between maintaining a communal tradition and adopting a new and individualized form of agriculture. Much depended on the attitude of customary village leaders. Most of these have tried to retain the old ways of doing things but, without their continued intervention, it seemed the systems might fossilise and eventually disappear in this area.

Change also occurs when population densities rise and land becomes short. This is currently taking place on many smaller Pacific Islands. Clarke and Thaman (1993) have coined the term 'agro-deforestation' to describe the conversion of complex agroforests to more simplified forms of agriculture. A combination of increasing population growth, poverty, landlessness cause by changes in usufruct rights, increasing wildfires and shifting aspirations have all contributed to these changes. There is also a shift in attitudes regarding the trade-offs between polycultures and monocultures and a growing view that the old insurance value of polycultures is no longer needed. Some young people also think that the former agroforests are unsophisticated 'bush' systems not worth supporting. One consequence of all this is that many food items are now being imported. Another is that many genetic resources (plant varieties as well as species) are being lost together with knowledge about establishing and managing agroforests.

Changes in attitudes amongst younger people have also contributed to the decline of some of the more isolated tea forests in northern Thailand. In these cases young people have left the village and fewer people are left who are interested in managing the tea forest (Sasaki et al. 2007). The market for these teas may have also declined in recent years following an earlier boom period prior to the 1980s when road access improved. Some growers have now begun switched to growing other crops, including drinking tea, but most are still cautious about a wholesale change because of the risks involved and their uncertainty over future prices. Interestingly, these declines in profitability may about to be reversed because a growing interest among tourists who wish to see these unusual tea forests (Sasaki et al. 2007).

The situation with 'jungle' rubber in Sumatra is more complex. Many rubber trees were planted in the first half of the twentieth century and their productivity is now declining Growers have several choices. One would be to fell the forests and replace it with high-yield clonal rubber material that is now available. This could lead to a conversion of the multi-species rubber agroforests to simple monoculture rubber plantations like those grown in Malaysia and elsewhere. The cost of a wholesale conversion would be high and possibly too much for many smallholders.

A second option, therefore, might be to fell only small patches of rubber agroforest at a time so that the conversion was done over a period. This would probably lead to the development of a new agroforest not much different to that now present. A third choice could be to replace the rubber by oil palm. It is unclear just which of these alternatives might develop. One indication of farmers views come from a study carried out by Williams et al. (2001). The intent was to test the effectiveness of various weeding treatments when introducing new high-yield rubber clones into species-rich agroforests. Disregarding the researcher's experimental designs, farmers tended to do a complete weeding in the newly established plantations because they regarded the clonal material as valuable. This suggested they were willing to make the switch from agroforests to monocultures. But the frequency and intensity of weeding differed amongst farmers with the wealthier farmers doing more than the poorer farmers who had greater demands on their time. The trial only lasted 21 months but suggests some richer famers may switch to monocultures but others may, by default, allow 'jungle rubber' to develop once more.

Conclusions

Forests are often able to regenerate naturally at degraded sites. These so-called secondary forests now occupy very large areas in many countries and the areas are increasing. The composition and structure of these forests depend on the site history and the landscape context. Sites where the intensity of the prior disturbance was limited and located near intact primary forest are likely to recover their floristic composition and structure more rapidly than those where the intensity of degradation is greater and located some distance away from intact primary forest. In these situations recovery may be rather more limited and an equilibrium may be reached where the forest structure is similar to that of the original forest but the species composition differs. In all cases the development of secondary forests at degraded sites is greatly enhanced by excluding further disturbances from the area.

In situations where populations are increasing and agricultural intensification is underway many forests are under threat and secondary forests are often under more threat than most. Those at greatest risk of being cleared are younger secondary forests on flatter land or near roads irrespective of the biodiversity they may contain or the successional trajectory on which they are set. The best way of protecting these forests is to increase their immediate value to their de jure or de facto owners. There are several ways in which this might be done. One is to manipulate the canopy cover to favour the growth of more valuable species already present. Another is to enrich the forest by planting more trees of these preferred species, including species that are not already present. These species might be timber trees or species providing foods and NTFPs. A number of methods for doing this have been developed. Secondary forests that have been modified in these ways may not be as biologically rich as undisturbed natural forest but are more likely to be protected.

The conservation of secondary forests may also be assisted by two other factors. One could be the development of a market for the ecosystem services they provide such as watershed protection or carbon sequestration. Another factor that may assist in their development is the process of urbanisation that is now underway in various locations and which reduces rural population pressures. This is unlikely to occur everywhere but may be especially important in the conservation of older secondary forests in steeper and more distant locations. For either of these factors to have any effect, appropriate institutional arrangements and policy settings will be needed that foster the protection of both public and privately owned secondary forests. These issues are discussed further in Chapter 12.

Secondary forests may not develop on highly disturbed or degraded lands. In these situations reforestation is only possible if seedlings are planted. There are a variety of ways in which this can be done and the next chapter discusses the most commonly used practice which is to establish plantation monocultures.

References

Adjers G, Hadengganan S, Kuusipalo J, Nuryanto K, Vesa L (1995) Enrichment planting of dipterocarps in logged-over secondary forests – effect of width, direction and maintenance method of planting line on selected Shorea species. Forest Ecol Manage 73:259–270

Aide M, Cavelier J (1994) Barriers to lowland tropical forest restoration in the Sierra Nevada de Santa Marta. Restoration Ecol 2:219–229

Alther C, Castella J-C, Novosad P, Rousseau R, Tran TH (2002) Impact of accessibility on the range of livelihood options available to farm households in the mountainous areas of northern Vietnam. In: Castella JC, Dang DQ (eds) Doi Moi in the mountains: Land use changes and farmers livelihood strategies in Bac Kan Province, Vietnam, The Agricultural Publishing House, Hanoi, pp 121–146

Apel U, Sturm K (2003) The use of forest succession for establishment of production forest in north-eastern Vietnam. In: Sim HC, Appanah S, Durst PB (eds) Bringing back the forests: Policies and practices for degraded lands and forests. Food and Agriculture Organisation of the United Nations, Proceedings of an International Conference, 7–10 October 2002, Kuala Lumpur, pp 39–49

Appanah S, Weinland G (1993) Planting quality timber trees in peninsular Malaysia: A review. Forest Research Institute, Malaysia, Kepong

Ashton MS, Gunatilleke CVS, Singhakumara BMP, Gunatilleke IAUN (2001) Restoration pathways for rain forest in southwest Sri Lanka: a review of concepts and models. Forest Ecol Manage 154:409–430

Barlow J, Gardner TA, Araujo IS, Avila-Pires TC, Bonaldo AB, Costa JE, Esposito MC, Ferreira LV, Hawes J, Hernandez MM, Hoogmoed MS, Leite RN, Lo-Man-Hung NF, Malcolm JR, Martins MB, Mestre LAM, Miranda-Santos R, Nunes-Gutjahr AL, Overal WL, Parry L, Peters SL, Ribeiro-Junior MA, da Silva MNF, Motta CD, Peres CA (2007) Quantifying the biodiversity value of tropical primary, secondary, and plantation forests. Proc Natl Acad Sci 104:18555–18560

Baur GN (1964) The ecological basis of rainforest management. Forestry Commission of New South Wales, Sydney

Bawa KS, Seidler R (1998) Natural forest management and conservation of biodiversity in tropical forests. Conserv Biol 12:46–55

Bebber B, Brown N, Speight M, Moura-Costas P, Yap SW (2002) Spatial structure of light and dipterocarp seedling growth in a tropical secondary forest. Forest Ecol Manage 157:65–75

Beccari O (1904) Wanderings in the Great Forests of Borneo, Archibald Constable & Co (Reprinted Oxford University Press, Oxford, 1986)
Bennett JA (ed) (2000) Pacific forest: A history of resource control and contest in Solomon Island, c. 1800-1997. Brill, Leiden
Boserup E (1993) The conditions of agricultural growth: The economics of Agrarian change under population pressure. Earthscan, London
Boucher DH, Vandermeer JH, de la Cerda I, Moallona MA, Perfecto I, Zamora N (2000) Post-agriculture versus post-hurricane succession in southeast Nicaragua rain forest. Plant Ecol 156:131–137
Bourke RM (1997) Management of fallow species composition with tree planting in Papua New Guinea. Research School for Pacific and Asian Studies. The Australian National University, Canberra, Canberra
Bowman DMJ (2000) Australian Rainforests: Islands of green in a land of fire. Cambridge University Press, Cambridge
Brewer SW (2001) Predation and dispersal of large and small seeds of a tropical palm. Oikos 92
Brook BW, Bradshaw CJA, Koh LP, Sodhi NS (2006) Momentum drives the crash: mass extinction in the tropics. Biotropica 36:302–305
Brown S, Lugo A (1990) Tropical secondary forests. J Tropic Ecol 6:1–32
Brown N, Press M, Bebber D (2000) Growth and survivorship of dipterocarp seedlings: differences in shade persistence create a special case of dispersal limitation. In: Newberry DM, Clutton-Brock TH, Prance GT (eds) Changes and disturbance in tropical rainforest in South-East Asia. Imperial College Press, London, pp 121–131
Bruenig EF (1996) Conservation and management of tropical rainforest: An integrated approach to sustainability. CAB International, Wallingford
Burslem D, Whitmore TC, Brown GC (2000) Short-term effects of cyclone impact and long-term recovery of tropical rain forest on Kolombangara, Solomon Islands. J Ecol 88
Cannon CH, Peart DR, Leighton M (1998) Tree species diversity in commercially logged Bornean rainforest. Science 281:1366–1368
Chazdon RL (2003) Tropical forest recovery: Legacies of human impact and natural disturbances. Perspect Plant Ecol Evol Syst 6:51–57
Chazdon RL, Coe FG (1999) Ethnobotany of woody species in second-growth, old-growth and selectively logged forests in northeastern Costa Rica. Conserv Biol 13:1312–1322
Chazdon RL, Peres CA, Dent D, Sheil D, Lugo AE, Lamb D, Stork N, Miller SE (2009) The potential for species conservation in tropical secondary forests. Conserv Biol 23:1406–1417
Chokkalingam U, Smith J, De Jong W (2001) A conceptual framework for the assessment of tropical secondary forest dynamics and sustainable development potential in Asia. J Tropic Forest Sci 13:577–600
Clarke WC (1971) Place and people: An ecology of a New Guinean community. Australian National University Press, Canberra
Clarke WC, Thaman R (1993) Agroforestry in the Pacific Islands: Systems for Sustainability. United Nations University Press, Tokyo, New York, Paris
Cohen AL, Singhakkumara BPM, Ashton PMS (1995) Releasing rain forest succession: a case study in the Dicranopteris linearis fernlands of Sri Lanka. Restoration Ecol 3:261–270
Connell JH (1978) Diversity in tropical rain forests and coral reefs. Science 199:1302–1310
Connell J, Green P (2000) Seedling dynamics over thirty-two years in a tropical rain forest tree. Ecology 81:568–584
Corlett R (1994) What is secondary forest? J Tropic Ecol 10:44–47
Corlett R (1998) Frugivory and seed dispersal by vertebratres in the Oriental (Indomalayan) region. Biol Rev 73:413–448
Corlett R (2002) Frugivory and seed dispersal in degraded tropical East Asian landscapes. In: Levey DJ, Silva WR, Galetti M (eds) Seed dispersal and frugivory: Ecology, Evolution and Conservation. CAB International, Wallingford
Corlett R (2009) Seed dispersal distances and plant migration potential in tropical East Asia. Biotropica 41:592–598

Dao TH, Tran CT, Le TC (2001) Agroecology. In: Le TC, Rambo AT (eds) Bright peaks, dark valleys a comparative analysis of environmental and social conditions and development trends in five communities in Vietnam's northern mountain region. The National Political Publishing House, Hanoi

Dawkins HC, Philip MS (1998) Tropical moist forest silviculture and management: A history of success and failure. CAB International, Wallingford

de Foresta H, Michon G (1993) Creation and management of rural agroforests in Indonesia; potential applications in Africa. In: Hladik CM, Hladik A, Linares OF, Pagezy H, Semple A, Hadley M (eds) Tropical forests people and food: Biocultural interactions and applications to development. UNESCO and Parthenon Publishing Group, Paris, pp 709–724

de Jong W (2002) Forest products and local forest management in West Kalimantan, Indonesia: Implications for conservation and development. Tropenbos International, Wageningen

de Jong W, Chokkalingam U, Smith J, Sabogal C (2001) Tropical secondary forests in Asia: introduction and synthesis. J Tropic Forest Sci 13:563–576

Dennis R, Hoffman A, Applegate G, von Gemmingen G, Katrtawinata K (2001) Large-scale fire: Creator and destroyer of secondary forests in Western Indonesia. J Tropic Forest Sci 13:786–799

Dirzo R, Mendoza E, Oriz P (2007) Size-related differential seed predation in a heavily defaunated neotropical rain forest. Biotropica 39:355–362

Doust SJ (2004) Seed and seedling ecology in the early stages of rainforest restoration. PhD thesis, School of Integrative Biology, University of Queensland, Brisbane

Dove MR (1993) Smallholder rubber and swidden agriculture in Borneo: a sustainable adaption to the ecology and economy of the tropical forest. Econ Bot 47:136–147

Driscoll P (1984) The effects of logging on bird populations in lowland New Guinea rainforest. PhD thesis, Zoology Department University of Queensland, Brisbane

Driscoll PV, Kikkawa J (1989) Bird species diversity of lowland tropical rainforests of New Guinea and Australia. In: Hamelin-Vivien ML, Bourliere F (eds) Vertebrates in Complex Tropical Systems. Springer-Verlag, New York, pp 123–152

Durno J, Deetes T, Rajchaprasit J (2007) Natural forest regeneration from an Imperata fallow: The case of Pakhasukjai. In: Cairns M (ed) Voices from the forest. Resources for the future, Washington, DC, pp 122–136

Elliott S, Blakesley D, Maxwell JF, Doust S, Suwannaratana S (2006) How to plant a forest: The principles and practice of restoring tropical forests. Biology Department, University of Chiang Mai, Chiang Mai

Erskine PD, Catterall CP, Lamb D, Kanowski J (2007) Patterns and processes of old-field reforestation in Australian rain forest landscapes. In: Cramer VA, Hobbs RJ (eds) Old fields: Dynamics and restoration of Abandoned Farmland. Island Press, Washington, DC, pp 119–144

Finegan B (1992) The management potential of neotropical secondary lowland rain forest. Forest Ecol Manage 47:295–332

Fox JED (1976) Constraints on the natural regeneration of tropical moist forest. Forest Ecol Manage 1:37–65

Friday K, Drilling ME, Garrity DP (1999) Imperata grassland rehabilitation using agroforestry and assisted natural regeneration. International Center for Research in Agroforestry, Bogor

Fujisaka S, Wollenberg E (1991) From forest to agroforest and logger to agroforester: a case study. Agroforestry Syst 14:113–129

Garcia-Fernandez C, Casado MA (2005) Forest recovery in managed agroforestry systems: The case of benzoin and rattan gardens in Indonesia. Forest Ecol Manage 214:158–169

Gardner TA, Barlow J, Parry LW, Peres CA (2007) Predicting the uncertain future of tropical forest species in a data vacuum. Biotropica 39:25–30

Garwood NC (1989) Tropical soil seed banks, a review. In: Leck MA, Parker VT (eds) Ecology of soil seed banks. Academic, San Diego, CA, pp 149–209

Geddes WR (1976) Migrants of the mountains: The cultural ecology of the Blue Miao (Hmong Njua) of Thailand. Clarendon Press, Oxford

Giambelluca TW (2002) Hydrology of altered tropical forest. Hydrol Process 16:1665–1669
Gibson CC, Williams J, Ostrom E (2004) Local enforcement and better forests. World Dev 33:273–284
Gilmour DA, Fisher RJ (1991) Villagers, forests and foresters. Sahayogi Press, Kaṭhmandu
Gouyon A, de Forests H, Levang P (1993) Does 'jungle rubber' deserve its name? An analysis of rubber agroforestry systems in southeast Sumatra. Agroforestry Syst 22:181–206
Hardwick K, Healy J, Elliott S, Garwood NC, Anusarnsunthorn V (1997) Understanding and assisting natural regeneration in northern Thailand. Forest Ecol Manage 99:203–214
Hau CH (1997) Tree seed predation in degraded hillsides in Hong Kong. Forest Ecol Manage 99:215–221
Heinimann A, Messerli P, Schmid-Vogt D, Wiesmann U (2007) The dynamics of secondary forest landscapes in the lower Mekong basin: a regional-scale analysis. Mountain Res Dev 27:232–241
Hoare P (2004) A process for community and government cooperation to reduce the forest fire and smoke in Thailand. Agricult, Ecosyst Environ 104:35–46
Hobbs RJ, Higgs E, Harris JA (2009) Novel ecosystems: implications for conservation and restoration. Trends Ecol Evol 24:599–605
Hobley M (1996) Participatory forestry: The process of change in India and Nepal. Overseas Development Institute, London
Holl KD (2002) Effect of shrubs on tree seedling establishment in an abandoned tropical pasture. J Ecol 90:179–187
Holl KD, Lulow ME (1997) Effect of species, habitat, and distance from edge on post-dispersal seed predation in a tropical rainforest. Biotropica 29:459–468
Holscher D, Mackensen J, Roberts J-M (2005) Forest recovery in the humid tropics: changes in vegetation structure, nutrient pools and the hydrologic cycle. In: Bonell M, Bruijnzeel LA (eds) Forests, water and people in the humid tropics: Past, present and future hydrological research for integrated land and water management. UNESCO and Cambridge University Press, Cambridge, pp 598–621
Hughes CE, Jones RJ (1998) Environmental hazards of leucaena. In: Shelton HM, Gutteridge RC, McMullen BF, Bray RA (eds) Leucaena: Adaptation, quality and farming systems. Australian Center for International Agricultural Research, Canberra, pp 61–70
Hviding E, Bayliss-Smith T (2000) Islands of rainforest: Agroforestry, logging and eco-tourism in Solomon Islands. Ashgate, Aldershot
Hynes RA, Chase AK (1982) Plants, sites and domiculture: aboriginal influence upon plant communities in Cape York Penisular. Archeol Oceania 17:38–50
Isager L, Ivarsson S (2002) Contesting landscapes in Thailand: Tree ordination as counter-territorialization. Crit Asian Stud 34:395–417
ITTO (2002) ITTO guidelines for the restoration, management and rehabilitation of degraded and secondary tropical forests. ITTO Policy Development Series No 13. International Tropical Timbers Organization, Yokohama
Jankowska-Błaszczuk M, Grubb PJ (2006) Changing perspectives on the role of the soil seed bank in northern temperate deciduous forests and in tropical lowland rain forests: parallels and contrasts. Perspect Plant Ecol Evol Syst 8:3–21
Jobbagy EG, Jackson RB (2000) The vertical distribution of soil organic carbon and its relation to climate and vegetation. Ecol Appl 10:423–436
Katz E, Garcia C, Goloubinoff M (2002) Sumatran benzoin (Styrax spp.). In: Shanley P, Pierce AR, Laird SA, Guillen A (eds) Tapping the green market: Certification and management of non-timber forest products. London, Earthscan, pp 246–256
Kuusipalo J, Hadengganan S, Adjers G, Sagala APS (1997) Effect of gap liberation on the performance and growth of dipterocarp trees in a logged-over rainforest. Forest Ecol Manage 92:209–219
Lamb AFA (1969) Artifical regeneration within the humid tropical lowland forest. Commonw Forestry Rev 48:41–53
Lamb D (1990) Exploiting the tropical rainforest: An account of pulpwood logging in Papua New Guinea. UNESCO and the Parthenon Publishing Group, Paris and Carnforth

Lasco RD, Visco R, Puhlin J (2001) Secondary forests in the Philippines: Formation and transformation in the 20th century. J Tropic Forest Sci 13:652–670

Laurance WF (2007) Have we overstated the tropical biodiversity crisis? Trends Ecol Evol 22:65–70

Leslie AJ (1989) Review of forest management systems of tropical Asia: Case studies of natural forest managment for timber production in India, Malaysia and the Philippines. Food Agric Organ United Nations, Bangkok

Letcher S, Chazdon RL (2009) Lianas and self-supporting plants during tropical forest succession. Forest Ecol Manage 257:2150–2156

Li Y, Xu M, Zou X, Shi P, Zhang Y (2005) Comparing soil organic carbon dynamics in plantation and secondary forest in wet tropics in Puerto Rico. Global Change Biol 11:239–248

Lugo AE (2009) The emerging era of novel tropical forests. Biotropica 41:589–591

Lugo AE, Brown S (1993) Management of tropical soils as sinks or sources of atmospheric carbon. Plant Soil 149:27–41

Maneeratana B, Hoare P (2007) When shifting cultivators migrate to the city, how can the forest be rehabilitated? In: Cairns M (ed) Voices from the Forest: Integrating Indigenous Knowledge into Sustainable Upland Farming. Resources Future, Washington, DC, pp 137–141

Marin-Spiotta E, Cusack D, Ostertag R, Silver W (2008) Trends in above and below ground carbon with forest regrowth after agricultural abandonment in the neotropics. In: Myster RW (ed) Post-agricultural succession in the Neotropics. Springer, New York, pp 22–72

Mayaux P, Holmgren P, Achard F, Eva H, Stibig H, Branthomme A (2005) Tropical forest cover change in the 1990s and options for future monitoring. Phil Trans R Soc B Biol Sci 360:373–384

Meehan HJ, McConkey KR, Drake D (2002) Potential disruptions to seed dispersal mutualisms in Tonga, Western Polynesia. J Biogeogr 29:695–712

Mesquita RCG, Ickes K, Ganade G, Williamson GB (2001) Alternative successional pathways in the Amazon Basin. J Ecol 89:528–537

Michon G (2005) Domesticating forests: How farmers manage forest resources. Institut de Recherchepour le Developpement, Center for International Forestry Research, World Agroforestry Center, Paris, Bogor

Mohamed AH, Othman AR (2003) Rehabilitation of Malaysian forests: perspectives and delimination of planting bamboo as a commercial species. In: Sim HC, Appanah S, Durst PB (eds) Bring back the forests: Policies and practices for degraded lands and forests. Food and Agriculture Organisation of United Nations, Bangkok, pp 99–106

Momberg F, Puri R, Jessup T (2000) Exploitation of Gagaru, and forest conservation efforts in the Kayan Mentarang National park, East Kalimantan. In: Zerner C (ed) People, plants and justice: The politics of nature conservation. Columbia University Press, New York, pp 259–284

Mouhot H (1966) Diary: travels in the central parts of Siam, Cambodia and Laos during the years 1858-61. Oxford University Press, Kuala Lumpur

Moura-Costa P, Yap SW, Ong CL, Ganing A, Nussbaum R, Mojium T (1996) Large scale enrichment planting with dipterocarps as an alternative for carbon offset - methods and preliminary results. In: Appanah S, Khoo KC (eds) Proceedings of the 5th round table conference on dipterocarps. Forest Research Institute, Malaysia, Kuala Lumpur, pp 386–396

Murphy PG, Lugo AE (1986) Ecology of tropical dry forest. Ann Rev Ecol Syst 17:67–88

Muscarella R, Fleming TH (2007) The role of frugivorous bats in tropical forest succession. Biol Rev 82:573–590

Nepstad D, Uhl C, Serrão EAS (1991) Recuperation of a degraded Amazonian landscape: forest recovery and agricultural restoration. Ambio 20:248–255

Ng FSP (1996) High quality planting stock – has research made a difference? Center for International Forestry Research, Bogor

Norden N, Chazdon R, Chau A, Jiang Y-H, Vilchez-Alvarado B (2009) Resilience of tropical rain forests: Tree community assembly in secondary forests. Ecol Lett 12:385–394

Norton DA (2009) Species invasions and the limits to restoration: learning from the new Zealand experience. Science 325:569–571

Osunkoya OO (1994) Post dispersal survivorship of north Queensland rainforest seeds and fruits: Effects of forest, habitats and species. Austr J Ecol 19:52–64

Pinard MA, Putz FE (1996) Retaining forest biomass by reducing logging damage. Biotropica 28:278–295

Post WM, Kwon KC (2000) Soil carbon sequestration and land-use change: processes and potential. Global Change Biol 6:317–328

Potter LM (1997) The dynamics of Imperata: Historical overview and current farmer perspectives, with special reference to South Kalimantan, Indonesia. Agroforestry Syst 36:31–51

Potter LM (2001) Agricultural intensification in Indonesia: Outside pressures and indigenous strategies. Asia Pacific Viewpoint 42:305–324

Potter L, Badcock S (2006) Can Indonesia's complex agroforests survive globalisation and decentralisation? In: Connell J, Wadell E (eds) Environment, development and change in rural Asia-Pacific: between local and global. Routledge, London and New York, pp 167–185

Richards PW (1952) The tropical rain forest. Cambridge University Press, Cambridge

Roder W, Phengchanh S, Keoboualapha B, Maniphone S (1995) *Chromolaena odorata* in slash-and-burn rice systems of Northern Laos. Agroforestry Syst 31:79–92

Romell E, Hallsby G, Karlsson A, Garcia C (2008) Artificial canopy gaps in a Macaranga spp. dominated secondary tropical rain forest – effects on survival and above ground increment of four under-planted dipterocarp species. Forest Ecol Manage 255:1452–1460

Santasomsat Y (2003) Biodiversity, local knowledge and sustainable development. Regional center for social sciences and sustainable development. Chiang Mai University, Chiang Mai

Sasaki A, Takeda S, Kanzaki M, Ohta S, Preechapanya P (2007) Population dynamics and land use changes in a Miang (chewing tea) village in northern Thailand. Tropics 16:75–85

Saunders A, Norton DA (2001) Ecological restoration at Mainland Islands in New Zealand. Biol Conserv 99:109–119

Schroth G, Harvey C, Vincent G (2004) Complex agroforests: Their structure, diversity and potential role in landscape conservation. In: Schroth G, da Fonseca G, Harvey C, Gascon C, Vasconcelas H, Isac A-M (eds) Agroforestry and biodiversity conservation in tropical landscapes. Island Press, Washington, DC, pp 227–260

Scott R, Pattanakaew P, Maxwell JF, Elliott S, Gale G (2000) The effect of artificial perches and local vegetation on bird dispersing seed deposition into regenerating sites. In: Elliott S, Kerby J, Blakesley D, Harwick K, Woods K, Anusarnsunthorn V (eds) Forest restoration for wildlife conservation. University of Chiang Mai, Chiang Mai, pp 326–337

Sheil D, Burslem DFRP (2003) Disturbing hypotheses in tropical forests. Trends Ecol Evol 18:18–26

Shilton LA, Altringham JD, Compton SG, Whittaker RJ (1999) Old world fruit bats can be long distance seed dispersers through extended retention of viable seeds in the gut. Proc R Soc London B Biol Sci 266:219–223

Silver WL, Ostertag R, Lugo AE (2000) The potential for carbon sequestration through reforestation of abandoned tropical agricltural and pasture lands. Restoration Ecol 8:394–407

Slik JWF, Bernard CS, Van Beek M, Bremen FC, Eichorn KAO (2008) Tree diversity, composition, forest structure and aboveground biomass dynamics after single and repeated fire in a Bornean rain forest. Oecologia 158:579–588

Stocker GC (1981) Regeneration of a North Queensland rain forest following felling and burning. Biotropica 13:86–92

Stocker GC (1988) Tree species diversity in rainforest – establishment and maintenance. Proc Ecol Soc Aust 15:39–47

Sunderland TCH, Dransfield J (2002) Rattan (various species). In: Shanley P, Pierce AR, Laird SA, Guillen A (eds) Tapping the green market: Certification and management of non-timber forest products. Earthscan, London

Taylor RH (1957) Plant succession on recent volcanoes in Papua. J Ecol 45:233–243

Thiollay JM (1995) The role of traditional agroforests in the conservation of rain forest bird diversity in Sumatra. Conserv Biol 9:335–353

Toh I, Gillespie M, Lamb D (1999) The role of isolated trees in facilitating tree seedling recruitment at a degraded sub-tropical rainforest site. Restoration Ecol 7:288–297

Toma T, Ishida A, Matius P (2005) Long-term monitoring of post-fire above ground biomass recovery in a lowland dipterocarp forest in East Kalimantan, Indonesia. Nutr Cycl Agroecosyst 71:63–72

Tracey JG (1982) The vegetation of the humid tropical region of Queensland. CSIRO, Melbourne

Tuomela K, Kuusipalo J, Vesa L, Nuryanto K, Sagala APS, Ådjers G (1996) Growth of dipterocarp seedlings in artificial gaps: An experiment in a logged-over rainforest in South Kalimantan, Indonesia. Forest Ecol Manage 81:95–100

van Nieuwstadt MGL, Sheil D, Kartawinata K (2001) The ecological consequences of logging in burned forest in East Kalimantan, Indonesia. Conserv Biol 15:1183–1186

van Ooschot G, van Winkel I, Moura-Costa P (1996) The use of GIS to study the influence of site factors in enrichment planting with dipterocarps. In: Appanah S, Khoo KC (eds) Proceedings of the 5th Round Table Conference On Dipterocarps Forest Research Institute, Malaysia, Kuala Lumpur, pp 267–278

Vest PA, Westoby M (2004) Sprouting ability across diverse disturbances and vegetation types worldwide. J Ecol 92:310–320

Viera DLM, Scariot A (2006) Principles of natural regeneration of tropical dry forest for restoration. Restoration Ecol 14:11–20

Webb LJ (1958) Cyclones as an ecological factor in lowland tropical rain-forests, North Queensland. Austr J Bot 6:220–228

Webb LJ, Tracey GT, Williams WT (1972) Regeneration and pattern in the subtropical rainforest. J Ecol 60:675–695

Weinstock JA (1983) Rattan: Ecological balance in a Borneo rainforest swidden. Econ Bot 37:58–68

White KJ (1976) Notes on enrichment planting in lowland rain forests of Papua New Guinea. Tropical Forest Research Notes. Office of Forests, Papua New Guinea, Port Moresby

Whitmore TC (1984) Tropical rain forests of the far East. Clarendon Press, Oxford

Wibowo A, Suharti M, Sagala APS, Hibani H, van Noordwijk M (1997) Fire managment in Imperata grasslands as part of agroforestry development in Indonesia. Agroforestry Syst 36:203–217

Williams SE, van Noordwijk M, Penot E, Healey JR, Sinclair JR, Wibawa G (2001) On-farm evaluation of the establishment of clonal rubber in multistrata agroforests in Jambi, Indonesia. Agroforestry Syst 53:227–237

Wollenberg E, Ingles A (1998) Incomes from the forest. Center for International Forestry Research and the International Union for the Conservation of Nature, Bogor

Woods P (1989) Effects of logging, drought, and fire on structure and composition of tropical forests in Sabah, Malaysia. Biotropica 21:290–298

Wright SJ, Muller-Landau HC (2006) The future of tropical forest species. Biotropica 38:287–301

Wunderle J (1997) The role of animal seed dispersal in accelerating native forest regeneration on degraded tropical lands. Forest Ecol Manage 99:223–235

Zimmermann B, Elsenbeer H, De Moraes JM (2006) The influence of land-use changes on soil hydraulic properties: Implications for runoff generation. Forest Ecol Manage 222:29–38

Chapter 6
Monocultural Plantings

> Prospects for further development in plantation forestry are most promising. Incredible growth rates, particularly in tropical regions, have been obtained, sometimes with species which have done poorly in their native habitat.
>
> Westoby 1961 (in Westoby 1987, p. 37)

> Unless the world is more or less to go without choice cabinet hardwoods these will have to be planted on a much greater scale than at present, as they are currently not favoured by currently practiced silvicultural systems.
>
> (Whitmore 1984, p. 285)

Introduction

Natural forest regrowth is an attractive form of reforestation but it is not always possible to achieve. This is especially true where forests have been cleared and replaced by large expanses of grasslands that are regularly burned. It is even more difficult when lands have been heavily degraded and soil conditions have changed or where any remaining patches of natural forest are so scarce so that tree seeds must be dispersed over long distances. Under these circumstances tree-planting is a more reliable form of reforestation. Tree plantations do have some advantages and one of these is their capacity to generate greater financial returns than natural regeneration. There are a number of reasons for this. One is because the species used can be chosen to meet specific needs or markets. Likewise, plantations can be established in locations that reduce transport and other costs (e.g. flatter ground at sites close to good roads, markets or ports). Finally, the productivity of plantations is usually much greater than most regenerating natural forests because of site preparation and fertilizer use. These advantages mean that plantations can be very efficient and profitable producers of goods such as timber and therefore be attractive ways in which to carry out reforestation.

However, there are many types of plantations that might be used. These differ in the types of species utilised, the numbers of species planted, the length of time they

D. Lamb, *Regreening the Bare Hills*, World Forests 8,

DOI 10.1007/978-90-481-9870-2_6,

are grown before being felled and the way the trees are managed during this period. Not surprisingly, these various types of plantations differ in their capacity to produce particular products or specific ecosystem services. This chapter begins the discussion of planted forests by considering monocultural plantations. These are the most commonly used form of plantation and, as the name suggests, are essentially the same as most agricultural crops in that just a single species is planted. The identity of this species is obviously critical – but which one should be chosen from the many that are usually available and what factors should influence this choice? This chapter explores the question of species choice, the ways that simple monocultures might be managed and the capacity of these plantations to generate various goods and ecosystem services.

Reasons for Establishing Plantations

Before discussing monocultural plantations it is useful to consider the reasons why trees are grown and what types of plantation are likely to appeal to different growers. There are, broadly speaking, four groups of prospective tree-planters. All of them may have decided to reforest their land but they are likely to have contrasting reasons for doing so and, therefore, differing views on how it should be done. Only some will be interested in plantation monocultures.

Private Industrial Growers

Plantation monocultures are usually favoured by privately-owned timber companies. These companies grow trees for industrial purposes and do so to create a new timber resource at the cheapest possible cost. They usually have a specific and clearly defined market in mind and this market often requires timber with certain prescribed properties whether the trees are being grown for pulpwood, veneer or for sawlogs. Consequently, most of these growers are usually only interested in growing a single tree species in a plantation monoculture. Ideally, logs should be uniform in size so that machinery can be designed to handle this specific size. Again, this requirement usually limits the number of species grown in a particular plantation estate. Industrial efficiency demands that the productivity (timber volume produced per hectare per year) be maximized and, for this reason, species that can be grown on shorter rotations are likely to be far more attractive than those grown on long rotations. Operational efficiency requires that timber coming from the plantations should be produced in a regular and predictable way, since industrial mills cannot handle erratic or episodic supplies. This means plantations should be relatively large in area. These growers usually prefer a few large contiguous areas of plantation rather than many small scattered plantations, since harvesting costs will then be lower. The capacity of industrial plantations to produce ecosystem services as well as timber is not usually a major consideration although most industrial plantation managers would aim at limiting erosion and protecting watersheds.

State Forestry Agencies

The objectives of state forestry agencies are often similar to those of many industrial growers since most state-owned plantations are established to create a new industrial timber resource. These plantations are almost invariably monocultures and, at least in the past, have been grown on long rotations to produce sawlogs. However, unlike industrial companies, state-owned plantations are sometimes established without a precise market in mind. In such cases, the intention has been to create a timber resource in the expectation that it will attract an industrial user to the area once the plantation is mature. That is, the objective of reforestation has often had less to do with maximizing a return on the funds committed and rather more to do with fostering rural development and employment opportunities. Government owned plantations may be large or small, depending on the availability of land, since most government forestry agencies are rarely able to access good quality land (these being allocated to agricultural users). Some state forestry agencies also plant trees to protect degraded watersheds but these generally use the same species and silvicultural methods as used in commercial plantations simply because the techniques are known to them.

Smallholders and Community Forestry Groups

Smallholders are likely to have a rather more diverse set of objectives than industrial growers or state forestry agencies. Some grow trees in plantation monocultures, especially if there is an established market for the timber of a particular species. Those with land near the plantations of a large industrial company may also be able to act as out-growers for that company using the same species and growing their plantations on fixed rotations following company prescriptions. But other smallholders not in this situation may choose to grow a wider variety of species including timber trees and species able to produce foods or medicines. These growers may favour either monocultural plantings or plantations involving species mixtures. And, unlike industrial growers, many of these smallholders may grow their trees without a fixed rotation period in mind. Instead, they may fell their trees on an opportunistic basis when cash is needed such as for a wedding or some other family occasion. In this sense, trees represent a form of savings. The time of harvest is also likely to be dictated by the size of individual trees rather than the overall stand productivity since the markets smallholders sell into usually pay more for bigger trees.

Community groups undertaking reforestation will often have similar attitudes to those of most individual smallholders although these groups may also create new forests for religious reasons, to protect watersheds, or to consolidate the group's landownership claims (something that may be also done by individual households).

Special Purpose Groups

Finally, there are a number of organizations or groups who are interested in reforestation for purposes other than timber production. Rather than having a fixed rotation in mind these growers are likely to prefer to establish new forests that are self-regenerating. Most of these groups will also prefer to use a form of reforestation other than a simple monoculture because they are interested in generating ecosystem services rather than forest products and believe more diverse plantings are a better way of achieving this. Some groups may wish to 'rehabilitate' degraded land such as former mine sites or to stabilise soils to protect watersheds. The environmental conditions at these sites often mean that exotic species dominate these plantings. Others, such as conservation groups, might be primarily interested in Ecological Restoration and use mostly native species because their aim is to restore, as far as possible, the original biodiversity.

The consequence of these differing objectives is that quite different silvicultural techniques and species may be needed by different growers. There are now established forms of silviculture for large-scale industrial plantations (Evans and Turnbull 2004; Krishnapillay 2002; Weinland 1998) but rather fewer for small-scale timber growers or restoration ecologists (though see Bristow et al. (2005) and Elliott et al. (2006)).

Implementing Reforestation on Degraded Lands

Many sites available for reforestation have impediments that limit plant growth. Soil conditions are one of the most common constraints. Constraints imposed by soil physical properties are usually easily identified and can be corrected by ripping, ploughing or mounding soils to improve penetrability or drainage. Such tasks are easily carried out by industrial forestry organizations but are obviously harder for smallholders lacking heavy machinery. Nonetheless, difficult soil conditions have been overcome by simply digging large planting holes and filling these by scraping in surrounding topsoils. Whisenant (1999) reviews some of the other methods by which soil physical conditions can be improved at degraded sites.

Soil chemical deficiencies are usually more difficult to remedy. The first problem is to identify which nutrient is deficient (or present in toxic concentrations). Sometimes visual symptoms give an indication. For example, Webb et al. (2001) provide a number of colour plates showing the symptoms of various nutrient deficiencies in *Swietenia macrophylla* and *Cedrela odorata* while Dell et al. (2001) have done the same for eucalypts. These can be useful first steps in identifying deficiencies but the absence of visual symptoms does not necessarily mean there are no nutrient constraints. Visual symptoms are also less useful when there are multiple nutrient deficiencies.

Apart from visual symptoms the most common approach to identifying deficiencies is to assess the chemical properties of successive soil horizons. Since large numbers of samples are needed, both across the site and at several depths, this can be an expensive undertaking. Such analyses often reveal that degraded soils lack organic matter, have low cation exchange capacities and are very acid. This means the soils are deficient in many nutrients, are less able to retain nutrients supplied in fertilizers and that elements such as phosphorus may be limiting while the concentrations of others, such as aluminium and manganese, may be approaching toxic levels.

Soil chemical analyses are useful in indicating potential nutritional problems but can be difficult to interpret because tree species differ markedly in their tolerance of various soil conditions. Soil phosphorus is an especially difficult problem because there is a difference between total soil phosphorus and that which is actually available to different plant species. The availability of phosphorus depends on soil pH and on the properties of the root rhizosphere. This means soil analyses provide a useful guide but need to be complemented by other techniques that identify which nutrient is actually limiting plant growth.

In addition to visual symptoms and soil analyses, two other approaches are commonly used to identify deficiencies. One is to analyse foliar samples to determine which elements appear to be present in concentrations lower than recognized threshold levels. This approach has been widely used (Dreschel and Zech 1991; Miller 1986; Reuter and Peterson 1997; Specht and Turner 2006) but its success depends on samples being collected according to strict protocols. These protocols include the requirement that only young but fully expanded leaves are sampled; that these samples be taken from upper parts of the crown and preferably involve sun leaves; and that the samples be dried (not more than 70°C) as soon as possible after collection. Although there are tables with suggested thresholds or critical levels below which deficiencies may be occurring (e.g. Miller 1986; Reuter and Robinson 1997) these are not absolute and can vary with species. Because comparatively little is known about many tropical tree species it can be useful to supplement the notional deficiency levels in these tables by sampling and comparing unhealthy and seemingly healthy trees of the same species.

An example of the first approach is shown in Table 6.1 which contains foliar analyses of a number of species growing in young plantation in northern Vietnam. All sample leaves were young, fully expanded sun leaves. These data show large variations in the levels of foliar nutrients. Using published thresholds as indicators it seems that there are multiple nutrient deficiencies affecting trees in these plantations. Every species has foliar phosphorous concentrations below those normally considered adequate while a number of species appeared to also have low nitrogen, potassium, calcium, copper and boron concentrations at some sites. Most of the soils in the region have very acid pH levels and this may have been the reason why some species appeared to have toxic levels of manganese. On the other hand,

Table 6.1 Range of foliar nutrient levels (minimum and maximum values) found in species in established plantations in northern Vietnam. Figures in *bold* are below nominal critical levels or, in the case of manganese, exceed the supposed toxic level (Lamb and Huynh 2006)

	N (wt%)		P (wt%)		K (wt%)		Ca (wt%)	
	Min	Max	Min	Max	Min	Max	Min	Max
Acacia auriculiformis	2.29	2.29	**0.07**	0.07	0.82	0.824	1.42	1.42
Acacia mangium	2.99	3.63	**0.10**	0.19	0.59	2.376	**0.17**	0.37
Canarium album	1.83	2.53	**0.12**	0.20	0.83	1.480	0.24	0.57
Castanopsis fissa	**1.45**	1.89	**0.06**	0.09	0.48	0.713	**0.15**	0.56
Chuckrasia tabularis	**1.38**	2.43	**0.10**	0.23	**0.37**	1.235	0.44	0.96
Cinnamomum cassia	**1.10**	1.44	**0.06**	0.10	0.40	1.129	0.27	1.43
Cinnamomum iners	**0.81**	1.73	**0.05**	0.13	**0.16**	0.702	0.48	0.76
Dracontomelum duperraneum	2.05	2.32	**0.15**	0.17	1.01	1.075	0.41	2.32
Endospermum chinensis	2.23	2.23	**0.17**	0.17	1.59	1.592	0.52	0.52
Erythrophloem fordii	2.08	6.30	**0.09**	0.41	**0.26**	1.225	**0.07**	2.50
Mangletia glauca	**1.13**	2.46	**0.08**	0.15	0.70	2.172	0.35	0.91
Michelia mediocris	2.49	3.75	**0.11**	0.18	0.95	1.494	0.36	0.64
Peltophorum tonkinenis	1.92	4.02	**0.08**	0.14	**0.39**	1.226	**0.12**	0.43
Pinus massoniana	1.22	1.60	**0.08**	0.09	**0.38**	1.137	0.37	0.69
Limiting values	1.8		0.18		0.40		0.20	
	Mn (mg/kg)		Cu (mg/kg)		B (mg/kg)		Zn (mg/kg)	
	Min	Max	Min	Max	Min	Max	Min	Max
Acacia auriculiformis	120	119	14	14	45	45	15	15
Acacia mangium	87	668	10	24	**10**	55	26	99
Canarium album	**13**	38	7	23	**13**	37	25	100
Castanopsis fissa	225	***3,042***	**4**	12	**12**	57	22	89
Chuckrasia tabularis	31	1,556	**4**	33	**1**	28	20	221
Cinnamomum cassia	470	1,861	**4**	11	19	43	16	56

Cinnamomum iners	455	***3,393***	**3**	37	20	66	13	140
Dracontomelum duperraneum	168	190	14	41	13	31	32	71
Endospermum chinensis	3,746	***3,746***	**5**	5	22	22	25	25
Erythrophloem fordii	40	426	**6**	34	**12**	92	**10**	169
Mangletia glauca	206	***3,547***	13	30	15	48	18	221
Michelia mediocris	113	2,332	12	31	22	95	16	475
Peltophorum tonkinenis	78	1,088	9	14	15	36	22	108
Pinus massoniana	250	336	**5**	13	**9**	33	48	329
Limiting values	20	*1,200*	2–5		15		10	

aluminium concentrations were not excessive even though these, too, can be affected by acid pH levels. The table also shows that species differed in their ability to utilize these soils and that some were more tolerant than others. For example, many species had multiple deficiencies including *Acacia mangium*, *Canarium*, *Castanospermum*, *Chukrasia*, *Cinnamum cassia*, *C. iners*, *Erythrophloeum*, *Peltophorum* and *Pinus* but others such as *Acacia auriculiformis* and *Michelia mediocris* were, seemingly, only limited by phosphorus. Some caution is needed in using foliar analyses because so little is known about the nutritional requirements of these particular species.

A second approach sometimes used to confirm such preliminary diagnoses are nutrient omission pot trials (Asher et al. 2002). In this case seedlings of the plantation species are grown in pots containing soils from the plantation area. Several fertilizer treatments are applied to these pots. One treatment adds a complete nutrient solution to the soil (the 'all on' treatment). A second adds a complete nutrient solution but one that lacks nitrogen ('all on – N'). A third adds a complete solution lacking phosphorus ('all on – P'). In this way successive treatments can work through each essential element. Finally, the trial is balanced by a control treatment that has no added nutrient solution. Deficiencies can be identified by comparing the growth of seedlings in each treatment. Nitrogen is unlikely to be limiting plant growth if the seedlings in the 'all on – N' treatment grow as well as those in the complete ('all on') treatment. This suggests the soil alone was able to supply sufficient for the plant's needs. On the other hand, if plants in the 'all on – P' treatment did not grow as well as those in the 'all on' treatment there may be a phosphorus deficiency since the soil alone was unable to make up the difference. Figure 6.1 shows results from a pot trial using soils from the same area from which the foliar samples in Table 6.1 were collected. Seedlings of *Eucalyptus urophylla* were grown in pots for 3 months. The seedlings were given an 'all on' treatment as well as selective omissions of the main nutrients. The results support the suggestion that *Eucalyptus urophylla* is limited by nutrient deficiencies (compare all-on and no nutrients) when grown in these soils. They also support the suggestion that phosphorus is deficient but that the seedlings are able to obtain sufficient nitrogen from these soils.

Taken together, soil analyses, visual symptoms, foliar analyses and nutrient omission pot trials can suggest which nutrients are likely to be limiting for plant growth in the field. However, they will not reveal how much of each nutrient will be needed to overcome the deficiency, the frequency at which this should be applied nor the best form of fertilizer in which the nutrient should be applied. Such important details require carefully designed field fertilizer trials. Industrial plantation companies and state forestry agencies can afford such research if they are to embark on a large scale reforestation program but small holders cannot. In most cases the best that can be done is to apply farm yard manure or compost at the time seedlings are planted or to use an arbitrary dose of an all-purpose NPK fertilizer in the hope that this overcomes the most commonly occurring deficiencies such as nitrogen and phosphorus.

Acid soils are widespread in many tropical areas and are especially difficult to deal with (Marcar and Khanna 1997; Myers and De Pauw 1995). Not only are they

Fig. 6.1 Omission pot trial using *Eucalyptus urophylla* to identify soil nutrient deficiencies. The seedling on the far left has been grown in soil from the site and been given a fertilizer application that includes all essential elements. The seedling second from the left has received no additional fertilizers and is growing less well. The seedling on the far right has received all fertilizers except nitrogen and is clearly able to get sufficient nitrogen from the soil alone. The seedling second from the right has received all fertilizers but phosphorus and is not able to get enough phosphorus from the soil. See text for details

associated with low levels of available phosphorus, calcium and magnesium but also with high concentrations of forms of aluminium and manganese that are toxic to plants. Additions of lime can change the soil pH and so reduce some of these problems. However, it is difficult to add the lime in a way where it is able to move beyond the top few centimetres of soil. Under these circumstances the best alternative for many tree growers may be to use a species that can tolerate the particular soil conditions at their site (i.e. change the species to suit the site rather than trying to modify the site to suit a particular species). Certain tolerant exotic species (especially putative nitrogen fixers such as *Acacia* spp.) may have particular advantages here (Marcar and Khanna 1997). Over time litterfall and nutrient cycling in plantations of such species may help improve topsoil organic matter thereby overcoming some fertility problems making it possible to grow a wider range of species.

Of the biological changes that may develop in degraded soils the most likely to occur is a loss of mycorrhiza. This might occur when topsoil is eroded or because of changes induced by the agricultural crops previously growing at the site. Many mycorrhizae are lost when the host trees are removed (Whisenant 1999). In most cases these changes can be overcome by ensuring tree seedlings are inoculated in the nursery prior to being planted in the field. This is usually done by simply adding a little natural forest topsoil to potting mixes, or topsoil from successful plantations of the particular species. There has been speculation that some species of the family Dipterocarpaceae are particularly sensitive to mycorrhizal problems when they are being replanted at deforested sites but this may depend on the degree of degradation that has occurred at particular sites as well as the species of dipterocarp being planted and the situation needs clarification (Lee 1998).

Degradation can also affect the populations of nodule-forming bacteria such as *Rhizobium* and so reduce the capacity of nitrogen-fixing plants to improve soil nitrogen levels. In some cases it may be sufficient to inoculate seedlings in the nursery, as with mycorrhiza, prior to them being planted in the field. In cases where acid soils are common, it may be necessary to identify strains that are tolerant of these pH levels (Kahindi et al. 1997; Myers and De Pauw 1995).

Most nutrient problems are evident from the earliest stages of plantation development but some can also develop at a later stage. An example of this is illustrated in Fig. 6.2 where regular collections of leaves and twigs from the forest floor can cause a drain of nutrient-rich material from the site. If this is carried out over a long period the losses may lead to a reduction in tree growth. Nutrient losses can also occur during the transition from the first to the second rotation. These can occur in two ways. The first is in timber and bark removed from the site in logs. Sapwood usually has a higher nutrient concentration than heartwood so the sapwood-rich young logs removed in pulpwood harvesting can result in a very large nutrient drain. The second cause of nutrient loss occurs because of site preparation when logging debris such

Fig. 6.2 Constant gathering of twigs and leaves from under a eucalypt plantation in Vietnam is imposing a long-term nutrient drain on this plantation ecosystem and is limiting tree growth

as leaves and branches are burned to clear an area prior to replanting. Some nutrients such as phosphorus remain on the site in ash but nitrogen is usually volatilized by fire and is lost from the site. Second rotation productivity declines are often induced by such losses and fertilizer applications are needed to compensate for the losses (Evans and Turnbull 2004; Nambiar and Brown 1997).

The Particular Case of Mine Site Rehabilitation

Old mine areas represent a rather special type of degraded site. Plant establishment and growth can be limited by unstable topography as well as by the physical and chemical properties of soils at these sites (Table 6.2). Some sites have mine wastes left piled in unstable steep mounds. These new topographic arrangements can be more prone to erosion and also lead to changes in the way water flows across the site. However, the main problems usually concern the soils that are left behind. Prior to mining the soil at a site will be represented by a surface mantle (say the top 5 m) of pedogenic material including an organically enriched surface (A) horizon overlying a sub-soil B horizon containing relatively little organic matter. This mantle lies above a zone of weathered rock (the C horizon). Open cut mines remove and stockpile all this soil and rock material (often referred to as the 'overburden')

Table 6.2 Common problems at former mine sites

Factor	Reasons
Slope	Tailings or other mine wastes left with steep slopes are unstable and prone to erosion.
Topography	Cause changes in drainage patterns and surface hydrology.
Soil texture	Coarser textured soils (e.g. with crushed rock) more easily leached and have lower water storage capacity. Fine textured soils may have lower water infiltration capacities.
Soil compaction and crusting	Extent depends on texture and movement of heavy machinery. Will impede plant growth. Crusting can develop when sodic materials are stored.
Surface soil temperatures	Likely to be higher in absence of leaf litter.
Soil organic matter	Low organic matter levels reduce nutrient supply but also affect mycorrhiza and levels of root nodule bacteria (e.g. *Rhizobium*).
Soil nutrient deficiencies	A variety of possible deficiencies including N and P. Worsened when topsoils are diluted when mixed with sub-soils.
Extreme pH	Acid pH created by oxidation of sulphides such as by iron pyrites. Likewise, some minesites have high and alkalinities.
Soil toxicities	Heavy metals may be common in tailings and mobilized by acidity.
Salinity and sodicity	Especially common in former coal mines and metalliferous tailings.

to access the mineral body (which may be bauxite, iron, gold, tin, coal, etc.). The ore body is then removed and the mineral extracted by crushing and treating ore to extract the relevant mineral. The resulting material is referred to as 'tailings' and often still contains a relatively high level of these minerals (e.g. as metal sulphides). Underground mines don't have an overburden but do have material removed in tunnelling as well as crushed rock material that once contained the ore.

Those revegetating former minesites must then deal with a new 'soil' material made up of the original soil surface mantle (i.e. the A and B horizons) and crushed rock. When these are mixed together the properties of the new soil will depend on the relative proportions of the different components. But even when the original soil material is kept separate, the more fertile A horizon material can be mixed through the other B horizon material and diluted. As a result there may be major changes to nutrient availability, soil texture and compaction. Depending on how long the soil was stockpiled before being replanted there may also be reductions in the populations of mycorrhiza and *Rhizobium* bacteria.

Tailings can be especially problematic. Nutrient levels are likely to be low and the material coarse-textured. Those tailings containing iron pyrites (FeS_2) can produce leachates that are highly acid (<pH 4) especially if the material is exposed to air and has small particle sizes. Acid leachates can mobilize heavy metals remaining in these tailings (e.g. copper, lead, zinc, nickel and chromium) which together can inhibit plant growth and seriously affect the quality of any water draining through the mine site. Some minewastes may have pH values >9 and this will affect the availability of phosphorus and some trace elements. Finally, salinity can be a problem in soils produced following coal mining and in some metalliferous tailings (Bell 1996; Bradshaw and Chadwick 1980; Ripley et al. 1996).

The best way of dealing with these problems is to design the mine beforehand with the rehabilitation program in mind and ensure that the new topsoil is favourable for root growth and that more difficult material (e.g. that having pyrites) is buried well below the rooting layer. If problematic material can be identified in the design stage the costs of rehabilitation will be much less because double-handling can be avoided. Other methods of dealing with acidic and toxic tailings by burying, capping or immobilizing this material are described in more detail by Ripley et al. (1996) and Bell (1996).

Current best-practice requires that a 0–30 cm topsoil layer is removed and stockpiled prior to mining (Bell 2001; Grigg et al. 1998). This can then be respread over any overburden or tailings prior to revegetation ensuring that soil organic matter, seedbanks, mycorrhiza and nutrients are saved and replaced in their original stratigraphic position. Topsoil storage should not exceed 6 months since seed viability and mycorrhizal populations appear to decline after this time. This topsoil should not be respread more than about 10 cm deep because seedlings cannot emerge from depths greater than this.

The type of soil seed bank present in topsoil may depend on the month in which the topsoil is removed. Thornton and Dahl (1996) worked at a mine site in the seasonally dry woodlands of tropical Australia and found the best time to remove and stockpile topsoil was in the month just prior to the wet season because the population of grass seeds was lowest then. This meant there were fewer weed problems once revegetation commenced. Respreading topsoil does not guarantee that plants grown in this new soil will prosper and it may be necessary to rip the site before planting to reduce compaction induced by the topsoil spreading operation. Likewise, it may be necessary to fertilise the area to overcome nutrient deficiencies. Revegetation is then normally carried out immediately by planting seedlings or by direct seeding into the newly spread topsoil. The scheduling of the different operations has to take account of seasonal climatic conditions so that planting is carried out at the optimum time.

If topsoil is not stockpiled and respread after mining revegetation can be rather more difficult even when no toxic materials are involved (Fig. 6.3). Organic matter and nutrient levels in former subsoils will be low and there will be no seed store, rhizobia or mycorrhiza. Most native plants will have difficulty in growing at these sites and much of the nutrients applied in fertilizer may be lost by leaching. In some situations it may even be necessary to import topsoil from elsewhere to cover the mine wastes prior to beginning a revegetation program that relies heavily on a few tolerant exotic species.

In some cases the objective of those seeking to establish some plant cover at former mine sites is to simply stabilise and protect the site. This can be done using the plantations of the type described below. More commonly, however, various kinds of mixed-species plantings are used. In terms of the classification outlined in Chapter 4, these represent either Rehabilitation or Ecological Restoration plantings

Fig. 6.3 Reforesting land after tin mining in Malaysia is made more difficult because the former topsoil at this site was not saved. The sandy soils now lack organic matter and only a few exotic species such as *Acacia* spp. are now able to tolerate the site conditions and grow

and are preferred because they are more likely to be self-sustaining over a longer period. These types of reforestation will be considered in later chapters.

The Standard Plantation Model

The most common form of planting used by state forestry agencies and by industrial timber growers is to establish even-aged plantation of trees in regularly spaced monocultures (Evans and Turnbull 2004). The species used are generally those with known timber properties that can be grown using well-established silvicultural techniques. They are also species whose seed can be easily obtained and from which seedlings can be easily grown in nurseries. Many industrial growers have undertaken sophisticated tree-breeding programs and developed species hybrids and polyploid plants. Some are beginning to use clonal planting material of especially productive individuals.

They major problem for all those engaged in reforestation of this kind is the time trees take to grow until they can generate a financial return. The tyranny of compound interest means the debt incurred in preparing a site and planting trees accumulates over a rotation. Thus, with an interest rate of five percent, the debt on an investment of one unit of cash will increase to 2.6 units within 20 years, seven units within 40 years and climb to 18.7 units by 60 years. The grower must therefore assume the trees they plant will survive and grow, that there will be a market for the trees when they reach the optimal size for harvesting and, finally, that the net profits from harvesting and selling the trees will cover these mounting debts. There are several ways to reduce these problems. One is to use fast-growing species to shorten the time until harvesting. This is the most attractive option for those able to sell pulpwood and, in this case, rotations of <10 years are common. The other way is to only grow species able to produce sawn timber attracting a high market price. These types of trees usually take longer to reach harvestable sizes (often >20 years) but the assumption is that the higher-quality timbers will compensate for the longer rotation period.

In both types of plantation the planting site is usually prepared by clearing and burning all the existing vegetation. It is then ripped and ploughed using heavy machinery. Seedlings are raised in plastic polybags or semi-rigid tubes and planted when they are around 30 cm tall although some are planted as open-rooted stock and teak is usually planted as 'stumps' (or trimmed seedlings). Depending on the site, fertilizer may be applied shortly after planting as a ring around the seedlings and about 30–50 cm away. The type and amount of fertilizer are usually determined by soil and foliar analyses, pot trials and field fertilizer trials. Trees are invariably planted in rows but the spacing between trees (within and between rows) varies considerably. An overall density of between 800 and 1,100 trees per hectare (tph) is common and this can lead to canopy closure in 3–5 years depending on growth rates and planting density. Weeds will grow between the trees and the greater the tree density the sooner it is before canopy closure occurs and weed control becomes unnecessary (but higher planting densities mean the costs of raising and planting seedlings are also higher).

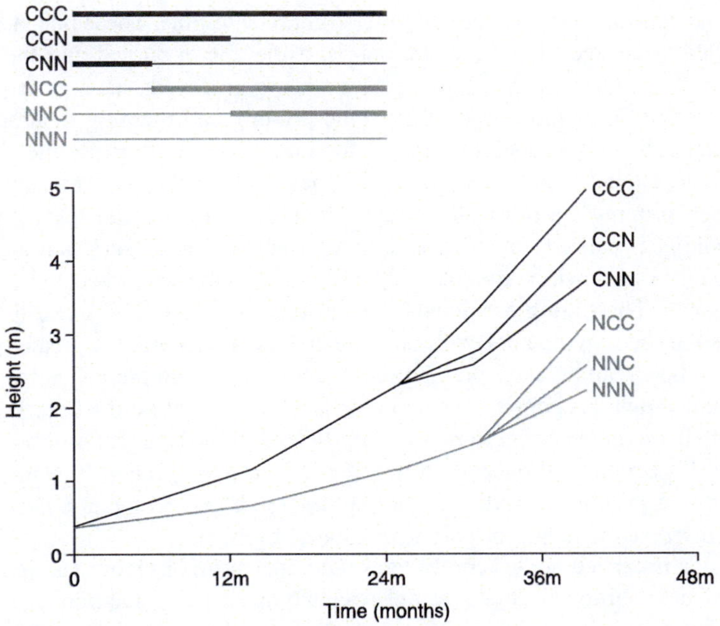

Fig. 6.4 The timing and duration of weed control affects tree growth rates. In this experiment using *Arauacaria cunninghamii* in sub-tropical Australia weed control was carried out at various times over a 24-month period. Treatments included weed control for the entire period (CCC), weed control for 12 months only (CCN), weed control for 6 months only (CNN), no weed control for the first 6 moths but then control for the remainder of the period (NCC), no weed control for 12 months and then control for the remaining 12 months (NNC) and finally, no weed control over the entire period (NNN). Heavy lines in the upper diagram show periods of weed control while the thin lines show no weed control (Paul Ryan, unpublished data)

Weed control is important and it is imperative that weeding is carried out from the time of planting. This is illustrated in Fig. 6.4 which shows that seedlings weeded during the first 12 months after planting grew much faster than those where weeding was delayed for 12 or 24 months. The growth of trees regularly weeded for 24 months was nearly double those that were not weeded. However, there are some situations where seedlings appear to derive a temporary benefit from being temporarily shaded by taller grasses or shrubs that are left along planting lines. In northern Thailand Hardwick et al. (1997) observed that tree survival was significantly greater at the end of the dry season (8 months after planting) in unweeded sites. They attributed the difference to the weeds sheltering the young tree seedlings from heat stress and transpiration losses. This benefit is unlikely to persist once the seedlings are established. After this time the competitive effects of weeds are likely to outweigh such benefits. Climbers and vines can represent an especially serious weed problem because they can swamp seedlings or pull down and distort seedling shoots unless they are removed (Kimura and Nishiyama 1999; Krishnapillay 2002; Neil 1984).

Subsequent management of the plantation depends on the management objective. In pulpwood plantations little further input is needed once the tree canopies close since most weeds likely to affect tree growth are excluded by shade although some minor vine control might still be necessary from time to time. These plantations are clear-felled at an age of 7–10 years, debris at the site is burned and the area is replanted. If, on the other hand, the trees are being grown for sawlogs or ply logs on rotations of 20 or more years then some pruning and thinning treatments are usually applied. This is done to improve log quality and to increase the size, and hence value, of individual trees. Some species such as those of *Eucalyptus* are largely self-pruning but most others retain branches on the lower trunk for some time. Pruning is carried out to remove these branches and ensure that any wood subsequently laid down on the tree trunk will be of high quality because it will be free of knots. The timing and intensity of pruning is crucial. If it is too late the branches will be big, the knots will be large and these will affect a greater proportion of the bole. But an early pruning may remove too much photosynthetic tissue and so slow growth. A common prescription is to prune trees before branches reach more than 3 cm diameter but to remove less than 50% of the green crown. The age at which this occurs will depend on the species and tree spacing. In some plantations a further pruning extending higher up the tree is carried out at a later date in order to increase the length of bole with knot-free timber.

Thinning is carried out to remove trees with poor form or vigour and reduce the effects of competition on the best trees remaining in the plantation. It can also remove dying trees and reduce the potential for future insect or pathogen problems. The obvious legacy of thinning is that the stand will have a smaller number of trees. On the other hand, more of these will have bigger diameters (Fig. 6.5). Many of the thinned trees have their crowns in sub-canopy positions. Thinning is often carried out several times during the rotation to progressively reduce the density of trees down to 100–200 trees per ha. Thus teak initially planted at 2.6 m × 2.6 m spacings (equivalent to 1,600 tph) in Myanmar are thinned at age 7, 14 and 21 years to leave a final stocking of 111 trees per hectare (Htwe 2000). In the case of *Swietenia macrophylla* plantations in Indonesia with an initial stocking of 1,070 tph, one

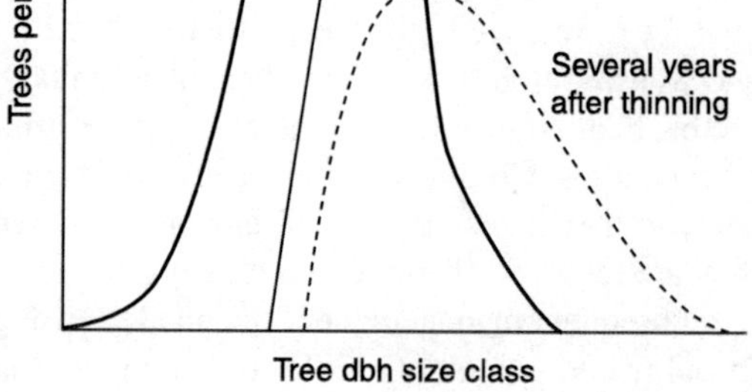

Fig. 6.5 A hypothetical illustration showing the effect of thinning is to reduce the number of trees per hectare but to increase both the number and the proportion of larger diameter trees in the stand. Thinning reduces competition and allows more larger trees to develop over time

prescription specified thinning at 10-year intervals, beginning at age 20, until the stocking was reduced to 140 tph at age 60 years (Pancel 1993). Thinning prescriptions once specified the tree densities or stand basal areas that should be attained at certain ages. These were then achieved by field staff assessing and marking individual trees for removal. Over time, as labour costs have increased, thinning has become less frequent and more intensive in an effort to shorten rotation lengths. Many modern industrial plantations are now initially thinned by systematically removing all trees in alternate rows. This can be done by with little damage to the residual trees.

Thinning might seem paradoxical. Why not simply plant fewer trees in the first place? The reason is that a high initial planting density ensures that the tree crowns merge at an early age and exclude weeds. The high density also encourages early height growth and limits branching. However, this advantage quickly disappears once the trees begin to compete with each other for light and soil resources. Thereafter the high tree density slows growth of individual trees. Thinning does not change the overall volume of timber produced by the plantation but it can have marked effect on the size class distribution of trees and improve the growth rates of the larger trees.

In practice the rates of thinning often differ from the recommended schedules. A common reason is that it is often difficult to find a market for the small logs produced by early thinnings. Some managers deal with this problem by carrying out non-commercial thinnings (i.e. simply felling unwanted trees and leaving them to decompose on the ground) but many growers may find this difficult to do. Either they cannot afford it or they regard such a practice as 'wasteful'. But if thinnings are delayed the stand will stagnate and individual trees will remain thin and spindly. If such stands are then heavily thinned to 'catch up' with the prescription the residual trees may become susceptible to windthrow under conditions of strong wind and wet soils. Most industrial growers try to resolve these dilemmas in ways that maximize their overall profitability by balancing expenditure on thinning treatments against the type of material required by their markets and whether there are compensatory price premiums for larger logs. There needs to be careful coordination between thinning and pruning operations since there is no point in pruning trees that are to be shortly removed in thinning.

Sawlog plantations are usually felled some time between 30 and 60 years age. This is well before the trees are physiologically mature. In industrial sawlog plantations the precise length of the rotation depends on variations in tree growth rates over time and on financial considerations such as interest rates and market prices. Rotations can also be set by operational considerations such as the annual capacity of mills or preferred log sizes. After clear-felling, logging debris on the sites is often burned to improve access and the areas are replanted. Some managers now avoid burning because it leads to a large loss of nitrogen. Instead they crush logging debris using heavy machinery and then replant the site.

Considerable research has been undertaken to develop methods for raising seedlings and establishing plantation of well-known species of *Pinus*, *Eucalyptus*,

Box 6.1 Why plantations sometimes fail?

Reforestation is not always easy and many plantations fail. Records of failure are rarely kept but the most common reasons are:

- Species are planted at unsuitable sites (where site conditions don't match the environmental tolerances of those species).
- Poor quality seedlings are used (these have been kept too long in the nursery so that seedlings are pot-bound or twisted and the shoot-to-root ratio is imbalanced).
- Planting is carried out at the wrong time of the year (e.g. too late in the rainy season so no follow-up rains).
- Seedlings are planted too deep or too shallow.
- The plantation soils are too compacted.
- The availability of soil nutrients is too low for the species planted.
- Soil chemical present in toxic concentrations (e.g. at former mine sites).
- There is insufficient weed control (too infrequent and for too little time).
- Sites are not protected from fire or grazing.
- There are severe insect and/or disease problems.

Seedlings that fail to establish in the first month or so can be replaced if the numbers are not too great. But plantations that persist with low overall survival rates are difficult to manage. Grasses often flourish at such sites increasing the risk of wildfires that may destroy the whole plantation.

Acacia and teak (Evans and Turnbull 2004; Nambiar et al. 1999). Prescriptions have been developed to guide pruning and thinning and specially-bred genotypes are being developed to produce uniform trees of these species that have particular timber properties. The site requirements of these species are reasonably well-known and the growth rates and productivities that can be expected at such sites are also established. Because of this monocultural plantations of these systems are popular and comparatively easy to manage. On the other hand, it is important to note that not all newly established plantations succeed and Box 6.1 outlines some common problems.

Limitations of This Standard Model

While this model is widely used, including by many smallholders, it has some disadvantages and does not necessarily suit the objectives of all of those likely to be undertaking reforestation. One disadvantage is that growers must choose a single tree species. In doing so they must assume that the goods it produces will still be wanted several years (often decades) in the future. Industrial growers are able to reach an informed decision more easily than smaller growers because they usually have better access to market information and they are able to achieve economies of scale. However the types of species favoured by industrial operators may not be those suited to smallholders wanting trees able to provide subsistence benefits as well as commercial returns. Moreover, many farmers have land that is some distance from industrial mills or markets. This means transport costs will be high. In such cases copying the reforestation choices of governments or industrial growers may not be the most appropriate course of action.

A second disadvantage of the industrial model is the delayed timing of financial returns. Again, this problem faces all tree-growers whether they are industrial operators or smallholders. As already noted it can be overcome, to some extent, if there are intermediate returns from the sale of thinnings. But small volumes of small diameter logs might be difficult for a farmer to sell. Industrial growers deal with the problem by planting fast-growing trees that have short rotations. But growers producing only small volumes may not be able to sell such timber unless their farm plantation is close to a market or has a special contract with a mill.

A third problem is the rigidity of the system; the spacing and density of trees is specified as is the time for thinning and the time of the final harvest. This is necessary for industrial growers that require a fixed daily or weekly intake into their mill. But it may be less suited in the more fluid circumstances in which many smallholders and communities live. Opportunities for selling produce may occur episodically (e.g. a visiting buyer with a truck makes an offer) or plantation owners may suddenly require funds for particular purpose unrelated to the optimal rotation length. Of course the model, in theory, is sufficiently flexible to handle such contingencies but government extension officers and professional advisers sometimes insist that the prescriptions be strictly followed in order to 'protect' forest lands and optimize productivity.

Fourthly, the plantations created by the industrial model may not be able to generate all of the ecosystem services that society is beginning to require. This depends on the species used and the rotation lengths adopted. Plantations using short rotations are clearly less suited for watershed protection than those using long rotations because of the frequent disturbances that can generate erosion. Plantations grown on longer rotations may offer better protection because many acquire a shrubby and protective understorey over time. But some don't. Teak is a well known example and erosion in teak monocultures can be severe because the tree canopy cover is sufficiently dense to exclude understorey plants and because the trees are deciduous (Siswamartana 2000). Heavy rain falling before a new canopy has developed can be especially damaging and Fig. 6.6 shows erosion occurring in a young teak plantation in northern Laos. Nor do even-aged plantation monocultures provide habitats for many of the wildlife found in tropical forests. As noted earlier, these simple plantations may be preferable to degraded grasslands but may not be appropriate in situations where watershed protection or conservation issues are especially important.

Finally, there is a broader ecological disadvantage arising from using a simple form of reforestation on a large scale. The model is now being widely applied and plantations exceeding 100,000 ha are becoming more common. The establishment of large monocultural plantations using a handful of carefully selected genotypes

Fig. 6.6 Substantial soil erosion can occur under teak monocultures. This is because the high levels of shade limit understorey development and because the trees are deciduous for part of the year (Photo: Sean McNamara)

inevitably leads to permanent landscape simplification. The same is happening, of course, with agricultural technologies such as oil palm and rubber plantations which are expanding across many parts of the region (this corresponds with the process of 'agro-deforestation' now occurring across the Pacific and described by Clarke and Thaman (1993). Collectively, these changes are leading to a significant degree of genetic impoverishment. The result of these changes is that some of the most diverse landscapes on earth are becoming some of the most homogenous. As shown in Chapter 1 it is already evident that this is leading to significant biodiversity losses. It is also likely to be producing ecosystems that are more vulnerable to biological and economic risks.

The Hazards of Monocultures

Plantation monocultures are less resilient than natural forests because they do not have the diversity of species and functional types present in the latter. These species, plus the complex food webs in which they are embedded, buffer the system against environmental changes. Some natural forests have very few tree species and the occurrence of small areas of natural monocultures was discussed in Chapter 1. Nonetheless, the increased commercial productivity that might be produced by most plantation monoculture is likely to be associated with an increased level of risk.

One of the risks is that of increased damage from pests and diseases. All of the tree species commonly used in plantations in the Asia-Pacific region are subject to a variety of pests and diseases. Some of the more important of these pests have been discussed by Speight and Wiley (2001) and Nair (2007) while Lee (1999) and Wingfield et al. (2001) have reviewed some of the more common plantation diseases. Insects are usually kept in check in natural ecosystems by parasites or predators. But, in the absence of these controls, significant amounts of damage can occur. Sometimes this can be enough to limit the usefulness of that particular tree species as a plantation species. Among these are members of the Meliaceae family (e.g. *Cedrela*, *Toona*, *Swietenia*, *Khaya*, *Chukrasia*) which can be affected by shoot tip borers belonging to the genus *Hypsipyla*, the widely planted exotic *Leucaena leucocephala* which is badly damaged by a psyllid and *Agathis robusta* which has been phased out of plantation programs in northern Australia because of a coccid moth. Other potential plantation species are sometimes affected by insect damage but can still be used. For example, teak (*Tectona grandis*) plantations can be defoliated by the moth *Hyblaea puera* although the trees can recover in the subsequent growing season.

Many of these tree species are grown as exotics and it is important to separate the 'exotic' effect from the 'monoculture' effect. Exotics are species that are not native to the area in which they are being planted. They may still be indigenous to the country but are now being grown outside their natural range. One view holds that exotics do better grown outside their natural range because they escape their natural predators and parasites. However they may also be more prone to damage because they are growing outside their ecological niche. Nair (2007) examined the difference between the 'monoculture' effect and the 'exotic' effect for some of the more common plantation species in Asia (*Acacia mangium*, *Acacia crassicarpa*, various *Eucalyptus* spp., *Falcataria moluccana*, *Gmelina arborea*, *Hevea brasiliensis*, *Leucaena leucocephala*, *Pinus caribaea*, *Swietenia macrophylla*, *Tectona grandis*). There did indeed appear to be evidence of a 'monoculture' effect with more damage being found in monocultures than in mixed-species plantations. However, the difficulty with such simple comparisons is that there are considerable differences in the nature of the non-monoculture plantations. Some have many species while others have few and some have species planted in alternate rows while others are planted in more intimate mixtures. This issue will be discussed further in Chapter 7. In the case of the 'exotic' effect, the empirical evidence was that neither the intensity of pest damage nor the number of insect species associated with a particular plantation species was consistently determined by its exotic status. Most species had less damage when grown as exotics although some had more (*Leucaena leucocephala, Pinus caribaea*). Nair (2007, p. 141) concluded that 'While [monoculture] plantations are at greater risk of pest attack than natural forests, plantations of exotics are at no greater risk than plantations of indigenous tree species'.

Several factors determine the risk of pest damage in monocultural plantations of exotic species. Some are to do with the presence of closely related tree species growing in the same location and some to do with the size of the new plantation.

There is also evidence that pests build up with time since a plantation species was first introduced. Particularly serious damage may result when insect pests from the home of the introduced tree species are accidentally introduced to a country and eventually reach the plantation. Damage can be substantial if their natural enemies are not able to colonize at the same time. An interesting example of the complexity of these issues comes from Fiji. The valuable hardwood *Swietenia macrophyll* from Central America was introduced to Fiji in the expectation this isolated location would allow it to escape from the shoot borer *Hypsipyla grandella* that attacks it in its native habitat. The species grew well in the absence of *Hypsipyla* but the log quality was subsequent damaged by boreholes caused by several species of ambrosia beetle. At least one of these is endemic to Fiji and would have had no previous relationship with *Swietenia* (Roberts 1978).

Similar patterns have been observed with tree diseases. Lee (1999) describes a number of stem cankers, leaf and shoot blight diseases, rusts and root rots affecting a variety of tree species across the region. One important problem is a stem rot causing damage to some older *Acacia mangium* plantations. This is serious because of the enthusiasm many growers have for this *Acacia*. The rot appears to become progressively more damaging once trees exceed 6–8 years and effectively limits the age to which plantations of this species can be grown (although a *mangium* x *auriculiformis* hybrid seems less affected). *Acacia mangium* is also affected by severe root rots (*Ganoderma* spp., *Phellinus* spp.). These have killed up to 40% of trees in 10–14-year old plantations in Malaysia and there is evidence that the problem becomes more damaging in some second and third rotations (Eyles et al. 2008). Another potentially serious plantation disease is yet to reach the region. This is the rust *Puccinia psidii*, also known as guava rust fungus, that has been found affecting a variety of native Myrtaceae in South America. Eucalypts are members of the Myrtaceae family and although the disease has not reached Australia it has already affected *Eucalyptus* plantations in South America. The disease is a potentially serious problem for natural eucalypt forests as well as plantations across the Asia-Pacific region (Coutinho et al. 1998).

It is important to recognize that these pests and diseases do not affect every plantation containing susceptible species and it is still difficult to assess the risks that pests and diseases pose for timber tree plantations grown in monocultures. Part of the problem is that tree plantations are still a relatively new form of land use and many plantation areas are still less than 30 years old. Insect pests and diseases may take several years to develop in such areas because the new environment is unfavourable to that insect or potential pathogen. Alternatively, it may be that a disease has to undergo a form of incubation during which more virulent genotypes of the pathogen eventually evolve (Wingfield et al. 2001). If true this could mean that the full impact of diseases and insect pests on tree plantations in the region is yet to be felt. On the other hand, perhaps there are reasons to be more optimistic. Lee (1999) notes that rubber (*Hevea brasiliensis*) has been successfully grown in monocultural plantations in southeast Asia for many years and diseases and insect pests have been kept under control by active monitoring and control measures. But this might simply reflect the level of scrutiny applied. Rubber has been sufficiently profitable to support a considerable degree of monitoring and it remains to be seen whether the same will be true of timber plantations.

The other hazard faced by those growing monocultures is an economic one. Markets and market prices can change over time and sometimes dramatically so. The problem of relying on a single market was discussed earlier in Box 4.3 which described how the demand for some products can decline over time and how an over-supply of others can reduce the price being offered. General purpose or utility timbers may be especially susceptible to the latter problem.

Species Choices

The extraordinary diversity of tree species in tropical forests means there are a large number of species from the region that might be grown in plantations (Appanah and Weinland 1993; Cameron and Jermyn 1991; Do and Nguyen 2003; Lemmens et al. 1995; Russell et al. 1993; Soerianegara and Lemmens 1993). Of course, not all are of high commercial value but, as natural forests shrink, the attractiveness of the more highly priced specialty timbers in particular should increase. This being the case, there is a certain irony in the fact that Brazil, the home of perhaps the richest tropical forests on earth, has chosen to use eucalypts for many of its largest timber plantations while Australia, the home of eucalypts, has until recently, used exotic pines for most of its plantations.

Tropical foresters have carried out experimental plantings of some of the more attractive timber species for perhaps 100 years (and even earlier in the case of teak). In Australia, trials in the early 1900s tested the highly sought-after red cedar *Toona ciliata* (syn. *australis*) and the indigenous conifer *Araucaria cunninghamii.* In Malaysia, early trials used *Neobalanocarpus* (syn. *Balanocarpus*) *heimii.* But plantations were rarely attractive while natural forests could supply much larger diameter logs at low cost and the silvicultural methodologies needed to grow these same species in plantations were not fully understood. Plantations only began to attract attention across much of the Asia-Pacific region after the 1960s when it became clear that timber supplies from natural forests were beginning to decline. The dilemma then was what species to plant? Although a variety of species were tested there was also a tendency for certain species to become 'fashionable'. Many of these were exotics and, over time, as the problems with each species became evident, a new exotic species was seized upon. This was despite the wealth of valuable indigenous tree species present in the region.

The history of plantation development in Malaysia is representative of that experienced in much of the region. Ng (1996) describes how the Malaysian plantation program initially aimed to create a timber resource for a future pulp industry. It also sought to compensate for the difficulties then being experienced in developing enrichment planting techniques in logged-over native forests. In common with many tropical countries at this time, a number of exotic pines were tested (including *Pinus caribaea* and *P. merkusii*) but these had to be abandoned in Malaysia because of the high frequency of foxtailing and difficulties in producing seed. The pines were then followed by *Gmelina arborea*, *Eucalyptus deglupta* and *Falcataria moluccana*. Plantations of each eventually ran into problems. Either the species were too sensitive to site conditions, had heartrot or it proved difficult to get a sufficient quantity of high quality seed. A large number of other, mostly exotic, species were also tested. None of these were successful until *Acacia mangium* was stumbled upon. By chance this was planted in some highly degraded sites, including *Imperata* grasslands, and grew extremely well. Since then it has been picked up and used widely across the region. Although *Acacia mangium* was subsequently found to suffer from heartrot it is still regarded as a valuable pulpwood species because it can be grown on short rotations where the heartrot problem is tolerable.

This search for a single miracle tree in Malaysia has been replicated by other forestry agencies across the region. In addition to *Acacia* several species of *Eucalyptus* have also proved able to grow very rapidly at sites with relatively poor soils (Turnbull 2003, 1991). The ability of these species to grow rapidly sends a powerful visual signal to prospective growers and makes them obvious candidates for future plantation programs. In some cases this is undoubtedly true; *Acacia* and *Eucalyptus* plantations now cover large areas of previously bare and badly degraded hillsides in many parts of the world that would have been difficult to reforest otherwise. These plantations have been highly profitable for their growers. The problem is that such species can easily become seen as the only species to use in plantations and this is not true. Smallholders are likely to be as impressed as government foresters by species that are able to grow quickly. But their economic circumstances, the location of their landholdings and transport costs means it may not be possible for them to grow these species profitably. Besides, many farmers expect their trees to provide more than large volumes of cellulose and the circumstances of many is such that they may be better advised to grow multi-purpose trees or slower-growing trees producing more valuable sawlog timbers.

Fast Growing or Slow Growing Species?

One of the first decisions to be made by a grower is whether to use fast-growing species and manage these on short rotations (<10 years) or whether to use slower-growing species grown on a much longer rotation (>20 years). Many plantations now use species such as *Eucalyptus urophylla, Acacia mangium, Styrax tonkinensis, Falcataria moluccana* and *Gmelina arborea* because of their fast growth rates. These have replaced the slower growing native species once harvested from natural forests and for which there was once a local or international market.

The obvious advantage of using fast-growing species is that short rotations mean there is an earlier financial return and establishment costs can be recouped more quickly. Some fast-growing exotic species are also useful because they grow well on poor soils and because they come as a 'silvicultural package'. In these cases seed is available from previous tree-breeding programs and there are established nursery techniques and plantation management systems that make them easier to grow in plantations. The chief disadvantage of using most fast-growing species is that they are mostly used to produce pulpwood, firewood or utility timbers for which growers receive a relatively low price. In the case of eucalypts this is because the timber of young trees is subject to various growth stresses that often results in serious distortion and quality degradation in sawn material unless particular care is taken (Hillis and Brown 1978). This means logs from young trees are difficult to use as sawlogs although they can be used for veneer. *Acacia* do not suffer from this problem. On the other hand, as noted above, some such as *Acacia mangium* can be subject to a heartrot from an early age and may need to be harvested when log diameters are not especially large (Old et al. 2000).

Fast-growing species have some other problems as well. They tend to transpire more water than slower growing species and large plantations may change local hydrological flows. In addition, more nutrients are also removed from the site during successive short rotations and can cause productivity declines in subsequent rotations if these losses are not corrected by fertilizers.

Plantations of slower-growing species (whether natives or exotics) are less attractive because of the time that must pass before there is any financial return. These species are also less attractive because, with the exception of a handful of exotics, less is generally known about their silvicultural attributes and site requirements. But, as Durst and Brown (2000) note, growth rates are only part of the equation for determining financial competitiveness. Factors such as investment costs, interest rates, market prices and rates of return are also important. If tree growth rates alone were the only factor there would be no Nordic timber industry. For an industrial grower the best financial results are usually obtained if the grower also owns the processing plant and the whole operation is close to the final market, irrespective of the type of tree being grown. But many slower-growing species are particularly attractive because of the quality of their timbers or because they provide other goods such as fruits, resins or medicines. The high market value of these goods means these species can be profitably grown at sites some distance from markets.

Changes in natural forest cover and in markets may be creating a new niche for plantations of these slower growing species. Leslie (1999) reviewed the international timber markets and suggested there was likely to be a 'wave of timber' coming from the very large areas of plantations that were then being established in the southern hemisphere. Most of these plantations are based on *Pinus radiata* and eucalypts and produce what Leslie referred to as 'commodity' timbers and pulpwood. He cautioned that his forecast depended on assumptions about future planting rates, plantation productivities and supplies from natural forests. However, he suggested there was likely to be a market niche for 'genuine specialty and decorative timbers' or what was referred to above as 'high-value' species. Natural forests are the ideal source of such timbers but future supplies from these forests are being threatened by poor logging, deforestation and because many are placed in reserves of one kind or another in response to public pressures. In short, he believed there should be an opportunity to meet an emerging deficit by growing these species in plantations. Since that time the areas of productive natural forest have continued to decline suggesting his forecast is still valid. Not all of these species need be natives and there are a number of well-known, high value exotic species in addition to teak such

Table 6.3 Advantages and disadvantages to farmers of fast-growing species (e.g. exotics such as *Acacia* and *Eucalyptus*) producing pulpwood and utility timbers and slower growing species able to produce specialty timbers or higher value goods

	Fast growing species	Slow growing species
Advantages	Usually able to tolerate poor soils	Timbers may attract higher prices.
	High quality seed readily available	Likelihood future market will strengthen as supplies from natural forests decline.
	Silvicultural methods established	Generate more diverse range of goods.
	Faster cash flow (and hence higher financial return?)	More flexible harvesting schedule.
	Well-known in marketplace	Suitable for local markets as well as international markets.
	Less risky because of shorter rotation	Isolated plantations viable since can bear higher transport costs.
Disadvantages	Produce a limited range of products (young eucalypts suffer from growth stress and some *Acacia* from heartrot)	Slower cash flow (unless thinnings or NTFPs can be sold).
	Logs mostly attract a lower price	Often less tolerant of poorer soils.
	Lower price means must be grown near markets to be profitable	Often difficult to get good quality seed.
	Have high water use	Less silvicultural knowledge available.
	Frequent harvesting and large sapwood content impose higher nutrient drain on site	More risky because of longer rotations.

as *Swietenia macrophylla* and *Khaya* spp. which may also suit the circumstances of some growers. Some of the advantages and disadvantages of fast- and slow-growing species are summarized in Table 6.3.

Sources of Information on Species Choices

What are the alternatives to fast-growing exotic species and where might one find information about these alternatives? There are several sources of information.

Biogeographic Distribution and Knowledge of Silvicultural Attributes

An obvious starting point for those interested in exploring the use of more native species is to identify those within whose natural distribution the proposed planting site is located. These species will be adapted to the climates and soils of the region (although not necessarily adapted to the degraded soils now available for reforestation). Some of these species might be distributed over a large geographic area and be very common throughout their range. Others may be widely distributed while being only found at particular sites within this distributional area. With proper management, both types may be useful. This species list can then be fine-tuned by excluding those with unsuitable timber properties or silvicultural attributes. Less attractive species would be those with timber that is difficult to season or saw and which is not decay-resistant. Likewise species with poor apical dominance, poor form and large branches as well as species with a low capacity for self-pruning would also be less attractive. Species with more open canopies may be less suited at sites with aggressive weeds since they may need constant tending. Knowles and Parrotta (1995) describe a system developed for evaluating species for reforestation that used silvicultural criteria such as ease of collecting seeds, producing planting stock and the degree to which a species needed some initial overhead shade to be established.

Traditional Knowledge and Farmer Preferences

Another source of information is traditional knowledge. Many farmers will have planted trees on their land. Some of these may be in home gardens but others may be in small woodlots. The numbers of species found in such plantings can be high. Clarke and Thaman (1993) recorded over 400 species that have been used in agro-forests across the Pacific. These include food trees as well as timber species. Large numbers are also found in more localized surveys. For example, a survey of 37 small (mostly <5 ha) farms in northern Vietnam found a surprising total of 64 tree species had been planted (Lamb and Huynh 2006). Many were only found at one or two farms but nine species occurred at more than 20% of the farms. Many of these species are grown to produce fruit, nuts or resins for subsistence use and are not likely to be suitable for commercial plantings since the market for these goods is probably limited. On the other hand, many farm plantings also contained commercially attractive timber species. These included species that have been traditionally used as well as those planted following recommendations from government extension officers. All of these plantings represent a valuable source of information about the site preferences of different species and an indication of the growth rates that may be possible. The diversity of species being used also points to the fact that farmers are not inherently conservative when it comes to their use of tree species but are willing to test new species and experiment with plantings if they believe they might benefit from doing so.

Foresters tend to assume people grow trees to produce timber or other goods but this is not always the case. Some of the difficulties in exploring the reason for farmer preferences are described in Box 6.2. This means some care is needed when interviewing farmers to ensure that both the interviewer and the interviewee under-stand what is being discussed. It must also be acknowledged that male and female farmers may have differing views on all of these matters.

Box 6.2 The best species to plant

Some villages in northern Thailand have established small forests on degraded lands near their village. These forests contain a variety of mostly native species. Some were planted by villagers as seedlings and some had colonized the site from residual natural forests. Researchers from the University of Chiang Mai investigated the biological diversity now present in one of these forests near the village of Pakhasukjai (S. Elliott, personal communication, 2009). The researchers did fieldwork in the forest during the day and spent their evenings after dinner discussing with villagers the uses made of different species. What was used for roof thatching? What was best for flooring? Which species had medicinal values? After three days the field work was drawing to an end and they had accumulated an impressive list of species and their local uses and knew which had been most widely planted and which were less common. On the last evening, almost as an afterthought, one of the researchers asked their hosts what species they would take with them to start a new forest if they ever had to move the location of their village. They assumed the most widely planted species were among the most valued. However, they were taken aback to hear of five completely new species that, until then, had not been mentioned in the evening discussions. One was a species used to provide the post for tethering buffaloes being ritually slaughtered during ceremonies to mark the funeral of the village leader. Another was a species the leaves of which were scattered on the ground to promote soil fertility during ceremonies to mark the start of the agricultural year. The other three species were also of spiritual and cultural importance. The villages obviously valued the trees providing timber and building materials but, in their view, some of the less obvious species were of considerable importance as well.

Evidence from experimental field trials

Biogeographic and traditional knowledge provide useful starting points but field trials established at known dates and with known histories are crucial for evaluating the potential value of different species for use in plantations and to understand their site requirements. Such trials are especially useful if they use seed of known provenance, have been properly maintained and include more than, say, 20 individual trees of each species. Such trials can provide information on the growth rates to be expected at different types of sites as well as survival rates, branching patterns and the form of trees. Indeed, a prudent investor might demand to have this knowledge before they made a commitment to invest land, funds and time in reforestation. Unfortunately there are relatively few places in the tropics where plantation trials have been established long enough (>20 years) to generate reliable data on volume increments although there are many reports describing early height and diameter growth (e.g. Appanah and Weinland 1993; Cameron and Jermyn 1991; Otsamo et al. 1997). Most studies involving tree species able to produce higher value or specialty timbers have found they have mean annual increments (MAIs) of less than 10 m^3 ha^{-1} $year^{-1}$ (Appanah and Weinland 1993; Russell et al. 1993). This is well below the values to be expected in well-managed plantations of fast-growing species such as many *Eucalyptus* and *Acacia* species (often well in excess of 20 m^3 ha^{-1} $year^{-1}$) although the regional survey commissioned by ITTO (STCP 2009) showed many plantations of these species do not always achieve the productivity levels they are capable of producing (Table 3.4).

Growth rates of well-established plantings of some better known species are shown in Table 6.4. Some of these represent data from a single site while others represent values coming from a range of sites. In most cases little is known of the management history of these plantations so it is difficult to know whether these are

Table 6.4 Productivity of some specialty timber tree species grown in the Asia-Pacific region

Species	Mean annual volume inc. m^3 ha^{-1} $year^{-1}$	Age	Reference
Agathis spp.	22–28	50	Soerianegara and Lemmens 1993
Araucaria cunninghamii	15	50	Dale and Johnson 1991
Araucaria cunninghamii	10–24	?	Varmola and Del Lungo 2003
Cedrela odorata	2–3	9	Lemmens et al. 1995
Cedrela odorata	4	23	Cameron and Jermyn 1991
Chukrasia tabularis	7	20	Do and Nguyen 2003
Dipterocarpus baudii	7	26	Appanah and Weinland 1993
Drybalanops aromatica	8	46	Appanah and Weinland 1993
Elaeocarpus grandis	3	23	Cameron and Jermyn 1991
Endospermum peltatum	12	20	Soerianegara and Lemmens 1993
Erythrophloeum fordii	6	20	Do and Nguyen 2003
Eucalyptus deglupta	25	19.5	Lamb 1990
Flindersia brayleyana	4	22	Cameron and Jermyn 1991
Gmelina arborea	18	20	Do and Nguyen 2003
Gmelina arborea	25	12	Oliver 1999
Grevillea robusta	3	15	Cameron and Jermyn 1991
Michelia mediocris	4–7	44	Do and Nguyen 2003
Peronema canescens	10	20	Soerianegara and Lemmens 1993
Pinus caribaea var. caribaea	10–23	15	Varmola and Del Lungo 2003
Pterocarpus indicus	7[a]	?	Harrison et al. 2005
Shorea leprosula	8	35	Appanah and Weinland 1993
Shorea parvifolia	15	32	Appanah and Weinland 1993
Swietenia macrophylla	3–11	?	Hammond 2002
Swietenia macrophylla	8–13	?	Varmola and Del Lungo 2003
Tectona grandis	7–13	?	Varmola and Del Lungo 2003
Toona australis	8	56	Cameron and Jermyn 1991

[a] Estimated

truly representative of these species or if they are under-estimates of what might be achieved with better management. It is well known, for example, that careful site preparation, weed control, fertilizers and thinning can dramatically improve growth so these MAI values may be conservative. But site matching is also important. For example, the productivity of *Swietenia macrohylla* in the extensive plantations in Fiji range from stands classed as 'poor' and having MAIs of 3 m^3 ha^{-1} $year^{-1}$ to those classed as 'very good' and having MAIs of 11 m^3 ha^{-1} $year^{-1}$ (Hammond 2002). Similarly, both *Cedrela* and *Elaeocarpus* are known to be relatively fast-growing and these MAI values may not be indicative of these species on better sites or with better management. In short, there is a definite need to develop improved data bases containing information about growth of key species. Permanent mensurational plots in a number of different farm forests would be one way of achieving this.

Field trials are also important because they can provide the first indication of insect or disease problems. Likewise they can reveal how different species are able to withstand episodic environmental hazards such as cyclones (Fig. 6.7). In some cases infrequent events like these may substantially alter the choices of species that might be used in a planting program. For example, Neil and Barrance (1987) describe how species in plantations on Vanuatu differed quite markedly in their wind-firmness and tendency to snap during cyclones. Wind-firm species included *Agathis macrophylla, Castanospermum australe, Intsia bijuga, Nauclea diderichi* and *Swietenia macrophylla.* The more susceptible species included *Eucalyptus camaldulensis, E. deglupta, and Khaya ivorensis.* Observations in natural forests suggested *Endospermum medullosum* and *Antiaris toxicaria* were very wind-firm suggesting that they, too, should be included in future plantation trials even though the timbers are not especially high quality hardwoods.

Field trials can be difficult to establish and it may be many years before they start providing useful information. They can be easily compromised if survival rates are <90% or if they are not maintained properly. Other difficulties sometimes occur during long-term field trials (Box 6.3).

Evidence from Markets

The final source of evidence comes from the markets in which forest products might be sold. There is no point in growing a species if it can not be sold or if its market price is less than the cost of production (unless it is being grown solely for subsistence purposes or environmental reasons). Many of the species first tested in plantation trials were those whose high market value had already been defined by the prices paid for logs harvested from natural forests. International market prices for particular types of timber are one indicator but another that may be more relevant for smaller producers is the local timber market. This issue will be discussed more fully in Chapter 9. It is sufficient to point out here that this market may not be especially well informed since many small rural sawmillers may not know of

Fig. 6.7 Storms and cyclones can badly damage plantations in lowland areas. A plantation area in the wet tropics of north eastern Australia (**a**) before and (**b**) after a cyclone (Photo: Mila Bristow)

market prices in more distant locations. Nor might current prices be a guide to the prices that might be offered in, say, 20 years time. Nonetheless, such information should be part of the decision-making process when species choices are being made. It should be added that price surveys like these often reveal that rural saw-millers do not necessarily want only the most expensive timbers. Some of their customers will not be able to afford high-priced goods and many millers may prefer to buy several different types of timber to suit the needs of different customers (e.g. furniture makers, house builders etc.).

Box 6.3 Hazards of long-term silviculture experimentation

All land in Papua New Guinea is owned by communities and government foresters must arrange for a 'loan' of land if they wish to establish field experiments. A trial had been established in a grassland area in the highlands to test the effects of applying various fertilizers on the growth of trees at that site. The trial was designed as a randomized block with each plot containing 49 trees. There is nothing remarkable about this design but the field layout can be confusing unless the trees are well-marked. The researcher decided to mark each tree with its own aluminium tag. The first inspection was carried out 6 months after the trial was established. Driving down the road to the site the researcher saw a cheerful old man with what seemed to be a silver necklace around his neck walking towards him. Flamboyant body decorations on men are well-known in the highlands of Papua New Guinea but silverish decorations are unusual. Mild curiosity turned to horror when he recognized his tree tags had been converted into a necklace! Since the old man probably belonged to the clan from whom the site had been borrowed and since the research had taken the precaution of also making a map of the tree layout it was difficult to not agree with the old man that yes, indeed, it was a very good day and that tomorrow would probably be a good day too if it did not rain. The moral of the story is that when working outside the safe confines of an experimental field station, it is always prudent to have an alternative in case things go wrong.

Problems Needing Resolution Before Using a Wider Range of Species in Reforestation Programs

There are, of course, many plantations across the region that use a much wider range of species than the few now commonly used in the larger industrial plantings. These have been established by government agencies as well as by communities and individuals using both scientific and traditional knowledge. Nonetheless, it is hardly surprising that the sheer diversity of species present in tropical forests means that much more remains to be known about how to profitably grow these and other native species on deforested and degraded sites. The discussion below focuses primarily on growing higher-value or specialty timber trees for commercial purposes.

How to Get Seeds and High Quality Seedlings?

One of the problems in using a wider variety of species in plantations is that of getting sufficient seed. Unlike the species currently used in large industrial plantations it is commonly difficult to get good quality seed in large quantities and, in many cases, poor seed sources are used. These include collections made from just one or two trees because they are easily accessible. Likewise seeds are sometimes collected from short, stunted trees (i.e. genetically inferior individuals) because they are easier to collect from. It is far better to collect from at least 15–20 trees representative of the optimal growth of the species at a particular site. Ideally, these should be trees with preferred traits such as straight stems, strong apical dominance and small diameter branches. In the case of species with broad geographic distributions, seed should be collected from provenances throughout their range and tested in field trials to identify the best source.

Seed collections can be difficult and expensive. Those collecting seed need to know when fruiting is occurring to ensure the timing of visits coincides with seed being neither unripe nor over-mature. It is usually preferable to collect seed while it is still on the tree rather than waiting till it falls to the ground where insect predation increases and viability rapidly declines. However, this can be difficult. Some species produce seed at similar times each year but other species may only produce seed at infrequent and unpredictable intervals (e.g. many dipterocarps). Further, while some species are present in natural forests at relatively high densities, others may only occur as scattered trees that are hard to locate. Even when this is the case it is important that seeds are collected from a number of trees and not just one.

These collection difficulties are compounded by the fact that the seed of some species lose viability quickly and must be germinated immediately after being collected. By contrast, the seed of other species will remain dormant and may not germinate for over a year even after it has been planted. The magnitude of the dilemma is illustrated by data from Malaysia and Thailand. In a study of 335 Malaysian tree species Ng (1983) found the seed of 50% of species had completely lost viability within 6 weeks. A similar study using 262 species from northern Thailand found the seed of 50% of species had completely lost viability after about 3 months (Elliott et al. 2006). All of these factors make it difficult for those needing to raise a specified number of seedlings of a prescribed size for a particular planting date.

There are several possible solutions. One is to collect and replant wildlings (i.e. recently germinated seedlings already growing in the forest understorey). This can work if enough young seedlings can be found and survival after transplanting into the nursery is acceptable. The technique is quite appropriate for small farm plantings (and is one traditionally used by many shifting cultivators and by those establishing agroforests) but it may not be as useful for those wishing to establish larger plantation areas. Another way of dealing with limited or episodic seed supplies is to use various forms of vegetative propagation such as cuttings. Species differ in their capacity to strike from cuttings but there is mounting experience concerning ways of doing this including the maximum age of material from which cuttings can be taken and the hormonal treatments that can promote rooting (Elliott et al. 2006; Krishnapillay 2002).

A good deal remains to be learned about how long seedlings should be kept in nurseries and the best size to use when planting out into the field. Some species can be planted out when shoots are 20–30 cm tall but others may need to be taller.

Larger seedlings usually require larger pots which make transport to planting sites more difficult. Longer periods in nurseries also increase the risk that seedlings become pot-bound so that they fail when eventually planted in the field. Krishnapillay (2002) and Elliott et al. (2006) provide a good overview of the main methodologies currently available for dealing with these problems.

Do Some Species Need Early Shade?

For many years there has been anecdotal evidence that seedlings of some tropical tree species cannot be established in the open but need to be under-planted beneath an existing tree cover. When planted in the open their mortality is high or trees develop multiple leaders. One study carried out in Brazil found 47% of the 160 tree species tested grew poorly when planted in the open and only 37%grew well (Knowles and Parrotta 1995). The reason for these types of results is unclear but suggested explanations include that sensitive species suffer from (i) photo-inhibition or heat stress caused by high radiation levels, (ii) water stress caused by excessive transpiration or (iii) low humidity that affects transpiration. All factors may be involved although certain species may be more sensitive to one factor than another. In Southeast Asia the dipterocarps appear to be especially sensitive to being planted in open grassland sites (Otsamo et al. 1996).

There is accumulating evidence about which tropical tree species can tolerate early shade (Appanah and Weinland 1993; Kimura and Nishiyama 1999) but rather less evidence concerning which species actually require early shade to become established. One major study in Malaysia examined the growth of a number of dipterocarp species planted in strips of various width cut through young regrowth forest (Anonymous 1999). The purpose was to examine how much light was necessary for establishment and growth of each species in these conditions. Seedlings of many species failed when planted in either narrow strips or in the open. In these particular situations the best overall compromise between excessive competition and too much exposure appeared to be strips of 20 m width cut through the regrowth although there were considerable differences between species. Appanah and Weinland (1993) conclude most dipterocarps grow better with some side shading to enhance height growth. It is not only the dipterocarps that are affected and reports from Vietnam suggest the non-dipterocarps *Canarium album*, *Erythrophloeum fordii* and *Cinnamomum* spp. all required some shade for the first few years (Cameron 1995).

In contrast to these reports there is evidence from a number of extensive planting programs in other parts of the region in which a great variety of tropical tree species have been successfully established without the need for early shade (Elliott et al. 2006; Erskine et al. 2005). The reasons for these differences are not clear but may be associated with the microclimate at the time of planting, the physiology of the species used or the leafing/branching morphology of the species (bushier seedlings are likely to provide some protective shade for leaves in the inner crown). More work is clearly needed to clarify which tree species benefit from early shade, just how much shade is beneficial and how the length of time it should be maintained.

What Are the Preferred Sites of Different Species?

A third problem in using a wider variety of species is in knowing the particular site conditions in which the species will flourish. Climate matching programs can be useful when introducing species from overseas (Booth 1996) but these are less useful when reforesting degraded sites with native species. Even in undisturbed forests the realized niche or actual distribution of a species is not necessarily the same as the potential niche or area over which it is capable of growing (because of the effects of competition) while in disturbed or degraded sites a former habitat may no longer be suitable even though the climatic conditions are unchanged. Such changes may prevent some otherwise attractive species from being used unless the site can be ameliorated, perhaps with fertilizers. Informed judgements about the types of sites likely to be preferred by a particular species are useful (e.g. Appanah and Weinland 1993) but, ultimately, the only way of resolving the issue is by testing the species in small trial plantings at a number of sites having different microclimates and soils. These can indicate the site preferences of particular species but the approach can also reveal which species are sufficiently tolerant to be able to be planted at a wide variety of sites that have quite different conditions (Butterfield 1995).

What Are Appropriate Pruning and Thinning Schedules?

When trees are being grown for sawlogs, a fourth issue concerns tree form. This is particularly the case with high-value timber trees where poor form can substantially reduce the price a grower receives for a log. Seed of indigenous species collected from natural stands will usually produce trees that vary quite markedly in both form and vigour. Some seedlings are likely to be heavily branched or have multiple leading shoots even at an early age. A breeding program may eventually resolve many of these disadvantages but, in the short term, they mean that more care than usual will be needed to ensure high quality logs are produced. A simple form-pruning at the seedling stage may solve many problems. Likewise, slightly higher-than-usual planting densities may be useful in minimizing branching and accentuating stem straightness. Otherwise careful pruning will be needed once canopy closure occurs. Some smallholders may find pruned branches and leaves are useful as fuel or feed for stock.

Any variation in tree vigour will amplify the need for an early thinning to remove unthrifty trees or those with poor form that compete with better quality individuals for soil resources and light. There are two basic forms of thinning. One, referred to as 'thinning from above', removes the largest trees forming the upper canopy. In a monoculture the advantage of this is that these trees will have the largest value because of their size; the disadvantage is that it leaves behind a plantation composed of the least vigorous trees to grow through and form the final crop. The quality of seed subsequently collected from these residual trees is also likely to be poorer than seed collected from the more vigorous trees with better form. For this reason the more common form of thinning used is 'from below'. In this case sub-dominant or suppressed trees having poor stem form are removed in the first thinning to reduce the competitive pressure on better quality trees and allow these to take advantage of the increased space. As the trees grow and crowns expand, subsequent thinning might switch from tree form or canopy position to improving between-tree spacing to reduce competition.

Thinning should be carried out as soon as it is possible to identify the trees likely to form the final harvest. This might be when trees are around 6–8 m tall. The influence of tree density on growth is shown in Fig. 6.8. This shows how annual growth of *Flindersia brayleyana* dramatically slows as between-tree competition intensifies and the point where self-thinning commences is approached. The periodic mean annual diameter increment of 20-year old unthinned stands was about 0.2 cm $year^{-1}$ but was over 1.2 cm $year^{-1}$ in thinned stands (Keenan et al. 2005).

In the absence of empirical data from field trials it is possible to calculate appropriate tree spacings using crown ratios (ratio of crown diameter to stem diameter) and the expected stem diameter increments based on most recent growth (Keenan et al. 2005). The spacing between trees needed to avoid canopy closure (and hence between-tree competition) is indicated by the following formula:

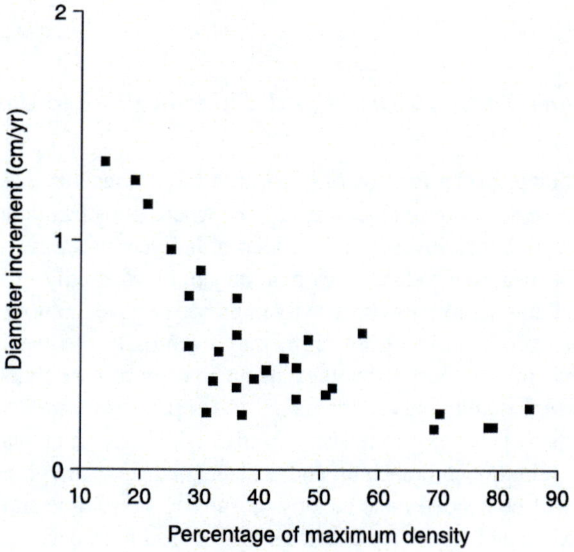

Fig. 6.8 Periodic mean annual diameter increment for 5 years after thinning in *Flindersia brayleyana* plantations in tropical Australia in relation to tree density (expressed as the percentage of the maximum density possible before self thinning occurs). The greatest tree growth rates were in stands having stocking rates less than 25% of the maximum stand density (Redrawn from Keenan et al. 2005)

$$S = CR(dbh + [Inc \times T])$$

where

S = spacing (m) needed
CR = crown ratio for the species.
Dbh = current stem diameter at breast height (cm)
Inc = expected annual diameter increment (cm $year^{-1}$)
T = time interval to next thinning (years)

Suppose a tree had a crown ratio of 20, a current dbh of 20 cm, an expected future diameter increment over the next few years of 1.0 cm $year^{-1}$ (based on recent growth measurements) and the next thinning was to be in 5 years time. Then the spacing between trees needed now to allow unimpeded growth for the next 5 years should be 5.0 m (equivalent to an area of 25 m^2 per tree or a density of 400 trees per ha). Similarly, a tree with a crown ratio of 25 but the same diameter and growth rate would require a spacing of 6.3 m. A more vigorously growing tree would achieve canopy closure earlier and need an earlier thinning.

Thinning can be especially attractive for some smallholders because it offers the possibility of an early financial return. But many small growers may be reluctant to do it unless there is a market for their trees. An equally important disincentive is that the idea of removing some trees in order to ultimately favour the growth of others may be counter-intuitive. This is despite the fact that substantial improvements in tree growth will be achieved by doing so. And the idea of first removing the worst trees may be even less appealing – surely one can get a better price today by selling the largest trees first? This may be true but it is at the expense of ultimately getting far better prices from the fastest growing trees with the best form. It also means any seed collected from the residual stand will be of a poor quality. This type of issue will probably be resolved over time as more demonstration plots, field evidence and technical advice becomes available to local growers.

One powerful way of demonstrating in the field the importance of tree spacing is to plant trees using a Nelder wheel design (Nelder 1962). This design involves using concentric rings of trees with the trees planted in spokes that intercept each of the rings (Fig. 6.9). The arrangement generates a pattern in which the mean space per tree is greater in trees growing in the larger diameter rings than it is for trees growing in smaller rings in the middle of the planting where the spokes are closer. These layouts can be designed to create tree densities equivalent to 3,600 tph down to about 50 tph (Lamb and Borschmann 1998). The adverse impact of close spacings on tree diameter usually becomes clear in a very short time.

Thinning may also be necessary for purposes other than improving timber production. For example, some tree-growers may wish to thin their plantations and increase the amount of light able to reach the forest floor in order to grow grass beneath their trees or underplant trees with other crops. And some growers may wish to encourage the expansion of crowns of fruit or nut trees (e.g. species of *Canarium, Illicium, Artocarpus, Durio, Terminalia*) to improve the harvest of these non-timber products. In both these cases tree densities of around 100 tph may be needed although local knowledge will often provide guidelines.

Fig. 6.9 A Nelder wheel layout for examining the effect of tree spacing on tree growth and morphology. The average area per tree is smaller (and hence the tree density is higher) in the inner rings of the trial than in the outer rings

What Are the Growth Rates?

Besides the type of timber produced, one of the key factors likely to influence the choice a landowner make about the species to plant is the potential growth rate. It is usually possible to predict how well many fast-growing exotics will perform because they have now been planted across a wide range of field conditions. By contrast, it is still difficult to say much about most indigenous species simply because there are too few field trials at too few sites that have been regularly measured over a sufficient period of time. The best rule-of-thumb available is that species with higher density timbers tend to be slower growing.

One way of generating a first estimate of the growth of these species in plantations is to measure their performance in the many, often scattered, plantings found on local farms. Knowing the size and age of a stand a crude measure of growth rates can sometimes be calculated. A modeling approach that used sparse data like this to develop preliminary stand models for plantation trees in Philippines was described by Venn and Harrison (2001) and a similar though less sophisticated study was carried out using growth data collected from a number of farm plantings and young species trials scattered across several provinces in northern Vietnam (Lamb and Huynh 2006). Most of the stands in this survey were less than 20 years old but the majority had similar tree densities and contained at least 20 individual trees. The planting dates at each site were verified and measurements were made of tree height and diameter. Growth curves were determined for each species by fitting a curve that approaches an asymptote representing the maximum diameter normally achieved by the species in natural forests (Fig. 6.10). The relationship between dbh and time (t) used a Chapman–Richards growth function:

$$\text{dbh} = dbh_{max} * (1 - b * e^{-k*t})^{c}$$

The variables required for parameterization of this equation are *dbh max* = maximum diameter at breast height for the species at maturity obtained from the local literature, and the co-efficients *k* and *c*, obtained through curve-fitting. The variable *b* is assumed to be one for all species studied. In this particular case a second curve was fitted approaching an asymptote equivalent to only 80% of the maximum tree diameter in order to generate a more conservative growth estimate. In this way a simple

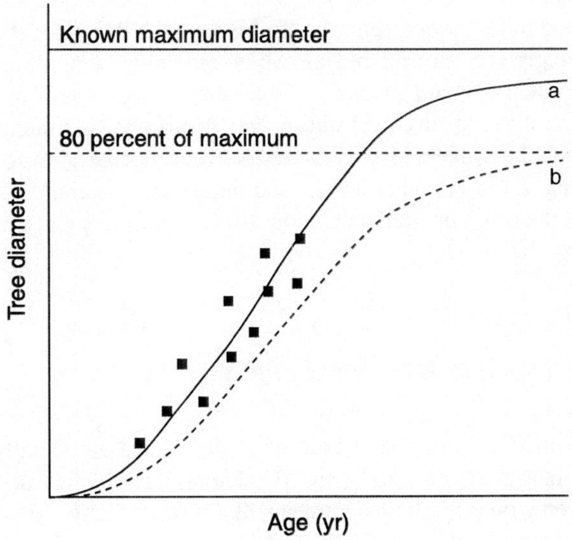

Fig. 6.10 Using growth trends in existing young plantations to obtain a first crude estimate of future growth rates of a particular species. (**a**) is a trend line based on these plantations and fitted to an asymptote representing the maximum recorded diameter for that species. (**b**) is fitted to an asymptote representing 80% of that maximum diameter to provide a more conservative estimate of future growth

Table 6.5 Estimated mean annual volume increments (MAI) of tree species grown in farm plantations in northern Vietnam assuming stand density of 800 tph and a 30-year rotation in comparison with MAI reported in published literature (see Table 6.4)

	Estimated MAI	Reported MAI
Species	$m^3\ ha^{-1}\ year^{-1}$	$m^3\ ha^{-1}\ year^{-1}$
Chukrasia tabularis	8.1	7.1
Erythrophloeum fordii	7.9	6.1
Michelia mediocris	7.4	4.5–6.8

first approximation could be made of the growth rates of various species. The results are shown in Table 6.5 together with other estimates of the productivity of these species derived from actual field trials. Overall, reasonably similar results were obtained giving reason to assume that a similar approach with other lesser-known species would provide a useful first approximation of growth and productivity.

Some judicious editing is usually called for. In the Vietnamese case, most sites were less than 20 years old but a few 80-year old stands established during the colonial era were also found. Their age made them attractive sites from a curve-fitting point of view but the data from these sites were well below the trend line generated by the young plots. One explanation is that trees of this species reach a growth plateau after around 20 years. However, a more likely explanation was that these older trees represent the residue of plantations that have been selectively felled from above leaving behind only the poorest and least vigorous trees. The poor form of these trees added weight to this view. Accordingly, data from these sites were deemed to be unrepresentative and were excluded from the analysis.

Of course, the growth rates of species can be manipulated and improved. Those with technical knowledge and financial resources can seek better quality seeds or seedlings and invest in improved plantation establishment and management. Many small farmers will be unable to do this and so may not be able to realize these potential advantages. On the other hand, such improvements could spread if extension services and learning networks develop to advise farmers and this is discussed further in Chapter 10.

What Are Appropriate Rotation Lengths?

Rotation lengths in government and industrial plantations are usually based upon detailed growth measurements and financial calculations. This is true whether fast-growing or slower-growing species are used. By contrast, many smallholders will probably have a rather more flexible attitude towards the timing of any fellings. They are unlikely to have detailed mensurational data other than a count of the total number of trees they own and perhaps a measurement, or an estimate, of the tree diameters (since diameter rather than volume is likely to be the main criterion used to assess the value of their trees at the time of sale). With the exception of those growing trees on contract to an industrial operator, many might opportunistically fell just one or two trees depending on the financial needs of the family. In such cases an even-aged plantation might gradually change into an uneven aged plantation.

However, a variety of other factors may also be important. Short rotations mean there is an earlier financial return to the grower, logs will be smaller and easier to physically carry to a road side and there is a shorter period during which theft might occur. On the other hand, long rotations mean the recovery rate is higher (i.e. proportionally less timber is wasted when larger logs are sawn) and there is more time for heartwood to develop. In species such as teak it is the heartwood that is most valued by the market. Other events determining the timing and intensity of harvesting may be changes in market access such as the development of new roads into the area, visits by log buyers or the arrival into the district of portable sawmills. Such practices will not maximize the financial returns to growers but probably suit their circumstances since, in the absence of rural banking services, many may regard their trees as a form of bank deposit to drawn upon when needed. Not all farm tree growers are in this situation of course and, over time, the management practices and rotation lengths of other small plantation owners will probably become similar to those used in many government plantations. Depending on species and market circumstances, these may vary between 20 and perhaps 60 years.

How Should Natural Regeneration Beneath the Plantation Canopy be Managed?

A factor that complicates the notion of a simple rotation is that some plantations gradually acquire a diverse understorey containing other species. Parrotta et al. (1997) summarize a number of studies that have examined the phenomenon. The extent to which this occurs depends on the type of landscape in which the plantation is located. If natural forest is nearby the new understorey may be mostly native species. If the surrounding landscape is largely an agricultural one then the understorey may be mostly weeds. The processes involved are essentially the same as those occurring in regrowth forests and were discussed in Chapter 5. Thus, small-seeded species are usually better represented than large-seeded species. Provided the extent of this understorey is modest there is probably little effect on the growth of the plantation trees. But, if left untended, the size of the understorey plants and diversity of species represented can become quite significant and pose a dilemma for managers.

An example of the extent to which colonization can occur is illustrated by a study in monocultural plantations in the humid tropics of northern Australia. In this case a survey using 151 plots (collective area totalling 1.2 ha) and located in mainly older (>40 years) plantations found 176 tree species were present (Keenan et al. 1997). The rates at which colonists were able to become established appeared to depend on the identity of the plantation species with plantations of hardwood species acquiring rather more diverse understories than conifer species (Firn et al. 2007; Keenan et al. 1997). Not surprisingly the species diversity and structural complexity of these understoreys increased with age. An illustration of this is shown in Fig. 6.11. Other studies have also found the identity of the plantation species affects the amount of understorey development with species having dense canopies having less understorey development than those with more open canopies. Thus teak monocultures in Thailand had much less understorey development than found in more open crowned polycultures (Kaewkrom et al. 2005).

The phenomenon has both disadvantages and advantages. From a silvicultural perspective the process might be seen as a problem because the newcomers slow the growth of the preferred plantation trees. In this case the colonists should be

Fig. 6.11 Colonization of a 60-year old plantation monoculture of *Flindersia brayleyana* by species from an adjacent patch of natural forest. Many of the colonists have now grown up and their crowns are occupying the upper canopy (Photo: J. Firn)

controlled or removed. On the other hand, the process might also be seen as having some benefits because the understorey is likely to protect the soils from erosion much more successfully than the structurally simple plantation monoculture. In addition, the increasing structural complexity of the plantation is likely to make it a more attractive habitat for some wildlife. Once such an understorey has developed the key question is how should such stands be managed? The obvious answer is that it depends on circumstances.

There are four possibilities. One is to fell the plantation trees (together with the colonists) as was the original intent. If desired, the site could then be replanted to grow a second rotation of trees. This option would be appropriate in circumstances where it was necessary to receive a return on the original financial investment. A second option might be to abandon any ideas of felling because the site is clearly beginning to acquire a significant conservation value. This might be the case where only small and scattered patches of natural forest remained in the area. The site would then be managed to conserve and enhance the newly acquired biodiversity. Such a choice might not be one that many private growers could afford but it might be attractive to some governments or non-government organizations. It might be especially attractive in parts of a plantation growing along water courses or in sections of a plantation that could form a firebreak through a plantation estate. A third option might be to selectively harvest the planted trees but leave the other species and allow the site to continue to regenerate. Provided the logging operation was done carefully this would have the advantage of repaying the investment costs and would leave the site as a secondary forest at which further successional development would be possible. This choice represents something of a mid-way choice between Option 1 and Option 2. Finally, and depending on the size and identity of the colonizing trees, the forest could be managed as a selection forest instead of being clear-felled. In this case planted trees and any commercially valuable colonists would be removed once they reached a suitable size class. Over time the proportion of trees belonging to the species that had been originally planted would probably decline. This situation might be attractive when the original forest had been surrounded by natural forest containing many commercially valuable species. This latter option might also be one with some attractions for private smallholders.

Monoculture Plantations, Biodiversity and Ecosystem Services

Most plantations are established to create a timber resource but many growers also expect them to provide certain ecological services as well. Plantations established using just one species can provide some of these services but not others. Unlike secondary forests, most plantations are eventually clear-felled and their capacity to provide various ecosystem services depends on the rotation length and the way this clear-felling and (subsequent re-planting) is managed.

Biodiversity

Monocultural plantations established for timber production might seem to offer few opportunities for protecting or conserving biodiversity. Most have been famously regarded as 'biological deserts' and young plantations of exotic species managed on short rotations of 7–10 years are unlikely to be very useful as wildlife habitats (although they may still be better than many agricultural crops). But the situation can be different in older plantations as the previous discussion concerning understorey development showed. Much depends on the plantation age and on the landscape context in which the plantation is located. A plantation established near a large area of otherwise undisturbed forest is almost certainly likely to acquire a species rich understorey of trees and shrubs brought in by seed dispersers (teak plantations with dense canopies being a notable exception). The structural complexity created by such colonists can often attract some wildlife colonists although many of them are likely to be habitat generalists rather than specialists from more undisturbed forests (Bell 1979; Kwok and Corlett 2000; Mitra and Sheldon 1993).

There are also some ways that even monocultural plantations can be made into more attractive habitats for other species. One way of doing this is to develop greater structural complexity across the plantation area. This will normally occur once thinnings and clear-fellings occur and the plantation estate becomes a more heterogeneous mosaic of young and old stands together with those that have been thinned to different degrees at various times. As well as increasing spatial heterogeneity these management interventions also modify litter layers and increase coarse, woody debris on the forest floor. All of these changes can improve habitat conditions for some species (Carnus et al. 2006; Hartley 2002).

A second way in which biodiversity might be conserved within the plantation estate is to avoid having large, contiguous plantation blocks. This can be done by separating different compartments with strips of natural vegetation, secondary forest or restored forest (Lamb 1998). As mentioned in Chapter 3, this approach is being developed in the large plantation program being developed by the Grand Perfect consortium of companies in Sarawak. The concession area covers 490,000 ha but just under half is to be used for plantations of mainly *Acacia mangium,* a third will be left for conservation purposes and the remainder for indigenous groups (Cyranoski 2007). A similar spatial arrangement is being developed in plantations in Sarawak (Wooff 2009) and Kalimantan (Marjokorpi and Otsamo 2006). The key issue with such schemes is to have a network of corridors to allow wildlife movement within the area as well as patches of natural (or restored) forest also within the area and able to provide habitat for particular species.

A third way in which biodiversity may accrue is if a number of different species, including native species, are grown in separate plantations. Thus the landscape becomes a mosaic of different monocultures (Lamb 1998). Although each remains comparatively simple they collectively create a more heterogenous landscape. Different wildlife may find certain plantations more useful than others such that the overall collective benefit is greater than if the whole area had been planted with just one species. These landscape issues are discussed further in subsequent chapters.

Monocultural plantations will never be able to provide the habitats needed to sustain the full range of species found in undisturbed forests but they will add heterogeneity to otherwise deforested landscapes and facilitate the movement of at least some species between residual patches of natural forest.

Watershed Protection

Reforestation is usually assumed to improve watershed protection by protecting soil surfaces and helping to mechanically re-enforce upper soil horizons. By doing so it helps reduce surface soil erosion and limits shallow (but not deep >3 m) mass wasting. The best form of reforestation is one that creates several leaf layers above ground and litter layers on the soil surface. This breaks the impact of raindrops and reduces soil movement. But even a simple forest cover can often reduce erosion. Zhou et al. (2002) measured erosion in a young (<14 years) eucalypt plantation in southern China over a period of 10 years. Litter and branches from the forest floor were regularly collected from the plantation for fuel. Even so, the annual erosion rate from the plantation averaged 9.1 kg ha^{-1} mm^{-1} of rainfall while that in a nearby site with bare soil and no trees was 43.7 kg ha^{-1} mm^{-1}. By comparison only 0.3 kg ha^{-1} mm^{-1} of soil was lost from a nearby mixed species plantation site that had a more complex canopy structure. Overland flow is likely to be greatest on steep slopes where there is little ground cover and the hydraulic conductivity of surface soils is low. There will be more sub-surface flow where slopes are gentler, there is denser ground cover and hydraulic conductivity is high.

Some might argue that the benefits of afforesting or reforesting well-established grasslands might be slight since the soil surface is already substantially protected. In these circumstances trees will simply shade the grass and reduce this cover. Leaving aside the effectiveness of different types of grass cover, evidence suggests that at least some grazed grasslands are not as effective as a good tree cover and that grasslands subjected to recurrent fires or grazing provide only modest cover (Sidle et al. 2006; Tomich et al. 1997). Tree plantations can be burned too, of course, but most are likely to be managed to exclude fires. Moreover, the transition from grassland to forest plus understorey is usually gradual enough to avoid the development of bare ground. Bare ground can be found beneath plantations when people collect litter and twigs for firewood but otherwise a good ground cover is usually present (compare Fig. 6.2 where litter is being collected with Fig. 4.2 where it is not).

In most tree plantations erosion is likely to be greatest at the time of logging. Logging means the upper canopy and much of the lower understorey are destroyed. There may also be considerable soil disturbance associated with log removal and replanting. Rain during these events is likely to cause erosion and this will be accentuated if logging debris is burned. Depending on site conditions it may take some time before sufficient plant cover regenerates and is able to protect the site once more. The shorter the rotation the more frequent these disturbances and the greater the hazard of erosion and sedimentation. Root systems help limit landslips but take some time to develop (Sidle et al. 2006). This means pulpwood plantations grown on short rotations are likely to be less effective than sawlog plantations that use longer rotations. Erosion from plantations can be reduced by leaving unlogged buffer strips along riparian areas and encouraging understorey development in these.

Water Flows

A common public perception is that reforestation of treeless sites will improve the yield of water from catchments and increase river flow in dry seasons. Unfortunately this is usually not true which means the consequences of reforestation may be quite different to those expected by many members of the community. A very large literature indicates that the large-scale reforestation of grasslands or former crop lands is more likely to reduce water flows because of the rainfall intercepted and evaporated from tree crowns and transpirational losses from the trees themselves. These are increased by the higher Leaf Area Indices of trees and perhaps by deeper root systems that give access to more soil water and groundwater resources (Bruijnzeel et al. 2005; Scott et al. 2005). The type of land being afforested/reforested is important. A global survey by Farley et al. (2005) involving a variety of species found that afforestation of grasslands reduced run-off by an average of 44% while reforesting shrublands reduced run-off by 31%. The decrease in run-off following reforestation was greater in higher rainfall areas although the proportional reduction was usually largest in drier locations. Likewise, the effect was especially pronounced in the dry season when the proportional reductions were greatest. A summary of some of the most important hydrological changes caused by afforestation or reforestation is given in Table 6.6.

The identity of the species used in afforestation/reforestation can make a difference to the extent of the changes in run-off. In their global survey Farley et al. (2005) found that, on average, the run-off from grasslands planted with pines decreased by 40% but those planted with eucalypts decreased by 75%. These between-species difference are likely to be affected by differences in root depth as well as leaf areas and canopy density (and hence the ability to intercept rainfall) with the importance of each factor depending on the overall rainfall and intensity of rain at a particular site. In general fast-growing species tend to use more water than slow-growing species and past controversies, especially in India, about the role of eucalypts in using water were probably a function of their often rapid growth.

Plantation age is also important. The reduction in run-off occurs soon after reforestation and can reach 10% within 2 or 3 years after planting although the full impact may not be observed for several decades. There is some evidence that run-off can eventually increase once plantations exceed 30 or more years (Bruijnzeel et al. 2005).

Peak flows and flooding can be affected by reforestation but the effect is inversely related to the size of the storm. The peak flow resulting from a small rainfall event may be reduced by reforestation (because there is more interception

Table 6.6 Generalised effects of deforestation and reforestation on forest hydrology in small catchments. See text for details and qualifications

Activity	Total water yield	Dry season flows	Local flooding
Deforestation	Water yield increases in proportion to amount of forest removed.	Flows reduced if topsoils degraded and infiltration capacity reduced but increased if soil infiltration capacity can be maintained.	Flooding may increase although rainfall intensity and catchment characteristics are probably more important determinants.
Reforestation	Water yield decreases in proportion to extent of catchment reforested but is also affected by planting density and tree age.	Flows usually decreased though may increase at seriously degraded sites if soil infiltration is improved by reforestation.	Can reduce flooding under certain conditions; effects in severely degraded sites unknown. Any effect reduced as size of catchment increases.

of rain by tree crowns and litter). In the case of larger storms the presence or absence of forests is likely to have little effect on watershed responses under conditions of saturated soils (van Dijk and Keenan 2007).

These generalizations may not apply to all sites and different patterns may occur when severely degraded sites are reforested. In these situations a new forest cover can improve soil structure and the rate at which water is able to infiltrate the topsoil thereby increasing the store of groundwater. Whether this changes the annual water yield or dry season flow depends on whether the improved rate of infiltration caused by reforestation is able to exceed the increased rate of evapotranspiration caused by the trees (Bruijnzeel 2004; Bruijnzeel et al. 2005).

Water is becoming an increasingly important resource in many agricultural areas meaning that the advantages of forest restoration will have to be balanced against the changes restoration may cause to run-off. However, all these hydrological impacts are dependent on scale and related to the proportion of the watershed covered by the new forests. In most cases this is small and the effects of vegetation on run-off are easily overwhelmed in larger storms (Van Dijk and Keenan 2007). Managers must then make trade-offs between reductions to overall or seasonal run-off, improvements in watershed protection and gains in other ecosystem services such as carbon sequestration that might arise from reforestation. There are some situations where the ability of trees to transpire water and thereby lower water tables is considered to be a good thing. Indeed, it may even be the primary motive for reforestation in places affected by salinisation or where swamps are to be reclaimed for agricultural purposes. But successes can be reversed when land managers fail to recognize the processes involved. This is illustrated by the case study from Papua New Guinea described in Box 6.4.

Box 6.4 The reversibility of hydrological changes

In the early 1970s some 1,500 ha of poorly drained lands near Mt Hagen in the Western Highlands of Papua New Guinea were planted with eucalypts to provide fuelwood for the tea factory being established in the district. *Eucalyptus robusta* was used because it is tolerant of badly drained sites although some drains were also dug to help get the trees established. All the land is owned by customary landholders but the Forestry Department was able to reforest it because the owners found it unsuitable for other purposes. The plantations prospered and the trees grew well causing a significant drop in the site's water table. Believing the area was now suitable for growing coffee some landowners began clearing blocks of trees. This caused the water table to rise once more and the new coffee plantings have subsequently failed. Trials are underway to determine whether a more diffuse tree clearing program might allow a tree-coffee combination that maintains the water table at a depth sufficient to allow coffee to grow (W. Amos, personal communication, 2009).

Carbon Sequestration and Storage

There is increasing global interest in the capacity of plantation forests to sequester carbon and this may become a means by which growers are able to derive additional income. There are two ways in which plantations can help sequester carbon. The most obvious of these is in the above-ground biomass. The more productive or older a plantation, the greater the amount of carbon immobilized. Estimates of the amount of carbon contained in the biomass of various types of plantation are shown in Table 6.7 together with estimates of biomass carbon in undisturbed forest and grassland. These plantations are located across a range of climatic conditions and soils and the data include direct measures as well as simply estimates of biomass carbon. Because of this they are simply indicative of a pattern without being definitive.

All the plantations have much lower carbon contents than the undisturbed tropical forest but higher contents than the grassland. On the other hand, the carbon content of older *Araucaria cunninghamii* plantations growing in a seasonally dry area of sub-tropical Australia were similar to that found in nearby undisturbed rainforests (Richards et al. 2007). They also show that fast growing species such as *Eucalyptus* and *Acacia* are able to sequester carbon more rapidly than the slower growing species managed on longer sawlog rotations. This means three 10-year rotations of a fast-growing species are likely to fix more carbon than, say, a single 30 year rotation of a slower growing species. But there is a difference between sequestration and

Table 6.7 Estimates of biomass carbon stored in natural forest, grassland and in plantations

Vegetation	Total tC/ha	Above ground biomass tC/ha	Roots and rhizome tC/ha	Reference
Undisturbed dipterocarp forest	295	235	60	Tomich et al. 1997
Imperata grassland	30	15	15	Tomich et al. 1997
Acacia mangium plantation (8 years)	135	105	30	Tomich et al. 1997
Eucalyptus grandis plantation (7 years)	164	140	24	Goncalves et al. 1999
Tectona grandis plantation (20 years)	120	104	16	Kraenzel et al. 2003
Araucaria cunninghamii (25 years)	–	110	–	Richards et al. 2007
Pinus elliottii plantation (29 years)	157	126	32	Simpson et al. 1999
Shorea robusta plantation (40 years)	–	270	–	Lugo et al. 1988
Araucaria cunninghamii plantation (50 years)	–	165	–	Richards et al. 2007
Araucaria cunninhamii plantation (63 years)	–	173	–	Richards et al. 2007

storage. The longevity of the sequestered carbon will be shorter in pulpwood plantations than sawlog plantations; fast grown timber is mostly used for short-lived products like paper or fuelwood and any fixed carbon soon re-enters the atmosphere. Timber produced in sawlog plantations with longer rotations is more likely to be used for products with a longer life such as construction materials or furniture. In this sense the carbon is stored for a longer period (including the period of plantation growth). Although these represent large amounts of sequestration the benefits a grower might receive depend not only on the market for carbon but on the abatement costs (the opportunity costs of entering the carbon market) and the transaction costs (monitoring and payment arrangements including payments when trees are felled). These issues will be discussed further in Chapter 9.

The second way plantations might assist sequester carbon is if they can accelerate the build-up of soil carbon. Carbon sequestered in this form is likely to be stored for much longer than that contained in biomass. Studies of the changes in topsoil carbon following reforestation show there are considerable differences in the rate at which this occurs. These differences between species are associated with difference in rates of litterfall and fine root turnover as well as in litter chemistry and decay rates. In some cases soil carbon increases rapidly following reforestation but in other cases it may decrease for a number of years before eventually increasing (Fig. 6.12). A number of factors appear to influence these patterns including the

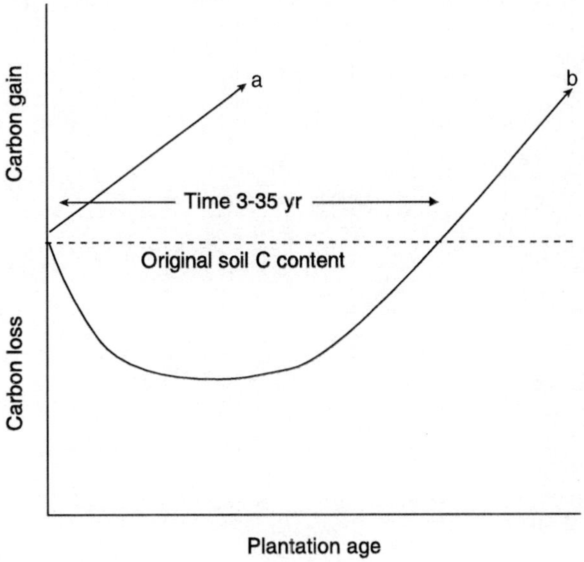

Fig. 6.12 The soil carbon content after plantation establishment. In some cases (**a**) the carbon content may increase immediately following reforestation but in many cases there is a decline in carbon content lasting up to 35 years before it begins to recover (**b**). Pattern **a** appears more likely to occur when broad-leaved species are used to reforest former croplands while **b** is more common when grasslands are reforested

previous land use, the species used in reforestation and the climate (Guo and Gifford 2002; Paul et al. 2002; Turner and Lambert 2000).

Rapid increases in soil carbon are most likely when reforestation takes place on former cropland and when broad-leave species are used in reforestation. The situation appears to be different when former pasture sites are reforested. In these cases broad-leaved species have less effect and soil carbon stores are sometimes reduced. Soil carbon also decreases when conifers are used to reforest pastures and the largest declines have been observed at sites with higher rainfall (>1,200 mm). In time soil carbon levels eventually rise again and exceed those at the time of planting but this may take 3–35 years (Turner and Lambert 2000; Paul et al. 2002). One study in the seasonally dry sub-tropics of Australia found soil carbon stocks had still not recovered 63 years after a pasture site had been reforested using the conifer *Araucaria cunninghamii* (Richards et al. 2008).

The reason for these contrasting patterns is not clear. Grasslands often have large root biomasses and carbon losses may result from the oxidation of these when sites are prepared for planting (Turner and Lambert 2000; Guo and Gifford 2002). Others suggest it is due to a low rate of litterfall in the early stages of plantation establishment which is unable to compensate for the continued decay of soil carbon from the original agricultural residues (Bashkin and Binkley 1998; Paul et al. 2002). As litterfall increases the net soil carbon also begins to increase. The reduction in soil carbon caused by reforestation with conifers may be due to their comparatively lower rate of root turnover meaning less carbon is accumulated in the topsoil. On the other hand, conifer plantations often accumulate much greater litter

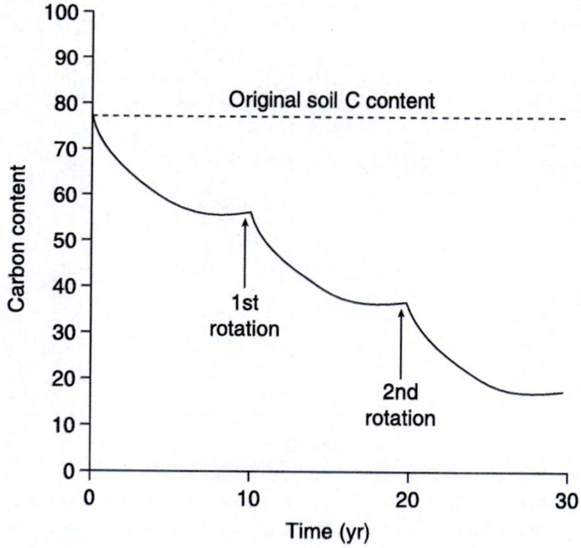

Fig. 6.13 Successive plantations grown using short rotations may cause a gradual decline in soil carbon over time

layers than do broad-leaved species and this may make up much of the difference (Paul et al. 2002).

Most of these studies have been carried out in temperate regions and more detailed work is needed in tropical environments. However, they suggest a major difference in the capacity of short and long-rotation plantations to sequester carbon. If rotation lengths are less than the recovery time then soil carbon levels may decline sharply after several felling cycles (Fig. 6.13). This, together with the ephemeral nature of carbon in the above ground biomass produced by such plantations means there is a clear difference between short and long-rotations plantations with the former being much less attractive than the latter.

Conclusions

Simple plantation monocultures are the most common form of reforestation now being carried out. Most involve a small number of exotic species that have been found to grow well in a variety of sites and produce timber for which there is a ready market. Plantations can be grown on short rotations for pulpwood or long rotations for sawlogs before being harvested. Most pulpwood plantations in the region use *Acacia* or *Eucalyptus* species while a small number of species such as *Gmelina arborea* and various species of *Pinus* dominate the sawlog market for utility timbers. Monoculture plantations suit industrial plantation owners because they can produce large amounts of a uniform material at a relatively low cost. Many industrial growers are now moving into a biotechnology phase where increasingly sophisticated forms of improved planting material are being developed for the desired species.

Some growers are also using species able to produce higher quality or specialty timbers. These include exotic species such as teak (*Tectona grandis*) and mahogany (*Swietenia macrophylla*) because they are well-known and popular in the international market. But there are a variety of indigenous species that might also be grown if the market demand generates prices able to compensate for their slower growth. At present surprisingly little is known about the silviculture of most of these species. This reflects the emphasis given in recent decades by government research bodies and industrial growers to finding ways of maximizing the plantation productivity using a handful of fast-growing exotics. By default these exotics have become regarded as the species to use when reforesting degraded or abandoned lands. In fact, they may be appropriate for this purpose in some cases but not in others.

Apart from this tendency to use a very small number of species, simple plantation monocultures do have some other disadvantages. Like all monocultures there are ecological risks associated with growing genetically similar plants across large contiguous areas and the young age of most plantation estates suggests the full range of biological hazards may not have yet been experienced. In addition, plantation monocultures are not able to provide a full range of goods and ecological services provided by natural forests. They provide habitats for only a limited number of wildlife species and may impose severe drain on water resources from a comparatively early age. On the other hand, some can sequester and store large amounts of carbon from an early age.

Plantation monocultures will suit some smallholders but not others. The rapid growth of some exotic tree species is obviously an advantage to those needing an early cashflow. However, rapid growth does not always mean there will be an equally large financial return and the value of any plantation to a landholder depends on much more than just growth rates. Some smallholders may wish their plantations to provide a variety of products and might not be satisfied with simple monocultures that commit them to a single product. And many smallholders may be less able to judge, let alone undertake, the ecological or economic risks accepted by large industrial growers of plantation monocultures. In short, the limited number of goods and services together with the risks associated with monocultures may mean that mixed-species plantations could be a more attractive alternative for some growers. These issues are discussed in the next chapter.

References

Anon (1999) Integrated report on the multi-storied forest management project in Malaysia (1991–1999), Forestry Department Penisular Malaysia. Perak State Forestry Department, Japan International Cooperation Agency, Kuala Lumpur

Appanah S, Weinland G (1993) Planting quality timber trees in peninsular Malaysia: a review. Forest Research Institute, Malaysia, Kepong

Asher C, Grundon NJ, Menzies N (2002) How to unravel and solve soil fertility problems. Australian Center for International Agricultural Research, Canberra

Bashkin MA, Binkley S (1998) Changes in soil carbon following afforestation in Hawaii. Ecology 79:828–833

Bell HL (1979) The effects on rainforest birds of planting teak, *Tectona grandis*, in Papua New Guinea. Austr Wildl Res 6:305–318

Bell LC (1996) Rehabilitation of disturbed land. In: Mulligan DR (ed) Environmental management in the Australian minerals and energy industries: principles and practices. University of New South Wales Press, Sydney, pp 227–264

Bell LC (2001) Establishment of ecosystems after mining - Australian experience across a diverse biogeographic zones. Ecol Eng 17:179–186

Booth TH (1996) Matching trees and sites: proceedings of an international workshop held in Bangkok, Thailand 27–30 March 1995. Australian Center for International Agricultural Research, Canberra

Bradshaw AD, Chadwick M (1980) The restoration of land. Blackwell Scientific Publications, Oxford

Bristow M, Annandale M, Bragg A (2005) Growing rainforest timber trees: a farm forestry manual for North Queensland. Rural Industries Research and Development Corporation, Canberra. https://rirdc.infoservices.com.au/items/03–010; accessed 16 September 2010

Bruijnzeel LA (2004) Hydrological functions of tropical forests: not seeing the soil for the trees? Agric Ecosyst Environ 104:185–228

Bruijnzeel LA, Bonell M, Gilmour DA, Lamb D (2005) Conclusion: forest, water and people in the humid tropics: an emerging view. In: Bonell M, Bruijnzeel LA (eds) Forests water and people in the humid tropics. Cambridge University Press and UNESCO, Cambridge, pp 906–925

Butterfield RP (1995) Promoting biodiversity: advances in evaluating native species for reforestation. For Ecol Manage 75:111–121

Cameron DM (1995) Evaluation of performance of indigenous high value species within the forest development area and proposals for further research and development to support plantings by farmers. Jaakko Poyry Consulting AB/Interforest, Sweden

Cameron DM, Jermyn D (1991) Review of plantation performance of high value rainforest species. Australian Center for International Agricultural Research, Canberra

Carnus J-M, Parrotta J, Brockerhoff E, Arbez M, Jactel H, Kremer A, Lamb D, O'Hara K, Walters B (2006) Planted forests and biodiversity. J Forest 104:65–77

Clarke WC, Thaman R (1993) Agroforestry in the Pacific Islands: systems for sustainability. United Nations University Press, Tokyo, New York, Paris

Coutinho TA, Wingfield MT, Alfenas AC, Crous PW (1998) Eucalyptus rust: a disease with potentially for serious international implications. Plant Dis 82:819–825

Cyranoski D (2007) Biodiversity: Logging: the new conservation. Nature 446:608–610

Dale JA, Johnson TJ (1991) Hoop pine forest management in Queensland. In: McKinnell FH, Hopkins ER, Fox JED (eds) Forest management in Australia. Surrey Beatty & Sons, Chipping Norton, pp 214–227

Dell B, Malajczuk N, Xu D, Grove TS (2001) Nutrient disorders in plantation eucalypts, 2nd edn. Australian Center for International Agricultural Research, Canberra

Do DS, Nguyen HN (2003) Use of indigenous tree species in reforestation in Vietnam. Agricultural Publishing House, Hanoi

Dreschel P, Zech W (1991) Foliar nutrient levels of broad-leaved tropical trees A tabular review. Plant Soil 131:29–46

Durst P, Brown C (2000) Current trends and development of plantation forestry in Asia Pacific countries. In: Proceedings of the international conference on timber plantation development Forest Management Bureau of the Philippines Department of Naturalk Resources, International Tropical Timbers Organisation, Food and Agriculture Organisation of the United Nations, Manila

Elliott S, Blakesley D, Maxwell JF, Doust S, Suwannaratana S (2006) How to plant a forest: the principles and practice of restoring tropical forests. Biology Department, University of Chiang Mai, Chiang Mai

Erskine P, Lamb D, Bristow M (2005) Reforestation in the tropics and subtropics of Australia using Rainforest Tree Species Rural Industries Research and Development Corporation, Canberra. https://rirdc.infoservices.com.au/items/05–087.pdf; accessed 16 September 2010

Evans J, Turnbull JW (2004) Plantation forestry in the tropics; the role, silviculture and use of planted forests for industrial, social, environmental and agroforestry purposes. Oxford University Press, Oxford

Eyles A, Beadle C, Barry K, Francis A, Glen M, Mohammed C (2008) Management of fungal root-rot pathogens in tropical *Acacia mangium* plantations. For Pathol 38:332–355

Farley KA, Esteban GJ, Jackson RB (2005) Effects of afforestation on water yield: a global synthesis with implications for policy. Global Change Biol 11:1565–1576

Firn J, Erskine PE, Lamb D (2007) Woody species diversity influences productivity and soil nutrient availability in tropical plantations. Oecologia 154:521–533

Goncalves JL, Poggiana F, Stape JL, Serrano M, Mell SL, Mendes K, Gava JL, Benedetti V (1999) Eucalypt plantations in the humid tropics: Sao Paulo, Brazil. In: Nambiar EKS, Cossalter C, Tiarks A (eds) Site management and productivity in tropical plantation forests. Center for International Forestry Research, Bogor, pp 5–12

Grigg A, Mulligan DR, Dahl NW (1998) Topsoil management and long-term reclamation success at Weipa, north Queensland, Australia. In: Fox HR, Moore HM, McIntosh AD (eds) Land reclamation, achieving sustainable benefits. Balkema, Rotterdam, pp 249–254

Guo LB, Gifford RM (2002) Soil carbon stocks and land use change: a meta analysis. Global Change Biol 8:345–360

Hammond D (2002) Hardwood programmes in Fiji, Solomon Islands and Papua New Guinea. Forest Plantations Working Paper No 21, Forest Resources Division, Food and Agriculture Organization of the United Nations, Rome

Hardwick K, Healy J, Elliott S, Garwood NC, Anusarnsunthorn V (1997) Understanding and assisting natural regeneration in northern Thailand. For Ecol Manage 99:203–214

Harrison SR, Venn TJ, Sales R, Mangaoang EO, Herbohn JF (2005) Estimating financial performance of exotic and indigenous tree species in smallholder plantations in Leyte Province. Ann Trop Res 27:67–80

Hartley M (2002) Rationale and methods for conserving biodiversity in plantation forests. For Ecol Manage 155:81–95

Hillis WE, Brown A (eds) (1978) Eucalypts for wood production. Commonwealth Scientific and Industrial Research Organisation, Canberra

Htwe UMM (2000) Teak plantations in Myanmar. In: Enters T, Nair CTS (eds) Site technology and productivity of teak plantations. Food and Agriculture Organisation of United Nations, Bangkok, pp 83–98

Kaewkrom P, Gajaseni J, Jordan CF, Gajaseni N (2005) Floristic regeneration in five types of teak plantation in Thailand. For Ecol Manage 210:351–361

Kahindi JHP, Woomer P, George T, de Souza Moreira FM, Karanja NK, Giller KE (1997) Agricultural intensification, soil biodiversity and ecosystem function in the tropics: the role of nitrogen-fixing bacteria. Appl Soil Ecol 6:55–76

Keenan R, Lamb D, Woldring O, Irvine T, Jensen R (1997) Restoration of plant diversity beneath tropical tree plantations in northern Australia. For Ecol Manage 99:117–132

Keenan RJ, Doley D, Lamb D (2005) Stand density management in rainforest plantations. In: Erskine P, Lamb D, Bristow M (eds) Reforestation in the tropics and subtropics of Australia using rainforest species. Rural Industries Research and Development Corporation, Canberra, https://rirdc.infoservices.com.au/items/05–087; accessed 16 September 2010

Keenan RJ, Sexton G, Lamb D (1999) Thinning studies in plantation grown Queensland maple (*Flindersia brayleyana*) in north east Queensland. Int For Rev 1:71–78

Kimura Y, Nishiyama Y (1999) Silvicultural manual for multi-storied forest management, Forestry Department Penisular Malaysia. Perak State Forestry Department, Japan International Cooperation Agency, Kuala Lumpur

Knowles OH, Parrotta JA (1995) Amazon forest restoration: an innovative system for native species selection based on phonological data and field performance indices. Commonw For Rev 74:230–243

Kraenzel M, Castillo A, Moore T, Potvin C (2003) Carbon storage of harvest-age teak (*Tectona grandis*) plantations, Panama. For Ecol Manage 173:213–225

Krishnapillay B (2002) A Manual for Forest Plantation Establishment in Malaysia. Forest Research Institute, Malaysia, Kepong

Kwok HK, Corlett RT (2000) The bird communities of a natural secondary forest and a *Lophostemon confertus* plantation in Hong Kong, South China. For Ecol Manage 130:227–234

Lamb D (1990) Exploiting the tropical rainforest: an account of pulpwood logging in Papua New Guinea. UNESCO and The Parthenon Publishing Group, Paris/Carnforth

Lamb D (1998) Large-scale ecological restoration of degraded tropical forest lands: the potential role of timber plantations. Restor Ecol 6:271–279

Lamb D, Borschmann G (1998) Agroforestry with high-value trees. Rural Industries Research and Development Corporation, Canberra, https://rirdc.infoservices.com.au/items/98–142; accessed 16 September 2010

Lamb D, Huynh DN (2006) Mixed species plantations of high-value trees for timber production and enhanced community services in Vietnam and Australia (unpublished report FST2000/003). Australian Center for International Agricultural Research, Canberra

Lee SS (1998) Root symbiosis and nutrition. In: Appanah S, Turnbull JM (eds) A review of the dipterocarps: taxonomy. Ecology and silviculture. Center for International Forestry Research, Bogor, pp 99–114

Lee SS (1999) Forest health in plantation forests in Southeast Asia. Australas Plant Pathol 28:283–291

Lemmens RHMJ, Soeriannegara I, Wong WC (eds) (1995) Plant resources of South-East Asia No 5(2) timber trees: minor commercial timbers. Backhuys Publishers, Leiden

Leslie AJ (1999) For whom the bell tolls. Trop Forest Update 9:13–15

Lugo AE, Brown S, Chapman J (1988) An analytical review of production rates and stemwood biomass of tropical forest plantations. For Ecol Manage 23:179–200
Marcar NE, Khanna PK (1997) Reforestation of salt affected and acid soils. In: Nambiar EKS, Brown AG (eds) Management of soil, nutrients and water in tropical plantation forests. Australian Center for International Agricultural Research, Canberra, pp 481–525
Marjokorpi A, Otsamo R (2006) Prioritization of target areas for rehabilitation: A case study from West Kalimantan, Indonesia. Restor Ecol 14:662–673
Miller HG (1986) Nutrient control of growth in temperate forests. ITE symposium No 20, Natural Environment Research Council, Institute of Terrestrial Ecology, Cambridge, UK, pp 147–152
Mitra SS, Sheldon FH (1993) Use of an exotic tree plantation by Bornean lowland forest birds. Auk 110:529–540
Myers RJK, De Pauw E (1995) Strategies for the management of soil acidity. In: Date RA (ed) Plant soil interactions at low pH: principles and management: Proceedings of the third international symposium on plant–soil interactions at low pH, 12–16 September 1993, Brisbane, Queensland, Australia. Kluwer, Dordrecht, The Netherlands
Nair KSS (2007) Tropical forest insect pests: ecology, impact and management. Cambridge University Press, Cambridge
Nambiar EKS, Cossalter C, Tiarks A (eds) (1999) Site management and productivity in tropical forest plantations. Center for International Forestry Research, Bogor
Nambiar S, Brown AG (1997) Management of soil, nutrients and water in tropical plantation forests. Australian Center for International Agricultural Research, Commonwealth Scientific and Industrial Research Organisation and Center for International Forestry Research, Canberra
Neil PE (1984) Climber problems in Solomon Islands forestry. Commonw For Rev 63:27–34
Neil PE, Barrance AJ (1987) Cyclone damage in Vanuatua. Commonw For Rev 66:255–264
Nelder JA (1962) New kinds of systematic designs for spacing experiments. Biometrics 18:283–307
Ng F.S.P (1983) Ecological principle of tropical lowland rainforest conservation. In: Sutton S, Whitmore TC, Chadwick AC (eds) Tropical rainforest ecology and management. Blackwell Scientific Publications, Oxford, pp 359–375
Ng F.S.P. (1996) High quality planting stock – has research made a difference? Center for International Forestry Research, Bogor
Old KM, Lee SS, Sharma JK, Zi QY (2000) A manual of diseases of tropical acacia in Australia, Southeast Asia and India. Center for International Forestry Research, Bogor
Oliver W (1999) An Update of Plantation Forestry in the South Pacific. SPC/UNDP/AusAID/FAO
Otsamo A, Adjers G, Hadi TS, Kuusipalo J, Vuokko R (1997) Evaluation of reforestation potential of 83 tree species planted on *Imperata cylindica* dominated grassland. New For 14:127–143
Otsamo R, Otsamo A, Adjers G (1996) Reforestation experience with dipterocarps on grassland. In: Schulte A, Schone D (eds) Dipterocarp forest ecosystems: towards sustainable management. World Scientific, Singapore, pp 464–477
Pancel L (1993) Forestation. In: Pancel L (ed) Tropical forest handbook. Springer Verlag, Berlin, pp 646–725
Parrotta J, Turnbull JM, Jones N (1997) Catalysing native forest regeneration on degraded tropical lands. For Ecol Manage 99:1–7
Paul KI, Polglase PJ, Nyakuengama JG, and P.K. K (2002) Change in soil carbon following afforestation. For Ecol Manage 168:241–257
Reuter DJ, Robinson JB (1997) Plant analysis: an interpretation manual. CSIRO Publishing, Canberra
Richards AE, Dalal R, Schmidt S (2007) Soil carbon turnover and sequestration in native subtropical tree plantations. Soil Biol Biochem 39:2078–2090
Ripley EA, Redman RE, Crowder AA (1996) Environmental effects of mining. St Lucie Press, Delray Beach, FL
Roberts H (1978) When ambrosia beetle attack mahogany trees in Fiji. Unasylva 29:25–28

Russell JS, Cameron DM, Whan IF, Beech DF, Prestwidge DB, Rance SJ (1993) Rainforest trees as a new crop for Australia. For Ecol Manage 60:41–58

Scott DF, Bruijnzeel LA, Mackensen J (2005) The hydrological and soil impacts of forestation in the tropics. In: Bonell M, Bruijnzeel LA (eds) Forests water and people in the humid tropics: past present and future hydrological research for integrated land and water management. Cambridge University Press, Cambridge, pp 622–651

Sidle RC, Ziegler AD, Negishi JN, Nik AR, Siew R, Turkelboom F (2006) Erosion processes in steep terrain - truths, myths, and uncertainties related to forest management in Southeast Asia. For Ecol Manage 224:199–225

Simpson JA, Osbourne DO, Xu ZH (1999) Pine plantations on the coastal lowlands of subtropical Queensland, Australia. In: Nambiar EKS, Cossalter C, Tiarks A (eds) Site management and productivity in tropical plantation forests. Center for International Forestry Research, Bogor, pp 61–67

Siswamartana S (2000) Productivity of teak plantations in Indonesia. In: Enters T, Nair CTS (eds) Site, technology and productivity of teak plantations. Food and Agriculture Organisation of the United Nations, Bangkok, pp 137–143

Soerianegara I, Lemmens RHMJ (eds) (1993) Plant resources of South-East Asia 5 (1): timber trees; major commercial timbers. Pudoc Scientific Publishers, Wageningen

Specht A, Turner J (2006) Foliar nutrient concentrations in mixed-species plantations of subtropical cabinet timber species and heir potential as a management tool. For Ecol Manage 233:324–337

Speight MR, Wiley FR (2001) Insect pests in tropical forestry. CABI Publishing, Wallingford

STCP (2009). Encouraging industrial forest plantations in the tropics: Report of a Global Study. ITTO Technical Series No 33. International Tropical Timbers Organization, Yokohama

Thornton NM, Dahl NW (1996) Use of native seed in the rehabilitation of mining in the monsoonal tropics. In: Bellairs SM, Osbourne JM (eds) Second Australian native seed biology for revegetation workshop. Australian Center for Minesite Rehabilitation Research, Kenmore, Brisbane, pp 63–69

Tomich TP, Kuusipalo J, Menz K, Byron N (1997) Imperata economics and policy. Agroforest Syst 36:233–261

Turnbull JM (2003) Eucalypts in Asia: Proceedings of an international conference held Zhanjiang, Guangdong, People's Republic of China, 7–11 April 2003. Australian Center for International Agricultural Research, Canberra

Turnbull JW (ed) (1991) Advances in tropical Acacia research: proceedings of an international workshop held in Bangkok, Thailand, 11–15 February 1991. Australian Centre for International Agricultural Research, Canberra

Turner J, Lambert M (2000) Change in organic carbon in forest plantation soils in eastern Australia. For Ecol Manage 133:231–247

van Dijk A, Keenan RJ (2007) Planted forests and water in perspective. For Ecol Manage 251:1–9

Varmola M, Del Lungo A (2003) Planted forests database (PFDB): structure and content. Planted forests and trees working paper FP/25. Food and Agriculture Organization of the United Nations

Venn TJ, Harrison SR (2001) Stand yield models for Australian Eucalypt and Acacia plantations in the Philippines. In: Harrison SR, Herbohn J (eds) Socio-economic evaluation of the potential for Australian tree species in the Philippines, RomeACIAR Monograph 75. Australian Center for International Agricultural Research, Canberra, pp 79–92

Webb MJ, Redell P, Grundon NJ (2001) A visual guide to nutritional disorders of tropical tree species: *Swietenia macrophylla* and *Cedrela odorata*. Australian Center for International Agricultural Research, Canberra

Weinland G (1998) Plantations. In: Appanah S, Turnbull JM (eds) A review of dipterocarps: taxonomy, ecology and silviculture. Center for International Forestry Research, Bogor, pp 151–186

Westoby J (1987) The purpose of forests: follies of development. Blackwells, Oxford

Whisenant SG (1999) Repairing damaged wildlands: a process-oriented, landscape scale approach. Cambridge University Press, Cambridge

Whitmore TC (1984) Tropical rain forests of the far east. Clarendon Press, Oxford

Wingfield MJ, Slippers B, Roux J, Wingfield BD (2001) Worldwide movement of exotic forest fungi, especially in the tropics and the Southern Hemisphere. BioScience 51:134–140
Wooff W (2009) Sabah Forest Industries experiences in plantation forestry. Conference on the Current State of Plantation Forestry in Malaysia: A Special Focus on Sabah, 18–20 Nov 2009, Forestry Department Headquarters, Sandakan
Zhou GY, Morris JD, Yan JH, Yu ZY, Peng SL (2002) Hydrological impacts of reafforestation with eucalypts and indigenous species: a case study in southern China. For Ecol Manage 167:209–222

Chapter 7
Mixed-Species Plantings

> In developing forest culture, rather than controlling the species genes and the population structure, farmers have chosen to retain and manage the diversity of species, structures and functions inherent to local forest ecosystems. To the question, 'Why did you choose this model rather than monocrop forestry?' (Indonesian) farmers respond by asking "Why should we choose monocrop forestry rather than our diversified model"
> Indonesian farmers discussing their agroforests;
>
> Michon (2005, p. 163)

> Endeavours to establish pure stands everywhere is based on an old and highly detrimental prejudice ...Since not all tree species utilize resources in the same manner, growth is more lively in mixed stands and neither insects nor storms can do as much damage; also, a wider range of timber will be available everywhere to satisfy demands ...
>
> German silviculturalist von Cotta (1828) quoted by Pretzsch (2005, p. 42)

Introduction

The third way in which forests might be re-established (and the second way in which degraded land might be replanted) is to use multi-species plantations or polycultures. As was the case with monocultural plantations, these plantings are usually undertaken when natural regeneration is thought to be unreliable or when species with having particular economic advantages are required. Mixed-species forests are not half-hearted attempts to mimic the diversity present in natural forests but they do seek to take advantage of some of the functional advantages of species-rich natural systems including their capacity to use resources more efficiently and to reduce nutrient losses from the system.

Landholders using mixed-species plantations are usually aiming to generate a wider range of goods or ecosystem services than provided by simple plantation monocultures and any increase in plant biodiversity is mostly incidental to this objective. Plantations established to provide goods are commonly felled at some point but those used to generate ecosystem services may remain undisturbed. Examples of the former are those that include, perhaps, timber trees and fruit trees

D. Lamb, *Regreening the Bare Hills*, World Forests 8,
DOI 10.1007/978-90-481-9870-2_7, © Springer Science+Business Media B.V. 2011

while examples of the latter are the multi-species plantings established to stabilize heavily degraded sites such as former minesites. Depending on circumstances, the numbers of tree species used in mixed-species plantations might be small or large and may include exotic species as well as native species. According to the terminology discussed earlier, the approach represents a form of forest rehabilitation.

Mixed-species plantings are not widely used in industrial plantations because they are more difficult to manage than simple monocultures. Nonetheless, foresters have long been interested in their potential advantages (Wormald 1992; Pretzsch 2005). Trials can be found throughout the tropics and continue to be established. Unlike industrial tree-growers, smallholders have always had a more pragmatic attitude to polycultures and their agroforestry systems commonly use a variety of tree species. Again, this is not because of any particular desire to conserve biodiversity but because mixtures suit their ecological and economic circumstances and provide the goods they need while reducing their vulnerability to ecological hazards.

In recent years ecological researchers have renewed their interest in polycultures. At first this was because of an interest in the functional consequences of biodiversity loss. More recently their interest has being driven by questions about how ecosystems are assembled and the interactions between species with differing traits. Much of this research has involved laboratory studies and short-term field trials and there is a need for ecologists and silviculturalist to find ways of converting the results of these studies into more robust forms of reforestation that are applicable at a farm or landscape level and that take account of the economic drivers of reforestation.

This chapter begins this process by examining the silviculture of mixed-species plantings. After reviewing some of the potential advantages and disadvantages of polycultures it discusses different types of multi-species plantings at particular sites as well as mixtures at a landscape scale formed by spatial mosaics of simple monocultures. It then reviews some of the implications these silvicultural designs have for the provision of ecosystem services.

Some Potential Advantages of Mixed Species Plantations

Mixed-species tree plantations offer a number of potential advantages. These include the possibility of increasing stand productivity, improving the nutrition status, improving resistance to pests and diseases and generating various important financial benefits. And, as more species are used, the ecosystem is likely to gain some degree of ecological and economic resilience. None of these benefits are assured and their development depends on the types of species and mixtures used and on the environmental and economic conditions present. In some situations mixtures may also be a way of ameliorating site conditions making it possible to establish preferred species at sites at which they might not otherwise have been able to grow (Table 7.1).

Table 7.1 Potential advantages of multi-species plantations

Advantage	Reason
Enhanced production	Greater niche complementarity between species; contrasting phenologies (separation in time) or root or canopy architecture (separation in space)
Improved nutrition	Greater ability to access and conserve nutrients leading to improved growth
Reduced damage from pests or disease	Susceptible tree species hidden in space or natural enemies of pests encouraged so damage is less
Improved financial benefits	Goods produced for several markets; greater flexibility in timing of cashflows
Site amelioration	Facilitator species modify site conditions to eradicate competitors and allow entry of preferred species

Enhanced Production in Multi-Species Plantations

There is increasing evidence that some multi-species plantations can have greater levels of productivity than monocultural plantations of their constituent species (Biot et al. 1995; Cannell et al. 1992; Forrester et al. 2006; Jones et al. 2005; Kelty 2006; Piotto 2008). The phenomenon has been recognized for some time and has been widely studied with grasses and agricultural crops (Harper 1977). Much of this early work addressed the nature of above- and below-ground competition and explored situations where inter-specific competition between two plant species might be less than any intra-specific competition among plants of the same species growing at the same density. These studies were done in so-called replacement series by exchanging plants of one species with those of another thereby creating a mixture. Harper described the Relative Yield as being the ratio of the yield of a species in a mixture to that of the species growing in a monoculture at the same density. In a 50:50 mixture a Relative Yield of 0.5 shows that the growth of that plant species was the same in the mixture and the monoculture. A Relative Yield of >0.5 showed it grew better in the mixture than in a monoculture while a value <0.5 showed it was adversely affected by growing in a mixture. The Relative Yield Total is the sum of the Relative Yields of all species in the mixture. When this is >1.0 the overall productivity of the mixture is better than that of either monoculture. The majority of these agricultural studies involved simple mixtures involving only two species. Joliffe (1997) reviewed the agricultural literature dealing with mixtures and found that, on average, mixed plantings had 12% more biomass than monocultures but gains of up to 30% were observed in some studies.

In recent years the topic has attracted additional attention because of the on-going reduction in global biodiversity. This renewed interest has focused on the question of the functional importance of this biodiversity loss. The problem has been addressed in two ways; some have explored the effect of progressive biodiversity losses from natural ecosystems while others have been more concerned with how ecological processes might change as more species are added to an ecosystem. Most of these studies have involved more than the two or three species commonly used in the earlier agricultural experiments. Much of this experimentation has been carried out in simple laboratory microcosms or in agricultural lands using grasses or herbs and very few studies have been done over long time periods using tree species. Nevertheless, some broad conclusions are beginning to emerge. These have been extensively reviewed by Hooper et al. (2005). Among these conclusions are (i) that certain combinations of species are complementary in their patterns of resource use and can increase the net primary productivity and degree of nutrient retention in a ecosystem over that provided by a single species, and (ii) that more species are required to ensure a stable supply of goods and services over the longer term and as larger areas are considered. This is because species respond differently to different environmental perturbations and stresses. However (iii), it is the functional characteristics of the species involved as well as the number of functional groups present that is important in influencing ecosystem properties rather than taxonomic diversity per se. The significance of differences in species diversity in higher trophic levels remains unclear.

Plant communities are often thought to be largely structured by inter-specific competition and the success of one species occurs at the expense of another. But two other mechanisms potentially able to improve the overall productivity of species mixtures are facilitation and complementarity. Facilitation occurs when one species increases the availability of a resource to others and thereby benefits the growth of those other species. An obvious example of this is when a nitrogen-fixing species is grown a mixture with other non-fixing species in an infertile soil. In these situations the nitrogen fixer is able to add nitrogen to the ecosystem and so reduce this limitation on the growth of the other species. Note that the beneficial role of this particular facilitator would disappear on more fertile soils where growth is not being limited by nitrogen availability. Another form of facilitation is when a tolerant species grows in the open and provides shelter or shade enabling another, less-tolerant, species to become established and grow at that site. Specific examples of facilitation will be discussed further below.

The second mechanism, complementarity, arises when species having different ecological niches are grown together. In this case niche partitioning reduces inter-specific competition and greater productivity is possible because, collectively, the species are able to use the resources at a site more efficiently than if only one species was involved. The significance of this is greatest when one or more of these resources are limiting for plant growth. So, for example, a productive mixture might be one which included a relatively shade-intolerant species with a sparse crown growing with a relatively shade tolerant species with a denser and deeper crown. The canopy would be stratified with the more shade intolerant species forming the upper canopy allowing sufficient light through to sustain the slower growing and shade-tolerant species in a sub-dominant position (Fig. 7.1). The mixture captures more light than either species would if growing in a monoculture. As noted in the previous chapter, many monocultures often colonized by dense understories

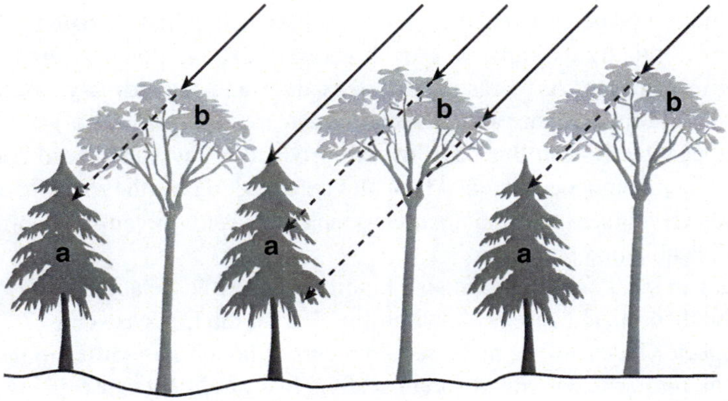

Fig. 7.1 Stable mixtures often have species with complementary attributes growing in stratified canopies. This mixture contains a shade tolerant species (**a**) and a shade intolerant and open-crowned species (**b**). Species (**a**) gets direct radiation in its upper crown as well as indirect radiation that penetrates he crowns of species (**b**). This allows **a** to persist in a sub-dominant position

because of the light able to penetrate the upper canopy layer. Although the shade-intolerant species is likely to be faster-growing than the shade tolerant species, at least initially, the growth rates should not be too different in order to prevent one species dominating and out-competing the other(s). In any case, Smith (1986) suggests the overstorey should not comprise more than 25% of the total number of trees. This allows space for the crowns of the dominant species to expand while the lower canopy trees help reduce branching in the stand dominants and provide what Smith (1986) refers to as a 'trainer' effect.

Other plant attributes enabling some complementarity would be differences in foliar phenology (e.g. deciduousness), root growth phenology or in rooting depths. A spatial separation in rooting depth with one species having only shallow roots while another has only deep roots is unlikely but examples have been found where one species in a has deeper roots than another which could allow some differentiation in patterns of soil resource usage (Ewel and Mazzarino 2008; Lamb and Lawrence 1993). Competition between species with differences in these attributes is reduced because they use resources at different times or from different spatial locations.

It is important to note that static measures of complementarity such as these can change over time and species once regarded as complementary can sometimes become competitive. Ewel and Mazzarino (2008) grew a palm (*Euterpe oleracea*) and a large herb (*Heliconia imbricata*) in mixtures with each of the trees *Hyeronima alchorneoides, Cedrela odorata* and *Cordia alliodora*. The first of these tree species is an evergreen but the other two are deciduous trees. The three tree species initially dominated their respective mixtures but over time a difference emerged in the way the two deciduous species and the evergreen species interacted with the palm. In the former the palm was able to take advantage of the period of deciduousness and grow up and join the canopy layer. The difference in leaf phenology gave the palm a competitive advantage even though all species had initially shared the site's resources. Eventually its intrusive growth allowed it to out-compete the trees and dominate the canopy. As a result, the trees died and net primary productivity declined. In the mixture involving the evergreen tree the palm also grew up and joined the canopy. But, in this case, overall productivity was enhanced compared with a nearby tree monoculture and the palm appeared to play the role of a complementary species rather than a competitive species, probably because of a difference in nutrient acquisition strategies.

Changes in site conditions can also modify competitive relationships (Pretzsch 2005). The theoretical cause is shown in Fig. 7.2. In the first case (Fig. 7.2 upper) the two species have similar niche requirements although they differ in their productivity at this particular site. A mixture of the two would not cause an increase in stand productivity because they would be competing for the same resources at the same time and A would out-compete B. A different situation prevails when the two species have different ecological niches (Fig. 7.2 lower). In this case the two species are mixed in stands at four different locations along an environmental gradient (e.g. in soil fertility or rainfall). The site conditions at location 1 are optimal for species A but are only marginally suitable for species B. In this case a mixture of the two would have no advantages because the growth of B would be poor. At location 2, however, the site conditions are suitable for both species and a mixture

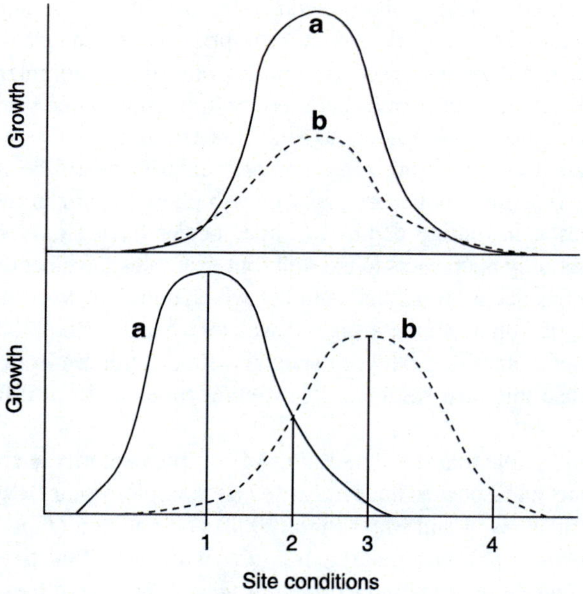

Fig. 7.2 Site conditions affect competitive interactions and the productivity of mixtures. In the upper diagram both species have similar site tolerances but **(a)** is more productive than **(b)** and the mixture would fail. The *lower diagram* shows the growth of species mixtures planted at four sites along an environmental gradient. Each species has different site tolerances. At site 1 species **(a)** grows well but conditions are marginal for **(b)**. At site 2 conditions allow **(a)** and **(b)** to grow. At site 3 species **(b)** grows well but conditions are marginal for **(a)**. At site 4 conditions are marginal for **(b)** but unsuitable for **(a)** (After Pretzsch 2005)

could be advantageous. At locations 3 conditions are optimal for species B but are marginal for species A; again a mixture here would have no advantages. At location 4 the conditions are only marginal for species B and are unsuitable for species A and it would soon disappear.

These models prompt two questions. The first is – what are the site preferences of potential plantation species? The second is – how much of each site type is available for reforestation and where might conditions be favourable for mixtures of particular species? Most tropical reforestation programs probably commence without enough knowledge to answer either question with any great confidence. Pretzsch (2005) argues that, in such cases, mixtures can be a way of redistributing risk and avoiding complete plantation failure.

Much of the research done on the relationships between diversity and productivity has been carried out in laboratories or in field experiments with grasses or shrubs that have only lasted a short period of 1 or 2 years. One of the few studies of the effects of increasing levels of tree diversity on productivity over a long period was that carried out by Erskine et al. (2006) who examined the growth of trees in 53 mostly multi-species plantations established on former farmland in the humid tropics of northern Australia. The plantations contained between one and eight species and were 6–9.5 years old when assessed. The composition of each plantation was largely a random assemblage although all species were commercially valuable timber trees. A total of 27 species were used across the various plantations and these included gymnosperms and angiosperms as well as some potential nitrogen fixers. Each plantation was at least 2 ha in size and had trees planted at densities of 600–800 tph. The study found evidence that increasing species richness (up to eight species) was associated with increased levels of productivity as measured by stand basal area or mean tree basal area (Fig. 7.3). A linear relationship was found between the number of tree species in each plantation and productivity but it was not clear that this would persist if more than eight species were used. Although the study design did not allow the underlying casual mechanisms to be identified there was some evidence that complementarity (associated with differences in canopy architecture) was involved but no evidence that any of the putative nitrogen fixers had improved productivity when included in the mixtures.

Contrasting results were obtained from another study in the same region suggesting that other mechanisms can also operate (Firn et al. 2007). This study involved mostly older (>65 years) plantations originally planted as monocultures. The species included well known native timber trees including two angiosperms, *Flindersia brayleyana* (Rutaceae) and *Toona ciliata* (Meliaceae) as well as two gymnosperms, *Araucaria cunninghamii* (Araucariaceae) and *Agathis robusta* (Araucariaceae). Each plantation was planted in close proximity to the others and all were on the same fertile, basalt-derived soil. The plantations were surrounded by nearby natural rainforest and, as a result, all were colonized by additional trees species. Sufficient time has passed for a number of colonists from the surrounding natural forests to grow up and join the canopy layer (see example in Fig. 6.11). At the time of the study, the richness of overstorey tree species (>10 cm dbh) in the former monocultures ranged from one to 17 species per 0.1 ha plot. Although additional

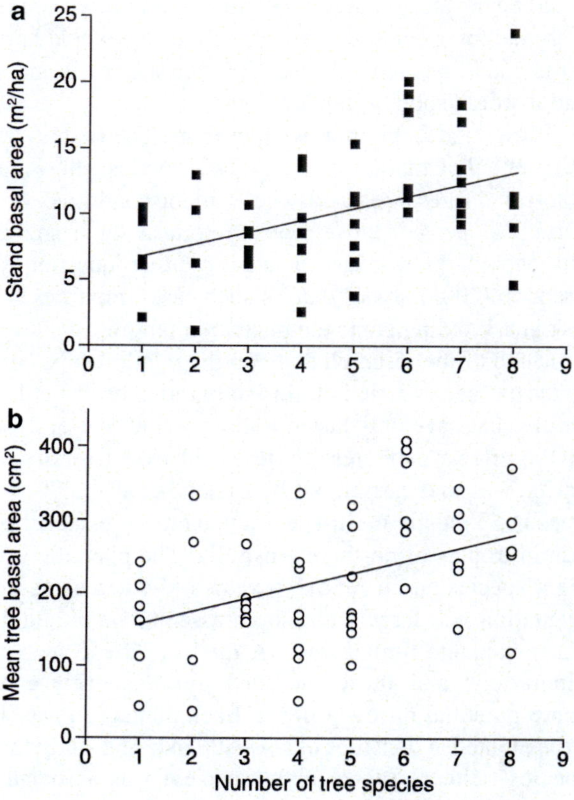

Fig. 7.3 Relationship between tree species richness and productivity (basal area) of (**a**) stands and (**b**) mean trees in stands of randomly assembled mixtures at age 6–9.5 years growing in northern Queensland, Australia (**a**: Stands: $N=53$; $r^2=0.21$; $p=0.001$; **b**: Mean tree: $N=53$; $r^2=0.18$; $p=0.001$). Each plantation >2 ha and stand densities were 600–800 tph (After Erskine et al. 2006)

trees were colonised the planted stands the overall tree density (trees >10 cm dbh) in the various sites was still broadly similar and ranged between 500 and 770 tph. The relationship between the species richness at the time of assessment and stand productivity is shown in Fig. 7.4. In this case, rather than production increasing with species richness, the reverse appears to be true. This is probably a result of the 'sampling effect'. This occurs in situations where one of the species included in a mixture happens to be particularly efficient in resource usage. In such cases most of the overall plantation productivity is derived from this species and not from the greater collective efficiency of the community. Each of the timber species originally established in these plantations were known to be highly productive plantation species and the additional species have not be able to contribute much additional productivity, at least until now.

Although the review by Hooper et al. (2005) highlighted the importance of functional groups in mixtures it is difficult to say much about the number of functional groups used in either of these two studies. Most can probably be classified as shade-intolerant long-lived secondary species although the variety of plant families represented suggest there may be some physiological and niche differences as well. The results suggest there is scope for increasing overall

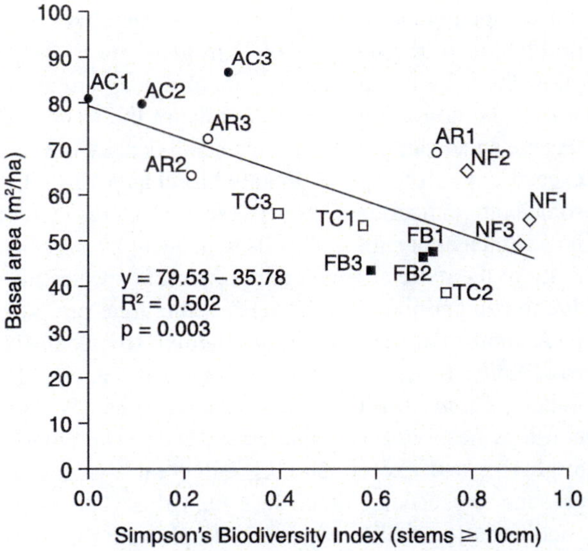

Fig. 7.4 Relationship between tree biodiversity in 0.1 ha plots and production (basal area) in old (>50 years) former monocultural forests in which natural regeneration from nearby secondary rainforest has added additional trees to the canopy layer. AC=*Araucaria cunninghamii*, AR=*Agathis robusta*, TC=*Toona ciliata*, FB=*Flindersia brayleyana* and NF=Natural Secondary Forest (Firn et al. 2007)

plantation productivity by using mixtures of several tree species with complementary traits. But they also show that great care is needed in deciding the composition of the mixture, the numbers of species to use and in the relative proportions of each species in the mix if any gains are to be worthwhile and of practical significance. It is also important to remember that species are not necessarily equally valuable and any gains in production need to be accompanied by gains in overall economic value if mixtures are to be attractive to forest growers. Some of these additional design issues will be discussed in more detail below.

Improved Nutrition

A second potential advantage of mixtures is that they may improve the nutritional status of a plantation beyond that occurring in monocultures. This might be done by increasing the amounts of nutrients entering the ecosystem, limiting impediments to nutrient cycling within the system or by reducing nutrient losses from the system. As before, facilitation and complementarity are important mechanisms by which these changes can occur.

One of the most widely studied situations is that where a nitrogen fixer is included in the mixture to increase the nitrogen stored in the ecosystem and improves the nitrogen nutrition of other species when their litter and roots decompose. There is now strong evidence that such mixtures can improve overall plantation productivity at sites with less fertile soils where nitrogen is limiting for plant growth (Binkley 1992; Forrester et al. 2006; Khanna 1998; Rothe and Binkley 2001). Some care is needed in mixing nitrogen fixers with other species to avoid one species from out-competing the other. Where the nitrogen fixer is the more vigorous species the problem may be avoided if the nitrogen-fixer is short-lived or it can be thinned to allow the more commercially valuable species to grow unhindered in the improved soil. If the non-nitrogen fixer dominates the stand and suppresses the fixer it may receive an advantage because the stand density has been reduced. This benefit may exceed any advantage attributable to changes in nitrogen nutrition (J. Ewel, personal communication, October 2009). Mixtures including nitrogen-fixers will have no advantages at sites with soils with adequate supplies of nitrogen and the productivity of the more commercially important species may even decrease at such sites due to competition if the nitrogen-fixing trees are growing vigorously.

A number of tree species are thought to able to fix nitrogen under appropriate conditions. In the Asia-Pacific region these include species of genera such as *Acacia, Falcataria, Gliricidia, Leucaena* and *Sesbania* as well as species of the non-legumes *Casuarina* and *Parasponia*. The amounts of nitrogen these species might fix can exceed 200 kg ha^{-1} $year^{-1}$ (Khanna 1998) although the actual amounts are usually much less than this. The rates of fixation depend on site conditions (particular the available phosphorus levels) as well as the degree of nodulation, tree density and age. Depending on the nutrient status of these other species, the additional nitrogen may enhance their growth, especially if the additional nitrogen is provided at an early stage of plantation growth. Or changes in the ratio of foliar N/P may accentuate any limitation on growth caused by phosphorus (Siddique et al. 2008). In such cases phosphorus fertilization may be needed to capture the full advantage from using the nitrogen fixer. Other ways mixtures might increase the supply of nutrients other than nitrogen to the active nutrient cycle is if one of the species has root systems able to access refractory or less-accessible soil nutrients such as phosphorus. This might be done with particular mycorrhizal associations, by changes induced by activity in the rhizosphere of certain species (Khanna 1998) or by roots able to explore deeper soil horizons and thus act as 'nutrient pumps' that sustain shallower-rooted species in the plantation mixture.

Evidence concerning the effect of mixtures on nutrient cycling within forests is equivocal. The quantities of nutrients cycled through litterfall can be higher in mixtures involving nitrogen-fixing trees than in monocultures of the non-nitrogen fixing species (Rothe and Binkley 2001; Forrester et al. 2006). However, the rates at which these litters subsequently decompose appear to vary a good deal. Some studies report enhanced decomposer activity in more diverse ecosystems (Balvanera et al. 2006) while others find no effect or even that decay rates are slowed (Rothe and Binkley 2001). It is generally agreed that the rate of litter decay is largely influenced by litter quality and by the types of soil fauna that are present. Both are consequences of the particular species producing the litter and any mixture effects may be more a result of these factors and not tree species richness. In the case where nitrogen fixers have been used, the subsequent rates of soil nitrogen mineralization appear to be greater in mixtures than in pure stands (Khanna 1998). The amounts of particular nutrients being cycled might also be affected by complementary differences in the timing of nutrient uptake.

The role of mixtures in limiting nutrient losses from new forests is barely studied. This is despite the fact that leaching losses can be high, especially in young plantations, because of the often high rainfall intensities experienced in tropical areas. Leaching losses are reduced and nutrient retention is increased by the presence of dense root systems that intercept and immobilize ions being moved through the soil, especially if these belong to fast-growing species. However, there have been surprisingly few studies of rooting systems in different types of young plantations or in regrowth forests. One study reported by Berish and Ewel (1988) compared root development in agricultural and forestry monocultures, natural successional vegetation, the same regrowth vegetation enriched with additional species and a diverse multi-species community that sought to mimic a natural succession. Fine root densities were much higher in the species-rich communities than in the short-lived agricultural monocultures but the tree monoculture quickly acquired a root density similar to that of the species-rich successional vegetation. Studies of nutrient movement in the soils supporting the various treatments showed nutrient losses were high under the agricultural crops but that nutrients were conserved with perennial vegetation irrespective of species richness (Ewel et al. 1991). In this case the addition of more species did not enhance nutrient retention. But even if the tree monoculture had been less effective, most plantations (with the exception of species like teak) soon acquire a diverse understorey containing a variety of life forms including grasses, herbs and shrubs, irrespective of whether a tree monoculture or mixture is established. Because of this, nutrient losses are probably limited after the first few years. Nutrient losses are also limited when overall plantation productivity is high since nutrients are rapidly taken up and immobilized in biomass.

In summary, plantation mixtures including nitrogen fixers may have improved supplies of nitrogen and improve the growth at sites where nitrogen would otherwise be limiting. There may be other nutritional advantages arising from mixtures but the circumstances under which these occur have been poorly studied.

Reduced Damage from Pests and Diseases

It is widely believed that monocultural crops are more susceptible to pests and diseases that diverse natural ecosystems, especially when these crops are even-aged and cover large areas. If so this could be because monocultures lack most of the trophic complexity and controls such as predators found in natural ecosystems. In addition they offer a large food resource to any insect and pathogen species adapted to use them. Their narrow genetic base and the closeness of plants are thought to allow rapid colonization and spread of pests or infection. Crops of single species also lack the physical or chemical barriers to insect movement often found between plants in natural ecosystems. Some of these problems were discussed in Chapter 5 in the context of the vulnerability of monocultures to pests and diseases.

There is some evidence supporting the view that diversity does, in fact, reduce pest and disease problems. Jactel et al. (2005) carried out a meta-analysis of over 50 field studies and found that insect pest damage is greater in single species stands than in mixtures containing these same species. They concluded this was because pests had poorer access to the host trees, there was a greater impact by natural enemies or the pests were diverted from less susceptible to more susceptible species. Similar conclusions were drawn by Nair (2007). In the case of diseases Pautasso et al. (2005) reported that tree diversity may also make forests less susceptible to fungal pathogens. A particularly striking example of how crop diversity and spatial patterns could significantly reduce damage from disease comes from southern China. This was a large-scale (3,300 ha) study carried out with rice in Yunnan province. Rice in this area is affected by a rice blast fungus (*Magnaporthe grisea*) and the use of fungicide is common. But Zhu et al. (2000) were able to demonstrate that disease-susceptible rice varieties planted in mixtures with resistant varieties had 89% greater yield and the disease was 94% less severe than when a simple monoculture was used. The result was so striking that farmers participating in the study were able to give up the use of fungicides. Wolfe (2000) suggested this result points to a need to rethink the trend in agriculture which is leading to not only a reduction in the number of crop varieties but also to a diminution in the genetic variation within these varieties. He suggested it might even be time for agriculturalists to explore using species mixtures in some situations as well. There is no reason to see why these same conclusions should not apply equally well to forest plantations.

But some qualifications are needed. Firstly, populations of polyphagous insects and generalist pathogens can first build up on a preferred host species and then spill over onto less palatable species growing nearby leading to contagion (Jactel et al. 2005). Blanton and Ewel (1985) observed a similar phenomenon with leaf cutting ants. This suggests simple mixtures may be only able to limit, but not prevent, insect damage. In some cases they may even confer associational susceptibility. Secondly, these types of studies are yet to generate specific silvicultural guidelines to improve plantation designs. Just how much plant diversity is needed? Would alternate row plantings of two species provide any benefit or must the host tree be 'hidden' amongst a much greater variety of other taxa? And at what spatial scale should the mixture occur? Might a patchwork mosaic of monocultures be sufficient or must a more intimate tree-by-tree mixture be used? Sometimes these questions can be answered when the biology and behavior of a specific insect pest is known but it is very difficult to develop generic prescriptions to deal with unknown future insect pests. For this reason few plantation managers are likely to move to mixed species stands simply to reduce insect damage although they may be happy to accept that mixtures may provide some insurance value, especially against indigenous pests or diseases.

Financial Benefits

Mixed-species plantings are more expensive to establish and manage than simple monocultures and are likely to be less attractive to large industrial growers for this reason. But they do have some distinct financial benefits and for many smallholders these financial benefits may be the most persuasive factor leading them to change from monocultures to mixtures. One advantage of mixtures is that they can offer a diversity of products such as different quality timbers suited for different markets or NTFPs as well as timbers. When future markets and prices are uncertain then diversifying products and income sources is a way of building economic resilience and is likely to appeal to smallholders aware of their commercial vulnerability. The disadvantage, of course, is that financial returns are reduced in situations where a large market develops for a particular product. In this case a mixed-species plantation reduces the amount of that particular good that can be sold by an individual producer. Some growers such as those acting as out-growers for large industrial enterprises are in the fortunate situation of being confident about the market they will supply and will willingly forego diversity. But many farmers are not in this position. In this respect it is interesting that households who acquired eucalypt plantations from village cooperatives during the doi moi period in Vietnam were often observed then inter-planting these monocultural plantations with species such as *Acacia auriculiformis, Styrax tonkinensis* and *Manglietia glauca* as a means of diversifying their income sources (Fahlen 2002). They evidently felt the advantages of diversity outweighed the disadvantage of not maximising their returns from growing a single crop of eucalypts. Other smallholders across the region have taken a similar view (Nibbering 1997; Pasicolan et al. 1997).

A second potential financial advantage of mixtures is that some designs can overcome one of the main disincentives to tree-growing by generating an early cashflow. There are several ways this might be done. One is by including faster-growing species that can be harvested at an early age with slower-growing species that make up most of the plantation. Another would be to include multi-purpose tree species able to supply, for example, fruit, nuts or resins from an early age. Or, finally, non-tree species supplying food or other products could be grown in the understorey as temporary or permanent components of the plantation. Complementarity is involved in both cases. But it is a form of economic and not ecological complementarity that generates these advantages.

Ameliorating Site Conditions at Cleared or Degraded Sites

The conditions at many degraded sites are such that only a small number of species – sometimes mostly exotic species – may be able to grow there. These conditions may be associated with infertile soils, aggressive weeds, high solar radiation levels or some other micro-environmental condition. By first planting a tolerant species these adverse environmental conditions are changed and the site can become suitable once more for a much wider range of species. Under ideal circumstances these species would have a modest market value and create a financial asset at the same time they were modifying the site's environmental conditions.

Amelioration might be achieved in two ways. One way would involve growing a species such as a nitrogen-fixer and then removing it once soil conditions had been improved. The facilitator would then be replaced by the preferred species.

Another way would be to use the facilitator as a nurse crop and growing the preferred species beneath it's canopy cover. In this case the benefit comes from the changed aerial micro-environment. Care is obviously needed to ensure the species in the mix are also complementary and that the nurse species does not act as an inhibitor. This could be achieved by choosing nurse trees having sparse crowns or by planting these at low densities. In many cases where this approach is used the nurse species is removed after a few years once the preferred species has become established. The approach has been used with agricultural crops such as coffee (hence 'shade coffee'). The idea of protective nurse trees also forms the basis of some silvicultural systems such as the Shelterwood method (Smith 1986).

The potential advantages of mixed species plantings come at a cost. The more species being used the more complicated the management and, until recently, few large industrial tree-growers have seen the advantages being sufficient to overcome the disadvantages. But this appears to be changing and it is interesting that some of the world's most efficient and best-managed timber companies in northern Europe are beginning to explore using simple mixtures in order to enhance production and biodiversity (Berqvist 1999; Fahlvik et al. 2005; Mönkkönen 1999; Pretzsch 2005).

Species Functional Types

What types of species should be used in mixed-species plantings? At least one of them must have a significant subsistence or commercial value to attract a land-owner's interest. And in many situations some may need to be capable of a facilitative role such as nitrogen fixing to ensure the plantings are established and grow. But, where complementarity is critical for the success of the mixture, the various species will need to belong to different functional types and have traits that complement each other. There is still some uncertainty concerning how different functional types might be classified. Wilson (1999) thought there were two basic types. The first were species that shared the same environmental conditions and hence were likely to be found together. The second were those that used the same resources and, thus, were unlikely to be found together because of competitive exclusion. Noble and Gitay (1996) identify five possible ways in which functional types might be classified:

- Phylogeny – groups of taxa with similar evolutionary histories.
- Life form or structure – groups of taxa with the same life form.
- Resource use – taxa using the same resource(s).
- Response to a defined perturbation – taxa that have a similar response to changed environmental conditions.
- Role in ecosystem function – taxa having similar patterns of resource use or biochemical function (e.g. nitrogen fixation).

Recent research on functional types has been largely carried out by those seeking to model the dynamics of existing forests. In these situations phylogeny appears not to have been highly valued although it may reflect important functional attributes such as basic physiological properties or resistance to certain pests. Instead, rather more effort has been undertaken exploring the other four categories. Thus Ewel and Bigelow (1996) argued function follows form and that the diversity of life forms present defines the way a forest can function. Likewise Köhler et al. (2000) used resource usage and differentiated species types according to their tolerance of shade tolerance and height at maturity. Ashton et al. (2001) developed a similar classification based on successional types that recognized pioneers of stand initiation, pioneers of stem exclusion, late successional dominants, late successional non-dominants, late successional sub-canopy and late successional understorey. Noble and Gitay (1996) found tolerance of fire was a useful way of classifying species while Gitay et al. (1999) and Gourlet-Fleury et al. (2005) explored the use of classifications based on growth rates and longevities. The variety of plant traits represented here include differences in stature, longevity, shade tolerance, growth rate and the architectural characteristics of canopies and roots. These, together with differences in growth phenology, offer scope for finding species combinations forming complementary partnerships.

In overcoming land degradation, one of the tasks is to develop ecosystems that are resilient. Students of resilience argue it is promoted by not only having species from a variety of functional types but also having multiple representatives from each type (Diaz and Cabido 2001; Elmqvist et al. 2003). But it may not be the diversity of functional types that is important. Rather, it may be that new forests need to have a certain combination of functional types to deal with the particular environmental conditions at a site. Perhaps a better way of addressing the issue might be to ask what functional types should be present that enable a particular new forest to cope with the ecological and economic circumstances that might be present over, say, the next 40 years? This period is the time frame in which managers might operate and during which various environmental stresses might develop. In this case the functional groups might include:

- Species for production: short and long-lived trees (all of which have some market value); shade-tolerant and shade-intolerant species able to occupy different positions in the forest canopy.
- Species able to reduce nutrient stress: species with high nutrient use efficiencies and using various mycorrhizal symbionts; deep-rooted species able to explore soil profiles as well as shallow-rooted species able to capture nutrients recently cycled through litter layers; some species capable of nitrogen fixation.
- Species able to tolerate occasional water stress: deep rooted species able to tolerate seasonal droughts and deciduous or semi-deciduous species.
- Species able to tolerate fire: species with thick, heat resistant barks and/or species able to resprout after fire.

The list includes those affecting ecosystem functioning (because of the way they access resources such as light, nutrient or water) as well as others that differ in the way they respond to disturbances (such as fire or drought). But the new forests should also play a role in sustaining regional biodiversity and so include species able to sustain mutualistic partnerships with wildlife, especially those able to disperse seed. Some species will have traits that place them in several of these categories. Not all of these will be equally relevant in every location and it should be possible to fine-tune the list and target the more appropriate traits according to the environmental conditions present now or likely to develop in future.

Although this suggests a way of identifying target species to use in mixtures commercial growers must also find ways of linking these functional types with the landowner's evolving economic circumstances. In some cases the combination might not be too difficult to imagine. For example, a fast-growing tree able to be sold after only a few years as a utility timber might be grown together with a slower-growing species able to produce a specialty timber but needing a longer growing period. Likewise it should be possible to find species providing NTFPs as well as timber amongst these various ecological groups. In developing mixtures for smallholder the over-riding task may have less to do with boosting productivity and more to do with satisfying the immediate and longer-term economic needs of the household. Various approaches to deal with this task are outlined below.

Designs for Mixed-Species Plantations

Some of the ways plantations might be designed to take advantage of the potential benefits of polycultures are outlined in Table 7.2. The Table is divided into two parts. The first part is concerned with four types of mixed-species planting that might be established at a particular site. Two of these are even-aged with all tree species being planted together at the start. The other two are uneven-aged with one or more of the species being planted below a nurse or cover crop once it has been established. Circumstances will dictate the actual numbers of species that might be involved in each mixture. The second part of the table concerns simple monocultures but is specifically concerned with the opportunities to foster diversity at a landscape scale by having a spatial mosaic of these different monocultures.

Cash Crop Grown Beneath a Timber Plantation

This is system is shown as Design 1 in Table 7.2. The purpose of the design is to generate an income from the site before the overstorey trees are ready to harvest. This helps reduce one of the major disincentives to tree growing experienced by smallholders. The mixture is initiated by first planting the overstorey trees. These might be a single preferred species or a mixture of several species. In either case the species chosen are

Table 7.2 Plantation designs involving mixtures

At individual sites				Across landscapes
Uneven-aged		Even-aged		Uniform age
Design 1	Design 2	Design 3	Design 4	Design 5
Trees and understorey crop	Trees only	Different rotation lengths	Single long rotation	Simple tree monoculture
NTFP species established beneath trees	Final crop trees under-planted beneath a temporary tree cover crop	Trees grown on short rotation mixed with others on long rotation	Permanent mixture of two or more species	Involving a single species or a spatial mosaic of monocultures with several species
Temporary mixtures			Permanent mixtures	

those needing rotations of more than 20 years before they can be logged for commercial purposes. Once the plantation trees are established and canopy closure has been achieved, the short-term cash crop is planted in the understorey between the rows of trees (unlike the well-known taungya system where the crops are planted at the same times as the trees and are abandoned once tree canopy closure occurs).

Some care is needed in selecting both the overstorey and understorey species to minimize competitive interactions and ensure they are complementary. The overstorey trees should have open-crowns that produce mottled light rather than a uniform shade. They should be able to produce useful timber or NTFPs, be able to develop straight unforked stems and be compatible with the crop (e.g. by having deeper roots). Other desirable attributes might include being capable of fixing nitrogen, being tolerant of heavy pruning or pollarding and being wind-firm. The understorey cash crop must be able to tolerate the environmental conditions in the understorey (both the light conditions and the degree of root competition) and be able to generate a financial return in a relatively short time. A perennial species able to produce successive harvests at frequent intervals would be especially attractive but one that could be harvested and then replanted might be attractive as well. The best species would also be one for which there is already some demand and an established market.

A number of such systems have evolved in agroforestry practices in various parts of Southeast Asia although few of these have been intensively studied. It is likely that some have been prompted by the declining availability of timber and NTFP supplies from natural forests while others are simply attempts to maximize the income from small areas of land using a system that requires little maintenance once it is established. These systems include food crops such as pineapples, bitter bamboo, cardamom, tea, ginger, lemongrass as well as traditional medicinal plants or rattans planted under plantations of timber species (Rao et al. 2004). In each case the amount of shade and numbers of trees vary according to the particular requirements of the understorey crop. Sometimes the trees are pruned or pollarded to help adjust light levels. Alternatively, light levels can be managed by planting several rows of trees and leaving a belt up to 10 m wide before the next rows of trees for the underplants.

Mixtures of trees and rattans require some particular forms of complementarity. The rattans should not be so heavy that they distort their host trees. At the same time the trees used should not be those with long branchless boles such as species like *Eucalyptus deglupta* that make it difficult for the canes to access the canopy (Weidelt 1996). The numbers of rattans per tree should also be limited to avoid the canes becoming entangled and difficult to harvest (Fig. 7.5).

This design has the advantage of flexibility and a large number of species combinations is possible. Provided the cash crops grown in the understorey are not too big they are unlikely to hinder the growth of the trees. And, depending on the species used, the structural complexity of the new plantation could be more attractive to some wildlife than a simple tree monoculture although the overall plant species diversity is not likely to be very great. The primary disadvantage of the design is the need to identify complementary species. However, this may not be as much of a problem as it seems because continuous testing by farmers appears to have already generating many working examples of systems that form ecologically and financially successful combinations. Of course any thinning of the plantations trees may have some consequences for the understorey cash crop either because these plants are physically damaged or because of subsequent changes in environmental

Fig. 7.5 Design 1 (trees and an understorey crop) represented by rattan established in a young plantation in Laos PDR. In time the rattans will grow up into the canopy

conditions (although the new environmental conditions after thinning might allow other species to be used).

Case Study 1: Shade coffee

An example of under-cropping that is common throughout the tropics is the growth of coffee (*Coffea* spp.) beneath a tree overstorey. In most cases the trees are arranged to provide around 30% shade and the coffee is then planted beneath them at spacings of 3 m or less. Greater amounts of shade have been beneficial at more marginal sites (DaMatta 2004). A large variety of woody species have been used to provide the overstorey cover including species of *Casuarina, Erythrina, Falcataria, Gliricidia, Grevillea* and *Leucaena*. There are a number of variations to the basic model (Somarriba et al. 2004). So, for example, a broad mixture of tree species have been used in some locations to provide the overstorey shade and Philpott et al. (2008) provide examples from southern Sumatra. In such cases, tree management for, say, fruit production, may not lead to optimal shade conditions for the coffee. And sometimes the system has involved other food crops as well as coffee. A system in the highlands of Papua New Guinea evolved from one that planted Arabica coffee beneath an established *Falcataria moluccana* or *Leucaena leucocephala* canopy to one where coffee and *Casuarina* were planted into food gardens. In this case farmers modified the initial system which mostly produced just subsistence foods to pass to one dominated by coffee, bananas and *Casuarina oligdon* and, then finally, to a simple coffee and *Casuarina* system. This system provides food, cash, fuelwood and timber (Bourke 1985).

Coffee is grown without an overstorey tree cover in some places (and is referred to as 'sun-coffee'). Production levels can be higher but the system has a number of disadvantages (Somarriba et al. 2004; DaMatta 2004). There is a greater need for agrochemical inputs, sites are more prone to soil degradation and there is evidence of a biennial production cycle with a 'good' year being followed by a 'poorer' year. The longevity of the coffee plants also tends to be shorter when grown in full sun rather than shade. Farmers using this monocultural system are more exposed to risks arising from international price fluctuations while coffee grown beneath a tree cover provides famers with lower cost and a greater diversity of products to buffer income fluctuations. In short, sun coffee might be best suited to larger growers able to support the more intensive management but smallholders might do better with coffee grown with a tree cover crop, especially at less fertile sites. In this case the trees form a complementary mixture with the coffee but also act as facilitators to improve economic productivity.

Uneven Aged Plantations Involving Only Trees

This silvicultural system is represented by Design 2 in Table 7.2. It is similar to Design 1 in that it involves underplanting beneath an existing tree canopy. However, it differs from Design 1 because the underplantings are also trees and these eventually go on to replace the overstorey species or 'nurse' trees that were initially planted. That is, the design represents only a temporary mixture. Its main purpose is to ameliorate otherwise unfavourable site conditions using a tolerant tree species to facilitate the establishment of a more preferred species. The nurse trees may exclude shade intolerant weeds (or at least make them easier to control), improve the soil or change some other properties of the micro-environment that are hindering the establishment of a preferred species.

The facilitator or nurse tree species should have certain attributes. They must be tolerant of a wide range of site conditions and be capable of rapid growth, at even degraded sites, such that a closed canopy is formed within a few years. The crown should be initially dense enough to hamper weeds but be not so dense that it also limits the growth of the preferred species. This means most nurse species will be from early successional stages and most underplanted species are likely to be from mid or later successional stages. Ideally, the nurse tree should acquire a commercial value at an early age so that its harvesting will yield a financial return. Otherwise it should be short-lived so that the preferred species can eventually grow up and replace it. If felling is done it should not damage the more valuable species growing in the understorey. This means that felling probably has to occur before the overstorey trees are very large.

The attributes of the under-planted species must have some degree of complementarity with the nurse species. In particular, they must be physiologically tolerant of some early shade and be not merely able to survive but to grow in height while the nurse trees are present. They must also be substantially more valuable than the nurse trees to persuade the landowner to bother with underplanting. Unless this is the case it would make more sense to simply use successive rotations of the nurse tree. Commercial worth would be one expression of value but ecological or conservational significance could be another.

Case Study 2: Improving conditions at a degraded site – Vietnam

Many coastal forest areas in central Vietnam have been highly degraded by past military activities. The Hai Van Pass area between the cities of Hue and Danang was affected in this way and became grasslands dominated by *Imperata cylindrica.* Although small scattered patches of natural forest remain nearby it was difficult for any of these species to regenerate within the degraded areas and attempts at plantings failed. The Phu Loc District Forest Protection Department has rehabilitated the area and established native species using a cover crop of *Acacia auriculiformis*. The first *Acacia* trees were planted in 1986 at densities ranging from 1,650 to 3,300 tph. These excluded the grasses and other weeds and probably added nitrogen to the soils (although the extent to which this occurred has not been evaluated). When these trees were 8 years old they were thinned by cutting 5 m wide strips (later reduced to 2.5 m) through the plantation. These timbers were sold in the local market and the revenue

Fig. 7.6 Design 2 (final crop trees under-planted beneath nurse trees) used at a degraded site in Vietnam. The *Acacia* overstorey has been under-planted with several native tree species able to produce high-quality timbers. The *Acacia* has facilitated the establishment and early growth of these but must be removed at some point when the advantages of facilitation are outweighed by the disadvantages of competition

was used to enlarge the *Acacia* plantation. Seedlings of a number of commercially valuable native species were planted under the *Acacia* canopy at densities of 200–500 tph (Fig. 7.6). These species included *Dipterocarpus alatus, Hopea odorata, Parashorea chinensis, P. stellata, Scaphium lychnophorum* and *Tarrietia javanica*. As these grew up more of the residual *Acacia* were gradually removed exposing the preferred species to full light. These trees could then also be sold to establish more of the plantation. Several hundred hectares of plantations have been established in this way.

One of the key silvicultural issues concerns the trade-off between the advantages and disadvantages of the nurse trees. These enable the preferred species to establish and survive but, at some point, they also begin to inhibit growth. This raises the question – just when should the overstorey cover be removed? This issue was addressed by McNamara et al. (2006) who related the mean annual increment of 8 year old trees of the under-planted species to the density of the overstorey *Acacia* trees amongst which they were planted. It was hypothesized that a strong relationship would imply the *Acacia* cover was affecting the growth of that species but any lack of a relationship would mean that growth was not yet being inhibited. The results showed that, at 8 years, high densities of *Acacia* trees had begun to inhibit the growth of the *Dipterocarpus alatus* while the height growth of other species such as *Hopea odorata* were still apparently unaffected (Fig. 7.7). Further monitoring

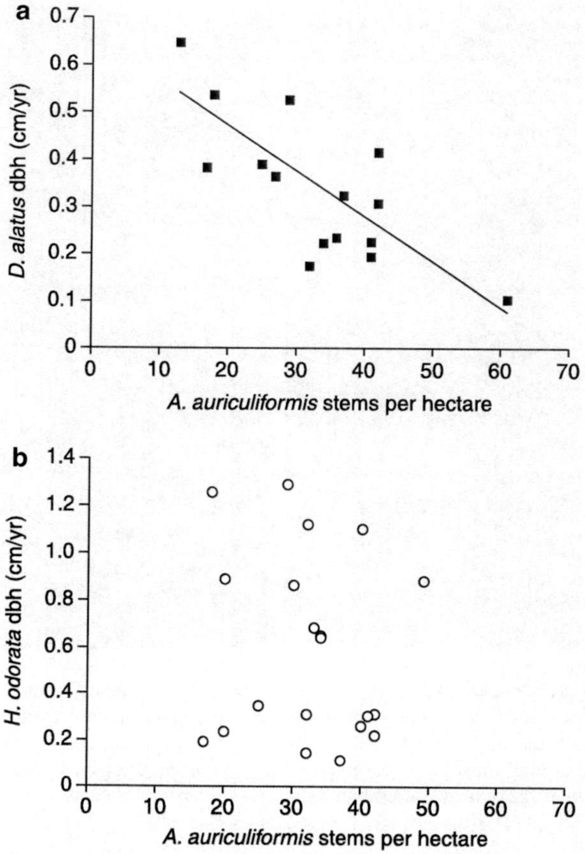

Fig. 7.7 Annual diameter growth increment (cm/year) of underplanted trees age 6–9 years planted beneath established *Acacia* overstories differing in tree density. (**a**) *Dipterocarpus alutus* – growth declines with increasing *Acacia* overstorey density ($p = 0.0022$, $r = -0.7075$, $n = 16$) and (**b**) *Hopea odorata* – no relationship between growth rate and overstorey density at this age (After McNamara et al. 2006)

will be needed to determine just when the final nurse trees should be removed. A similar approach has been outlined by Kuusipalo et al. (1995) for use at degraded sites in Indonesia.

Case Study 3: Improving conditions at a degraded site – Malaysia

Many dipterocarp species grow poorly when planted in open areas and appear to need some initial shade at the time of establishment. The Multi-storied Forest Management Project in Malaysia sought to establish whether various dipterocarp species could be established by planting seedling beneath a nurse crop (Anon 1999). The experimental plots covered 180 ha of *Acacia mangium* plantation which acted as a cover crop or nurse trees. At the time of underplanting these plantations were 4–5 years old and had a stocking of 900 tph. The *Acacia* were probably then more than 7 m tall (this height was not reported but is estimated on the basis of mean annual increment data). Five under-planting treatments were examined.

Each involved felling strips (i.e. rows of planted trees) through the *Acacia* plantation and planting the dipterocarps along these clearings. In treatment 1 alternative row of *Acacia* were cleared and replaced by a row of dipterocarp seedlings (1:1), treatment 2 removed and replaced every second two rows of *Acacia* (2:2), treatment 3 removed and replaced four rows leaving the next four rows (4:4), treatment 4 removed and replaced eight rows leaving the next eight rows (8:8) and treatment 5 removed and replaced 16 rows leaving the next 16 rows (16:16). These five treatments correspond with strips that were 3, 6, 12, 24 and 48 m wide. The single row (1:1) replacement left the new seedlings under canopy cover while the 8:8 and 16:16 strips essentially left the new dipterocarp seedlings as open plantings. A variety of dipterocarps were tested with each treatment involving measurement plots containing about 120 trees of each dipterocarp species. The strips were weeded and kept free of vines.

Most of the planted species had substantially lower survival levels (with many failing completely) as the width of the planting strips increased and seedlings were exposed to higher radiations levels. However, a few species (e.g. *Hopea odorata, Shorea leprosula*) were much less affected and were able to tolerate the more open conditions (Fig. 7.8). Tree height and stem diameters of all the surviving trees were much less affected by the strip width irrespective of species (Anon 1999).

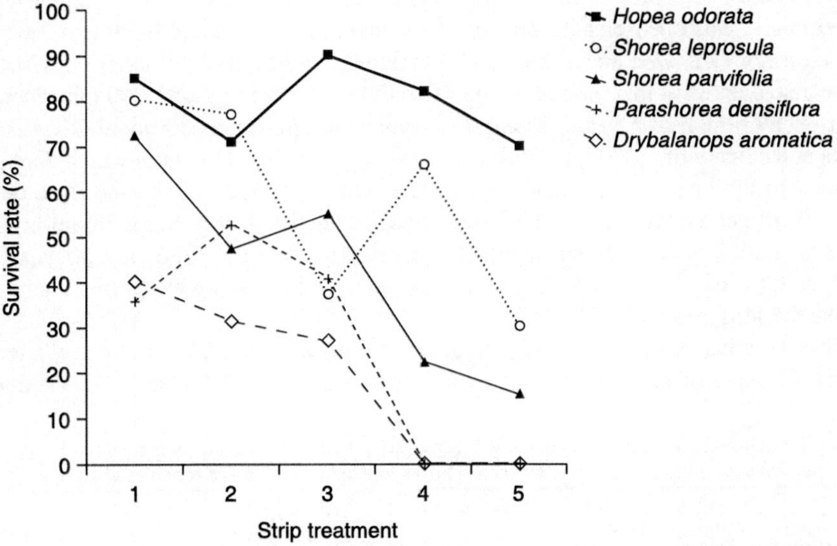

Fig. 7.8 Survival of trees planted in strips of various width cut through 3 year old *Acacia mangium* plantations in Malaysia. Treatment 1- strips were created by removing and replacing every alternate row of *Acacia* (1:1); Treatment 2 – remove and replace every alternative second two rows (2:2); Treatment 3 – remove and replace every alternative four rows while leaving four rows (4:4); Treatment 4 – remove and replacing every alternative eight rows while leaving eight rows (8:8); Treatment 5 – remove and replace every alternative16 rows while leaving 16 rows (16:16). Survival was measured 60 months after planting for *Shorea leprosula* and *S. parvifolia* and after 48 months for the other species (After Anon 1999)

The overstorey *Acacia* was removed after 8 year when these were 18 m tall. Care had to be taken in removing the *Acacia* to avoid damaging the dipterocarps. Because of this it was concluded that, at this age, underplantings in 12 m wide strips (the 4:4 treatment) represented an appropriate balance between providing enough overstorey cover to maintain survival while minimizing felling damage when the *Acacia* were removed.

Case Study 4: Reducing insect damage

The family Meliaceae includes some of the world's more valuable timber species but genera such as *Cedrela, Chukrasia, Khaya, Swietenia* and *Toona* spp. are frequently damaged by shoot borers from the genus *Hypsipyla* (Speight and Wiley 2001; Wormald 1992). In Australia, the growth of *Toona ciliata* (red cedar) is badly affected by *Hypsipyla robusta*. There have been many attempts to grow *Toona ciliata* (red cedar) in plantations because it's timber is so valuable but all have failed because of repeated insect damage to young shoots. Anecdotal evidence suggested damage is much less when the red cedar trees are grown in shade. Accordingly, a trial was initiated in north Queensland in 1941 using *Grevillea robusta* (another commercially attractive timber species) to create an overstorey or nurse crop for the *Toona ciliata* (Keenan et al. 1995). There were seven treatments in the trial. Treatment 1 was open-planted *Toona* established at 2,000 tph and treatment 2 was a mixture of *Grevillea* and *Toona* each planted at 1,000 tph. Additional *Grevillea* were also planted at this time in another five plots at a stocking of 1,000 tph. Then, over each of the next 5 years, *Toona* was planted among these additional *Grevillea* plots at a density of 1,000 tph. This led to a sequence of treatments in which *Toona* planted in the open, at the same time as *Grevillea* or beneath progressively older *Grevillea* trees creating a series of stands each with 2,000 tph. Some thinning of trees of both species was subsequently carried out over the following 50 years. Although the trial was unreplicated it has the advantage of being monitored over an unusually long period.

Insect damage was less and survival and tree form was best in *Toona* trees planted under the level of canopy cover provided by treatments 4–7 (Table 7.3). But this

Table 7.3 Growth of *Toona ciliata* at 54 years after being grown beneath a nurse crop of *Grevillea robusta* to reduce damage by the tip borer *Hypsipyla robusta* (Keenan et al. 1995)

Treatment	*Grevillea/Toona* planting dates	Age (year) of overstorey	Percent with multiple leaders	Average *Toona* stem volume m^3
1	–/1941	–	65	2.82
2	1941/1941	–	34	2.92
3	1941/1942	1	3	2.70
4	1941/1943	2	3	1.10
5	1941/1944	3	2	1.30
6	1941/1945	4	3	0.96
7	1941/1946	5	3	0.91

imposed a limitation on growth because of shading and individual *Toona* trees grew better as open-planted trees or if planted with the youngest *Grevillea* (treatments 1–3). For plantation managers the silvicultural compromise might be to create a temporary mixture by planting *Toona* beneath one year old *Grevillea* (treatment 3) and allowing the trees to grow tall enough to develop a commercially attractive tree bole. Thereafter the stand could be managed as a simple monoculture.

The obvious silvicultural question is how long to keep the beneficial overstorey before converting to a monoculture? There is some evidence that *Hypsipyla grandella* in Costa Rica mostly fly below 6 m (Grijpma and Gara 1970). This suggests any cover could be removed once the trees grew up and exceeded this height. But the behaviour of *Hypsipyla robusta* appears to be different and there is evidence that damage caused by this species also occurs in taller trees. One study of the effect of *Hypsipyla robusta* on *Toona australis* carried out in the Philippines, Thailand and Australia found damage continued as trees grew taller up to heights of 4.5 m and that height was a good predictor of damage (Cunningham and Floyd 2006). Although there was some indication that damage rates may have been then declining these authors also report other observations of damage being found on >35 m tall *Toona* trees. This suggests the best option for managers would be to retain cover trees until the *Toona* trees had reached a merchantable bole length.

The advantage of Design 2 is that it allows preferred species to be established at sites where they would fail if normal planting methods were used. But there are two disadvantages. The first is that the method requires a market price for small-sized logs of the facilitator or nurse species. The second is that there is usually a trade-off to be made between the ecological advantages provided by the facilitator and the inhibition it will eventually cause because of shading and root competition. The age at which this occurs will depend on the particular species involved and requires field trials with the species concerned to explore this trade-off.

The three case studies that used this design involved three different forms of facilitation including weed exclusion, the provision of shade for species unable to tolerate full sunlight at the seedling stage and changing the aerial micro-environments leading to reduce insect damage. Where *Acacia* were used there may have also been some improvement in soil nitrogen levels. In each case the facilitator species (*Acacia, Grevillea* spp.) might have been grown as a successful monoculture without the need for the complexities of mixtures but managers preferred to use these to enhance the quality of the goods being produced by the plantation. But the commercial value of the facilitator species also meant the mixture generated a cashflow before the final harvest of these preferred species.

Even-Aged Plantation Using Species Grown Together on Short and Long Rotations

This silvicultural system is represented by Design 3 in Table 7.1. Like Designs 1 and 2 it is also a temporary mixture and there are several versions of the design. In one the primary purpose is to increase the financial attractiveness of a plantation containing commercially valuable but slow-growing sawlog species. This is done by planting a fast-growing species with the slower-growing species. The faster-growing species is harvested at an early age thereby providing a financial return while trees of the slower-growing species are able to continue growing until they reach their financially optimal harvesting size. The first harvesting operation also acts as a thinning operation.

Thinning in a monoculture plantation could obviously achieve the same outcome if a market could be found for the small-sized logs but this is often difficult. This system avoids such marketing problems by deliberately using a species able to provide a marketable product within the desired time period. As with any thinning operation, care is needed to ensure that harvesting does not damage the residual trees. In this system even more care is needed because the trees being harvested are larger than the residual trees. However, if row plantings are used it should be possible to minimize any damage because trees can be felled and removed along the rows. Some ecological complementarity between the species is needed to ensure the faster-growing species do not overly inhibit the slower-growing trees. But since the mixture is only temporary the extent to which this is required is probably a little less than in a permanent mixture.

A second version of the system is more concerned about using the species grown on the shorter rotation as a facilitator that fixes nitrogen and thereby boosts production of the preferred species. Ideally this nitrogen fixer should have a shorter longevity and it would also be advantageous if it was commercially valuable although this is not essential. The growth rate of the nitrogen fixer determines how the system is managed. If it grows quickly and over-tops the commercially preferred species then it may need to be thinned or removed. This assumes most of the nitrogen fixation has been carried out at an early age. If, on the other hand, it grows more slowly than the commercially preferred species if might be left to eventually senesce if it appears it is not affecting the growth of the overstorey species. In this case the death of these trees would also act as a stand thinning.

Case Study 5: Early cashflow from trees grown on short rotation – Vietnam

There is a strong market for sawlogs in many parts of rural Vietnam. But there is also a market for poles. A trial testing mixtures of trees able to produce both products was established at Doan Hung in Phu Tho province by the Forest Research Center (Lamb and Huynh 2006). Three native sawlog species (*Michelia mediocris, Canarium album* and *Chukrasia tabularis*) and *Eucalyptus urophylla* (for poles) were grown in alternate rows as pair-wise mixture of each species combination. The overall tree density was 1,100 tph. The eucalypt grew much faster than the other three species and after 3 years many trees exceed 9 cm diameter and 12 m

Fig. 7.9 Design 3 (trees with differing rotation lengths) used to grow *Eucalyptus urophylla* planted in alternate rows with *Michelia mediocris* in Vietnam. The plantation is now 4 years old. The eucalypt will reach a merchantable size within another few years and can be removed to generate a cashflow and reduce stand density. This will enhance the growth of the *Michelia* which will be grown until it reaches a sawlog size

height. These growth rates suggest the eucalypts will be saleable within another 2 or 3 years. This would leave the residual stand of sawlog trees with a stocking of 550 tph. Over the same time period the slower-growing native species achieved heights of 2 or 3 m and these are likely to be several meters taller by the time of the pole harvest (Fig. 7.9). Sufficient light penetrated the eucalypt canopy to reach their crowns and their growth was comparable with that when they were grown in simple monocultures. This particular trial is still too young to assess how the other species combinations will develop.

There was some indication that mixtures of the sawlog species with the eucalypt can improve tree form even at this early age. A sample of 20+ trees of each sawlog species in monocultures and mixtures found the mixtures induced reductions in the proportion of trees with forks (lateral branches >1 cm width arising below the main crown) or bends (a curve in main stem displacing it >4 cm within a vertical distance of 10 cm) and caused a small decrease in the average number of leading shoots in *Michelia* though not other species (Table 7.4). Of course many of these problems would normally be resolved by pruning or thinning the trees with poorer form but, nonetheless, the results indicate the potential power of the 'trainer' effect in these mixtures.

Table 7.4 Proportions or numbers of trees with forks and bends and average number of leading shoots in 4 year old trees of *Michalia mediocris, Chukrasia tabularis* and *Canarium album* when grown in mixtures with *Eucalyptus urophylla* of same age (Lamb and Huynh 2006)

	Michelia	*Chukrasia*	*Canarium*
Forks (%)			
Monoculture	21	35	51
Mixture	5	22	42
Bends (%)			
Monoculture	32	13	46
Mixture	14	7	33
Leading shoots (No.)			
Monoculture	1.13	1.26	1.55
Mixture	1.01	1.25	1.50

Case Study 6: Use of a nitrogen fixer to improve tree nutrition on infertile soils

Many eucalypt plantations in Australia are limited by nitrogen deficiencies. To test whether increasing proportions of a nitrogen fixer in a mixture might be able to improve productivity *Acacia mearnsii* was planted with *Eucalyptus globulus* in monocultures and in a replacement series in which the proportion of the two species varied but density remained constant (Bauhus et al. 2004; Forrester et al. 2004). The five stands included 100% eucalypt, 75% eucalypt + 25% *Acacia*, 50% eucalypt + 50% *Acacia*, 25% eucalypt + 75% *Acacia* and 100% *Acacia*. In the mixed-species stands seedlings of the two species were planted together in an intimate mixture along rows. Height growth was comparable for the first 6 years but thereafter growth of the *Acacia* slowed and the eucalypt became the dominant canopy species. The *Acacia* appears to have increased the height and diameter growth of the eucalypt in mixtures compared with those in the monoculture by increasing nitrogen availability through fixation and by increasing nitrogen cycling. There was no evidence that the improved nitrogen levels also improved the photosynthetic abilities of the eucalypt. Productivity also appears to have been enhanced because of the development of a stratified canopy with the eucalypt in the dominant position and the *Acacia* in the sub-canopy (Bauhus et al. 2004; Forrester et al. 2004). Best stand growth was recorded in the 50:50 mixture which, at 11 years, had more than double the biomass found in the eucalypt monoculture. Although the study is still comparatively young there are several silvicultural options open depending on markets. One of these would be to thin or remove all of the *Acacia* and leave the eucalypt to grow longer. If there were no markets for small trees the *Acacia* could be left and allowed to continue adding nitrogen to the system as the eucalypts continue to grow although it is not clear just how long the *Acacia* will continue to do this.

Even-Aged Plantation with All Species Grown Together in a Single Long Rotation

This silvicultural system is represented by Design 4 in Table 7.1. Unlike the previous designs this is a permanent mixture (Fig. 7.10). This design might serve several purposes. One might be to capture the potential for multi-species plantations to have higher levels of productivity than monocultures. It is not clear how many species should be used although the optimum number is unlikely to be more than three or four in plantations established primarily for timber production. Larger numbers than these would be more difficult to manage and the overall financial worth of the plantations is likely to be reduced as additional less-valuable species are included.

A second situation where this design might be used is when there is a need to increase the variety of products to diversify income sources or subsistence goods. Thus a plantation might include different types of timber trees as well as species able to supply various NTFPs (e.g., fruits, nuts, spices, medicines). It might even

Fig. 7.10 Design 4 (permanent tree mixtures) represented by a 60 year old mixture of hoop pine (*Araucaria cunninghamii*) and Queensland maple (*Flindersia brayleyana*) in sub-tropical Australia. The two species have similar growth rates but the hoop pine is more shade tolerant and has a deep crown while the maple is less shade tolerant and has a shallow crown

be possible to design such a plantation to include species able to supply such goods from a relatively early age thereby improving the timing of any revenue stream. These plantations could include only a few, or perhaps many, species.

Finally, a third situation where the design would be attractive is where certain ecological services are needed as well as goods. These might be to improve watershed protection (hence some deep-rooted species able to stabilize hillslopes might be included) or to sustain local wildlife (in which case food trees might be added to stands of commercial timber species) or even to increase the local populations of certain threatened plant species. In some of these cases the diversity of species used could be quite high even though some non-native species able to tolerate existing site conditions might have to be used.

A variety of spatial layouts could be used depending on the attributes of the species and the purpose of the mixture. In the case of commercial plantings of canopy species these might involve might alternative rows or belts of each species (Fig. 7.11a) or more intimate tree-by-tree mixtures along planting lines of two or more species (Fig. 7.11b). Mixtures might also be achieved by planting trees in small groups (Fig. 7.11c) or blocks (Fig. 7.11d). Over time such a block planting might be thinned to leave just one or two trees. Alternatively, where the objective of reforestation is the provision of ecological services and no future logging is expected, less care might need to be taken with spatial locations and trees of each species might be randomly located throughout the stand. In this case the emphasis might be more concerned with ensuring canopy trees are surrounded by several sub-canopy species.

The more complex the mixture, the more scope there is to involve a larger range of species and functional types. But, at the same time, the greater the complexity then the more uncertain the silvicultural outcome is likely to be. One advantage that multi-species plantings might have is in allowing growers to quickly screen a large number of possibly useful species to identify those able to tolerate site conditions and form stable mixtures.

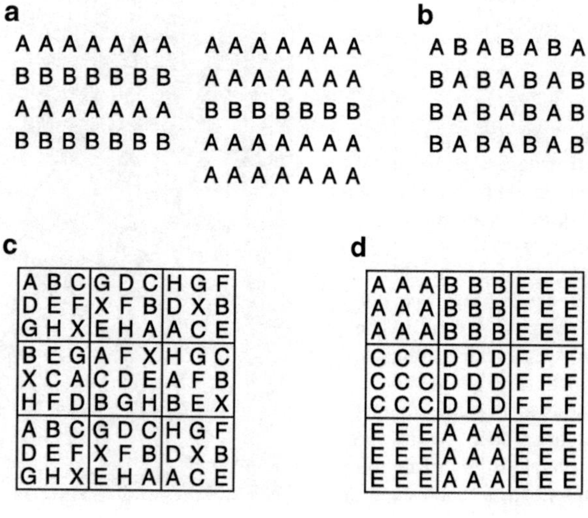

Fig. 7.11 Different types of spatial arrangements for growing trees in mixed species plantings (**a**) two species planted in alternative single or multiple rows, (**b**) two species planted in more intimate mixtures (**c**) random mixtures of multiple species within small block plantings and (**d**) block plantings of a single species. Trees of different species are indicated by (**A**), (**B**), (**C**), etc.

Case Study 7: Mixtures involving Sandalwood and a host species

Sandalwood *(Santalum spp.)* is a hemi-parasite with around 16 species that are distributed across the Asia-Pacific region. It produces highly valuable timber with an aromatic heartwood although there are large differences in the oil content of timbers of the various species and genotypes. Natural stands have been heavily exploited for more than 100 years and many have been cleared or degraded. Natural regeneration can be assisted by broadcasting seed into natural stands where a variety of potential hosts already exist but less than 1% of these seed may produce seedlings. Because of this limited success rate alternative approaches are being explored.

The problem for those wishing to grow sandalwood is in finding an appropriate host (or facilitator) able to sustain the seedlings in the nursery and then later in the field. Several approaches have been developed. One is to grow the sandalwood in the nursery and to introduce a 'pot host'. This is a host plant grown in the same pot or seedling tube. The pot host must be able to grow with the sandalwood and sustain it for as long as is needed (if the pot host dies the sandalwood will also perish) but not dominate it. The two must also grow together long enough in the nursery for the haustoria and root connection to be sufficiently robust to tolerate being transferred to the field and planted out. A variety of potential pot host have been tested but species of *Alternanthera* have often been used (Radomiljac 1998; Robson 2004).

Once in the field a new host is needed because the pot host, even if it was sufficiently long-lived, would be too close to the sandalwood and might overshadow it. This has been approached in several ways. One is to plant the sandalwood seedlings into old gardens sites or regrowth forest and allow the sandalwood to find a new host among the species already present. An alternative is to plant seedlings and a new host in alternate rows (or as alternative plantings along rows) such that the trees of the two species are within 2–3 m of each other. Under these circumstances the host may be parasitized within a year. Hosts used in field plantings have included *Casuarina equesitifolia* as well as species of *Acacia, Sesbania, Cassia, Dalbergia* and *Cathormium.* While a host with commercially attractive timber would be preferable some care is needed to ensure a balance is maintained between the sandalwood and host. Hosts with large spreading crowns may suppress the sandalwood (e.g. *Falcataria, Khaya* spp.). On the other hand, less vigorous hosts may succumb to competition from the sandalwood. It may even be necessary to use a third host to sustain the sandalwood through the rotation if the second host succumbs to such competition. Considerable skill is needed to select appropriate hosts and determine the numbers of these that should be used at any one time as well as the tree spacings to use. Some of the methodology being used to grow sandalwood in a large-scale industrial plantation in the dry tropics of northern Australia is outlined by Done et al. (2004) while current methods being used in smallholder operations in the Pacific are outlined by Ehrhart (1998) and Robson (2004).

Case Study 8: Mixtures involving pairs of commercially attractive tree species

A replicated trial was established in north Queensland to examine how several types of species pairs might interact (Huynh 2002). The species chosen were all local species and included *Eucalyptus pellita, Elaeocarpus grandis, Acacia aulacocarpa* and *Flindersia brayleyana*. The first two species are relatively fast growing and shade-intolerant species while *Flindersia brayleyana* grows less quickly and is relatively more shade tolerant. The *Acacia* usually grows well and is a putative nitrogen fixer under appropriate field conditions and is usually regarded as intolerant of shade. These four species were grown in pair-wise mixtures with trees of each species planted in alternative rows as well as in monoculture plantings where each species was planted was bounded by rows of itself. Overall tree density in both the mixtures and monocultures was 1,100 tph. The site was an old sugar cane farm and receives about 2,200 mm rain. As expected, the fastest growth was observed in trees of *Eucalyptus pellita* and *Elaeocarpus grandis.* After 38 months these reached 13.5 and 12 m respectively in monocultures while the *Flindersia* reached 8 m and the *Acacia* reached 7 m.

Growth of each species was statistically similar in mixtures and monocultures except for *Eucalyptus* which had significantly better growth when mixed with *Flindersia* than when grown in a monoculture. These two species appear to have formed a stratified canopy and be a complementary pair. The *Eucalyptus* also had a larger leaf area per tree, probably a consequence of the greater space available to its trees in the upper canopy layer. The leaf area of *Flindersia* was similar in this mixture and in a monoculture. When expressed as volume there was evidence that the Relative Yield Total of the *Eucalyptus-Flindersia* mixture at this age was greater than could be obtained in the respective monocultures (Table 7.5).

By contrast *Eucalyptus pellita* had much poorer growth when mixed with *Elaeaocarpus*. The height growth rates were similar and there was no sign of complementarity in the canopy architecture of the two species or of canopy stratification.

Table 7.5 Relative Yield Total in stand volume of pair-wise mixtures of *Eucalyptus pellita (EP), Elaeocarpus grandis (EG), Flindersia brayleyana (FB)* and *Acacia aulacocarpa (AA)* at 38 months after planting. A Relative Yield Total >1.0 shows production is greater than would occur in a monoculture (Huynh 2002)

Mixture	Relative yield of		Relative yield total of mixture	Conclusion
	First spp.	Second spp.		
FB + EP	0.73	0.89	1.62*	Benefit
FB + AA	0.71	0.77	1.48	
FB + EG	0.51	0.66	1.17	
AA + EP	0.47	0.78	1.25	
AA + EG	0.47	0.63	1.10	
EP + EG	0.50	0.35	0.85	Failure

*Gain is significant ($P<0.05$)

The Relative Yield Total of this mixture was <1.0. There was some evidence that *Eucalyptus* might form complementary mixtures with *Acacia* because of the difference in growth rates allows some canopy stratification but *Acacia* is not tolerant of much shade and the growth differences (between the mixture and the monocultures of each species) were not statistically significant. There was also some indication that the *Flindersia – Acacia* mix might have some advantages but, again, the mixture was not significantly better than monocultures of its component species. Overall, the trial supports the idea that species with complementary crown structures forming stratified canopies may form productive mixtures.

Case Study 9: Multi-species plantings involving four species; Costa Rica

Many farmers in Costa Rica grow small timber plantations on their farms (Piotto et al. 2003a). Despite this there is little understanding of the most appropriate species to use or the silvicultural requirements of these species. A series of mixed species plantations were established to explore how a variety of native tree species grew in plantations and whether greater levels of production could be achieved in monocultures or in mixtures (Petit and Montagnini 2006; Piotto et al. 2003b). Three of these trials were established in the humid lowlands at La Selva and used four species grown in monocultures and together in mixtures. Each trial included a putative nitrogen-fixer, a relatively fast growing species and a relatively slower growing species. All of these species were chosen because they are widely used by small farmers in the region and because they provided a variety of branching patterns, sizes and crown shapes. Planting density was originally 2,500 tph (i.e. 2 m spacings) to accelerate interactions between trees. Half of each trial was thinned at 3 and 6 years while the other half was left unthinned. Each trial suffered some mortality reducing the number of species in Trials 1 and 2 to three species each.

By age 10–11 years different results were found in each trial as a consequence of the particular species used. In Trial 1 two of the three species grew better in the mixture than they did in the monoculture (the exception being *Calophyllum brasiliense* which is a relatively slow-growing species normally found in mature forests). In addition, the survival rate of all species was also better in the mixed. In Trial 2 there was no difference in the growth of trees of the various species planted in monocultures or mixtures although, once again, survival rates were higher in the mixed stand than in the monocultures. In Trial 3 there was no difference in the growth of species planted in the mixtures or in monocultures except for *Genipa americana* which was suppressed in the mixture and grew much better in monocultures.

The trials highlighted the superior performance of *Vochysia guatemalensis, Terminalia amazonia, Jacaranda copaia* all of which are relatively fast growing species representative of early or mid successional stages. The two slower growing species from later successional stages (*Calophyllum brasiliense* and *Dipteryx panamensis*) could grow in mixed plantation mixtures but tended to be suppressed by the faster-growing species. Tree mortality and the loss of several species complicates these analyses. Although there is some evidence for enhanced productivity in Trial 1 it is difficult to know the nature of the mechanisms that have produced it and it may be that the better growth of the faster growing trees (*Jacaranda copaia* and *Vochysia guatemalensis*) was at the expense of the slower growing trees (*Callophyllum brasiliensis*). If so the gain was largely due to competition causing a reduction in the effective density of these trees rather than any degree of complementarity between the species remaining in the mixture.

Case Study 10: Multi-species planting involving 16 species

This trial was established to explore the growth in plantations of a variety of rainforest tree species in sub-tropical Australia at a time when little was known of the silvicultural characteristics and site requirements of most of these species. The 16 species included in the trial all had high-quality timbers but there was considerable variation in their properties and their timber densities ranged from 400 to 800 kg m^{-3}. The species included shade-intolerant secondary species as well as shade tolerant species representative of more mature successional stages. They also included hardwoods and softwoods. Most species came from nearby rainforests although several exotic species were also used (e.g. *Khaya nyasica, Cedrela odorata*). The plantation was established at a sub-tropical location in southern Queensland that has moderately fertile soils and receives around 1,500 mm of rain. Prior to being planted the site was used as a cattle pasture.

The layout of the trial ensured that trees of each of the 16 species were surrounded by those of at least three, and more commonly four, other species. This was done by planting a single tree of each species at random within a block of 16 trees (4 × 4) and each block was surrounded by similar adjoining blocks (Erskine et al. 2005b). The layout resembles that shown in Fig. 7.11c. Trees were spaced 3 m apart and all trees were planted at the same time. Most species grew well in the first few years and canopy closure occurred after 4–5 years.

At 5 years of age the trees ranged from 2.8 to 9.0 m height and there was evidence of several canopy layers developing. Given the differing ecological characteristics of the various species it was expected that species believed to be from early successional stages would form the upper canopy while those thought to be representative of more mature successional stages would form a subordinate canopy layer. But the limited knowledge of the growth rates of the various species in a plantation setting meant it was hard to predict how each species would fit within the evolving plantation structure. Several hypotheses were developed. The first was that height growth is closely related to timber density and all those species forming the upper canopy should have low timber densities (Whitmore 1984). This was indeed the case although the r^2 value was only 0.487 (Fig. 7.12). None-the-less, the relationship provides an indication of what to expect in terms of relative growth rates.

The second hypothesis was that the slower growing and shade tolerant species would form a sub-ordinate canopy layer beneath that produced by the fast-growing

Fig. 7.12 Relationship between height (at 5 years) and timber density for 15 tropical tree species growing in a mixed-species plantation in sub-tropical Queensland

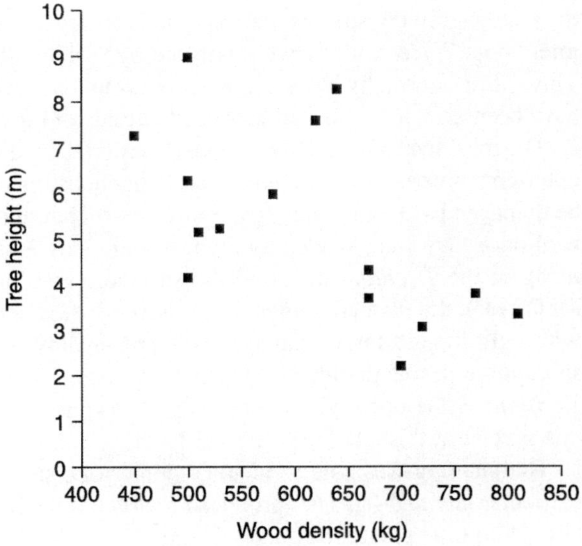

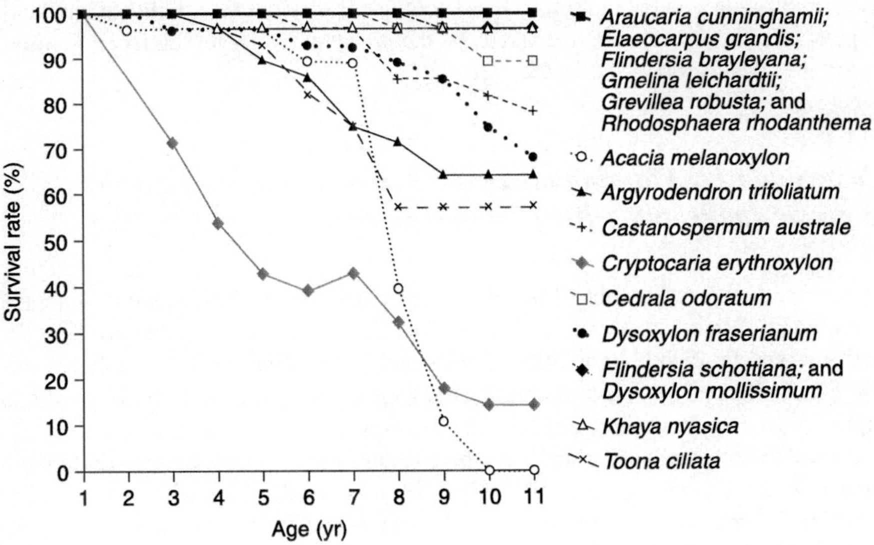

Fig. 7.13 Survival over time of species planted in the even-aged mixed-species plantation shown in Fig. 7.12. The mixture includes mostly native species from early and late successional stages although the exotics *Cedrela odorata* and *Khaya nyasica* are also included

secondary species. This also occurred but not nearly to the extent expected. In fact the survival rates of some supposedly shade tolerant species was poor (e.g. *Cryptocarya erythroxylon*) although they appeared to have stabilized after 10 years (Fig. 7.13). The reason for this is not entirely clear but it may be that in some locations the overall light levels in the unthinned plantation are simply too low to sustain these species. Even more surprising was that *Acacia melanoxylon*, one of the fastest-growing species that occupied the upper canopy layer, also began dying after about 7 years. Many were infected by various stem borers and bracket fungi. This species normally has a longevity of around 40–50 years in this region but may have been growing below its preferred altitudinal range.

The third hypothesis, based on result reported by Keenan et al. (1995), was that open-grown trees of *Toona ciliata*, an indigenous member of the Meliaceae, would be damaged by the tip borer *Hypsipyla robusta* but that surviving trees would grow well once they were shaded by the faster-growing early secondary species. In fact many of the *T. ciliata* died and the survivors grew poorly although other exotic Meliacea in the plantation including *Khaya nyasica* and *Cedrela odorata* grew well with only limited insect damage. This presumably reflects the fact these are less susceptible to this particular *Hypsipyla* species. This result confirms that planting *T. ciliata* in the open within a species mixture is not a useful method of growing this species at this site (c.f. Case Study 4).

The planting was able to identify timber tree species able to grow well at this particular site and has also produced a structurally complex multi-species plantation. With time plantations like this might provide habitats for a variety of species and so contribute to local biodiversity conservation. But the ability of mixed plantations to provide goods (e.g. timber) and ecological services (e.g. wildlife habitats) depends on how they are managed as the trees grow larger and competition increases. This is discussed further below.

Case Study 11: Planting mixtures of non-commercial species to create stable and self-sustaining new forests

As mentioned earlier, not all mixed-species plantings are established to generate commercial outcomes. In some degraded sites the motive is simply to develop resilient, species-rich communities of trees and shrubs that are able to tolerate the site conditions and be self-sustaining. This approach might be especially useful at highly degraded sites where changes to environmental conditions make it difficult to restore the original vegetation and where some form of rehabilitation (as defined earlier) is a more appropriate choice. These types of plantings can be especially useful at former minesites. Hobbs et al. (2009) referred to these plantings as 'novel ecosystems' because they involve species combinations and relative species abundances that have not occurred before in that particular biome.

An example of the approach is at the tailings dump at the Kidston gold mine in northern Australia. The area is occupied by tropical woodlands and has a strongly seasonal climate; the total rainfall is 720 mm but 80% falls between November and April meaning there is a long dry period. The topsoils at this site were mostly alkaline with pH levels ranging from 7.7 to 9.7 (c.f. 6.9 in nearby unmined areas). Because of this the objective of revegetation was to establish a self-sustaining community of trees, shrubs and ground cover with a structure resembling that of the pre-mining woodland rather than attempting to restore the original community (Roseby et al. 1998). Around 30 trees and shrub species were use together with a number of local and introduced grasses. All were selected on the basis of their capacity to tolerate the climate and soil conditions. The tree and shrub species used were all native species (though not necessarily species previously growing at the site) and these were initially introduced as seedlings. The species used include at least three families (Myrtaceae, Mimosaceae and Casuarinaceae or Proteaceae depending on the particular site being revegetated) and up to five genera. All species were planted in equal numbers. None of the species were commercially important and there was no intention to carry out any form of harvesting. Trees were planted at 5 m spacings (i.e. 400 tph). After 3 years all species had survived although survival rates varied from 17–95%. A number had also fruited and seedlings of several species had become established on the ground beneath the tree canopy. Having identified those species best able to tolerate the site conditions research has subsequently sought to develop methods by which these can be established by direct seeding rather than as seedlings in order to reduce costs. The amount of seed used was adjusted according to viability and germination success in order to establish roughly similar populations of each species. Monitoring is carried out to verify that composition, tree density and ground cover targets are being met. Most local species are poorly dispersed and none of those in the surrounding natural vegetation have yet colonized the site although this may occur in the longer term.

These five case studies using Design 4 fall along a gradient of increasing uncertainty. Past experience has given silviculturalists a good deal of information about the nature of the species mixes needed to grow sandalwood (Case Study 8). But the other four examples involve increasingly larger numbers of species and more uncertainties about outcomes. This might not matter in cases such as mine site rehabilitation where some losses can be tolerated provided sufficient species remain to provide a stable community (Case Study 11). But it is a wasteful approach in situations where a commercial benefit is required. The pair-wise comparison (Case Study 9) might be a useful approach to identifying complementary species when something is known about the attributes of each species while the mixes involving four and 16 species (Case Studies 10 and 11) might be useful when knowledge about site tolerances, growth rates and shade tolerances is more limited. But can these sometimes ad hoc assemblages be replaced by mixtures composed of species known to be complementary with each other because of the way they use resources?

Identifying Ecologically Complementary Species

As noted earlier, complementarity can come from species having different growth phenologies and using resources at different times or by having differing canopy or rooting architectures and so using resources from differing spatial locations. Two highly shade intolerant species are unlikely to form a complementary and commercially viable mix because one will eventually overtop the other and out-compete it. On the other hand, a mixture of a relatively shade tolerant species and a shade intolerant but open-crowned species, both of which are able to grow at much the same rate, could be complementary and grown together because of their contrasting tolerances and abilities to partition space. Examples of both outcomes were observed in Case Study 8.

Much depends on the longevity of the shade-intolerant species and the length of the proposed rotation. If the upper canopy species is not able to persist until the end of the rotation then care needs to be taken to accommodate the changing densities and spatial patterns that its disappearance will cause. Ashton et al. (2001) suggest shade intolerant species should be planted to surround tolerant species and short-lived pioneers should surround longer lived species to take account of complementarities in resource usage and in self-thinning among species. The spacings between trees should reflect known crown morphologies. Over time, the short-lived species will die and the soil and light resources they have used are taken over by slower growing shade tolerant species which ultimately form the canopy. The proposals depend heavily on knowledge of tree longevities, maximum tree sizes and shade tolerance.

Classifications such as shade-tolerant or shade-intolerant are necessarily superficial descriptors of species attributes and imply that a dichotomy exists when there is really a continuum, especially at the post-seedling stage. While acknowledging the idea of a continuum Poorter et al. (2006) identified four functional types among adult trees. In the case of trees of small stature these included (i) short-lived (<30 years) shade-intolerant pioneers and (ii) shade tolerant species able to establish and survive in the shade. Among trees with larger stature were (iii) long-lived (>30 years) pioneers that are shade intolerant and (iv) partially shade-tolerant species that can establish in the shade but need light to grow. Based on this classification the upper canopy layer should have the long-lived pioneers while the partially shade-tolerant species are candidates for the sub-dominant canopy positions. Shade intolerant species tend to have narrow and shallow crowns. They also have a rapid leaf turnover and are grow rapidly in height. More shade-tolerant species tend to have deeper and broader crowns with longer-lived leaves. They are better at capturing light in patchy or variable light environments but under appropriate conditions may be able to grow as fast in height as narrow crowned species. This led Poorter et al. (2006) to suggest that species with a range of crown allometries can probably co-exist in the upper canopy. As noted earlier, the maximization of timber productivity is not always the primary objective of mixed-species plantations. This means there may be considerable scope for assembling relatively stable mixtures of species able to occupy the sub-dominant canopy position based on crown attributes and, perhaps, wood density (as a surrogate index of growth rate).

Empirical field trials and further studies in natural forests will eventually generate knowledge of the species traits necessary to make more informed judgements about the species needed to create these types of plantations. In the meantime an alternative approach to identifying compatible species to use in the sub-dominant strata might be to identify those having a similar response to the competition they experience when planted in a mixture. In this case the hypothesis would be that two species with similar competitive abilities should form a stable mixture because both intra-and inter-specific competition is similar.

A large number of Competition Indices (CI) have been developed that might be used for this purpose (e.g. Biging and Dobberton 1992, 1995; Burton 1993; Vanclay 2006a). Some of these are distant-dependent indices meaning they take account of the spatial patterns of trees while others are distance independent indices that do not. Most assess the growth of the subject tree (e.g. height, diameter, crown development) in relation to the growth of surrounding trees. One simple version of such an index is shown in Fig. 7.14a. In this case the trees were established in a mixture using a regularly spaced, square planting grid and growth of the subject tree is related to the algebraic sum of the differences in it's height (h_i) and that of each of the four surrounding trees (h_j).

$$CI = \sum_{j=1}^{n} \left(h_i - h_j\right)$$

By assessing the CI of a number of trees of the subject species in a plantation mixture it is possible to examine how growth is affected by competition. Figure 7.14b, c show an example where *Erythrophloeum fordii* is much less sensitive to being overtopped by surrounding neighbours than *Canarium album* which suffers when surrounded by taller neighbours. The slope of the relationship is an index of the sensitivity of that species to competition from surrounding trees. Other things being equal, compatible species able to form stable mixtures are those with similar slopes. This approach was used to identify potentially compatible species for use in plantation mixtures in sub-tropical Queensland (Huynh 2002; Lamb et al. 2005). Such indices are necessarily simple measures of the competitive abilities of particular species because growing conditions and competitive relationships change as trees age and root systems extend. Other more complex measures have been used that take account of larger numbers of surrounding trees and their distances from the subject tree. For example Vanclay (2006a, b) used:

$$CI = \sum \left(H_j / H_i\right) / \text{Distance}_{ij}$$

Using this index and the SIMILE modeling environment he was able to explore inter-actions in a young plantation containing a mixture of *Eucalyptus pellita* and *Acacia peregrina*. The index helped identify the optimal proportions of the two species in the mixture, the spacings that might be used and the spatial layout. This analysis suggested it did not matter whether the species were planted in alternative rows or as alternative trees along rows. Vanclay (2006a, b) argued the approach has the advantage of overcoming problems of unexpected tree mortality and any differential effects that tree density might have on species in the mixture.

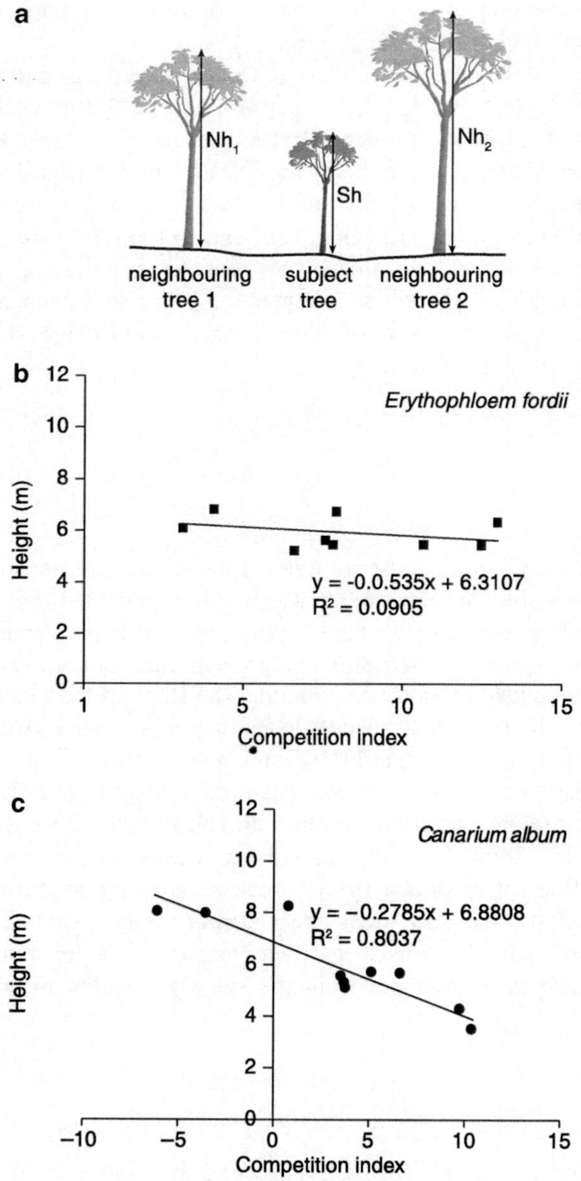

Fig. 7.14 Competition indices can be used to assess the relative competitive abilities of different species. (**a**) the competition experienced by a tree is represented by the algebraic sum of the differences in height between the target tree and each of the four neighbouring trees (**b**) the heights of *Erythrophoeum fordii* trees are not related to differences in the value of the competition index suggesting it is tolerant of some competition (**c**) the height growth of *Canarium album* declines sharply as the competition index increases suggesting it is more sensitive to inter-specific competition (Lamb and Huynh, unpublished data)

Some Management Issues

Mixtures are inherently more difficult to manage than simple monocultures and this is one of the reasons they are less popular amongst large-scale forestry enterprises even though some designs offer the possibility of increased productivity. But these additional management inputs may not be a major disincentive for other land managers, including smaller growers, and many of these may have already used mixtures in previous agroforestry practices on their farms. For such landowners the need for additional management inputs may be outweighed by the advantages of mixtures including the greater product diversity, improved timing of cashflows and reduced risk. Nonetheless there are a number of management issues such farmers must confront depending on the type of mixture that is being used.

The Number and Type of Species to Use

One early management decision that must be made is the number of species to be used. Some mixtures may involve only a small number of species such as when a NTFP crop is grown in the plantation understorey (Design 1) or where one tree species is grown on a short rotation and mixed with another growing on a longer rotation (Design 3). In such cases the mixture may only involve two species. But other mixtures may use rather more species such as when a single nurse species is used to facilitate the establishment of a number of other higher-value species planted in the understorey (Design 2) or when a permanent mixed-species stand is established (Design 4). In these cases the number and type of species could be substantially greater. The choice of just how many species to use clearly depends on the circumstances and objectives of the grower establishing the plantation. Only a relatively small number of carefully selected species might be used where the objective is enhanced production or income diversification but rather a large number may be needed to provide certain ecological services.

The type of species is also important. In addition to being able to tolerate site conditions they must be able to facilitate or complement each other and some must be able to produce certain goods (e.g. timber, fruit, resins) for which there is a known market. But, they should also contribute to building resilience. Diaz and Cabido (2001) and Elmqvist et al. (2003) suggest there is a benefit from including a variety of functional types as well as some diversity within these types but there may be a limit to just how much diversity plantation managers can cope with. Perhaps the best that might be done is to aim to use species able to tolerate the environmental stresses most likely to occur in the lifetime of the plantation (e.g. nutritional stress, drought, wildfire). There may be rather more scope for adding diversity and building resilience at a landscape scale rather than at every site and this will be discussed further in Chapter 10.

A second issue arising from this first one concerns the relative proportions of each species or species-type that should be used. Apart from Smith's (1986) suggestion that overstorey trees should not exceed 25% of the total stand there are few other guidelines for practitioners. In most cases the decision is one that must also take some account of the financial implications of the choice and most landowners will favour the most commercially attractive species. But functional effectiveness does not necessarily mean species must be present in large numbers and just a few trees of ecologically attractive species such as wildlife food or even rare or endangered species added to a new plantation might provide significant functional or conservation value at a relatively low cost to production.

Thinning

All closely spaced plantations eventually require thinning if they are not to stagnate. As noted in Chapter 5, thinning can be done 'from above' or 'from below'. Thinning from above is normally not practiced in monocultures since it leaves behind a genetically impoverished stand to grow over the duration of the rotation. But this disadvantage does not necessarily apply in mixtures and a form of 'thinning from above' is integral to Design 2 involving the temporary nurse trees and also to Design 3 where the faster growing species is removed to generate an early cash-flow while leaving trees of the more valuable species to grow through and form the final crop. All of these thinnings reduce stand density, alter competitive relationships and leave more space for the crowns of the residual trees. The chief problem with this type of operation is that it can damage the smaller residual trees. This damage can be reduced if trees are planted in rows so that trees are felled and successively removed along a row. Removal of canopy trees also means that substantially more light is able to reach the ground so that care may needed once more to control weed growth until the residual trees are able to close their canopies.

The more commonly practiced 'thinning from below' removes smaller, less vigorous trees as well as those with poor form. Again, it alters competitive relationships and the trees left behind are then those best-suited to take advantage of the improved growing conditions (e.g. more light and less root competition) provided by thinning and to grow into high-value logs. In principle it is equally applicable in mixed species plantations such as those created using Design 4. But, putting aside the question of whether or not the timber produced by thinning can be sold, the issue for the manager is whether to allow thinning to remove only slower growing trees, irrespective of their identity, or whether to retain some trees because of their future market value or even for the sake of maintaining a certain level of biodiversity? The answer clearly depends on the grower's objectives. Some landholders establish mixed-species plantations for production and 'conservation' purposes without being entirely clear about which is most important. This dilemma highlights the need to have quite explicit management objectives.

An illustration of some of the trade-offs that might be made comes from a desktop study using data from the plantation described above in Case Study 9. In this case 16 species had been planted at the same time in a random mixture in an endeavor to create a plantation that might provide some conservation benefits as well as sawlogs although all the trees of one species had died and been lost from the mixture. by age 10 years the overall planting density was 1,100 tph. Over a period of 15 years some species had grown rapidly and become canopy dominants, others became sub-dominants while others had been suppressed or died. By this time the canopy height of this unthinned stand was around 20 m and some of the larger trees had diameters exceeding 30 cm dbh.

A variety of thinning prescriptions were explored to investigate how thinning might affect species richness but no attempt was made to anticipate the financial impact of thinning (Erskine et al., 2005a). The prescriptions included (i) a simple mechanic thinning that removed every second (diagonal) row of trees thereby halving the density but maintaining an even spacing between all tree, (ii) removing trees with poor form irrespective of the impact on stand species richness, (iii) removing all trees unable to provide a straight bole exceeding 5 m length irrespective of the impact on species richness and (iv) removing all trees with small stem diameters, again irrespective of the impact on species richness. A final treatment (v) involved testing the effect of removing the canopy dominant *Elaeocarpus grandis* which has a large crown and which is beginning to suppress adjoining trees (i.e. a thinning from above).

The effect of these different prescriptions on residual species richness, stand basal area, residual stocking and overall stem form is shown in Table 7.6. All treatments reduced the basal area although the greatest reduction came from the mechanical thinning of every second row and the loss of the dominant *Elaeocarpus* had only a modest impact. The greatest effect on residual stocking was thinning to remove trees without a >5 m bole while tree form was, unsurprisingly, most improved by the treatment that sought to do so. Both of these prescriptions also reduced the basal area by more than 40% meaning that competition has probably been reduced to the point which should allow an appreciable increase in the rate of growth. Perhaps surprisingly, species richness was not greatly affected. The greatest impact came from the treatment removing the smaller trees but, even then, 12 of the original 15 species still remained (although the magnitude of any reduction

Table 7.6 Effect of differing thinning prescriptions on the attributes of a 15 year old mixed species plantation (Erskine et al. 2005a)

Prescription	Basal area $m^2\ ha^{-1}$	Residual stocking tph[a]	Average tree form[b]	Species richness
Pre-thinning	32.1	843	6.2	15
Remove every second row	15.5	417	6.2	15
Remove trees of poor form	19.4	440	7.8	15
Remove trees without straight bole length >5 m	17.0	330	7.3	12
Remove trees with small dbh	29.5	538	7.2	13
Remove the canopy dominant	23.8	773	6.1	14

[a] Trees per ha

[b] Tree form based on a ten point scale (one poor and ten good)

obviously depends on the threshold size used). Removing the dominant *Elaeocarpus* would reduce overstorey competition but the treatment had less immediate impact on basal area or on the average form of the residual trees than might have been expected.

In short, it appeared it is probably possible to develop thinning prescription which may involve several of these strategies which boost tree quality, stand growth and still retain much of the original species richness. It remains to be seen how the various types of species in such a plantation might be able to respond to such treatments over time and what the overall financial consequences of the treatments could be. Of course different forest growers will have different objectives and so make different trade-offs. For many smallholders the primary value of species richness is in the diversity of goods it generates and therefore in its insurance value. Such growers are likely to strike a balance that is different from those more interested in conservation issues and the provision of ecosystem services.

Rotation Length

The longevity of the plantation rotation can be based on the biological rotation (the time needed to maximize the timber yield before growth rates begin to decline) or the financial rotation (the time needed to optimize the present net return on capital invested). In most industrial monocultural plantations all trees are felled at the end of the rotation and the site is replanted. But having a pre-determined rotation length is probably less important for growers attracted to using mixed-species plantings. Some may follow industrial growers and clear-fell after a certain time. Others may adopt an opportunistic attitude whereby individual trees are removed according to the needs of the moment or as market opportunities present themselves. Regeneration might be carried out after clear-felling to produce a second even-aged mixed-species planting. Alternatively, some growers may decide to plant in the canopy gaps left after a tree is removed creating an uneven-aged stand. The situation can be complicated by natural regeneration. Wormald (1992) quotes Muttiah (1965) who describes a plantation established in Sri Lanka early in the twentieth century containing an even-aged mixture of *Swietenia macrophylla*, *Tectona grandis* and *Artocarpus integrifolius*. This was established to facilitate the establishment of the *Swietenia* which would have otherwise been damaged by insects. After 60 years many of these trees had reached >78 cm diameter and were felled. But large numbers of *Swietenia* seedlings regenerated beneath the residual canopy and the forest was gradually converted to an uneven-aged, mixed-species stand. Of the trees in the new forest only 20% were *Tectona grandis* or *Artocarpus integrifolius*.

Rotations are important when goods such as timber are being produced but are less relevant when the objective is to produce ecosystem services. In cases where payments are made for ecosystem services the advantages of clear-felling may disappear and mixed-species stands may be managed using some kind of selection system or logging may even be abandoned (e.g. Case Study 11).

Mixtures at a Landscape Scale – a Mosaic of Monocultures

The discussion hitherto has concerned mixtures at particular sites. But diversity can occur at a variety of scales ranging from individual sites to landscapes. Mixtures at this larger scale may be represented by a number of small monocultural patches of each species scattered in a mosaic across the landscape. The obvious advantage of this patch mosaic is that the establishment and management of these monocultures relatively simple. On the other hand, the variety of environmental conditions present across the landscape allows different species to be carefully matched to their most appropriate sites. Taken together, the overall plantation area can then generate a wider variety of forest products than a simple monoculture.

The ability to create these types of mosaic depends on land ownership patterns. In landscapes containing large numbers of smallholders there will be a natural tendency for the development of many small forest plantations within an agricultural matrix although landowners with poorer soils and steeper sites are more likely to plant trees than those with uniformly good soils. Once a species becomes popular it is likely to dominate plantings throughout a district. But, even so, many farmers continue to plant a variety of species. One survey involving 45 households in the Philippines found farmers had planted 126 premium species and a further 36 non-premium species on their farms as well as 74 species of fruit trees (Emtage 2004). Another survey in northern Vietnam that found 64 tree species had been planted at 37 farms (Lamb and Huynh 2006). Nibbering (1999) and Pasicolan et al. (1997) have reported similar findings. In the longer term this diversity of species may have some economic as well as ecological advantages.

The situation is different when landholdings are larger. In this case the prospective plantation owner(s) may decide to have a mosaic of monocultures across their plantation estate rather than an extensive monoculture using one species because it will allow particular species to be matched with their most appropriate sites. But several questions need to be resolved. First, what proportion of the landscape should be reforested? Secondly, how might the non-plantation matrix be managed? And, thirdly, what should be the size of individual forest patches? The type of landscape being reforested and its spatial heterogeneity will determine the answers to all of these questions. In a mostly agricultural landscape the trees might be established in areas unsuited to cropping. In a badly degraded forest area the plantations might be located on the more accessible areas with the remaining land being allowed to develop natural forest regrowth. In both cases the spatial distribution of land suitable for crops or plantations determines the proportions reforested and the area of forest patches. But other issues might influence these decisions as well. Spatial mosaics can be attractive for conservation reasons, especially when steep lands are reforested and riparian strips are protected. In such cases the extent of plantings and the way areas outside the plantation are managed may be strongly influenced by the need to protect watersheds or provide conservation corridors for wildlife.

An example of a deliberate attempt to create a spatial mosaic of different monocultures comes from Shandong province in China where foresters are

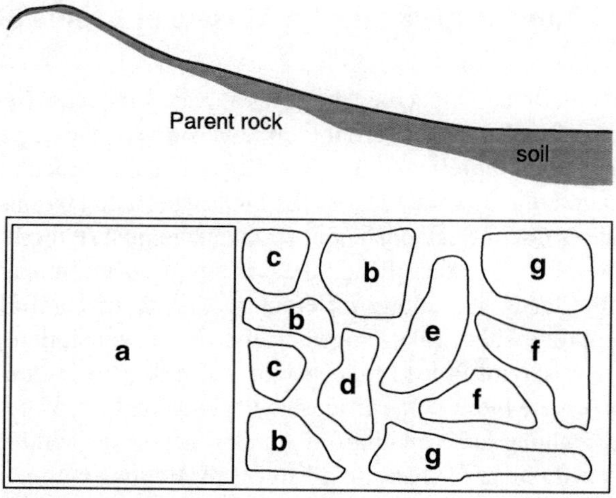

Fig. 7.15 Design 5 (mosaic of monocultures) showing a spatial mosaic of monocultural plantings on a degraded hillslope. Only a few species (**a**) can tolerate sites on ridgetops with shallow soils. Soil depth increases at lower topographic positions widening the number of species that might be planted. Species (**b**) and (**c**) can be used at mid-slope positions but (**d**), (**e**), (**f**) and (**g**) are used at locations with deeper soils and more favourable conditions. These monocultures are planted in patches embedded in a matrix of natural regeneration

establishing 'eco-forests' on severely degraded hills. There is considerable variability in environmental conditions across these landscapes. Soil depths are shallow at the tops of hills and only a few species can tolerate the exposed site conditions. Soils are deeper on lower slopes and environmental conditions are more favourable allowing a wider variety of species to be used. Extensive monocultures are used on the most exposed hilltop sites using the few species able to tolerate these sites. On lower slopes patches of different species up to a maximum size of 2 ha are used. These patches are separated by belts of other species to form a complex spatial mosaic. An example of the approach is shown in Fig. 7.15 and an illustration of mosaic plantings being applied in practice is shown in Fig. 7.16.

The issue of how to foster tree planting at a landscape scale and design spatial layouts to improve the environmental outcomes is discussed in more detail in Chapter 11.

Providing Ecosystem Services

The variety of silvicultural designs used in mixed species plantings mean that some are not very different from simple monocultures in their ability to conserve biodiversity and supply various ecological services while others are substantially better. The larger the number of tree species involved then the greater the difference is likely to be.

Fig. 7.16 Design 5 (mosaic of monocultures). A landscape mosaic achieved by growing small monocultural patches of each species at appropriate sites. In this case about six tree and shrub species are represented on the landscape

Biodiversity

Multi-species plantations are likely to have greater structural complexity than monocultures and so make the plantations more attractive to some wildlife species. One study of multi-species timber plantations in tropical Australia compared the biodiversity present in young (age 5–15 years) monocultures and mixed-species plantations and found both types of plantations attracted a variety of birds, lizards and mites but most were habitat generalists and only a small proportion of these were 'rainforest' species (Kanowski et al. 2005). Nearby 'restoration' plantings aged 6–22 years had higher numbers of birds, including rainforest specialists and lizards, though mite species numbers were similar to those in the timber plantations. Part of the reason for these differences is almost certainly the fact that the mixed species timber plantations used less than 20 tree species while the 'restoration' plantings had up to 100 tree species. This would have affected structural complexity and possibly also food resources. The value of this study is constrained, however, by the relatively young age of the trees and a different pattern emerges as the plantations become older.

Much greater levels of plant diversity were found in older (40–70 years) timber plantations growing in the same area. These were originally established as monocultures but had been colonized by additional tree species from nearby natural forest. This converted simple monocultures to structurally complex and species-rich mixtures. Keenan et al. (1997) found as many as ten woody species in could be present in 78.5 m^2 plots. Overall there were 155 tree species and a total of 350 plant species across the 151 plots sampled in the survey. This diversity created habitats and conditions making the plantations more attractive to wildlife and Kanowski et al. (2005) found they had almost 75% of the birds found in intact rainforest and comparable numbers of lizard and mite species.

One of the more widely studied forms of mixed-species planting are the shaded coffee plantations in which coffee is grown as an understorey beneath a tree cover. A variety of different planting designs have been used and some in Central America contain significant numbers of vertebrates and invertebrates (Perfecto et al. 1996; Somarriba et al. 2004). Again it appears that the more structurally and floristically diverse these are, then the greater their value for biodiversity conservation. The few studies of shaded coffee plantations in Asia suggest these support rather less biodiversity than those in Latin America (Philpott et al. 2008). This may be because of their simpler structure and much lower levels of overstorey shade or because the particular canopy trees used provided little food for native birds.

The complex agroforest systems of southeast Asia represent a particular type of multi-species planting. In many of these a small number of tree species dominate the canopy layer but there can be a species-rich understorey. Thiollay (1995) studied so-called jungle rubber (*Hevea brasiliensis*), damar (*Shorea javanica*) and durian (*Durio zibethinus*) agroforests in Sumatra and found all had large numbers of bird species present although fewer than were found in nearby undisturbed natural forest. Overall about 40% of the birds recorded in natural forests were missing from the agroforests. Those absent were the larger frugivores and insectivores and forest interior specialists. Similar results were reported by Beukema et al. (2007) who found the species richness of terrestrial pteridophytes was higher in rubber agroforests than in nearby natural forest while the numbers of bird species was comparable and the numbers of epiphytic pteridophytes and vascular plants were lower. There were fewer species of birds and pteridophytes with more specialized habitat requirements in the agroforests than in natural forests.

These observations suggest mixed-species plantations have the capacity to improve regional conservation outcomes in landscapes where patches of residual natural forest remain. Their ability to do so will depend on factors such as the overall canopy architecture and on the food resources available. The wildlife species most favoured will be the habitat generalists rather than the habitat specialists although the longer the plantations remain and the more plant colonists they acquire the more attractive the habitat conditions are likely to be for these habitat specialists. But the landscape context is important and plantings close to residual forest patches are likely to more effective in supporting the conservation of biodiversity than more isolated plantings. Mixed-species plantations may be especially useful in providing a buffer zone around natural forest patches and, for some wildlife species at least, corridors between such patches. Mixed species plantations can also add heterogeneity to landscapes and so help mobile seed dispersers. Any plantation logging will, of course, substantially alter these new habitats but the impact this has on the newly acquired biodiversity will depend on the scale and location at which it is done. These issues will be discussed further in Chapter 11.

Soil Protection and Hydrological Flows

The capacity of plantation monocultures and secondary forests to protect watersheds has been discussed in previous chapters. In principle, the ability of mixed species plantations to conserve soils and protect watersheds is likely to be better than most monocultural plantations and may, in some cases, approach the levels of protection found in secondary forests. This is because of the multiple canopy layers present in most mixed-species plantations in comparisons with the single canopy layer in monocultures. This was confirmed in studies reported by Zhou et al. (2002) which found substantially less erosion in structurally complex mixed-species plantation in China than from simple monocultures. Note, however, that in many cases it is the herbs, shrubs and small trees growing close to the ground in the understorey that are important rather than the diversity of canopy trees.

In the case of hydrological flows the difference between monocultures and mixed-species plantations may be smaller. Hydrological flows depend on rates of evapo-transpiration and the infiltration capacity of surface soils. Simple mixtures may not differ very much from monocultures although species-rich and structurally complex mixtures may intercept more rainfall and so generate less run-off. A study by Zhou et al. (2002) found this to be the case in a comparison of interception and stemflow in a eucalypt plantation and a nearby mixed species plantation and run off and stormflow were less in the mixture than in the monoculture.

As with all reforestation programs, the success of any mixed-species plantings in improving soil protection and hydrological functioning depend on the landscape context in which it is carried out and this will be discussed further in Chapter 11.

Carbon Sequestration

Mixed species forests, like monocultures, can absorb and sequester large amounts of carbon both above and below ground. The carbon content of above-ground biomass is closely related to stand biomass so that mixtures with enhanced productivity are likely to also have enhanced carbon uptake. Most comparative studies have been carried out at relatively young ages when small differences in plantation age lead to large differences in the amounts of carbon sequestered. Perhaps the more interesting question is whether *older* mixed-species plantings sequester more carbon than monocultures of the same age. At these ages the relative contribution of species with low-density and high-density timbers is likely to be changing depending on their proportions in the original mixtures and on their longevities. Large differences between mature monocultures and mixtures might emerge if these species with high timber densities have been advantaged by being grown in a mixture. But differences in the way the two types of plantations are likely to be managed – thinning schedules, clear-felling or selective logging – make it difficult to know if older mixed-species plantings do have any real advantage over monocultures of the same age. For all practical purposes, the differences between the two systems may be small.

There is some evidence that larger amounts of carbon are sequestered in soils by mixed-species plantings than by monocultures. One study found enhanced levels of soil carbon accumulated over time in mixtures involving three tree species and two monocots compared with the amounts found in monocultures of the individual tree species (Russell et al. 2004). Root growth tended to be greater in mixed species plantings than in monocultures suggesting this was the source of the additional carbon. However, there was also evidence that beyond a certain production threshold, it was the chemistry of the detrital input, most especially that from roots, that was responsible for increased levels of soil carbon rather than the amount of root material.

Other studies have shown that more soil carbon accumulates in plantations involving nitrogen fixing species (Kaye et al. 2000; Resh et al. 2002). This appears to be due to a slower rate of decomposition of older recalcitrant soil carbon as more nitrogen is fixed. Mixtures of a nitrogen fixing species with other non-nitrogen fixing species can sometimes yield higher rates of carbon sequestration both above- and below-ground than monocultures of either species (Kaye et al. 2000). Further work is needed to explore just how mixtures, but especially those involving nitrogen fixers, might be designed and managed in order to optimize above-ground and soil carbon sequestration.

Conclusions

Mixed-species plantings offer landholders some potentially significant advantages over plantation monocultures despite their greater management complexity. Depending on the type of mixture used, these advantages include their capacity to produce a variety of goods rather than just one product and their ability to generate a wider variety of ecosystem services and conservation values than monocultures. In addition they are sometimes able to facilitate the establishment of commercially preferred species at highly degraded sites and to generate an early cashflow thereby making tree-growing a more attractive land use option for many smallholders. Although much of the research carried out on mixtures has concerned whether or not they have higher levels of productivity than monocultures it is these other issues that are more likely to induce many landholders to adopt mixed-species plantings.

But random species assemblages and *ad hoc* plantings are unlikely to generate these benefits and there are, as yet, no simple recipes or blueprints to follow. The two key elements underpinning most successful mixtures are facilitation and complementarity. In some situations enough will be known of the attributes of commercially attractive species to make preliminary judgements about their complementarity with other species. Differences in crown architecture and growth phenology are useful starting points. In such cases pair-wise comparisons may be a useful way of testing these combinations. But otherwise it may be useful to use larger numbers of species in mixtures and learn from experience, accepting that some failures will certainly occur.

Most commercially-focused growers will probably use only small numbers of species in mixtures. This is to prevent management becoming too complex and to avoid the contribution from the most valuable species being diminished by other, less economically attractive species. Simple mixtures can probably be managed using some form of thinning followed by a clear-felling and replanting. More complex mixtures may end up being selectively logged and eventually becoming uneven-aged stands.

All mixed-species plantations involve making some form of trade-offs. These might involve the numbers or proportions of different types of species used, the balance between managing for production or managing for conservation or the ways in which thinning is carried out. In some situations the final outcome may be one that suits no one and it may be that some trade-offs are easier to manage at a landscape scale rather than at every site. This will be discussed further in Chapter 11.

A distinction was drawn earlier in Chapter 4 between Rehabilitation and Ecological Restoration. Mixed-species plantings that involve some (tolerant) exotic species may be the best way of reforesting some highly degraded sites but there will always be some situations where the preferred choice is to try to re-establish only native plant species and to attempt to restore the original ecosystems. This is discussed in the next chapter.

References

Anon (1999) Integrated report on the multi-storied forest management project in Malaysia 1991–1999. Forestry Department Peninsular Malaysia, Perak State Forestry Department, Japan International Cooperation Agency, Kuala Lumpur

Ashton MS, Gunatilleke CVS, Singhakumara BMP, Gunatilleke IAUN (2001) Restoration pathways for rain forest in southwest Sri Lanka: a review of concepts and models. For Ecol Manage 154:409–430

Balvanera P, Pfisterer AB, Buchman N, He J-H, Nakashizuka T, Raffaelli D, Schmid B (2006) Quantifying the evidence for biodiversity effects on ecosystem functioning and services. Ecol Lett 9:1146–1156

Bauhus J, van Winden AP, Nicotra AB (2004) Above ground interactions and productivity in mixed species plantations of *Acacia mearnsii* and *Eucalyptus globulus*. Can J Forest Res 34:686–694

Berish C, Ewel JJ (1988) Root development in simple and complex tropical successional ecosystems. Plant Soil 106:73–84

Berqvist G (1999) Wood volume yield and stand structure in Norway spruce understorey depending on birch shelterwood density. For Ecol Manage 122:221–229

Beukema H, Danielsen F, Vincent G, Hardiwinoto S, van Andel J (2007) Plant and bird diversity in rubber agroforests in the lowlands of Sumatra, Indonesia. Agroforest Syst 70:217–242

Biging G, Dobberton M (1992) A comparison of distance-dependent competition measures for height and basal area growth of individual conifer trees. Forest Sci 38:695–720
Biging G, Dobberton M (1995) Evaluation of competition indices in individual tree growth models. Forest Sci 41:360–377
Binkley D (1992) Mixtures of nitrogen-fixing trees and non-nitrogen-fixing tree species. In: Cannell MGR, Malcolm DC, Robertson PA (eds) The ecology of mixed-species stands of trees. Blackwell Scientific Publications, Oxford, pp 99–123
Biot Y, Blaikie PM, Jackson C, Palmer-Jones R (1995) Rethinking research on land degradation in developing countries. World Bank, Washington, DC
Blanton CM, Ewel JJ (1985) Leaf-cutting ant herbiverory in successional and agricultural tropical ecosystems. Ecology 66:861–869
Bourke RM (1985) Food, coffee and casuarinas: an agroforestry system from the Papua New Guinea highlands. Agroforest Syst 2:273–279
Burton PJ (1993) Some limitations inherent to static indices of plant competition. Can J For Res 23:2141–2152
Cannell MGR, Malcolm DC, Robertson PA (1992) The ecology of mixed-species stands of trees. Blackwell Scientific Publications, Oxford
Cunningham S, Floyd RB (2006) *Toona ciliata* that suffer frequent height-reducing herbivore damage by a shoot-boring moth (*Hypsipyla robusta*) are taller. For Ecol Manage 225:400–403
DaMatta FM (2004) Ecophysiological constraints on the production of shaded and unshaded coffee: a review. Field Crops Res 86:99–114
Diaz S, Cabido M (2001) Vive la difference: plant functional diversity mattters to ecosystem processes. Trends Ecol Evol 16:646–655
Done C, Kimber P, Underwood R (2004) Development of the Indian sandalwood industry on the Ord river irrigation area. Prospects for high-value hardwood timber plantations in the 'dry' tropics of northern Australia. Conference on Private Forestry North Queensland, Mareeba, Queensland, 19–21 Oct 2004. http://www.plantations2020.com.au/reports/pfnq/index.html; accessed 20 September 2010
Ehrhart Y (1998) Descriptions of some sandal tree populations in the south west Pacific: consequences for the silviculture of these species and provenances. In: Radomiljac AM, Ananthapadmanabho HS, Welbourn RM, Rao KS (eds) Sandal and its products. Australian Center for International Agricultural Research, Canberra, pp 105–112
Elmqvist T, Folke C, Nyström M, Peterson G, Bengtsson J, Walker B, Norberg J (2003) Response diversity, ecosystem change and resilience. Front Ecol 1:488–494
Emtage N (2004) Typologies of landholders in Leyte, Philippines and the implications for development of policies for smallholder and community forestry. In: Baumgartner DM (ed) Proceedings of human dimensions of family, farm and community forestry international symposium. Washington State University Extension Washington State University, Washington, DC, pp 81–88
Erskine P, Lamb D, Huynh DN (2005a) Managing competition in a mixed species plantation. Improving Productivity in Mixed-Species Plantations; Held in association with XXII IUFRO World Congress, Ballina, NSW, Australia, 5–7 Aug 2005
Erskine PD, Lamb D, Borschmann G (2005b) Growth performance and management of a mixed rainforest tree plantation. New For 29:117–134
Erskine PD, Lamb D, Bristow M (2006) Tree species and ecosystem function:can tropical multi-species plantations generate greater productivity? For Ecol Manage 233:205–210
Ewel JJ, Bigelow SW (1996) Plant life-forms and tropical ecosystem functioning. In: Orians GH, Dirzo R, Cushman JH (eds) Biodiversity and ecosystem processes in tropical forests. Springer, Berlin, pp 101–126
Ewel JJ, Mazzarino MJ (2008) Competition from below for light and nutrients shifts productivity amongst tropical species. Proc Natl Acad Sci USA 105:18836–18841
Ewel J, Mazarino MJ, Berish CW (1991) Tropical soil fertility changes under monocultures and secondary communities of different structure. Ecol Appl 1:289–302
Fahlen A (2002) Mixed tree-vegetative barrier designs: experiences from project works in northern Vietnam. Land Degrad Dev 13:307–329

Fahlvik N, Agestam E, Nilsson U, Nyström K (2005) Simulating the influence of initial stand structure on the development of young mixtures of Norway spruce and birch. For Ecol Manage 213:297–311

Firn J, Erskine PE, Lamb D (2007) Woody species diversity influences productivity and soil nutrient availability in tropical plantations. Oecologia 154:521–533

Forrester DI, Bauhus J, Khanna PK (2004) Growth dynamics in a mixed-species plantation of *Eucalyptus globulus* and *Acacia mearnsii*. For Ecol Manage 193:81–95

Forrester DI, Bauhus J, Cowie AL, Vanclay JK (2006) Mixed species plantations of *Eucalyptus* with nitrogen-fixing trees: a review. For Ecol Manage 233:211–230

Gitay H, Noble IR, Connell JH (1999) Deriving functional types for rain-forest trees. J Veg Sci 10:641–650

Gourlet-Fleury S, Blanc L, Picard N, Sist P, Dick J, Nasi R, Swaine M, Forni E (2005) Grouping species for predicting mixed tropical dynamics: looking for a strategy. Ann For Sci 62:785–796

Grijpma P, Gara R (1970) Studies on the shoot borer *Hypsipyla grandella* (Zeller). 1. Host selection behaviour. In: Grijpma P (ed) Studies on the shoot borer *Hypsipyla grandella* (Zeller) Lep Pyralidae. Inter-American Institute of Agricultural Science Miscellaneous Publication 101, Turrialba, pp 23–33

Harper JL (1977) Population biology of plants. Academic, London

Hobbs RJ, Higgs E, Harris JA (2009) Novel ecosystems: implications for conservation and restoration. Trends Ecol Evol 24:599–605

Hooper DU, Chapin FS, Ewel JJ, Hector A, Inchausti P, Lavorel S, Lawton JH, Lodge DM, Loreau M, Naeem S, Schmid B, Setala H, Symstad AJ, Vandermeer J, Wardle DA (2005) Effects of biodiversity on ecosystem functioning: a consensus of current knowledge. Ecol Monogr 75:3–35

Huynh DN (2002) The ecology of mixed species plantations of rainforest tree species. PhD Thesis, Department of Botany, University of Queensland, Brisbane

Jactel H, Brockerhoff E, Duelli P (2005) A test of the biodiversity-stability theory: meta analysis of tree species diversity effects on insect pest infestations, and re-examination of responsible factors. In: Scherer-Lorenzen M, Körner C, Schulze E-D (eds) Forest diversity and function: temperate and boreal systems. Springer, Berlin, pp 235–262

Joliffe P (1997) Are mixed populations of plant species more productive than pure stands? Oikos 80:595–602

Jones HE, McNamara N, Mason WL (2005) Functioning of mixed species stands: evidence from a long term experiment. In: Scherer-Lorenzen M, Körner C, Schulze E-D (eds) Forest diversity and function: temperate and boreal systems. Springer, Berlin, pp 111–130

Kanowski J, Catterall CP, Procter H, Reis T, Tucker N, Wardell-Johnson G (2005) Biodiversity values of timber plantations and restoration plantings for rainforest fauna in tropical and subtropical Australia. In: Erskine PD, Lamb D, Bristow M (eds) Reforestation in the tropics and subtropics of Australia using rainforest tree species. Rural Industries Research and Development Corporation, Canberra, pp 183–205, https://rirdc.infoservices.com.au/items/05–087; accessed 20 September 2010

Kaye JP, Resh SC, Kaye MW, Chimner RA (2000) Nutrient and carbon dynamics in a replacement series of *Eucalyptus* and *Albizia*. Ecology 81:3267–3273

Keenan R, Lamb D, Sexton G (1995) Experience with mixed species rainforest tree plantations in North Queensland. Commonw For Rev 74:315–321

Keenan R, Lamb D, Woldring O, Irvine T, Jensen R (1997) Restoration of plant diversity beneath tropical tree plantations in northern Australia. For Ecol Manage 99:117–132

Kelty MJ (2006) The role of species mixtures in plantation forestry. For Ecol Manage 233:195–204

Khanna PK (1998) Nutrient cycling under mixed-species tree systems in southeast Asia. Agroforest Syst 38:99–120

Köhler P, Ditzer T, Huth A (2000) Concepts for the aggregation of tropical tree species into functional types and the application to Sabah's lowland rain forests. J Trop Ecol 16:591–602

Kuusipalo J, Adjers G, Jafarsidik Y, Otsamo A, Tuomela K, Vuokko R (1995) Restoration of natural vegetation in degraded *Imperata cyclindrica* grassland: understorey development in forest plantations. J Veg Sci 6:205–210

Lamb D, Huynh DN (2006) Mixed species plantations of high-value trees for timber production and enhanced community services in Vietnam and Australia (unpublished report FST2000/003). Australian Center for International Agricultural Research, Canberra
Lamb D, Lawrence P (1993) Mixed species plantations using high value rainforest trees in Australia. In: Leith H, Lohman M (eds) Restoration of tropical forest ecosystems. Kluwer, Dordrecht
Lamb D, Huynh DN, Erskine P (2005) Designing mixed species plantations: progress to date. In: Erskine P, Lamb D, Bristow M (eds) Reforestation in the tropics and sub-tropics of Australia using rainforest tree species. Rural Industries Research and Development Corporation, Canberra, pp 129–140, https://rirdc.infoservices.com.au/items/05–087; accessed 20 September 2010
McNamara S, Tinh DV, Erskine PD, Lamb D, Yates D, Brown S (2006) Rehabilitating degraded forest land in central Vietnam with mixed native species plantings. For Ecol Manage 233:358–365
Michon G (2005) Domesticating forests: how farmers manage forest resources. Center for International Forestry Research, The World Agroforestry Centre, Bogor
Mönkkönen M (1999) Managing Nordic boreal forest landscapes for biodiversity: ecological and economic perspectives. Biodivers Conserv 8:85–99
Nair KSS (2007) Tropical forest insect pests: ecology, impact and management. Cambridge University Press, Cambridge
Nibbering JV (1997) Upland cultivation and soil conservation in limestone regions of Java's south coast. In: Boomgaard P, Colombijn F, Henley D (eds) Paper landscapes: explorations in the environmental history of Indonesia. KITLV Press, Leiden, pp 153–184
Nibbering JW (1999) Tree planting on deforested farmlands, Sewu Hills, Java, Indonesia: impact of economic and institutional changes. Agroforest Syst 46:65–82
Noble IR, Gitay H (1996) A functional classification for predicting the dynamics of landscapes. J Veg Sci 7:329–336
Pasicolan PN, de Haes HAU, Sajise PE (1997) Farm forestry: an alternative to government driven reforestation in the Philippines. For Ecol Manage 99:261–274
Pautasso M, Holdenrieder O, Stenlid J (2005) Susceptibility to fungal pathogens of forests differing in tree diversity. In: Scherer-Lorenzen M, Körner C, Schulze E-D (eds) Forest diversity and function: temperate and boreal systems. Springer, Berlin, pp 263–289
Perfecto I, Rice RA, Greenberg R, VanderVoort ME (1996) Shade coffee: a disappearing refuge for biodiversity. Bioscience 46:598–608
Petit B, Montagnini F (2006) Growth in pure and mixed plantations of tree species used in reforesting rural areas of the humid region of Costa Rica, Central America. For Ecol Manage 233:338–343
Philpott S, Bichier P, Rice RA, Greenberg R (2008) Biodiversity conservation, yield and alternative products in coffee agroforestry systems in Sumatra, Indonesia. Biodivers Conserv 17:1805–1820
Piotto D (2008) A meta-analysis comparing tree growth in monocultures and mixed plantations. For Ecol Manage 255:781–786
Piotto D, Montagnini F, Ugalde L, Kanninen M (2003a) Performance of forest plantations in small and medium-sized farms in the Atlantic lowlands of Costa Rica. For Ecol Manage 175:195–204
Piotto D, Montagnini F, Ugalde L, Kanninen M (2003b) Growth and effects of thinning of mixed and pure plantations with native trees in humid tropical Costa Rica. For Ecol Manage 177:427–439
Poorter L, Bongers L, Bongers F (2006) Architecture of 54 moist-forest tree species: traits, tradeoffs, and functional groups. Ecology 87:1289–1301
Pretzsch H (2005) Diversity and productivity in forests: evidence from long-term experimental plots. In: Scherer-Lorenzen M, Körner C, Schulze E-D (eds) Forest diversity and function: temperate and boreal systems. Springer, Berlin, pp 41–64
Radomiljac AM (1998) The influence of pot host species, seedling age and supplementary nursery nutrition on *Santalum album* Linn. (Indian sandalwood) plantation establishment within the Ord river irrigation area, Western Australia. For Ecol Manage 102:193–201
Rao M, Palada MC, Becker BN (2004) Medicinal and aromatic plants in agroforestry systems. Agroforest Syst 61:107–122
Resh SC, Binkley D, Parrotta JA (2002) Greater soil carbon sequestration under nitrogen-fixing trees compared with Eucalyptus species. Ecosystems 5:217–231

Robson K (2004) Experiences with sandalwood in plantations in the South Pacific and north Queensland. Conference on Prospects for high-value hardwood timber plantations in the 'dry' tropics of northern Australia. Private Forestry North Queensland, Mareeba, Queensland. http://www.plantations2020.com.au/reports/pfnq/index.html; accessed 20 September 2010

Roseby SJ, Mulligan DR, Menzies NW, Ritchie PJ, Currey NA (1998) Ecosystem development on tailings at Kidson gold mine, north Queensland, Australia. In: Fox HR, Moore HM, McIntosh AD (eds) Land reclamation: achieving sustainable benefits. Balkema Press, Rotterdam, pp 137–142

Rothe A, Binkley D (2001) Nutritional interactions in mixed species forests: a synthesis. Can J For Res 31:1855–1870

Russell AE, Cambardella CA, Ewel JJ, Parkin TB (2004) Species, rotation, and life form diversity on soil carbon in experimental tropical ecosystems. Ecol Appl 14:47–60

Siddique I, Engel VL, Parrotta J, Lamb D, Nardoto G, Ometto J, Martinelli L, Schmidt S (2008) Dominance of legume trees alters nutrient relations in mixed species forest restoration plantings within seven years. Biogeochemistry 88:89–101

Smith DM (1986) The practice of silviculture. Wiley, New York

Somarriba E, Harvey CA, Samper M, Anthony F, Gonzalez J, Staver C, Rice RA (2004) Biodiversity conservation in Neotropical coffee (*Coffea arabica*) plantations. In: Schroth G, da Fonseca GAB, Harvey CA, Gascon C, Vasconcelas HL, Izac A-M (eds) Agroforestry and biodiversity conservation in tropical landscapes. Island Press, Washington, DC, pp 198–226

Speight MR, Wiley FR (2001) Insect pests in tropical forestry. CABI, Wallingford

Thiollay JM (1995) The role of traditional agroforests in the conservation of rain forest bird diversity in Sumatra. Conserv Biol 9:335–353

Vanclay J (2006a) Spatially-explicit competition indices and the analysis of mixed species plantings with the Simile modeling environment. For Ecol Manage 233:295–302

Vanclay J (2006b) Experimental designs to evaluate inter- and intra-specific interactions in mixed plantings of forest trees. For Ecol Manage 233:366–374

Weidelt H-J (1996) Rattan – distribution, morphology, use and ecologically well adapted cultivation. In: Schulte A, Schone D (eds) Dipterocarp forest ecosystems: towards sustainable management. World Scientific, Singapore, pp 627–647

Whitmore TC (1984) Tropical rain forests of the far east. Clarendon, Oxford

Wilson JB (1999) Guilds, functional types and ecological groups. Oikos 8:507–522

Wolfe M (2000) Crop strength through diversity. Nature 406:681–682

Wormald TJ (1992) Mixed and pure forest plantations in the tropics and subtropics. Food and Agriculture Organisation of the United Nations, Rome

Zhou GY, Morris JD, Yan JH, Yu ZY, Peng SL (2002) Hydrological impacts of reafforestation with eucalypts and indigenous species: a case study in southern China. For Ecol Manage 167:209–222

Zhu YY, Chen HR, Fan JH, Wang YY, Li Y, Chen JB, Fan JX, Yang SS, Hu LP, Leung H, Mew TW, Teng PS, Wang ZH, Mundt CC (2000) Genetic diversity and disease control in rice. Nature 406:718–722

Chapter 8
Ecological Restoration

> The differences among outcomes of successional events in seemingly similar assemblages or ecosystems may well follow broadly interpretable patterns, but the itineraries are not easily predictable at the onset of the journey.
>
> Hobbs and Morton (1999, p. 46)

Introduction

Ecological Restoration is the process of assisting the recovery of an ecosystem that has been degraded, damaged or destroyed (SER 2004). Different practitioners interpret this in various ways. Some think it necessitates re-establishing the original ecosystem or trying to replicate the communities present in some supposed reference ecosystem. Others take a rather more pragmatic approach and seek to restore as many of the original species as possible and develop a functionally effective and self-sustaining system even if it is, ultimately, one with a slightly different composition than the original ecosystem. In practice, there is probably little difference between these approaches. Both rely on re-establishing native species representative of the site and both recognise that the process will take many years to reach maturity.

Ecological Restoration is an obviously attractive approach in a world with increasing areas of degraded land. But it is difficult to do at the most badly degraded sites because of the environmental changes that have occurred. It is also the most demanding form of reforestation because of the numbers of species that must be managed and because, in most cases, little is known about many of these and how they interact. Ecological Restoration is usually less able to generate livelihood benefits that other forms of reforestation. And when livelihood benefits do develop they are often indirect and occur after some years have passed. It is also comparatively expensive to carry out. Even so, some promising examples of restoration have also been found at locations where vegetation has been destroyed and soils damaged, including at former minesites. Most of these new ecosystems are still relatively young and recovery is by no means complete but promising results are being observed.

D. Lamb, *Regreening the Bare Hills*, World Forests 8,
DOI 10.1007/978-90-481-9870-2_8, © Springer Science+Business Media B.V. 2011

For those interested in Ecological Restoration there are two key questions. One is the ecological question – how to initiate a succession and re-assemble a new forest ecosystem? The task involves commencing as well as managing and directing successional trajectories since restoration takes place over much longer time scales than commercial timber plantations. The second question concerns the social context in which ecological restoration might be carried out. Under what conditions might it be attempted and how might it be encouraged? This chapter seeks to address both these sets of questions. It deals primarily with situations where little residual forest vegetation remains and where tree species must be brought in from outside. Situations are rarely as simple as this and many deforested sites still retain seed stores or old root systems able to regenerate if the site is protected from further disturbances.

Re-Assembling Forest Ecosystems

There has been a long and sometimes intense debate amongst ecologists concerning the ways forest ecosystems develop. At issue are questions like – How do successions commence? What regulates their composition? And, how do these successional communities change over time? The debate links theories about successional development with those concerning community assemblage. The former is mostly concerned with the ways species populations change over time while the latter is more concerned with the interactions between these species that give rise to particular successional trajectories (Nuttle et al. 2004).

Broadly speaking, there have been three views about the way forest ecosystems develop. Some say it is a deterministic process regulated simply by physical and biotic factors. Others view it as a stochastic process depending entirely on the species able to colonise and the order in which they do so. A third view lies mid-way between these and suggests that ecosystems are structured and restricted to some extent but can develop alternative stable states because of the randomness inherent in all ecosystems (Temperton and Hobbs 2004). The weight of empirical evidence favours this last hypothesis, especially in tropical forests. For example, Webb et al. (1972) monitored succession development in rainforest over 12 years following a small but intense disturbance and concluded that certain changes were determinate and predictable (e.g. a structural framework controlled by the environment) but that there remained a probabilistic element in the final species composition governed by microsite variation and species interactions.

In recent years there has been some discussion about whether there are rules governing how these new ecosystems are assembled (Temperton et al. 2004; Weiher and Keddy 1999). Such 'assembly rules' are taken to mean the principles that determine how communities within ecosystems come together and function within a particular set of environmental conditions. Some 'rules' are self evident and trivial. For example, shade intolerant pioneer species must not be planted beneath the canopy of species representative of more mature successional stages, epiphytes can only enter a succession when trees have become established, predators need prey species etc. But more meaningful 'rules' are less evident. Is the arrival sequence of species important? Must the successional sequence that generated the original target community be repeated or can one follow a different successional trajectory and omit some intermediate stages? And how important are differences in the relative abundances of the various species in the community?

Recent work has focused on two aspects of the problem. One of these concerns the filters that regulate the species able to take part in the new community. The other concerns the interactions between these new species once they have re-colonised the site. In both cases the 'rule' being sought is how to overcome impediments to community assemblage.

Filters

Filters represent the bottlenecks that regulate the plant species able to colonize the site and form the new ecosystem. Two types of filter exist. One is a constraint on dispersal limiting certain species already present in a regional species pool from reaching the site. This may be because their seed have no dispersal mechanism, their dispersal agents are absent, or these agents are not able to cross the intervening landscape to reach the site. The most likely early colonists will be species that are tall, long-lived and which regularly produce large amounts of seed that is dispersed by wind or by wildlife that are habitat generalists able to use a variety of ecosystems within a landscape. A second filter is imposed on those species that are able to reach the site. This is the constraint imposed by the site's environmental conditions which might prevent the colonizers from becoming established. These constraints might include the presence of tall grasses or vines, the degree to which soils are compacted or the presence of a large herbivore population (Fig. 8.1). The identity of the species that do pass through these filters and their time of arrival will influence the ecosystem's future successional trajectory.

Neither of these filters is static but may change over time as a result of feedback. As more tree species become established the habitats become more complex enabling new wildlife species to visit or use the site. Some of these visitors will bring seed of other plant species which further modify the site. The new conditions may also facilitate the arrival of some species (e.g. because the site is more sheltered) but inhibit the establishment of others (e.g. because the site is too shady).

There is scope for managers able to modify both filters. Most obviously they can do so by bringing seedlings or seed to a site. The species most in need of assistance are those with large seed that are poorly dispersed or species with no particular dispersal mechanism other than gravity. But managers can also modify the site by ripping or ploughing soils or adding fertilizers. Likewise they can change light conditions by thinning stands to create canopy gaps.

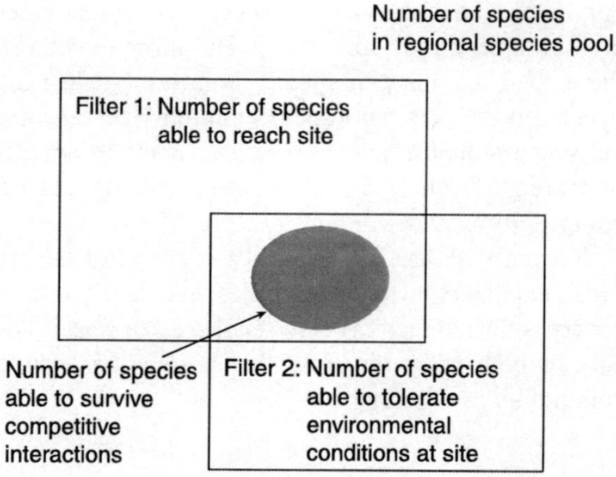

Fig. 8.1 The species pools and ecological filters that determine the species able to reach a site and colonise it. The species able to persist at the site and take part in successional change then depends on competitive interactions between the colonists

Interactions Between Species

The fate of those species able to reach and colonise a site then depends on the interactions they have with other members of the new ecosystem. The nature of these interactions will change over time as the system evolves and the two filters themselves change. The initial interactions are largely a function of the attributes of the early colonists. This was the case with in a restoration program that following mining (Box 8.1). In this case the forest was in a temperate region and had a simply tree flora but a species-rich understorey. Even 30 years after restoration had been being initiated, few additional plant species were able to enter and colonize the new community and evidence suggests the overall successional trajectory was determined by the identity of the initial colonists. But an open forest in a dry Mediterranean climate is not necessarily a model for what may happen in a tropical forest and there are a number of factors that could generate quite different outcomes. One is the greater role of seed-dispersing wildlife in most tropical regions. Another is that light rather than water or temperature is likely to be the main limiting factor and that light conditions can change over time as short-lived, early colonists die and create gaps in the forest canopy. Some species are favoured by gaps and full sunlight while others can tolerate shade. These differences mean there is usually a considerable turnover of species in tropical forest successions.

A distinction can be made between species interactions that facilitate the population growth of new colonists and those that inhibit any population growth. An example of facilitation is when a fast growing pioneer tolerant of high radiation levels acts as a nurse trees for seedlings of later successional species unable to tolerate full sunlight. Similarly, fruit-bearing trees can provide perches and food for seed-dispersers while nitrogen fixers can improve nutrient resources. Inhibition can occur when the first colonist at a site commands most of the site's resources and so

Box 8.1 An Apparently Successful Forest Restoration Program

One of the longest and most intensive attempts at forest restoration has taken place in Western Australian forests following bauxite mining. The alumina company Alcoa commenced mining in the 1960s and now restores about 550 ha each year. The region has a rainfall of 1,000–1,200 mm and a Mediterranean climate with a warm dry summers and cool wet winters. The region has around 784 plant species and is recognized as a biodiversity hotspot. The sites being mined are occupied by open *Eucalyptus marginata* forests with a small number of tree species but a very diverse shrub and ground flora. Mining removes bauxite ore to an average depth of 5 m from small scattered patches of forest each of about 10 ha. As a result around 50% of the landscape remains unmined (Koch 2007a). Surveys carried out before mining identify the plant and animal communities present and these are used as targets for restoration.

Considerable effort is used to develop an appropriate post-mining habitat (i.e. reduce the habitat filter). Before mining the topsoil and overburden are removed in two layers (0–15 cm and 15–80 cm) and replaced in the same sequence after mining. This is done once the mined area has been deep-ripped to break up the compacted pit floor. The sites are ripped a second time after the topsoil has been replaced to remove compaction caused by the soil return operation and to provide contour furrows to limit erosion. Logs and rocks are also added to provide fauna habitat.

Successional development in these forests is thought to match the intial floristics model in which the composition of vegetation in the first few years after a disturbance controls the long term floristics and functioning of the ecosystem (Koch 2007b). Because of this rehabilitation managers seek to reintroduce the complete plant species complement at the same time (thereby overcoming the dispersal filter). Around 140 species are commonly involved. Some (around 28 species) are brought back in topsoils, some as seed (78–113 species) and the remainder as seedlings. These latter are mostly species with scarce seed or that are difficult to germinate and need special methods of propagation. Sufficient seed is applied to achieve a tree density of around 1,300 tph after the first summer. Some species that are prone to early herbivore damage and these are protected by temporary screens but otherwise the survival rates are generally high. The newly established communities are given a single fertilizer treatment. Most species in these forests are poorly dispersed and so there is little change in composition over time. This means it is difficult to change the vegetation composition once it is established although opportunities to do so probably occur after event such as fires.

Koch and Hobbs (2007) have reviewed the outcomes of the methods being used and conclude that productivity in the restored forest is high, structural attributes are developing appropriately and that all ecosystem functions

(continued)

Box 8.1 (continued)

including nutrient cycling appear to be recovering. The numbers of both plant and animal species (including herbivores, detritivores, nectivores, insectivores and carnivores) and their respective diversities now match those in unmined sites although some wildlife are still limited by certain habitat features such as rotting logs or tree hollows that will take more time to develop. The sites are also able to tolerate fire and the fire-adapted species are able to regenerate normally. In short, the methods used appear likely to have established an appropriate successional trajectory and will eventually restore a forest very similar to the original forest ecosystems present before mining.

prevents later arrivals from becoming established. Thus fast-growing trees with dense crowns will prevent shade intolerant species from colonizing or growing and vines can smother or pull-down new tree seedlings. Both processes can occur simultaneously. Moreover, a species initially acting as a facilitator (e.g. because it fixes nitrogen) might later become an inhibitor (because it has a dense crown). Likewise some wildlife can be facilitators because they disperse seed but also act as inhibitors because they consume a high proportion of these seeds.

Putting Theory into Practice

One way of putting theory into practice and accelerating successions would be by limiting the role of inhibitors and enhancing the role of facilitators (assuming, of course, that their identity is known). This might be done by removing or thinning tree species acting as inhibitors or by adding seeds or seedlings of species able to act as facilitators. An example of the use of facilitator species has already been discussed in Chapter 7 when the role of nurse trees was considered. In that case the nurse tree excluded weeds and provided shelter thereby allowing target species to be established. But managing facilitation can be difficult because, over time, facilitators can sometimes become inhibitors. In the previous example the nurse trees were eventually removed to allow the more valuable species to grow on independently.

The transition is more difficult to arrange if no commercial harvesting option is available. An illustration of this was provided by the experience of those seeking to influence successional development at a former mine site in subtropical Queensland, Australia. In this case 30–40 native species were established using seedlings and direct sowing of seed. The site had sandy, infertile soils and 700 gm of seed of *Acacia concurrens* (a putative nitrogen-fixer) were added to the seed mix being broadcast across the site in order to improve soil nitrogen levels and so facilitate the growth of the other species. All species grew well but the *Acacia* proved to be especially fast-growing. It soon over-topped the other species and began shading them out. At this point the number of species at the site declined sharply. When, after about 10 years, the *Acacia* began to senesce there were few other species still remaining beneath its canopy. The large amount of woody material produced by the *Acacia* created a major fire hazard. A wildfire eventually burned through the area and triggered the germination of large numbers of *Acacia* seed that had entered the soil seed bank. This produced an even denser population of *Acacia* seedlings. The managers tried decreasing the amount of *Acacia* seed used in the seed mix but eventually removed it entirely in order to ensure that adequate populations of the other species were able to persist. In this case the presumed nutritional benefits of the facilitator were outweighed by the disadvantage of its faster height growth and extended canopy cover.

This experience does not mean that facilitation is not worth attempting but it does mean that the ecological attributes of the facilitator and the species taking part in the succession must be clearly understood. It also leaves open the question of just what proportions of seeds or seedlings of each species can be used in the initial founder community. Should the ratios reflect the proportions of species in the original (or a reference) forest or should certain species be favoured over others? There is insufficient experience to date to provide answers to these questions.

A second way in which restorationists might manipulate interactions is by changing the sequence in which species are introduced into a site. For example, one might ensure facilitators are early arrivals but species that act as inhibitors are only added once the successional trajectory is established. Laboratory studies suggest assembly histories are especially influential in small ecosystems (Futami 2004). An example of just how significant this arrival sequence can be is provided by the contrasting rates of successional development beneath old (>50 years) monoculture plantations in north Queensland (Firn et al. 2007; Keenan et al. 1997). Much faster rates of colonization occurred when broad-leaved species (*Flindersia brayleyana, Toona australis*) were planted (i.e. were the first colonists) than when conifers (*Araucaria cunninghamii, Agathis robusta*) were planted. After 60 years around 17 species per 0.1 ha plot had colonized the hardwood plantations but less than nine species had colonized the conifer plantations (Firn et al. 2007). These differences may be due to differences in canopy architecture and light environments on the forest floor but may also be due to differences in litter quality.

Similar observations were made by Kuusipalo et al. (1995) who noted quite different levels of species colonization under the canopies of monocultural plantations in Kalimantan, Indonesia. More species were found colonizing the understorey of *Acacia mangium* plantations than under *Eucalyptus deglupta* or *Paraserianthese falcataria*. They attributed this difference to the canopy architectures of the various species. More open canopies (*Eucalyptus, Paraserianthes*) allowed more grasses or other species to persist making colonization harder than where trees with dense crowns (*Acacia*) excluded the grasses. Likewise, Finegan and Delgado (2000) observed differing successional trajectories developing on old pasture sites in Costa Rica initially colonized by either *Vochysia ferruginea* or *Codia alliodora* both of which are fast growing and long-lived pioneer tree species. They concluded that the successional differences at the various sites were probably due to spatial and temporal variations in seed rain linked to characteristics of these two initially dominant trees species. The long-term outcomes of these interactions are unclear. Once the canopy of the original plantation monoculture begins to break up and colonists continue to reach the sites and enter the canopy layers, these founder effects may begin to diminish.

Too little is known about most wildlife to be able to deliberately manipulate restoration programs apart from ensuring a variety of plant food resources are included in the mix of tree species that are planted. Obligate mutualists will be especially difficult to accommodate. For example, in the case of nectivores occupying seasonally dry forests, it might be necessary to have a number of plant species proving a sequence of nectar sources throughout the year. Each species might also need to be represented by a sufficient number of individual trees to ensure at least several flowering trees are available in every month.

Examples of Ecological Restoration of Tropical Forests

A variety of ways of initiating successions to restore tropical forests have been suggested including transporting blocks of soil and/or litter containing seed, building artificial bird perches or planting clumps of trees in grasslands to attract seed-dispersing birds (Sampaio et al. 2007). More promising results have been achieved when seedlings of a variety of species have been directly planted using relatively high planting densities. It is probably true to say that few of these plantings have been informed by ecological theories concerning facilitation or inhibition. Nor, in most cases, have practitioners explicitly sought to restore the original forest since few of them had a complete list of species or access to seed of these species. Rather, they have been willing to define 'success' as the establishment of a new forest community dominated by mainly native species that appears to be self-sustaining and on a successional trajectory that will allow more species to enter the site in the longer-term. It is useful to examine some of these plantings in order to try to draw out some generalisations. All of the following Case Studies cover areas exceeding a few hectares and all have been monitored over 10 years or more.

Case Study 1: Hong Kong

Reforestation of bare and degraded hills in in Hong Kong is an early example of reforestation being carried out for largely environmental reasons (Corlett 1999; Nicholson 1996). It is also an example of a pragmatic response to a field situation where it is unrealistic to try to restore the original forest ecosystems because so little is known about them and because many of the original species are probably now extinct. Most forest cover was lost in Hong Kong before 1600 and only remnants have persisted since then. Nonetheless, the present vascular flora probably totals around 1,550 species of which 390 species are thought to be native tree species (Corlett 1999; Zhuang 1997). Many of the original fauna have also become extinct although there are still a number of species able to disperse seed (Dudgeon and Corlett 2004).

The sites available for reforestation in Hong Kong are difficult to restore. Most are occupied by grasses and are found on steep slopes with thin, acidic soils having low levels of organic matter. The rainfall totals 2,200 mm but it is strongly seasonal and fires are common during the dry season. The initial plantings used timber trees but were undertaken by the government for health and aesthetic reasons. Over time the emphasis changed to reforestation for timber production and then to watershed protection. Most recently, conservation has become the primary motive and the aim now is to develop a species-rich and self-sustaining forest community.

Initial plantings in the 1870s sought to use a variety of species but *Pinus massoniana* dominated the early plantings because it could tolerate the site conditions. It was established using seedlings and by directly sowing seed. Subsequently, a large number of exotics and native species were tested. Most of these new forests were cleared during the Second World War. After the war reforestation recommenced and, over time, conservation became the primary objective. *Pinus massoniana* was used again but the pinewood nematode (*Bursaphelenchus xylophilus*) reached Hong Kong in the 1970s and wiped it out within 10 years. Other exotics used included *Acacia confusa, A. auriculiformis, A. mangium, Casuarina equisetifolia, Eucalyptus citriodora, E. robusta, E. tereticornis, E. torrelliana, Lophostemon confertus, Melaleuca quinquenervia and Pinus elliottii.* Native species were tested but could not survive in open plantings although they did prosper when planted beneath a nurse tree canopy. These natives included species such as *Castanopsis fissa* and *Cinnamomum camphora.*

Over time the nurse trees have facilitated the development of a diverse forest community although the canopy layer has become dominated by *Machilus* and a handful of other species. Most of these are light-demanding trees and are not found as seedlings on the forest floor. Potentially large shade-tolerant tree species are also rare amongst understorey seedlings. One plant family still present in some old forest patches in Hong Kong but which is noticeably under-represented in most reforested sites is the Fagaceae. Many of the species in this family are shade tolerant and were probably an important component of the original forests but their seed are large and dispersal is limited presumably because any original dispersal agents are now extinct (Dudgeon and Corlett 2004). A number of wildlife species now use the restored forest including an estimated 35–45 bird species (Nicholson 1996).

Case Study 2: Amazonia, Brazil

One of the largest tropical forest restoration projects is that being carried out after bauxite mining in Para state in the Brazilian Amazon (Parrotta et al. 1997; Parrotta and Knowles 1999, 2001). The site has a rainfall of 2,200 mm and is occupied by evergreen equatorial rainforest. Over 100 ha per year of mined land have been treated over a period of more than 15 years. The project managers have sought to address both the habitat and dispersal filters that were discussed earlier. Topsoil (to a depth of 15 cm) was removed before mining and then replaced once mining was completed. The area was deep-ripped and planted with seedlings of 80–100 species at a density of 2,500 tph. These grew well and soon formed a closed canopy. After 10 years the trees were 11 m tall. Most of the species used were from later successional stages and some of these were found to perform poorly when planted in the open (Knowles and Parrotta 1995). New species soon appeared in the community in addition to those that had been planted. These species came from seed pools that had been present in the topsoil as well as from the undisturbed rainforest surrounding the site. At the time of the assessment (10 years) these new colonists made up 70–83% of the species present. Most of the colonizers had small seed sizes since the birds, mammals and primates able to carry larger seed were unwilling to venture into the still young, new community. Unsurprisingly, most colonists were found at sites nearest to intact forests and the density and diversity of these species declined with increasing distance from intact forest although there was still abundant colonization by woody plants at sites more than 600 m away. At the time of the study the structure and overall richness of species at restored sites was still less than in nearby undisturbed forest. However surveys found the restored forests contained 141 species from 38 families in comparison with 157 species from 39 families in the natural forest and the species-area curves for the two were becoming similar.

Case Study 3: North Queensland, Australia

A series of rainforest restoration plantings on former farmland in the humid tropics of northern Queensland also show promising results can be obtained if good site preparation is carried out and vigorous seedling stock is used (Erskine et al. 2007; Freebody 2007; Goosem and Tucker 1995; Kanowski et al. 2005; Lamb et al. 1997; Tucker and Murphy 1997; Wardell-Johnson et al. 2005). The area has an annual rainfall ranging from 1,400 to 3,600 mm and the dominant natural vegetation is evergreen rainforest. Reforestation was carried out using a series of mostly small (<10 ha) plantings. These were established on former pastures with the specific intent of restoring species-rich rainforest ecosystems resembling the nearby natural forests.

A variety of methods were initially used although most involved planting around 30–40 tree species at the same time at densities of about 2,500–300 tph. In some cases care has been taken to 'seal' edges by planting boundaries with species having deep and dense crowns to reduce so-called edge effects. The total area of restored forest is around 1,300 ha (Vize et al. 2005).

Ordination studies showed that, at 15 years, the plant assemblages at these restored forests were still different from natural forests but that substantial numbers of additional species had begun to colonise these sites (Tucker et al. 2004). Studies in other plantings in the same region found substantial colonisation of rainforest plantings could begin from as early as age 7 years. These colonisers include a variety of life forms (canopy and sub-canopy trees, shrubs, vines, ferns) most of which are dispersed by birds, bats and mammals (Tucker and Murphy 1997). As in Case Study 2 above, the extent of colonisation depends on the distance the new plantings are from intact forest and White et al. (2004) found restoration plantings surrounded by pasture and more than 600 m from natural forest were largely colonised by weed species rather than rainforest trees. Even where tree seedlings did colonise these isolated plantings the new seedling pool could be dominated by just one or two species. However, it is still too early to determine if these early colonists will skew the successional trajectory or whether this is a temporary phenomenon reflecting episodic changes in fruiting patterns in the natural forest (Tucker and Murphy 1997).

Once a canopy cover is established then species from other trophic levels begin to occupy the new forests (Tucker 2000). Analyses of the birds, lizards and mites populations have found the species richness was almost half that in nearby natural forest in less than 22 years. On the other hand, most of these new species were generalists representative of more open habitats rather than rainforest specialists (Tucker et al. 2004; Kanowski et al. (2005). Apart from the age of the new forest, one of the key variables influencing wildlife diversity was the distance of the planted site from intact natural forests and, as was the case with plant colonists, more isolated sites had less diversity. But, taken collectively, the findings indicate the plant and animal communities being established are representative of the nearby natural forests. In addition, they also suggest that appropriate successional trajectories and feedback processes have been established within a comparatively short period.

Case Study 4: Chiang Mai, Thailand

Restoration plantings have been carried out on former croplands in the hills near Chiang Mai in northern Thailand (Elliott et al. 2004; Elliott et al. 2006; Elliott et al. 2003). The area has a rainfall of 2,100 mm but a pronounced dry season (from November to April when the mean monthly rainfall is <100 mm) and the surrounding forests are evergreen rainforests. The areas planted lie within the Doi Suthep-Pui National Park and had been illegally cleared by farmers some 20 years earlier to grow cabbages, corn and other cash crops. The purpose of reforestation was to re-establish the former biodiversity on these now degraded sites to help restore the integrity of the Park. Planting commenced in 1997 and new areas have been added each year since then. Restoration is carried out by removing weeds with weedicides and planting around 20–30 local tree species on a single occasion early in the wet season. The species used are chosen to represent early and later successional stages and the planting density used is around 3,100 tph. Fertiliser is applied at the time of planting and twice more in the first rainy season. The sites are tended for another 2 or 3 years by which time canopy closure takes place.

Seedlings of colonists often begin appearing on the forest floor within a few years of planting. Surveys carried out when the plantings were 6 years old found

70 species represented in the seedling populations. Only nine of these were species that had been originally planted and the remainder species were new colonists. This was double the richness of tree species found as seedlings in an unplanted control site. The density of seedlings under the new forest canopy was about 1,600 per hectare. Some of these species had been carried a considerable distance. For example, the nearest tree able to supply seeds of one species found in the faeces of Large Indian Civet were at least 5 km away.

Birds began visiting the sites once trees were more than a few meters tall. Most of these were species normally found in open habitats but several were frugivores able to disperse forest seed. After canopy closure more specialized forest bird species began to appear and, by the seventh year, a total of 55 bird species had been observed in the oldest plantations with trees of *Erythrina subumbrans* and *Melia toosendan* both attracting 28 species (Elliott et al. 2004). A total of 87 bird species have been observed. These were observed either in the planted plots or in the unplanted control areas and 45 of these species were also recorded in nearby intact natural forests (located about 2 km away although there are scattered patches of secondary forest that lie between the planted and natural forest). This indicates that after 7 years of plantings the two areas share about 63% of the bird species. Nests of some of these birds were observed in 2 year old trees although most tree species only began to acquire nests after age 5 years. A number of medium sized and larger animals have also been sighted within the plantings including the Common Barking Deer, Leopard Cat, Malayan Pangolin, Hog Badger, Large Indian Civet, Burmese Ferret Badger Siamese Hare, Javan Mongoose, Hoary Bamboo Rat, wild pigs and fruit bats. A number of these have brought seed of new species into the sites. Restoration appeared to reduce the abundance of mice and rat species (potential seed predators) although there may have been a slight increase in the richness of these species in the early stages of forest development.

Case Study 5: Khao Phaeng Ma, Thailand

This hilly site adjoins Khao Yai National Park in central Thailand and was once state forest under the control of the Royal Forestry Department. The Dry Evergreen Forest occupying the sites was cleared in the early 1970s when settlers were encouraged by another government agency to move in and grow corn. This was despite the fact the area was classed as an important watershed protection forest. By 1983 soil fertility had dropped (as had the price of corn), and many farmers abandoned their farms. The land was then largely grassland with a few scattered residual trees although middlemen from whom the farmers had previously borrowed money still claimed ownership of the land.

In 1994 a national reforestation program began in throughout Thailand to honour the King. Reforestation at Khao Phaeng Ma was initiated by an NGO, the Wildlife Foundation-Thailand. Initial funding came from a donation made by the private sector and a management committee made up of representatives of the local community, the Royal Forestry Department, local government agencies and the Wildlife Fund Thailand. The subsequent success of the project was undoubtedly due to the good working relationship that developed between the local community and the NGO.

The nominal landowners were persuaded to bequeath 'their' land to the project (in exchange for their name being placed on a sign acknowledging their donation). In this way a total of 800 ha became available to the project. Several hundred hectares were planted in the first year using seedlings supplied by the Royal Forestry Department. Because of the rapidity of events only a small number of species were available for planting and most of these were exotics. One of these was *Leuceana leucocephala* which can become invasive and dominate successions because of its capacity to create a large soil seed pool. However, this problem has not eventuated since fire has been successfully controlled. In subsequent years seedlings of a much larger variety of mostly local species were used. There was no knowledge of how this type of reforestation should be done and, in the absence of technical advice, the species chosen depended on what was available at the time and traditional local knowledge. Likewise, the planting methods were somewhat ad hoc. Initial planting done by local villagers who were paid but subsequent plantings was done by volunteers and the total area was covered in a few years. Tourists now visiting the area are encouraged to buy a seedling from the project's own nursery and plant this in designated locations to help enrich the new forest with a more diverse range of species.

Ecological succession was rapid (Fig. 8.2). Once canopy closure occurred new species began colonizing from the nearby National Park and enriching the understorey.

Fig. 8.2 Restored forest (15 years old) at a former grassland site, Khao Phaeng Ma, central Thailand. The forest was restored using planted seedlings and some direct sowing. Additional species have colonized the site from outside. Note the gap which contains a salt lick used by a herd of gaur (*Bos gauris*) that, along with other large vertebrates including bear and deer, have migrated into the site from a nearby national park

The forest now contains 232 plant species of which only 69 were planted. About 174 species of wildlife have also been recorded including large animals such as gaur (*Bos gauris*, a large wild oxen species), Asian black bear (*Ursus thibetanus*), Serow (*Capricornis* sp.) and marbled cat (*Pardofelis marmorata*). A herd of 100 gaur are now established in the new forest and viewing these has become a popular activity for tourists visiting the site. Although established primarily as a conservation forest the site has made a significant contribution to the local economy. Local communities collect NTFPs from the area and have raised money from organizing educational and study tours for visitors (Rawee Thaworn and Boripat Sunthorn; personal communication).

Some Tentative Principles Governing the Ways in Which Forest Ecosystems Might Be Restored

With the exception of the Hong Kong example these Case Studies have several things in common. All have covered comparatively large areas and all were carried out in landscapes where natural forests remained nearby. In most cases, they used relatively large numbers of species to initiate their respective successions and planted seedlings of these on a single occasion and at high planting densities. In each case additional species from these natural forests began to colonize the newly planted sites once canopy closure occurred and weeds such as grasses were excluded. In no case did exotic woody weeds become a problem. Wildlife appear to have begun to colonize and use the new forests from a comparatively early age. The Hong Kong case study differs because it had a much longer lead time and suffered from various disturbances. But, there too, successional development has been important in enriching the initial planted forests.

Overall the results are promising and, although all the evidence comes from successions that are still young, the Case Studies suggest it is possible to establish an appropriate successional trajectory using these types of techniques. On the other hand, the functional attributes of these new forests are still unclear and the problem of re-introducing and sustaining endangered, vulnerable or sparsely distributed and rare species – wildlife as well as plants – remains to be addressed.

In each case a considerable amount of knowledge had been acquired about the biology of most of the species used including flowering and fruiting times, seed storage and nursery techniques as well as knowledge about their successional status. But it is still premature to propose anything as sophisticated as 'assembly rules' that might be used to restore tropical forests. This is because too few sites have been examined in any detail and because most of those that have been studied are still comparatively young. However, these case studies together with observations by others (Ashton et al. 2001; Freebody 2007; Goosem and Tucker 1995; Lamb et al. 1997; Mansourian et al. 2005; Rodrigues et al. 2009) mean it is possible to suggest a series of draft principles or propositions that might guide those seeking to undertake Ecological Restoration of tropical forests. These are offered here simply as a starting point that can be modified as further experience accumulates.

1. *It is possible to move a community across a threshold from one state condition to another by protecting the sites from future disturbances, eliminating impediments to colonization (i.e. removing the dispersal filter) and changing habitat conditions (i.e. modifying the habitat filters).* Plant species needing particular assistance to establish are those that are poorly dispersed (e.g. those with large fruit or seed), those that fruit infrequently and those that are endangered, vulnerable or rare. The removal of grasses and other weeds is probably the key habitat impediment although changes in soil physical and chemical properties may also be needed. Certain wildlife species are likely to face similar constraints on their establishment and will also need assistance to become established.
2. *The future state of any restored forest is heavily dependent upon the starting point.* Species used to initiate a succession will have a strong influence upon the way it develops, at least in the short term. Amongst those species used there should be species able to quickly 'capture' the site and exclude weeds and initiate key successional processes as well as those able to attract seed-dispersers and so foster successional development.
3. *Successional development is likely to be slower at more heavily degraded sites and in more strongly seasonal environments.* The more degraded a site has become the fewer will be the original species able to still occupy and tolerate site conditions. In some cases it may even be necessary to use exotic species to facilitate the entry of native species to such sites.
4. *Many successions can be initiated by planting species from different successional stages at the same time.* That is, it is not necessary to mimic natural successions although it can be useful to include some short-lived pioneers in these initial plantings to ensure that canopy closure is rapid and weeds such as grasses are excluded.
5. *Successional development will be aided by the presence of species with a range of longevities since this will ensure the continued development of canopy gaps.* Such canopy gaps prevent stagnation and allow species in seedling pools to grow up and reach the canopy layer. Canopy gaps can also be created by allowing certain disturbances (e.g. fire in the case of some seasonal forests) once the new forest has reached a stage where the species present can tolerate these.
6. *It is better to use a number of species in the initial planting mix rather than only a few.* This is because too little is known about the ecology of most species and the way they interact and because some species are likely to fail, especially when planted at degraded sites. The impact of such losses will be less in more diverse plantings. A larger number of species is also likely to produce a more structurally complex forest which is more attractive to a wider range of seed-dispersing wildlife. In order to be self-sustaining and resilient the mixture should include plant species from several functional groups such as pioneers, longer-lived secondary species, species representative of more mature successional stages, fleshy-fruited species and nitrogen-fixers. Some species may belong to more than one of these groups.
7. *Some species will not be able to enter a succession until others are already present and able to facilitate their entry (conversely, the early entry of some species may limit the colonization by others).* Some species from later successional

stages need some shade to become established and will fail when planted in the open. Fast growing species with dense crowns and heavy litter layers may out-compete and exclude some slower growing species. Likewise the early development of large populations of seed predators and herbivores may prevent the subsequent development of many later plant colonists. Some form of adaptive management will be needed to deal with these events because too little is usually known about the ways species interact.

8. *The landscape context is important and additional native species will only colonize a new restoration site when there is relatively intact natural forest nearby.* The actual rate of any post-planting colonization will depend on both the distances involved and on the presence, as well as the ability, of wildlife to travel across the intervening landscape but, in general, the more distant this natural forest the slower any colonization will be.
9. *New plantings are likely to be more attractive to seed-dispersing wildlife when they offer perch trees, are structurally complex and provide a food resource.* Carefully designed mixtures that include species with these attributes are likely to be more effective than random mixtures.

In Practice

Guidelines such as these can be useful as a way of focusing attention on critical issues but implementing them in the field can be difficult. In some cases little might be known of the biology of the key species such as growth rates, shade tolerances or longevities. Likewise little might be known about how to get enough seed to raise large numbers of seedlings of various species to plant at a specified future date. There are also likely to be financial constraints limiting resources available for establishment and maintenance.

Given these limitations, there are two stages in implementing restoration in the field. The first concerns site preparation to avoid the habitat filter. Like commercial plantation establishment, sites must be protected from disturbances such as fires and be free of weeds and pests. Some woody plants that develop large soil seed stores (e.g. *Mimosa pigra, Leuceana leucocephala*) can be especially problematic. Likewise fencing or poisoning might be needed to exclude some animal pests. The second stage involves overcoming the dispersal filter. There appear to be three broad approaches that can be used with the choice being determined by the site conditions and landscape context as well as by the resources available.

Nurse Tree Method

The habitat filter can be a serious problem at exposed sites or degraded sites with heavy weed cover and reduced soil fertility. Under these conditions, facilitators or nurse trees can be used to initiate ecological restoration by changing environmental conditions. Provide colonists can reach the site (i.e. there is no dispersal filter) successional development can then proceed. The simplest approach is to use a single nurse species able to tolerate conditions at the site, exclude weeds and improve environmental conditions. A variety of species can be used as nurse trees including natives and exotic species but their seedlings should be easy to raise in a nursery and they should be able to grow quickly and form closed canopy forests within a short period. They should not be short-lived pioneer species because they must last long enough for grasses or other weeds to be excluded and tree colonists to become fully established.

Planting densities of around 1,100 tph might be used. The resulting canopies should be dense enough to reduce weed cover but not be so dense that seedlings of native species cannot colonize the understorey (Kuusipalo et al. 1995; Otsamo 2000a). Thus open-crowned eucalypts are probably not suitable as nurse species because they allow too many grasses to persist while *Gmelina arborea* has too dense a canopy in the wet season and too open a canopy in the dry season. But *Acacia* or *Falacataria* species can be useful. Some nurse trees can eventually become impediments to successional development and may need to be removed as thinnings or girdled and allowed to die in situ in order to allow enough light to reach the new colonists. Otsamo (2000b) found 260 m^2 gaps cut in an *Acacia mangium* plantation were also a useful way of encouraging natural regeneration.

The advantage of this technique is its simplicity and low cost. In fact nurse trees have sometimes been harvested and the proceeds used to reforest additional areas (McNamara et al. 2006). The disadvantage of the approach is that it is limited to situations where there are nearby natural forests able to supply seeds of a variety of colonists as well as wildlife able to act as dispersal agents. Nurse trees are less likely to be able to facilitate the establishment of wind-dispersed species which are more favoured by open conditions.

Framework Species Method

This second approach is another version of the nurse-tree technique but uses a larger number of species and avoids the problem of the initial facilitators possibly becoming inhibitors. The method uses 20–30 species to initiate the succession and provide a framework for future successional development (Elliott et al. 2006; Freebody 2007; Goosem and Tucker 1995; Lamb et al. 1997). Again, it is most suitable for sites close to existing natural forest able to provide plant and wildlife colonists.

The identity of the species used is important. All are trees and must be able to tolerate site conditions. They must also be able to grow quickly and close canopy within a few years so that weed are excluded. But, importantly, at least some of these fast-growing species (up to 30%) should also be short-lived so that canopy gaps are created enabling subsequent colonists to grow into the canopy. A mixture made up entirely of short-lived pioneers is likely to be less successful because they will die before the preferred species have become established and if too many of the planted seedlings die too quickly the site may be colonized by grasses.

The mixture must contain some species able to provide a fruit reward within a few years to attract seed-dispersers as well as species with a branching architecture that encourages these seed-dispersers to perch and deposit seed on the site. Planting densities of 2,500–3,000 tph are used but there are no guidelines concerning the proportion of each species within this total. More practical considerations are likely to influence this choice. Because large numbers of seedlings are needed the most favoured species are likely to be those with readily available seed and that are easy to raise in a nursery. Among the species that should be included in the mix are those that are poorly dispersed (e.g. large seeded species), plants that are the favoured food plants of important wildlife species or perhaps nitrogen fixers able to improve site conditions. In the seasonally drier tropics it would also be advantageous to include species tolerant of fire (or able to resprout from stumps).

The method is suitable for sites near existing natural forest remnants where seed dispersal is possible. It is not suited for highly degraded sites some distance from intact forest where seed dispersal rates are likely to be slow. Special care is also needed to monitor sites being restored using this method because, as with the Nurse tree method, sites can be colonized by weeds as well as native species. The primary advantage of the method is that critical species can be targeted for inclusion but that costs can be reduced by not needing to collect seeds from a large number of different species and raise these in a nursery.

Maximum Diversity Method

This third approach uses a much larger number of species, all planted at the same time, to initiate the succession (Goosem and Tucker 1995). The method might be more useful at sites distant from natural forests and where the rate at which new colonists brought in by seed dispersers is slow. It assumes that there are both dispersal and habitat filters operating. The number of species actually used depends on the capacity of local nurseries and the need for rapid biodiversity establishment but might reach 80–100 species (e.g. Knowles and Parrotta 1995). This allows for the failure of some species unable to tolerate site conditions (e.g. being planted in the open) but also caters for those that are only slowly dispersed. Although some pioneer species might be included to ensure some canopy gaps are periodically created most (90%) of the species should be from later successional stages. Special prominence might be given to species with large seed that are usually poorly dispersed. The planting mix might also include endangered, vulnerable or rare species. It might also include life forms other than trees. Rodrigues et al. (2009) report that using >50 species has generally provided better outcomes in restoration projects in Brazil although the reason for this is unclear.

Most species should be represented by at least 20–30 individuals per hectare to ensure sufficient of each species survive but otherwise there are no guidelines concerning the relative proportion of each species; those known to be functionally important or slow to reproduce might deserve to be planted in greater numbers. Site preparation, planting methods and early weed control are the same as used in the Framework Species Method and should aim to achieve canopy closure in 12–18 months. The advantage of the method is that it ensures species diversity is high from an early stage. It also ensures particular species are part of the succession thereby avoiding the risk they may not be brought to the site by seed-dispersers.

It is important to emphasize that there is not a single approach to ecological restoration and the choice of which of these various approaches to use depends on the degree of degradation that has occurred and on the landscape context. Where natural regeneration from soil seed pools or old root systems is possible or where the distances to natural forest are short then the Nurse Tree approach might be quite sufficient. On the other hand, more degraded sites and sites more distant from intact forest will probably require the Maximum Diversity Method. Some judgement is called for in making these decisions. Not all patches of residual forest are the same since some retain a wide range of plant species while others may be small or degraded and be dominated by one or two secondary or weed species. Similarly, seed dispersal from a forest patch retaining many wildlife species may be rapid but poor from one where few wildlife species remain. In practice restorationists must often assume all residual forest patches are the same.

The rates at which species richness is recovered differ according to which of these approaches are used. This is illustrated in Fig. 8.3. The Maximum Diversity method provides a fast initial response although subsequent successional development may be slow if this approach is used in more isolated or difficult sites where opportunities for

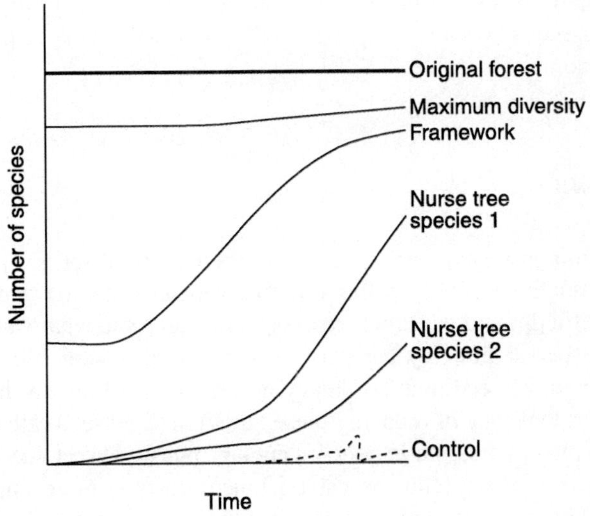

Fig. 8.3 Expected recovery rates at a degraded site using various restoration techniques. Recovery is likely to be more rapid using the Maximum Diversity method although the actual rate depends on the landscape context. The recovery rate is slower using the Framework Species method, again depending on the landscape context. Recovery rates using single species of nurse tree vary with the attributes of the particular species used. Recovery will be much slower if no planting is carried out and may not occur if the site is subject to frequent disturbances

further colonization are limited. The Framework Species method may allow rapid successional development once canopy closure occurs, again depending on the landscape context. Recovery is slowest using the Nurse Tree Method although the rate will vary according to the identity of the species used. All methods generate faster outcomes than if no intervention occurs. In that case some woody species may eventually colonise the site but these can be destroyed by periodic wildfires or by grazing.

All three methods may require some follow-up including replanting after excessive early seedling deaths and intensive weed control until canopy closure occurs. Depending on the planting density this may take several years. Weed control will be particularly necessary at the margins of planted sites where the 'edge effect' operates. Monitoring will also be necessary to ensure exotic trees are excluded, the populations of planted species whose numbers have declined over time are maintained or to introduce new species not colonizing sufficiently rapidly. But perhaps the key problem is that of ensuring that the full range of species originally present at the site have an opportunity to re-establish at the new forests. The sheer diversity of tropical forests means that most species occur at low densities and are sparsely distributed. It will always be difficult to locate these trees in natural forests, collect their seed and raise seedlings. Because of this, all three methods depend on natural dispersal mechanisms to cope with the problem of uncommon species. The dilemma is that the populations of such species are likely to decline even further as deforestation and fragmentation increase. Eventually, a point may be reached where it is impossible for those remaining to reproduce and disperse seeds in sufficient numbers to establish in a new regenerating forest. In most cases managers will have no idea when this threshold is crossed. The same may be true of certain wildlife species when their populations become too sparse.

Direct Seeding

Restoration plantings are expensive because of the variety of species involved and because of the relatively high seedling planting densities that are generally used. It can also be difficult, and therefore expensive, to overcome what was referred to above as the dispersal filter by carrying seedlings into distant sites or to steep areas. Potted or tubed seedlings are heavy and there is a limit to the number a planter can carry. One way of reducing these costs is to sow seed rather than plant seedlings. This also avoids the expensive nursery phase. Direct seeding may be especially attractive where reducing establishment costs is more important than achieving high levels of productivity (Woods and Elliott 2004).

Direct seeding has been widely used as a silvicultural tool to regenerate forests in temperate regions (Mergen et al. 1981). In these situations seed are often introduced following a burn that removes competing vegetation. In most of these cases only a single species is sown. But direct seeding using a large number of species is routinely used to restore vegetative cover at former minesites. A variety of readily available agricultural tools and machines have been used to apply the seed (Dalton 1993) and, in some situations, it has been done from the air (Mergen et al. 1981). It is difficult to distribute small seed by aircraft because they can blow away. To avoid this some form of clay pelleting is often used to add weight and ensure they fall within the target area. The pellets may also include insecticides to prevent predation by insects and a dye so the distribution of the seed is more apparent to those on the ground.

Direct seeding has not been as widely used in the tropics although trials were carried out in Hong Kong as early as the nineteenth century (Hau 2000). In that case *Pinus massoniana* was directly sown into degraded grasslands. The trials were successful and direct seeding became the primary method of reforestation with this species until the 1940s but it was later abandoned because of insect pests and increased fire risks which destroyed the new seedlings. Some years ago, degraded lands in Indonesia occupied by *Imperata cylindrica* and *Mimosa* were also been successfully reforested by sowing seed of *Sesbania grandiflora, Leucaena leucocephala, Calliandra calothyrsus* and *Acacia auriculiformis.* In this case the best results were achieved after the sites had been ploughed or burned (Hadipernomo 1979 quoted by Mergen et al. 1981). But the technique was not followed-up and neither was it widely adopted elsewhere. Presumably this was because of other unsuccessful trials that have not been reported.

In recent years researchers have begun to explore the technique again. Parrotta and Knowles (1999) used a mix of 27 native species, most of which were short-lived native pioneer species from nearby rainforest, to restore forest to a 17 ha block of land following bauxite mining in Amazonia, Brazil. Seeds were planted after 15 cm of topsoil had been replaced across the site and the area had been deep-ripped. The seeds germinated and grew well. At age 10 years the trees had achieved heights and basal areas comparable with those planted as seedlings. Like other areas at the mine site established using seedlings, the newly regenerated forests attracted large numbers of new colonists from the nearby natural forest. Although seed of 27 species were used the site became dominated by a single species (*Sclerolobium paniculatum*) which, by 10 years, had >50% of the stand basal area. There was also evidence that grasses had begun to colonise the site at age 10 years because the canopy had begun to open as some of the *Sclerolobium* began to senesce.

Another trial in Brazil involving a mixture of species had more equivocal results. The seed of five native pioneers were planted (to 5 cm depth) at a grassland site in the Atlantic forest region following weedicide treatment to remove the grasses and fencing to exclude predators (Engel and Parrotta 2001). Two of the five species (*Schizolobium parahyba and Enterolobium contorstisiliquum*) had high germination and survival rates and also grew quickly. The other species failed possibly because of poor seed quality or because of a dry period that followed sowing. None-the-less, a stocking rate of 1,000–1,800 tph was achieved within several years and there was an indication that the young trees had begun to alter site conditions enabling outside colonists to establish at the site at a faster rate than in nearby unplanted control sites.

The survival rate of directly sown seed depends on the types of communities into which seed are sown. Camargo et al. (2002) tested how well the seed of 11 species were able to establish in a variety of habitats included bare soil, grassland and several types of forests. The seed of each species was sown at a time corresponding with time of each species shed seed and was lightly buried in the soil. The success of each species varied with habitat but, overall, most species were able to establish at the site with bare soil where there was no predation or herbivory. Even so, the overall field germination rates were less than 5%. Much lower establishment rates were observed in grassland sites or in secondary or primary forest. The species that survived and grew best were those with larger seed.

One situation where direct seeding might not be expected to succeed is where sites are occupied by dense bracken fern. These provide dense shade and strong root competition. However, working in Mexico, Douterlungne et al. (2010) found that simply broadcasting seeds of the fast-growing native pioneer, balsa (*Ochroma pyramidale*), coupled with a short period of slashing (<4 months) to control the bracken enabled the trees to establish and shade out the bracken. The rate of seedling establishment was lower than if seedlings were used but in this case balsa seed was easy to collect in large numbers thereby outweighing this disadvantage. Once the balsa canopy was established other species could colonise and become established.

Finally, a simple field trial covering 30 ha in the highlands of northern Laos used seed of *Pinus kesiya, Keteleeria davidiana, Schima wallichii* and *Quercus serrata* (Lehmann 2002). These were sown after water buffalo ploughed the site to remove weeds and create a roughened soil surface. Seed were broadcast by hand after the first rains of the wet season. Good germination and seedling establishment was observed although the best results were obtained when sites had been ploughed and harrowed immediately prior to sowing.

Perhaps the most extensive use of direct seeding in restoration activities has been at former minesites. In these cases seed are typically sown into recently spread topsoils and where the problems of weeds and predation by wildlife are much less. An example of current practice is the restoration activities carried out after bauxite mining in seasonal environments in northern Australia dominated by open monsoonal forests and woodlands. Topsoils are removed and stockpiled prior to mining. Once mining finishes these topsoils are respread and about 50, mostly native, species representing a variety of life forms are directly sown onto the site (Bell 2001). The amount of seed of each species is adjusted to take account of differing viabilities and establishment rates. Seeding is carried out early in the wet season and sites are monitored and supplementary sowing or seedling planting is carried out to fill gaps or supplement populations of particular species.

Limitations on the Use of Direct Seeding

These experiences suggest direct seeding may be a useful technique to employ in Ecological Restoration but it does have certain limitations. One of these is that it needs large amounts of seed meaning that in most situations only a small number of species can be used. A second is that the actual establishment rates (i.e. the proportion of seed that germinate and produce a seedling) can be low and erratic. This means some directly sown areas may acquire only a patchy cover of trees. This may be caused by seed predation, low seed viability or by unsuitable environmental conditions. Seed size also influences the success rate.

Seed predation: Seed predation is an obvious problem when large numbers of seed are being broadcast. However, the rates at which predation occurs vary not only with habitats and the identity of the predators but the qualities of the particular plant species (e.g. size, type of seed coat, chemical protection etc.). The density of seeds present also affects the amount of seed predation with more seed escaping predation when large amounts are used. Larger vertebrate predators tend to consume larger seed or fruit and are more likely to be active at sites near forest margins or where the site has some shrub cover. These types of predators are less likely to be found in more open or grassy sites which are more likely to be occupied by smaller predators like rodents and insects which tend to feed on smaller seed. Given the variety of factors affecting predation it is not surprising that the rates found at degraded sites appear to vary considerably. Woods and Elliott (2004) found very little predation by rodents in an open former agricultural site occupied by herbaceous weeds in northern Thailand although ants removed some smaller seed. By contrast, Hau (1997) found an average of 74% of seeds from 12 species were removed with 60 days from degraded grassland and shrublands in Hong Kong. More seed were removed from shrubland sites than grassland sites with the most seed being removed within a week. Rats were the main predators but ants took smaller seed.

Seed viability: Seed viability also regulates the degree of germination success. Seeds of some species lose their viability within days while others are able to retain it for some time. Some of this difference is associated with the timing of seed shed. A study of 262 species in the seasonal forests of northern Thailand found most species shedding seed in the late dry and early wet season germinated very quickly but seed of many of the species that shed their seed later in the wet season or in the dry season remained dormant until germination was triggered by the rains of the following wet season (Elliott et al. 2006). Similar observations have been made in seasonally dry forests in South America (Viera and Scariot 2006). Most of those undertaking direct seeding trials have broadcast seed early in the wet season to ensure seedlings are well-established before drier conditions arrive. The implication of this is that care will needed in collecting seed and assembling species mixtures if seed with sufficiently high viabilities are to be used at the best sowing dates. Species normally shedding seed at the commencement of the wet season are most likely to be successful if sown at that time.

Rapid germination is also needed if seed are to avoid dessication and take advantage of favourable conditions. This was not the case in a direct seeding trial with seed of *Alphitonia petriei*, a pioneer species in north Queensland. Seeds were sown into sites where weeds had been removed but it took 6 weeks for field germination to occur (c.f. 3 weeks in a glasshouse) by which time grass and other weeds had grown up and began competing with the new seedlings (Sun et al. 1995). Higher germination rates can often be achieved if seeds are treated prior to being sown. A variety of treatments can be used although the two most common are to scarify seeds or to soak them in hot water to break the seed coat. A delayed germination also means seed are exposed to the risks of predation for a longer time.

Microsite suitability: One of the factors most affecting seed germination rates is the type of micro-sites into which the seed are placed. Using the earlier terminology, this might be thought of as one part of the environmental filter. Germination is likely to be lower when seed are placed on exposed sites on a hard soil surface than if they are deposited in surface crevices or buried. A study by Doust et al. (2006) in north Queensland tested the importance of these micro-site differences. This study involved sowing the seed of 18 tree species in various micro-sites in lowland (<100 m) and upland (1,030 m) locations in the early wet season. In both locations seed that were lightly buried had much greater success than seed spread onto bare soil or into mulch developed from slashed weeds (Table 8.1). That is, simply broadcasted the seed was relatively ineffective. Woods and Elliott (2004) also noted in trials carried out in northern Thailand that burial dramatically improved germination success to rates matching those achieved in the controlled conditions of the nursery. Some form of prior weed control is usually necessary before seeds are broadcast or sown except in cases where restoration is being carried out immediately after mining ceases. But there are situations in dry tropical forest areas where some limited vegetative cover can be at least temporarily beneficial because it protects the new seed and seedlings from dessication (Hardwick et al. 1997; Viera and Scariot 2006).

Seed size: The seed sizes of species often affect their survival and growth. Large seeded species (more than, say, 5 g) are usually most successful but especially if they are buried and able to avoid predators (Doust et al. 2006, 2008). The food reserves in these seed enabled them to tolerate adverse conditions and persist. Similar observations have been made by Camargo et al. (2002) who found field germination rates of less than 1% for seed weighing less than 1 g but higher values

Table 8.1 Percentage of viable seed established after 8 months when sown into various microsites at lowland and upland grassland areas in the early wet season (Doust et al. 2006)

Treatment	Establishment rate (% of viable seed added)	
	Lowland	Upland
Seed broadcast on top of mulch	3.3	1.8
Shallow soil burial, mulch removed	25.5	24.3
Shallow soil burial, below mulch	26.4	22.6
On cultivated soil surface, mulch removed	3.2	2.7
In furrow, no mulch	19.7	25.9
On top of furrow, no mulch	10.9	22.9

in heavier seed. On the other hand, Doust et al. (2008) found larger-seeded species tended to have slower growth which made their seedlings more susceptible to competition from weeds. Smaller seed have fewer food reserves but, once established, their seedlings can grow faster and may have a better chance of avoiding being over-topped by weeds.

Provided these micro-site and seed size factors are understood the net seedling density can still be reached by increasing the seed application rate. But this means that large amounts of seed of some species might be needed. It may be easy to collect large amounts of seed of some species but very difficult for others, especially those that flower and fruit only infrequently or which are normally only sparsely present in natural forests. This requirement means that it will probably not be possible to re-introduce many tropical tree species to degraded sites using direct sowing. Some other desirable seed attributes are listed in Table 8.2. Very few species will have all these attributes and some selection will be needed depending on the species available and the characteristics of the sites being restored.

In summary, direct seeding has some distinct advantages but also has some risks and may not be suitable in all situations (Table 8.3). It is most likely to be advantageous when large amounts of seed of the preferred species can be collected and quickly applied early in the rainy season in relatively weed-free sites. Seed burial will probably give better germination rates than broadcast sowing but the latter may be suitable if sites have been ripped or ploughed to provide a more heterogenous series of micro-sites. But even under these conditions it is a more risky method of initiating successional development simply

Table 8.2 Desirable seed attributes for species being used in direct seeding

Attribute	Reason
Seed readily available	Large amounts of seed are needed because establishment rates are sometimes low
High viability	High viability reduces the amount of seed needed
Rapid germination	Allows seedlings to take advantage of short-lived favourable conditions
Large seed size	Large seed often have higher rates of establishment and survival
Rapid seedling growth	Rapid seedling growth allows seedlings to escape competition
Tolerance of some shade and competition	Some tolerance enables seedlings to persist

Table 8.3 The advantages and disadvantages of direct seeding as a method for restoring forests

Advantages	Disadvantages
Lower costs	Less reliable than seedlings
Can treat large areas quickly to take advantage of good planting conditions	May have patchy distribution
	Not all species can be used
Plantings have a 'more natural' appearance	Large numbers of seed are needed

because the rates of predation, germination and seedling survival will vary from year to year and, on some occasions, plant cover may be very patchy. There are likely to be some situations where a combination of direct seeding and seedling planting is useful. In this case the more common and easily collected species are sown from seed while species with very small seed, or those that are difficult to collect, are raised in the nursery and then planted. Direct seeding might also be carried out once planted seedlings have achieved canopy closure. In this case the seed are broadcast into a weed-free environment. Provided this is done before many wildlife have begun to colonize the sites the predation rates may be still at acceptably low levels.

The Social Context

Ecological Restoration does not normally provide an immediate financial return meaning that there must be other benefits if a landowner is to be induced to undertake it and then continue to protect the developing forest into the future. In the five Case Studies described earlier these benefits included 'conservation', watershed protection and, in one case (Case Study 4; Chiang Mai), a belief that restoration would help those carrying it out to be granted conditional tenure on other land. The conservation motive is becoming more important over time as governments and communities recognize just how much deforestation has occurred and the magnitude of the dangers involved in losing biodiversity. But conservation is not something that many small landholders can afford to contemplate unless they have food security or are paid for their contribution. In Vietnam, smallholders are being paid to establish Protection Forests as part of the nation Five Million Hectares Reforestation Project. Watershed protection is a key purpose of these activities although conservation is also an explicit aim (De Jong et al. 2006; MARD 2001). Few of these plantings could currently be described as Ecological Restoration or are likely to create forests resembling those once present at these sites but they are still significant because large areas are being reforested for purposes other than simply production.

Conservation-oriented NGOs have sought to accelerate the pace of restoration by raising funds and assembling partnerships to drive restoration, especially on degraded lands that are under-used or have been abandoned by small farmers. The north Queensland activities in Australia (Case Study 3), the Chiang Mai project in Thailand (Case Study 4) and the Khao Phaeng Ma project also in Thailand (Case Study 5) are all examples of this. In each case significant areas have been restored. The importance of these projects is they often energize local communities and become educational tools showing just what can be done. In some cases, such as at Chiang Mai and Khao Phaeng Ma, they become the focal point of eco-tourism and so begin generating financial benefits. Some benefits may also develop if smallholders or communities can become seed collectors or establish nurseries to supply the large numbers of seedlings required.

NGOs are also becoming more effective in re-orienting government policies and persuading them of the need to restore degraded lands for conservation purposes. Events now underway in the Atlantic forest area of Brazil may be a portent of things to come elsewhere in the tropics (Rodrigues et al. 2009; Wuethrich 2007). In this case NGO activity and public pressure led to governments beginning to enforce environmental protection laws. But further pressure has led to the declaration of a one million hectare reforestation target and the allocation of significant state and federal government funds and tax concessions to fund the programme. Additional funds will be drawn from the World Bank and the Global Environmental Facility. In Sao Paulo state laws now mandate that each reforested hectare includes a minimum of 80 tree species with seed of each being gathered from at least 12 mother trees to ensure genetic diversity is maintained. These laws also specify that the species mix includes pioneers and species from later successional stages (Wuethrich 2007). But perhaps the most interesting aspect of this ambitious project is the efforts being undertaken to involve local landholders, including small farmers, in the project because managers recognize that little will be achievable without their participation. Efforts are being made to develop a PES scheme through which these landholders are compensated for their participation. A database is also being developed so landowners can register their riverine lands for restoration by volunteers (Rodrigues et al. 2009).

Although the Brazilian program is still at a relatively early stage the experience of reforestation in the temperate forests of Korea is instructive. In this case the original motive, when reforestation commenced in the 1950s, was to rehabilitate degraded landscapes using monocultures to produce fuelwood (Lee and Suh 2005). Over time the objective has completely changed and a rather more complex form of reforestation has developed in which ecological restoration has become far more important and 78 tree species are now being used. Forest restoration has become a significant national endeavour and a source of pride. Success often breeds success and it is conceivable the same thing could happen in Brazil and, indeed, in other tropical locations as more restoration projects are undertaken.

Monitoring and Adaptive Management

Restoration plantings, unlike commercial plantations, are established without a harvesting date in mind. The intention is that successional development should take place and that non-pioneer species should regenerate and pass through successive generations until the system begins to approach a state resembling that occupied by the original forest (or, in the case of an especially degraded site, a stable and persistent, species-rich community). It is rarely possible for restorationists to introduce wildlife or cryptic biota such as fungi, insects or soil organisms but it is assumed these will colonize sites once plants develop appropriate habitats. Some of the attributes of such a restored ecosystem are shown in Box 8.2.

Box 8.2 Attributes of Restored Ecosystems (After SER 2004)

1. The restored ecosystem has an assemblage of species characteristic of the original or reference ecosystem.
2. The restored ecosystem has mostly indigenous species.
3. All functional groups necessary for the continued development and/or stability of the restored ecosystem are present or have the potential to colonise.
4. The physical environment of the restored ecosystem is capable of sustaining reproductive populations of the species necessary for the continued successional development of the ecosystem.
5. The restored ecosystem apparently functions normally for its ecological stage of development.
6. The restored ecosystem is suitably integrated into the landscape with which it interacts through biotic and abiotic flows and exchanges.
7. Potential threats to the integrity of the restored ecosystem are eliminated or reduced as far as possible.
8. The restored ecosystem is sufficiently resilient to endure the normal range of stresses found in the local environment.
9. The restored ecosystem is self-sustaining to the same degree as a reference ecosystems and has the ability to persist indefinitely under existing environmental conditions.

But successions do not necessarily progress in this idealised fashion. Key species or functional groups may fail while others species may become excessively dominant. Hobbs and Norton (1996) argue that, instead of a continuum, ecosystem development often passes through a series of identifiable states and transitions. Some of these transitions are undesirable and may take the system across thresholds into states from where considerable resources are needed to re-instate the desired trajectory. An illustration of some of these states and transitions is shown in Fig. 8.4. In this case, restoration of a deforested area moves the ecosystem through a series of Desired States (although some are only short-lived and may be better described as stages) until the final mature successional state is reached. But the successional trajectory can be diverted by large-scale seedling deaths, sapling deaths or unbalanced competitive relationships leading to various Deviated States such as grasslands or forest communities dominated by a small number of species.

Such problems can only be dealt with using some form of adaptive management whereby the evolving ecosystem is monitored and management interventions are made when it is deemed necessary to maintain the desired successional trajectory. These interventions might take the form of supplementary or enrichment plantings. Alternatively, they may involve thinning dense stands to open closed canopies and allow further species development.

Two forms of monitoring might be used depending on the nature of the restoration program and the resources available. The simplest would be to carry out periodic

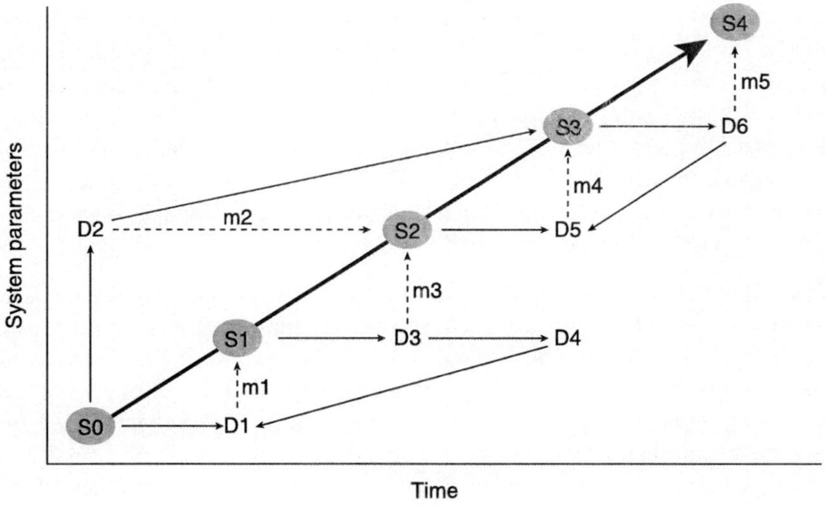

Fig. 8.4 Possible states and transitions taking place during ecological restoration. The Desired States (S), Deviated States (D) and possible Management Interventions (m) are outlined below. Desired State: S0: age 0–1 year, site cleared of weeds and seedlings planted. S1: age 5–7 years, species-rich community forming a closed canopy. S2: age 15–20 years, forest stand composed of pole sized trees of multiple species. S3: age 20+ years, canopy composed of multiple species and life forms with many becoming reproductive. S4: aged 30+ years, structurally complex and self sustaining forest with >100 tph and composed of most of the desired species and life forms; these species represented in seedling populations. Deviated States: D1: planted seedlings or directly sown seed fail to establish (because of droughts, herbivory, predation, weeds); site left bare or with only weeds D2: sowing or planting density too great and a few dominant pioneers exclude other species. D3 some trees or species fail and their deaths create many canopy gaps and allow grasses and weeds to flourish. D4: recurrent fires in grassy areas gradually destroy all residual trees species. D5: competitive interactions, pests or diseases allow one canopy species to dominate and begin excluding others. D6: established plants fail to regenerate (there is limited reproduction or there is excessive seed predation or seedling herbivory). Management Interventions: m1: replant or reseed failed areas. m2: thin canopy to allow suppressed species to grow. m3: replant canopy gaps with fast-growing species and improve fire management. m4: thin or girdle trees of dominant species to alter competitive relationships and undertake enrichment planting; m5: remove pest species, manipulate canopy density

visual inspection to assess the way the new forest is evolving. The first of these inspections should be done several months after planting to determine seedling survival. Subsequent inspections might reveal whether weeds are becoming a problem, whether an appropriate tree density or canopy cover is being maintained, whether the amount of soil erosion is acceptable and whether the site remains free of disturbances such as fires or grazing animals. A series of permanently marked photo-points could be used to provide a visual record. This type of bare-bones monitoring gives an indication of forest diversity, structure and whether important functional outcomes of restoration such as erosion control are being achieved.

A more detailed and quantitative monitoring program involving a network of permanently marked plots or transects may be needed in other situations where there are more precise restoration targets or legal obligations to meet certain revegetation standards, such as at minesites. In these cases the monitoring program should be designed to assess progress towards the benchmarks and provide answers to specific questions. The attributes of a restored ecosystem outlined in Box 8.2 provide an obvious starting point. The answers to these questions should confirm that the successional trajectory is still appropriate or, if it is not, suggest a course of action that might be followed to correct any adverse changes. In the case of plants the questions a monitoring program might pose could be:

- Does the ecosystem contain most of the characteristic species of the former forest ecosystem? If not, are these species continuing to colonize the site? (Perhaps all those species able to reach the site have arrived? Perhaps there are now too few wildlife able to carry out seed dispersal? If recruitment is insufficient action may be needed to accelerate the process. For example, seeding or enrichment planting).
- Is the canopy cover intact and continuing to exclude grasses and other weeds? (It may be necessary to plant additional trees in any new gaps to ensure an appropriate canopy cover is maintained).
- Is each species (or functional type) represented by an adequate population of individuals? (Additional seedlings might need to be planted if the density of key species is too low).
- Is there a full range of life forms present such as canopy and sub-canopy tree species, vines, shrubs, epiphytes, palms, tree ferns etc.? (If not, steps may be needed to facilitate their recruitment).
- Is the diversity of species representative of later successional stages increasing, stable or decreasing? (The reasons for any failures should be explored and replanting carried out if needed).
- Is there a seedling population on the forest floor? (The absence of a population may be due to seasonal effects, herbivory or a lack of seed).
- Are existing species regenerating? (If not it may be because they are still be too young but it may also be that seeds or seedlings of new colonists are failing to survive).
- Is the forest floor stable with a litter layer present or is erosion occurring? (Unchecked erosion could be damaging to the forest).

A similar range of questions might be developed to monitor wildlife to ensure the transition to a self-sustaining community has been achieved. Likewise, specific questions could be developed to monitor the development of endangered, vulnerable or rare plant or animal species in the new forests although this may be difficult to achieve in many tropical forests because of a shortage of biologists with the necessary specialist skills.

There is, finally, the question of whether or not to have a 'reference' site against which to assess progress. This depends on circumstances. It may be a useful practice if appropriate reference sites exist but these may not be present in highly degraded landscapes. Box 8.1 describes a long-term forest restoration project in which monitoring and adaptive management have been key components in devising what appears to have been a remarkably successful restoration technique.

Conclusion

Ecological Restoration is difficult and takes time. It is difficult because little is known about the ecological attributes of most tropical forest biota and because chance events can play an important role in the way systems develop. It takes time because most trees are long-lived and many take years to reproduce. Under these circumstances it is sensible to avoid defining 'success' as being the recreation of a forest identical to that which previously existed at a particular site. Instead a more modest short-term goal might be to develop a functionally effective, species-rich and self sustaining community of mostly native species.

Theory concerning assembly rules and successional processes has not been especially useful in providing guidance on how tropical forests should be restored. On the other hand, pragmatic approaches involving the establishment of many species at a single time appear to be generating surprisingly good results. There is some uncertainty concerning just how many species are needed to obtain a satisfactory outcome and much probably depends on the landscape context in which restoration is being practiced. Even the protective environment provided by a simple monoculture can be enough to initiate a succession when there is sufficient natural forest nearby.

The early patterns of successional development are often determined by the nature of the founder community but there is some evidence that colonists from the surrounding forest eventually overwhelm these differences and lead to a degree of convergence provided a canopy cover is maintained. The outcome may be less predictable when there is less natural forest remaining or it is more distant. Because successional development is so uncertain restoration projects should be closely monitored and some form of adaptive management should be practiced.

Biologically speaking, Ecological Restoration is the most difficult form of reforestation but it is one that is also difficult for socio-economic reasons. It may provide substantial benefits in the form of ecosystem services but these are rarely quantified or paid for. This means there will always be a constraint on the extent to which restoration will be practiced. And even when Ecological Restoration is undertaken there is always the risk they may be cleared again if economic circumstances change. Nonetheless, the scale of degradation, a rising demand for conservation and the protection of native biodiversity and the beginnings of markets for ecosystem services all mean that Ecological Restoration is likely to become more widely practiced in future.

References

Ashton MS, Gunatilleke CVS, Singhakumara BMP, Gunatilleke IAUN (2001) Restoration pathways for rain forest in southwest Sri Lanka: a review of concepts and models. Forest Ecol Manage 154:409–430

Bell LC (2001) Establishment of ecosystems after mining – Australian experience across a diverse biogeographic zones. Ecol Eng 17:179–186

Camargo JLC, Ferraz IDK, Imakawa AM (2002) Rehabilitation of degraded areas of Central Amazonia using direct sowing of forest tree seeds. Restor Ecol 10:636–644
Corlett R (1999) Environmental forestry in Hong Kong: 1971–1997. Forest Ecol Manage 116:93–105
Dalton G (1993) Direct seeding of trees and shrubs. a manual for Australian conditions. Department of Primary Industries, Adelaide
De Jong W, Sam DD, Hung TV (2006) Forest rehabilitation in Vietnam: history, realities and future. Center for International Forestry Research, Bogor
Doust SJ, Erskine PD, Lamb D (2006) Direct seeding to restore rainforest species: microsite effects on the early establishment and growth of rainforest tree seedlings on degraded land in the wet tropics of Australia. Forest Ecol Manage 234:333–343
Doust SJ, Erskine PD, Lamb D (2008) Restoring rainforest species by direct seeding: tree seedling establishment and growth performance on degraded land in the wet tropics of Australia. Forest Ecol Manage 256:1178–1188
Douterlungne D, Levy-Tacher SI, Golicher DJ, Danobeytia FR (2010) Applying indigenous knowledge to the restoration of degraded tropical forest clearings dominated by bracken fern. Restor Ecol 18:322–329
Dudgeon D, Corlett R (2004) The ecology and biodiversity of Hong Kong. Friends of the Country Parks, Hong Kong
Elliott S, Anusarnsunthorn V, Maxwell JF, Gale G, Toktank T, Kuarak C, Navakitbumrung P, Pakkad G, Tunjai P, Thaiyang J, Blakesley D (2004) How to plant a forest. Proceedings of the 8th Biodiversity Research and Training Annual Conference, Surat Thani, Thailand
Elliott S, Blakesley D, Maxwell JF, Doust S, Suwannaratana S (2006) How to plant a forest: the principles and practice of restoring tropical forests. Biology Department, University of Chiang Mai, Chiang Mai
Elliott S, Navakitbumrung P, Kuarak C, Zangkum S, Anusarnsunthorn V, Blakesley D (2003) Selecting framework tree species for restoring seasonally dry tropical forests in northern Thailand based on field performance. Forest Ecol Manage 184:177–191
Engel VL, Parrotta JA (2001) An evaluation of direct seeding for reforestation of degraded lands in central Sao Paulo state, Brazil. Forest Ecol Manage 152:169–181
Erskine PD, Catterall CP, Lamb D, Kanowski J (2007) Patterns and processes of old-field reforestation in Australian rain forest landscapes. In: Cramer VA, Hobbs RJ (eds) Old fields: dynamics and restoration of abandoned farmland. Island Press, Washington, DC, pp 119–144
Finegan B, Delgado D (2000) Structural and floristic heterogeneity in a 30-year-old rain forest restored on pasture through natural secondary succession. Restor Ecol 8:380–393
Firn J, Erskine PE, Lamb D (2007) Woody species diversity influences productivity and soil nutrient availability in tropical plantations. Oecologia 154:521–533
Freebody K (2007) Rainforest revegetation in the uplands of the Australian Wet Tropics: The Eacham Shire experience with planting models, outcomes and monitoring issues. Ecol Manage Restor 8:140–142
Futami T (2004) Assembly history interacts with ecosystem size to influence species biodiversity. Ecology 85:3234–3242
Goosem S, Tucker N (1995) Repairing the rainforest: theory and practice of rainforest re-establishment in North Queensland's wet tropics. Wet Tropics Management Authority, Cairns
Hardwick K, Healy J, Elliott S, Garwood NC, Anusarnsunthorn V (1997) Understanding and assisting natural regeneration in northern Thailand. Forest Ecol Manage 99:203–214
Hau BCH (2000) Promoting native tree seed in land rehabilitation in Hong Kong, China. In: Elliott S, Kerby J, Blakesley D, Hardwick K, Woods K, Anusarnsunthorn V (eds) Forest restoration for wildlife conservation. Forest restoration research unit. Chiang Mai University and International Tropical Timbers Organisation, Chiang Mai, pp 109–120
Hau CH (1997) Tree seed predation in degraded hillsides in Hong Kong. Forest Ecol Manage 99:215–221
Hobbs RJ, Morton S (1999) Moving from descriptive to predictive ecology. Agroforestr Syst 45:43–55

Hobbs RJ, Norton DA (1996) Towards a conceptual framework for restoration ecology. Restor Ecol 4:93–110

Kanowski J, Catterall CP, Procter H, Reis T, Tucker N, Wardell-Johnson G (2005) Biodiversity values of timber plantations and restoration plantings for rainforest fauna in tropical and subtropical Australia. In: Erskine PD, Lamb D, Bristow M (eds) Reforestation in the tropics and subtropics of Australia using rainforest tree species. Rural Industries Research and Development Corporation, Canberra, pp 183–205. https://rirdc.infoservices.com.au/items/05–087; accessed 20 September 2010

Keenan R, Lamb D, Woldring O, Irvine T, Jensen R (1997) Restoration of plant diversity beneath tropical tree plantations in northern Australia. Forest Ecol Manage 99:117–132

Knowles OH, Parrotta JA (1995) Amazon forest restoration: an innovative system for native species selection based on phonological data and field performance indices. Commonw Forest Rev 74:230–243

Koch JM (2007a) Alcoa's mining and restoration process in south Western Australia. Restor Ecol 15:S11–S16

Koch JM (2007b) Restoring a Jarrah understorey vegetation after bauxite mining in Western Australia. Restor Ecol 15:S26–S39

Koch JM, Hobbs RJ (2007) Synthesis: is Alcoa successfully restoring a Jarrah forest ecosystem after bauxite mining in Western Australia? Restor Ecol 15:S137–S144

Kuusipalo J, Adjers G, Jafarsidik Y, Otsamo A, Tuomela K, Vuokko R (1995) Restoration of natural vegetation in degraded *Imperata cyclindrica* grassland: understorey development in forest plantations. J Veg Sci 6:205–210

Lamb D, Parrotta J, Keenan R, Tucker N (1997) Rejoining habitat remnants: restoring degraded rainforest lands. In: Laurance WF, Bierregaard RO (eds) Tropical forest remnants: ecology. Management and Conservation of Fragmented Communities. University of Chicago Press, Chicago, pp 366–385

Lee DK, Suh SJ (2005) Forest restoration and rehabilitation in Republic of Korea. In: Stanturf JA, Madsen P (eds) Restoration of boreal and temperate forests. CRC Press, Boca Raton, pp 383–396

Lehmann L (2002) Direct sowing as an alternative technique for afforestation. Nam Ngum Watershed Management and Conservation Project, Laos PDR

Mansourian S, Vallauri D, Dudley N (2005) Forest restoration in landscapes: beyond planting trees. Springer, New York

MARD (2001) Five million hectare reforestation program partnership: synthesis report. International Cooperation Department, Ministry of Agriculture and Rural Development, Hanoi

McNamara S, Tinh DV, Erskine PD, Lamb D, Yates D, Brown S (2006) Rehabilitating degraded forest land in central Vietnam with mixed native species plantings. Forest Ecol Manage 233:358–365

Mergen F, Abbott HG, Mann WF, Moulds FR, Nordmeyer AH, Scott JD, Vietmeyer ND (1981) Sowing forests from the air. National Academy Press, Washington

Nicholson B (1996) Tai Po Kau nature reserve, new territories, Hong Kong: a reafforestation history. Asian J Environ Manage 4:103–119

Nuttle T, Hobbs RJ, Temperton V, Halle S (2004) Assembly rules and ecosystem restoration: where to from here? In: Temperton V, Hobbs RJ, Nuttle T, Halle S (eds) Assembly rules and restoration ecology: bridging the gap between theory and practice. Island Press, Washington, pp 410–422

Otsamo R (2000a) Secondary forest regeneration under fast-growing forest plantations on degraded *Imperata cylindrica* grasslands. New Forests 19:69–93

Otsamo R (2000b) Early development of three planted indigenous tree species and natural understorey vegetation in artificial gaps in an *Acacia mangium* stand on an *Imperata cylindrica* grassland site in South Kalimantan, Indonesia. New Forests 19:51–68

Parrotta J, Turnbull JM, Jones N (1997) Catalysing native forest regeneration on degraded tropical lands. Forest Ecol Manage 99:1–7

Parrotta JA, Knowles OH (1999) Restoration of tropical moist forest on bauxite-mined land in the Brazilian Amazon. Restor Ecol 7:103–116

Parrotta JA, Knowles OH (2001) Restoring tropical forest on lands mined for bauxite: examples from the Brazilian Amazon. Ecol Engi 17:219–239

Rodrigues RR, Lima R, Gandolfi S, Nave AG (2009) On the restoration of high diversity forests: 30 years of experience in the Brazilian Atlantic Forest. Biol Conserv 142:1242–1251

Sampaio AB, Holl KB, Scariot A (2007) Does restoration enhance regeneration of seasonally deciduous forests in pastures in central Brazil? Restor Ecol 15:462–471

SER (2004) The SER international primer on ecological restoration. Society for Ecological Restoration International, Tucson

Sun D, Dickinson GR, Bragg AL (1995) Direct seeding of *Alphitonia petrie* (Rhamnaceae) for gulley revegetation in tropical northern Australia. Forest Ecol Manage 73:249–257

Temperton V, Hobbs RJ (2004) The search for ecological assembly rules and its relevance to restoration ecology. In: Temperton V, Hobbs RJ, Nuttle T, Halle S (eds) Assembly rules and restoration ecology: bridging the gap between theory and practice. Island Press, Washington, pp 34–54

Temperton V, Hobbs RJ, Nuttle T, Halle S (2004) Assembly rules and restoration ecology. Island Press, Washington

Tucker N (2000) Wildlife colonisation on restored tropical lands: what can it do, how can we hasten it and what can we expect? In: Elliott C, Kerby J, Blakesley D, Hardwick K, Woods K, Anusarnsunthorn V (eds) Forest restoration for wildlife conservation. international tropical timbers organisation and forest restoration research unit. Chiang Mai University, Chiang Mai, pp 279–295

Tucker N, Murphy TM (1997) The effects of ecological rehabilitation on vegetation recruitment: some observations from the Wet Tropics of North Queensland. Forest Ecol Manage 99:133–152

Tucker N, Wardell-Johnson G, Catterall CP, Kanowski J (2004) Agroforestry and biodiversity: improving conservation outcomes in tropical northeast Australia. In: Schroth G, da Fonseca GAB, Harvey CA, Gascon C, Vasconcelas HL, Izac A-M (eds) Agroforestry and biodiversity conservation in tropical landscapes. Island Press, Washington, pp 431–452

Viera DLM, Scariot A (2006) Principles of natural regeneration of tropical dry forest for restoration. Restor Ecol 14:11–20

Vize S, Killin D, Sexton G (2005) The community rainforest reforestation program and other farm forestry programs based around the utilization of rainforest and tropical species. In: Erskine PD, Lamb D, Bristow M (eds) Reforestation in the tropics and subtropics of Australia using rainforest tree species. Rural Industries Research and Development Corporation, Canberra, pp 7–22. https://rirdc.infoservices.com.au/items/05–087; accessed 20 September 2010

Wardell-Johnson G, Kanowski J, Catterall CP, McKenna S, Piper S, Lamb D (2005) Rainforest timber plantations and the restoration of plant diversity in tropical and subtropical Australia In: Erskine P, Lamb D, Bristow M (eds) Reforestation in the tropics and subtropics of Australia using rainforest tree species. Rural Industries research and development Corporation, Canberra, pp 162–182. https://rirdc.infoservices.com.au/items/05–087

Webb LJ, Tracey GT, Williams WT (1972) Regeneration and pattern in the subtropical rainforest. J Ecol 60:675–695

Weiher E, Keddy P (1999) Ecological assembly rules: perspectives, advances, retreats. Cambridge University Press, Cambridge

White E, Tucker N, Meyers N, Wilson J (2004) Seed dispersal to revegetated isolated rainforest patches in north Queensland. Forest Ecol Manage 192:409–426

Woods K, Elliott S (2004) Direct seeding for forest restoration on abandoned agricultural land in northern Thailand. J Trop Forest Sci 16:248–259

Wuethrich B (2007) Reconstructing Brazil's Atlantic rainforest. Science 315:1070–1072

Zhuang X (1997) Rehabilitation and development of forest on degraded hills of Hong Kong. Forest Ecol Manage 99:197–201

Chapter 9
Plantation Finances

> Net present value ….has much to recommend it to large agencies seeking a single numerical index of how good or bad (profitable or not) a potential investment is. Like similar and related measures, internal rates of return and the benefit/costs ratios, it reduces complex, almost baffling streams of costs and returns occurring over long time periods, to just a single index, facilitating comparisons between alternatives by bureaucrats. [Nonetheless] … farmers may make their decisions on multi-dimensional data – not the single numerical index of the administrator.
>
> Byron (1991, p. 176)

Introduction

It is all very well for conservationists and foresters to recognize the need to reforest degraded lands but this does not mean that the owners of such lands will necessarily feel inclined to do so. Some may while others will not. Much will depend on their circumstances and especially whether they believe they have secure tenure over the land they are using. But a key influence is also likely to be their perception of the profitability of reforestation. Are timber trees likely to be more profitable than alternatives such as annual crops or other tree crops? And can they afford the time delay before there is any financial return from their trees? Commercial tree-growing is a relatively new land use for many farmers and some might conclude that the risks and opportunity costs of tree growing are simply too high. On the other hand, there is widespread empirical evidence showing that some private landowners do find tree-growing is worth doing. There are probably several reasons for this:

There are markets for forest products: widespread deforestation and the consequent decline in the supply of timber and various NTFPs means there is an increasing market for many of these products and the market price of some goods are increasing.

There are ways of generating earlier cashflows: the most obvious way of doing this is to use fast-growing species but multi-purpose trees, mixed-species plantings and plantation thinnings can also provide early financial returns.

D. Lamb, *Regreening the Bare Hills*, World Forests 8,
DOI 10.1007/978-90-481-9870-2_9,

Plantations act as financial buffers: farm tree plantations can diversify income sources and act as a financial reserve or bank deposit. Trees have no fixed harvest date and opportunistic fellings can smooth out seasonal flows in farm income or provide funds for unexpected expenses such as weddings or funerals.

Opportunity costs are not always high: tree plantations can utilise infertile or steep land not suited for other agricultural crops. Plantations can also be financially attractive when land is not needed for food production because the household receives other off-farm income.

Tree growing is not labour intensive: this means it may complement rather than compete with other farming activities once the trees are established. It may be especially attractive for households with only a few members able to carry out fieldwork.

Tree growing is a way of asserting land ownership claims: the act of planting trees is often recognized as a way of asserting land ownership and may be a simple way of doing so when tenure is uncertain.

Throughout the region large industrial groups have recognized the opportunities to invest in reforestation. But a number of farmers are still cautious. Just what markets are open to them and what forms of silviculture are likely to give them the greatest returns? This chapter explores these issues. It begins by examining some markets for forest products and the nature of their market chains. It then examines the financial value of several types of reforestation. It concludes by examining role that the emerging markets in ecological services might play in influencing the decisions made by landholders. The majority of examples in this chapter happen to come from Vietnam because of the recent surge in reforestation by smallholders in that country. These examples illustrate many of the financial issues confronting landholders and policy-makers concerned with the financial aspects of reforestation although the marketing environment and financial circumstances of farmers in other countries will obviously differ.

Markets for Forest Products – Examples from Vietnam

A major uncertainty facing all prospective tree growers concerns the types of markets likely to exist in future. What types of forest products will be in demand and what size of logs will be wanted? More to the point, what will be the future prices that buyers will be prepared to pay? A number of international bodies regularly grapple with these questions but the forecasting problem is even more acute for small growers in places like Vietnam where a significant proportion of the buyers of farmer's trees are local timber processors. Many of these buyers will have only the most superficial knowledge of regional or international trends.

Fuelwood in Vietnam

In many rural areas fuelwood is one of the more commonly used forest products. Even in areas connected to the electricity grid, fuelwood is still used for cooking, heating and, in some cases, for preparing food for livestock. Large amounts of timber are also used in brick kilns and potteries. The per capita amounts consumed each year in Vietnam vary widely and range between 510 and 1,825 kg depending on forest productivity, the ease of collection and the need for household heating (McElwee 2001). Fuelwood is more commonly used by rural households but some urban households also continue to use wood or charcoal. The majority of this wood probably still comes from natural forests, including nature reserves, but increasing amounts are being collected from gardens and scattered trees in agricultural landscapes.

Much fuelwood is collected by users and is seen as a free-good but large amounts of fuelwood are also sold in the market place. Fuelwood sellers in rural markets are often from poorer households because wood collection requires little more than a bicycle or carrying pole to carry fuel from the forest to the market. This also means that most of the fuel must come from local sources. This form of timber trading had the advantage of being flexible and is a way of making quick cash when agricultural labour is not needed. In central Vietnam market prices fluctuate with season. Prices are lowest in February when agricultural activities are less demanding and more women have time to gather wood and branches for the market. Prices are 25–50% higher in May–June when the weather is hotter but there are fewer sellers. In Central Vietnam McElwee (2001) found fuelwood sales represented an average of 11% of household incomes but this proportion ranged from 2% to 42%. Some fuelwood is marketed in urban areas but there is a tendency for other fuels (kerosene, gas, electricity) to replace timber as incomes rise (Mercer and Soussan 1992).

There is no incentive to grow trees for fuelwood as long as it can be freely gathered but alternatives must be found as these sources dwindle. Some of the more widely planted tree species can provide good firewood and in Vietnam the most favoured species include *Casuarina equisetifolia* and *Eucalyptus* spp. On the other hand, *Acacia*, though also widely planted, is not favoured because its timber has a low heat yield and it is a smoky fuel. But few landowners are likely to establish plantations of these species just for fuelwood. Why use trees for low-priced fuelwood when they could be grown on a little longer and sold for more valuable products like poles or sawlogs? In these circumstances it is likely to be thinnings and prunings that get used for fuelwood although it is difficult to assess just what price a particular plantation owner might receive for this material. There are, nonetheless, some situations where fuelwood plantations may be economically attractive. For example, Pasicolan et al. (1997) reported small fuelwood plantations have become financially viable in parts of the Philippines because of the complete absence of alternative energy sources. Likewise, in Papua New Guinea, some plantations have been developed in the highlands primarily to produce fuelwood for a large tea factory.

In summary, there is likely to be a continuing demand for fuelwood, especially in rural areas and the sale of fuelwood will remain an attractive income source for some poor people. However, fuelwood production is unlikely to be anything other than an incidental by-product of most reforestation projects except, in some localised situations.

Sawn Timber and Poles in Vietnam

The most obvious market for many small rural tree growers are local sawmills serving domestic markets (Fig. 9.1). A study of 90 mills in northern Vietnam in 2003–2004 by Bui et al. (2005) found these accepted logs from a wide variety of species and often supported a surprisingly large number of small factories making furniture, window frames and other timber products. Most of these mills had less than five workers and many employed these on a seasonal basis. The annual input of each mill was commonly less than 30 m^3 of logs and the timber they used came from timber merchants (i.e. middlemen who bought logs from farmers and transported these to the mill) as well as direct purchases from farmers with home gardens and tree plantations. Some timber also came from old house poles and beams.

Fig. 9.1 A small sawmill representative of many found in rural areas of Vietnam. A high proportion of these are associated with small furniture factories

Her survey found these sawmillers used over 29 tree species. However, there was evidence that a gradual change was underway in this area with a decline in the use of higher-value species including *Chukrasia tabularis*, *Erythrophloeum fordii*, *Fokenia hodginsii* and *Madhuca pasquieri* and an increase in the use of faster-growing species such as *Melia azedarach*, *Manglietia conifera*, *Artocarpus heterophyllus* and exotics such as eucalyptus and acacia. This change reflected a decline in the supplies of higher-valued species caused by bans on harvesting in natural forests and an increasing supply from plantation grown trees. Most millers required logs greater than 40-cm diameter but some could use logs with small-end diameters (under bark) as little as 5 cm (for chopsticks and chair legs). An example of the size of logs being used is shown in Fig. 9.2.

Given the variety of species used it is not surprising there was also a considerable variation in the prices paid for timber. These prices varied with species as well as log size and shape. The highest priced timber was *Fokenia hodginsii* (around VND 5.4 million or roughly US$340 per m^3 at this time) while *Styrax tonkinensis* was only VND 0.3 million (around US$19) per cubic meter. *Eucalyptus* and *Acacia* spp. timber prices were around VND 0.550 million (US$34) per cubic meter. These prices were strongly related to government-defined timber quality classes (Fig. 9.3a). The location of the mill also affected prices with mills in more isolated locations appearing to pay lower prices than those in more densely populated areas with better road access. There was evidence of a significant rise in timber prices between 2000

Fig. 9.2 An example of the small and variable size of logs often available to small rural sawmillers in Vietnam

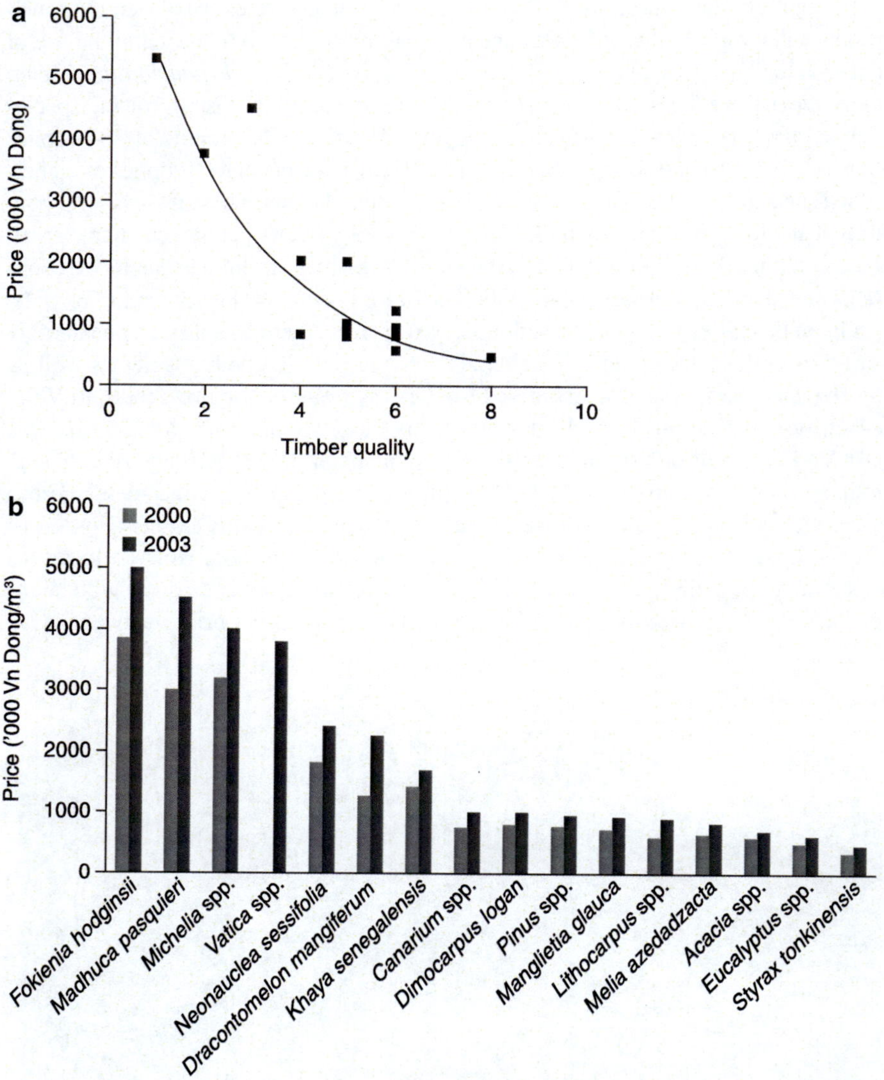

Fig. 9.3 Sawlog prices in 2000 and 2003 in northern provinces of Vietnam (**a**) there is a strong relationship between timber price and timber quality, (**b**) prices rose over the period 2000–2003 with the largest increases being found in higher quality timbers (Bui et al. 2005)

and 2003 with prices for the higher value species increasing more rapidly than for the lower valued species reflecting the changes in species availability (Fig. 9.3b). Again, this appears to have been caused by reductions in the supplies of logs entering the market. Even higher prices have been recorded in recent years (Phan Sy Hieu; personal communication, August 2010). A similar trend in the log price of native tree species was reported by Hines (1995) in central Vietnam.

When asked what species they would prefer in future, most sawmillers favoured premium quality species such as *Chukrasia tabularis* or *Erythrophloeum fordii*. However most sawmillers also had a realistic idea of what they were likely to be able to buy and also what their customers were likely to be able to afford. Small mills located in relatively remote areas mostly serve local customers who can only afford relatively low-priced furniture while larger enterprises in or near towns may have customers able to afford higher-priced products. These latter sawmillers are more likely to be prepared to pay for logs of higher-value species. As a consequence of these factors the species in greatest demand by most sawmillers in northern Vietnam were indigenous species of moderate quality such as *Artocarpus heterophyllus* (Jackfruit) which has been grown as a fruit tree in many homegardens as well as *Manglietia conifera*, *Melia azedarach* and *Acacia*. Only 34% of respondents were interested in purchasing eucalypt logs because the sawn timber produced from young trees tends to split and warp without careful sawing and seasoning treatments. All enterprises reported that it was becoming increasingly difficult to find suitable timber trees and some were despondent about their business prospects because of this. Overall the survey suggested there is a strong market for plantation-grown sawlogs in these provinces.

There was also a market for poles (logs that are 7–8 m in length and up to 20 cm in diameter) and many households used these for housing and other small structures. Pole prices in Phu Tho province in 2004 were around VND 16,000 each which means these logs were more valuable than pulpwood.

Pulpwood in Vietnam

The market for pulpwood timber is attractive to many growers because well-known species including eucalypts or *Acacia* can be grown on short rotations. In the northern province of Phu Ninh the Bai Bang paper mill buys pulpwood logs from locations sometimes more than 100 km from the mill (depending on the available transport network). It accepts eucalypts, *Acacia* and bamboo and draws on logs from State Forest Enterprise plantations as well as private growers. Growers in more distant locations may have difficulty in accessing this market. In 2006 the mill door price for eucalypts and *Acacia* grown by smallholders was around VND 550,000 (around US$34) per m^3. It is difficult to predict what impact future international pulpwood markets, especially the Chinese market, will have on these prices.

Forest Product Markets Elsewhere in the Asia–Pacific Region

Elsewhere in the Asia-Pacific region the market for timber products is quite different but the *relative* prices of sawlogs, poles and pulpwood are probably not too dissimilar to Vietnam, especially in those places where most natural forests have been removed or are now in protected areas. As in Vietnam, most smallholders producing sawlogs are likely to be supplying a domestic market although those growing higher-value timbers may be able to reach the international marketplace. The key determinant of price is usually the identity of the species but the actual price that growers receive for their trees depends on several other factors. One of these is the size of the trees at the time of sale. Many owners of small plantations may harvest trees well before the commercially optimal time because a buyer arrives and makes an offer or because they suddenly need the money (e.g. because prices for their agricultural crops have fallen that year or because they need funds for a wedding or funeral). Since log price is related to tree size an early harvest means they may not receive a high price. The price they receive will also be limited by log quality and the quality of logs produced by many smallholders may be reduced because the trees have not been pruned or the plantations have not been thinned. Logs with knots formed by old branch stubs or those that are bent or misshapen are usually substantially discounted. In some cases tree growers in some smaller island countries in the Pacific are not able to compete with the quality of imported timber and their poor-quality logs attract only low prices.

The price received by growers also depends on the amounts of timber that are available and these generally decline when large volumes of timber suddenly appear on the market at the same time. This occurred in parts of Vietnam and the Philippines when the growth performance of species such as eucalypts and *Gmelina arborea* led growers to plant large areas of these species. All matured and reached the market at the same time causing the price to fall (Pasicolan and Macandog 2007; see also Box 4.3). Some speculate that the same will happen when the large area of plantations (>700 ha) established in Malaysia and elsewhere in recent years to produce the aromatic timber known as agarwood or eaglewood (from species of *Aquilaria* and *Gyrinops*) begin to generate timber (Pang, 2009).

But, paradoxically, price declines can also occur when the market is under-supplied. This occurred when logging of native rainforests in Australia ceased in 1988. Some local landowners began establishing plantations of some of the high-value cabinet wood species in the expectation that the reduced supply would lead to even higher prices. But, in the absence of a continued log supply from the natural forests, market prices of these timbers actually declined. They now seem likely to remain depressed until a regular supply can be developed once more. Whether they will ever recover is a moot point. Markets can be fickle and may prefer to stay with readily available and well-known timbers such as teak in preference to the former high-quality native timbers.

The advantage held by internationally recognized species like teak is hard to over-emphasize. Unlike species such as eucalypts and *Gmelina arborea* that can be over-supplied in local markets, there appears to be an almost infinite international market for teak (or at least high quality teak). Laos is one place where teak grows naturally and there is currently something of a boom in teak growing now underway there (Box 9.1). Teak does not occur naturally in the Solomon Islands but logs of teak grown in plantations there are now fetching more than the average log price of trees from natural forests. The difference is accentuated because the company

Box 9.1 The Lao PDR Teak boom

Teak has been grown in Laos for many years but the planting rate increased sharply after the 1980s, especially around Luang Prabang in northern Laos where climatic conditions are favourable. In 1990 the total plantation area in the region was about 500 ha but by 2005 was estimated to have grown to 10,000 ha. This represents a 20-fold increase in 15 years. Most plantings have been done by smallholders rather than the government or companies and most are less than 1 ha in area (Midgely et al. 2007).

One of the key drivers has undoubtedly been the strong market for teak following a decline in the supply of other timbers from shrinking natural forests. This has been assisted by poorer markets for alternative land uses such as fruit trees or coffee, an expanding road network and supportive land tenurial arrangements (the government having undertaken a land allocation process at this time that was aimed at preventing shifting cultivation and making sedentary agriculture and tree-growing more attractive). Government extension efforts and the provision of planting material also helped overcome a lack of silvicultural knowledge by farmers (Hansen et al. 2007; Midgely et al. 2007). Once growers were seen to be harvesting trees and profiting from their plantings the popularity of teak increased even more. Financial analyses by Hansen et al. (2007) and Midgely et al. (2007) both show teak growing is more financially attractive than many other alternative land uses. Not all farmers are equally involved and the attractiveness of teak varies with household wealth. In wealthy households teak accounts for 52% of income but only 14% in poor households, especially those in hill areas. Roder et al. (1995) argued that poorer households had too little land to be able to take part in tree growing or could not acquire planting material.

The boom in teak planting has generated a series of other changes. One concerns changes in land use practices. In some cases good agricultural land along roads and rivers is being reforested and some farmers are now moving upslope to clear patches of natural forest to grow their food crops. In other cases steeper land is being planted and, under these circumstances, erosion can be severe (see earlier Fig. 6.6). A second set of changes involves land ownership. It appears many teak plantations are being sold after only a few years to town businessmen or government officials although the extent to which this occurs and whether it is trees or land (or both) that are sold is unclear. Hansen et al. (2007) report some extension staff saying that this may even be one of the main motives smallholders have for teak planting, at least in areas closer to town. It seems there is not only a teak boom but also a land boom. It also suggests that in this location, teak may be more attractive for town businessmen, wealthier farmers and government officials all of who appear to have been engaged in buying existing trees and all of whom can afford to take the longer term view on their investments than less-wealthy farmers for whom money-in-the-hand is more attractive. In short, teak plantations are improving the livelihoods of some poorer farmers but not in the way expected.

owning these plantations has received certification from the Forest Stewardship Council and logs are being sold at a significant premium over non-certified logs (V. Vigulu, personal communication, October 2008). Another well-known timber is sandalwood and it is attracting growers in Australia and the Pacific because of the decline in supplies from natural stands and the rise in sandalwood prices (L. Thompson, personal communication, October 2008). In both cases the price is set by the international market and not the local market as was the case with the Australian cabinet wood prices.

Finally, timber prices can change as demands change. An example of this occurring is the decline in the price of *Melaleuca* logs in the Mekong Delta region of Vietnam. This is occurring because people are beginning to use concrete for house posts rather than *Melaleuca* logs. Those growing *Melaleuca* must now sell smaller logs as poles and firewood and accept much lower prices (Fig. 9.4). Similar changes are likely to occur in other urban areas as some utility timbers are replaced by alternative materials.

In summary, the market for forest products in Vietnam and elsewhere in the Asia-Pacific region is changing as the area of natural forests decline and users of these products seek alternative supplies. Plantations are able to supply some of these products but the prices received by growers are often difficult to predict because market conditions vary with location. In some places there is strong evidence that prices are rising as supplies decline. In other places the market price

Fig. 9.4 *Melaleuca* logs being harvested in the Mekong Delta. They were once widely used for house posts but are being replaced in this market by concrete. Many small logs are now being used for poles, fish net posts and firewood

of forest products grown in plantations has remained stable or has even fallen when the local market has been saturated. Overall, higher quality timbers appear to have the best market prospects for smallholders not acting as out-growers for large pulp or woodchip mills.

Market Chains

The price that growers receive at the plantation gate is usually much less than that of a log at the mill door. The difference is made up of harvesting, transport costs and other fees. Many privately owned plantations are located in relatively inaccessible areas so that harvesting costs (felling trees and getting logs to the roadside) can be high. The subsequent transport costs can be a major element of market chains, especially for growers in more remote locations.

But many growers also face a variety of taxes and other charges imposed by local and provincial government bodies (Table 9.1). Le et al. (2004) examined some of these market chains in a detailed study carried out in northern Vietnam. In addition to taxes imposed by government bodies there were a variety of other 'informal' fees charged by communes and villages for trucks using local roads or passing through commune or village areas. There were also illegal demands for fees by police and minor officials. These were supposedly to check overloading or carry out 'forest inspections'. Fees such as these could be imposed at several checking points between the farm gate and the market. In one case a truckload of bamboo being transported over a 200 km distance attracted fees at 14 police and forestry inspection points before the bamboo reached its market. At each point the amounts charged were usually negotiable but there were no receipts. Depending on the travel distances and the locations through which forest products are transported, these charges can substantially diminish the price traders can afford to pay growers.

Table 9.1 Taxes and charges levied in 2004 on forest products in Quang Ninh Province, northern Vietnam (Le et al. 2004)

Tax or fee	Paid to
Formal	
Commune resource tax	Commune authorities by traders (fee per truck)
State Forest Enterprise	State Forest Enterprise by traders (variable amount)
Value added tax	District authorities by traders (5%)
Resource tax	District authorities by traders (planted forest products 13%)
Buy-from-afar tax	District authorities (10%)
Informal	
Commune road fee	"Guard station" by truck passing through each commune
Village fee	Village head by truck owners passing through each village
Forestry inspection fee	Forestry inspectors by traders
Police fee	Police by traders or truck owners

Not all market chains are as complex and the magnitude of the fees and taxes vary with location. Nonetheless, there are often other bureaucratic requirements that act as disincentives to tree growing by smallholders. Bui et al. (2005) reported that growers in some north western provinces could require permission to fell trees grown on their own land from the Communes Peoples Committee, the Forest Protection Unit and sometimes the Provincial Forest Service and national Ministry of Agriculture and Rural Development. Similar fees and regulations apply in other countries (Anyonge and Rochetko 2003; Bertomeu 2008). For example, in the Philippines all growers must register the trees on their land and obtain a permit before these can be felled. This registration is necessary before they can obtain a permit to transport the logs. In practice, many smallholders are unaware of the requirement or how to carry it out (Mangaoang et al. 2005) and Harrison (2003) quotes reports of farmers being imprisoned for harvesting trees on their own land that they had planted themselves. As he wryly notes, this provides an extremely negative signal about the merits of farm forestry.

The role of market agents or middlemen deserves some comment. In situations where timber plantations are scattered and marketing arrangements are still being developed timber merchants perform a useful role because they identify where the sometimes small volumes of sawlogs are located and help put log sellers in touch with log buyers. Studies in the Philippines by Herbohn and Harrison (2005) found some growers had only the vaguest ideas of where markets could be found beyond naming the nearest town and most had little knowledge of log or timber prices. Le et al. (2004) reported similar cases from Vietnam. Traders can assist by carrying out the actual harvesting operation, providing the transport to take logs from farms to mills and, where necessary, deal with the various fees and taxes. They are generally more efficient at this latter task than smallholders because they understand from experience how the system works. They can also provide growers with information on market prices since traders know which species or products are in demand and where that market is located. Traders operating over longer distances are likely to be better informed than those only working within a small area. Knowing the market price of timber puts traders in a powerful positions vis a vis growers and potentially enables them to capture a disproportionate share of the sale price. On the other hand, these agents would argue that, quite apart from transport costs and taxes, the large numbers of small growers, the isolation of many plantations, the often difficult access, the small volumes provided by each and the low log quality are the reasons why payments received by growers are sometimes low.

The complexity of the market chain mean it is difficult to say just what proportion of the mill door price an average tree-grower might receive. One estimate in northern Vietnam was that farm gate prices for pulpwood logs were around 50% of the price at the gates of the Bai Bang paper mill (Center for International Economics 1999). On the other hand, Nguyen (2004) estimates most growers who deal with traders supplying this mill receive only 25–30% of the mill door price.

Financial Models of Different Plantation Designs

Financial analyses are commonly carried out on proposed large-scale forestry projects to test whether they are credible investments. They are also useful when designing joint venture programs linking smallholders and private operators or for lobbying governments for financial and policy support for reforestation. These financial analyses take various forms. Some are simply financial analyses and evaluate cash costs and returns to the investor based largely on anticipated timber production and market prices. Other analyses try to quantify some of the social and environmental costs and benefits of the proposed project. These might include shadow prices for costs like family labour, the reduced access to former grazing lands being reforested by the project or benefits such as reduced fuel collection times, improved watershed protection or biodiversity conservation. Many of these costs and benefits are difficult to quantify in any meaningful sense. See, for example, the debate between McCauley (2006) and Maguire and Justus (2008) over the economic value of biodiversity.

A key variable in all of these financial analyses is the discount rate. This is the interest rate used to calculate the present value of a future asset such as income from selling plantation trees. The level at which this rate is set has been a subject of considerable debate amongst those who must deal with long-term land uses and investment decisions (Leslie 1987). High discount rates make the use of slow growing trees that need long rotations unattractive (they also encourage rapid rates of deforestation in natural forests because less value is then attached to future yields). Some have argued in favour of using low rates because they believe it is ethically indefensible for society to discount future assets and to limit activities that might create such assets. For this reason lower rates are sometimes used in social analyses than in private analyses (S. Harrison, personal communication, 2009). But most economists believe discount rates should match the rates used in the private sector. Alternative investment decisions have to compete for scarce savings and the cost of capital is measured by the current market interest rate. These rates fluctuate over time but rates used in cost benefit analyses have ranged between 7% and 12%. Leslie (1987) argues there is often a great deal of subjectivity in the choice.

Few smallholders contemplating tree-planting are in a position to do these sorts of evaluations. But, as noted by Byron (1991) in the quote at the head of this chapter, all smallholders will, nonetheless, make a careful judgement about whether they believe investing in tree-planting for commercial purposes is in their financial interest or whether the opportunity costs of doing so are too high. In making this judgement they will use whatever information is available to them and take account of a variety of non-financial factors and not just those that might be of interest to an accountant or banker. These factors are likely to include the timing of any cashflow as well as the likely overall financial benefit. Smallholders without much capital may be more concerned with the return on labour than the returns on capital. But, ultimately, the most compelling influence might be whether or not their neighbours are planting trees. Unless reassured by the experience of neighbours many farmers may be reluctant to take on what may seem to be a risky land use option.

If the decision to proceed is made then the next question is likely to be what *type* of reforestation to carry out? For example, would it be better to grow a high-value species for sawlogs on, say, a single 30-year rotation or to grow three 10-year rotations of a fast growing pulpwood species over the same period? The financial profitability of the three short rotations will almost certainly exceed that of the single sawlog rotation and the timing of the cashflow will probably be more attractive to many growers as well. But the final choice will depend on whether there is a local market for the pulpwood or whether growers prefer to establish slow-growing multi-purpose species suitable for a wider range of end uses and markets. Likewise, some growers may be willing to trade-off the overall profitability of pulpwood growing for the sake of more flexibility in the timing of harvesting from a sawlog plantation.

A Vietnam Case Study

A variety of species choices and silvicultural designs being used in smallholder plantations in Vietnam are shown in Table 9.2. The table includes monocultures as well as species mixtures that use some of the designs described in Chapter 7. Financial analyses of each of these types of plantation together with a sensitivity

Table 9.2 Silvicultural designs suitable for northern Vietnam used in financial analyses (Harrison et al. in press)

No.	Design	Management and reason for planting
1	Monocultures of fast growing species. A: *Acacia mangium* B: *Eucalyptus urophylla*	Use three successive 10 years rotations; sell as pulpwood
2	Monoculture of *Pinus merkusii*	A favoured species at some degraded sites. Can also be tapped for resin
3	Monoculture of *Chukrasia tabularis*	Provides a premium quality timber
4	Mixture of slow and fast-growing species; *C. tabularis* + *E. urophylla*	Provides an early cashflow as well as a large final income. Sell eucalypt for poles after 7 years and allow *Chukrasia* to grow for 30 years
5	Mixture of premium timber tree and NTFP tree *C. tabularis* + *Cinnamomum cassia*	Provides an early cashflow and a large final income. Harvest bark from *Cinnamomum* after 10 years and logs of *Chukrasia* after 30 years
6	Mixture of facilitator tree and premium timber tree; *Acacia mangium* and *C. tabularis*	Provides an early cashflow and allows the introduction of higher-value tree species. *Acacia* grown for 8 years and thinned to allow *Chukrasia* to be underplanted. Final *Acacia* trees removed at 16 years
7	Mixture of premium timber tree and understorey NTFP; *C. tabularis* and *Curcuma longa*	Provides an early cashflow from tumeric which begins after 7 years. *Chukrasia* felled after 30 years

analysis to test variations in market prices and discount rates has been carried out by Harrison (unpublished) using data gathered from field situations in northern Vietnam including timber prices from the sawmill survey described above. An updated analysis using more recent timber prices is given in Harrison et al. (in press). The plantation productivities are based on data from growth plots and in some cases on conservative predictions of future mean annual increments. The results are expressed as the Net Present Value (NPV) per hectare which is the absolute payoff and as the Internal Rate of Return (IRR) which is the rate of return on capital invested.

The different silvicultural designs involve several rotation lengths. With rotations of 20–30 years and a relatively high discount rate, any cashflow after the first rotation has little effect on the NPV. When rotations are much shorter there is a possibility of using the land for another purpose once felling is complete. Harrison overcame this problem by assuming there were three successive rotations using the same silvicultural design.

All designs presume an adequate level of site preparation, good species-site matching and good planting material. For the sake of simplicity no allowance was made for income from thinnings in monoculture plantations or for additional productivity of the plantation arising from thinning treatments. Other assumptions and inputs used in the analyses are outlined in Harrison et al. (in press)

The results of the initial analysis are summarised in Table 9.3. This shows:

- All plantation designs were profitable in the sense of having a positive NPV and an IRR exceeding 10% under the assumptions made and field conditions present in Vietnam.

Table 9.3 Financial profitability of alternative plantation designs outlined in Table 8.2 showing the Net Present Value (NPV) and Internal Rate of Return (IRR) (based on Harrison, unpublished)

No.	Design	NPV VND 1,000/ha	IRR %	NPV for pessimistic scenario
1A	*A. mangium* monoculture, 10 years rotation	15,440	32.1	+1,890
1B	*E. urophylla* monoculture, 10 years rotation	11,380	26.7	
2	*P. merkusii* monoculture, 30 years rotation	2,100	11.1	–22
3	*C. tabularis* monoculture, 30 years rotation	11,560	14.6	–4295
4	*C. tabularis* + *E. urophylla,* mixture slow (30 years) and fast (7 years)	13,330	15.4	+1,746
5	*C. tabularis* + *Cinnamomum cassia,* timber and NTFP	11,500	14.1	–2,529
6	*A. mangium* + *C. tabularis*, facilitator and premium spp.	14,900	22.0	+767
7	*C. tabularis* + *Curcuma longa,* timber and NTFP	27,100	26.7	+4950

- Plantations with short rotations had higher IRR than those with long rotations; among the monocultures it was not surprisingly the three rotations of *Acacia* grown for pulpwood had a higher IRR than a single rotation of the premium quality sawlog species *Chukrasia tabularis.*
- Premium timber species grown on longer rotations were more valuable when used in mixtures. Some of the mixtures analysed had levels of profitability that were comparable with the *Acacia* monoculture.
- The most profitable plantation (in terms of NPV) was the premium timber species, *Chukrasia*, underplanted with a NTFP (in this case, turmeric). This plantation also produces an early cashflow because of the sale of the tumeric although it requires rather more labour than the others involving only trees. The profitability of this type of plantation mixture obviously depends on the market price and productivity of the particular NTFP used.

These results are conditional on the assumptions made about stand productivities, farm gate prices and associated costs. Sensitivity analyses which explored the effects of higher or lower values for these parameters are described in Harrison et al. (in press). These changes had little effect on the overall conclusions. Even under the most pessimistic scenario when all parameters (labour costs, timber prices, growth rates) were set at marginal values, most silvicultural options remain positive although, under these circumstances, scenarios involving monocultures of slow-growing trees did generate negative returns.

Similar analyses were carried out by Hines (1995) using productivity and price data from sites in central Vietnam. Her overall conclusions were similar. She found that:

- Tree growing was profitable and achieved IRR of at least 14%.
- Mixtures of trees growing on rotations of different length (e.g., eucalypts grown for woodchips and *Acacia* grown for poles) could be more profitable than simple monocultures of eucalypts.
- Mixtures involving trees grown with crops such as pineapples were very profitable.
- Plantations containing trees producing NTFPs (e.g., *Cinnamomum cassia*) were very profitable.

The conclusions emerging from these financial analyses are matched by empirical evidence from Vietnam. Most farmers are unable to carry sophisticated analyses but soon become aware of new market opportunities. Many farmers with land near the Bai Bang paper mill are now growing simple monocultures of eucalypts or *Acacia* using rotations of less than 10 years. This is almost certainly due to the fact that these species grow well on poor soils, that the mill is a stable market and that these plantation designs are easy to manage. On the other hand, there is ample evidence that plantations using other timber species and other plantation designs like those in Table 8.2 are popular in many other locations as well. This simply means there is no single 'best' species choice or plantation design. The most profitable type of plantation depends on a farmer's location as well as their economic circumstances.

Finally, some mention needs to be made of the financial returns that might be made by households or communities retaining and managing their own secondary forests. Hines (1995) analysed of the income generated by selective harvesting in a regenerating secondary forest in central Vietnam. She found that the limited labour involved in harvesting about 4 m^3 $year^{-1}$ of fuelwood and poles each year could generate returns of 12.7% which was comparable with that of eucalypt monocultures in the same area. Though financially feasible, such a system might require considerable management inputs (e.g. to assess growth rates) if it was to remain truly sustainable.

The Financial Profitability of Tree-Growing Elsewhere in the Asia-Pacific Region

It is not possible to extrapolate the results of such analyses to other countries because of differences in market prices, plantation productivities (which are affected by site conditions and fertiliser applications), labour and management costs as well as discount rates. *Acacia* may be highly profitable plantation timber trees in parts of Asia but do not attract nearly as much attention in Fiji or from tree growers in north Queensland where many of these *Acacia* species originate. Current reforestation practices across the region show that the most popular form of industrial reforestation has involved eucalypts or *Acacia* spp. grown on short rotations for pulpwood. But it is also clear that across the region there are many growers, including smallholders, who have planted species that need longer rotations because they are slower growing. These species include utility timbers such as *Azadirachta excelsa*, *Eucalyptus deglupta*, *Falcataria moluccana*, *Gmelina arborea* and rubberwood (*Hevea brasiliensis*) as well as species producing premium timbers such as teak (*Tectona grandis*), mahogany (*Swietenia macrophylla*), hoop pine (*Araucaria cunninghamii*) and rosewood (*Pterocarpus indicus*).

Assessments of the IRR of plantations of these species when grown by smallholders are often well in excess of 15% and the NPV can also be high. Examples of such analyse are those carried out in the Philippines (Bertomeu 2006; Venn et al. 2000), Peninsular Malaysia (Fauzi and Noor 2002), Laos (Midgely et al. 2007) and Sabah (Lee 2008). Growers of these timber trees have clearly decided that they will eventually profit from either a higher market price for these particular timbers or because they offer some other advantages. One such advantage is that logs of these species are valuable enough to compensate for high transport costs from more isolated locations (or from areas where road systems are poor).

Trees take years to grow and financial circumstances may change of time. What might happen in future to alter these conclusions? Some trends are relatively clear. Road access is likely to improve as rural road networks expand and reduce the current isolation of many landholders. This suggests their transport costs will decline raising the returns received by growers (although this will improve the profitability of agricultural crops as well as timber meaning farmers might prefer the former to the latter). Further population drift from the country to the city is also likely meaning the opportunities for reforestation might increase (see alternative models described in Chapter 2). Finally, consumption patterns are tending to rise as living standards rise and timber supplies from natural tropical forests in the Asia-Pacific region are likely to diminish. These trends suggest a promising future for small timber producers. On the other hand, there are also likely to be increasing supplies of utility timbers from some of the very large plantations being established in many temperate countries (Leslie 2005). Small holders producing logs of only moderate quality may have trouble competing in such markets although they may be still able to sell into their local district markets. This crude analysis suggests there may be an improving market for the higher-quality timbers previously supplied from natural forests but perhaps a more stable market for utility timbers. A report by ITTO (2007) on recent trade and price trends suggesting an overall move towards higher value products (both sawnwood and logs) although this reported noted the importance of international economic conditions in the short term.

Reforestation Businesses

Reforestation can trigger considerable economic activity by providing timber for timber mills and furniture factories as well as business for transport companies.

But reforestation might also provide the basis for several other businesses. One of these could be privately owned nurseries and a large number of these appear to have spontaneously developed in parts of Vietnam. Some provide seedlings of only one or two species (e.g. eucalypts or *Acacia* spp.) while others raise a much greater variety of species (Fig. 9.5). These nurseries presumably are able to compete with government nurseries on price or on the variety of species on offer. Ideally, it would be preferable to have some form of quality assurance to ensure that only good quality seedlings are sold.

A second type of business that could develop might be one that carries out tree-planting. Many smallholders are able to do this themselves but some people could be interested in having someone else do it if the areas involved are large or if they are absentee land owners. There are several impediments to the development of such enterprises apart from the obvious need for an entrepreneur willing to take the initiative. One of these is the need to develop a way of providing year-round work for employees. In the absence of this skilled but part-time employees will drift away whenever a permanent job becomes available. A second impediment is the need to have a year-round cashflow to pay these employees. Neither of these problems may be insurmountable; although planting seasons are mostly short and defined by climate there are a variety of other tasks including seed collecting, nursery work and plantation management (weeding, pruning, thinning) that can be carried out at other times of the year and which might generate funds for a reforestation business. The greatest problem is likely to be that of getting started. This is where sponsorship

Fig. 9.5 Private nurseries can appear as more reforestation takes place. This nursery in the central highlands of Vietnam contained around 20 species most of which were native tree species

or partnerships with either governments or NGOs are likely to be crucial in nuturing silvicultural and management skills until the business are self-sustaining.

Payments for Ecosystem Services

In recent years there has been increased interest in the services that forest supply rather than in the goods they produce. This interest has increased as overall forest cover has declined. These services include watershed protection, improved water quality, carbon sequestration and biodiversity conservation. If land owners provide such services to the wider community through retaining forests or establishing new forests, should they not be paid for doing so?

The concept of payment for ecological services (PES) explicitly recognizes that trade-offs must be made as land use pressures intensify. As Wunder (2005) note, it is this conflict of interest that provides the rationale for PES. Such a payments scheme is likely to encourage some people to become tree-growers who might not otherwise have done so. The payment mechanism is also seen as a way of mobilising new, private sector funds to improve the livelihoods of rural communities. Depending on how a scheme was arranged, the payments could also be delivered from an early stage thereby overcoming one of the key disadvantages of most tree planting which is that financial benefits are only received some years after the initial investment is made.

In an ideal world a PES scheme would be a voluntary transaction in which a defined ecological service was bought by at least one buyer from at least one provider on the condition that the provider was able to secure and provide the service over time (Wunder 2005). In practice, a rather wider range of PES-like practices have occurred including a number where not all participants are volunteers (e.g. where land is granted to a farmer on the proviso that they reforest it) and where there are a variety of intermediaries between the seller and the buyer (e.g. in the form of various kinds of timber certification or eco-labelling). The particular complexities of the carbon market are outlined further below.

The PES concept has been widely discussed in recent years but, in the tropics, it has developed most strongly in Latin America (Wunder 2005). However the idea is spreading and a recent survey identified 30 watershed protection schemes in Asia that had been established or were in various stages of development for the purpose of paying landowners for the ecological services they provide (Huang and Upadhyaya 2007). Likewise, Suyanto et al. (2005) reviewed over 80 PES initiatives in Indonesia that covered biodiversity conservation, watershed protection, carbon sequestration and landscape aesthetics (although most of these were still at the formative stage). In Vietnam the government has begun organising pilot projects to test methodologies (Nguyen 2009; Peters et al. 2009).

Role of PES in Enhancing Conservation Outcomes

Although PES schemes are about the provision of ecosystem services the actual payments are often being made for land uses such as natural or planted forests that are assumed to generate that service rather than for a particular service itself. In such cases it is commonly assumed that all types of forests are equally able to provide the service. This is not necessarily true and the ability of different forms of reforestation to provide various ecosystem services is shown in Table 9.4. There is an added complexity in these relationships because the plantation age affects the degree to which various ecosystem services are provided. For example, a relatively young plantation with a well-developed understorey may be able to protect a hillslope from erosion but, by the same age, it will have sequestered only a small amount of carbon and would have little value as wildlife habitat (Fig. 9.6).

The ability of forests to supply ecosystem services also depends on the way these forests are being managed. Thus forests that are regularly clear-felled, burned and replanted will be less effective in controlling erosion than those that are selectively logged and remain unburned. This means it may be difficult to predict just how much of a specified service is likely to be delivered over time from a particular type of reforestation. Table 9.4 also shows that some trade-offs may be made between different services. For example, some types of reforestation improve biodiversity conservation but most reduce water yields. Finally, it should also be

Table 9.4 Summary of ecological services provide by different types of reforestation (based on Chapters 4–7)

Ecological service	Capacity of different types of reforestation to provide this service
Erosion and slope stabilisation	All forms of reforestation are able to provide some soil protection but those generating thick litter layers and/or understorey development are best. Structurally complex canopy layers are usually better than single canopy layers. Regrowth forests and plantations grown on long rotations are better than forests or plantations that are frequently disturbed.
Water yield	All forms of reforestation usually reduce the total water yield but the rate of decrease is greatest in plantations of fast-growing trees.
Dry season flow	Usually decreased by all forms of reforestation except in cases where severely degraded lands are reforested and the soil infiltration capacity is increased. Forests with significant understorey development or generating large litterfalls are more likely to ameliorate compacted soils.
Sedimentation and clean water	See erosion note above. However, water quality is also likely to be affected as much by the location of any reforestation (steep slopes, riverine strips) as by the type of reforestation.
Biodiversity conservation	Best supported by natural regrowth and structurally complex and species-rich plantations containing some food plants. Plantation monocultures are likely to be least effective.
Carbon sequestration	The rate depends on plantation productivity, rotation length and prior land use. Monocultural plantations grown on short rotations generally produce more biomass carbon over time than slower-growing trees grown on longer rotations. But, over time, such plantations may also cause a net reduction in soil carbon. There is some evidence of enhanced soil sequestration in species mixtures especially those including trees capable of N-fixation.

kept in mind that some forms of reforestation may be no more effective than other land uses in providing certain services. For example, grasslands may be able to stabilise soils and protect some watersheds just as well as many forms of tree cover. In such cases it may be hard to argue that reforestation is necessary (and the argument may be even harder to sustain if the type of reforestation used substantially reduces the overall yield of water).

In theory, any payments should be conditional on the buyer receiving the service they have paid for. Governments have been rather more tolerant about this than private buyers (Wunder et al. 2008). It is sometimes assumed the benefits will flow if soil conservation measures have been carried out on a farm or if a certain number of trees have been successfully planted and survive for, say, 5 years. A better approach would be to monitor the actual service being delivered. An obvious

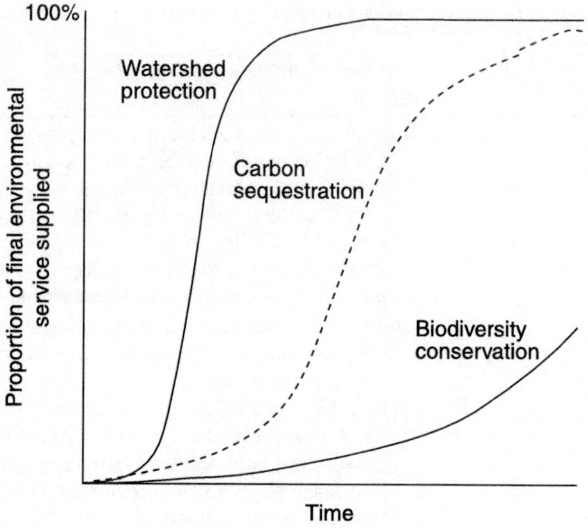

Fig. 9.6 The age of a plantation will strongly influence the degree to which it can supply ecological services. Young plantations with well-developed understories may be able to protect watersheds but will be much less able to provide habitats for wildlife

example might involve measuring the actual sediment levels in stream water draining from a farmer's land and paying accordingly.

But this process of quantification may be more easily carried out for some services than others. For example, biodiversity benefits are likely to be somewhat more difficult to assess than water quality (Which species to count? Should only native species be counted? Should one include only those species having a certain minimum population?). An additional dilemma is that some services depend on a group of suppliers acting together as might be the case with watershed protection. The problem comes if these have different areas and proportions of their land reforested or use different types of reforestation. How can one assess just how much of a particular service each farmer has supplied and so distribute the payments in a way that reflects these contributions?

Pagiola et al. (2007) dealt with both problems by using an 'ecosystem services index' to measure how different types of reforestation affected biodiversity and carbon sequestration on farms in Nicaragua. Thus a patch of primary forest scored a 1.0 on the biodiversity scale but a patch of improved pasture with trees scored only 0.3. Every hectare of land on a farm was rated according to an agreed index and the overall score for all land uses on the farm determined the payment received. The scheme was enthusiastically accepted by farmers and overall ratings on each farm increased rapidly as payments flowed and reforestation became more popular. To date, most PES schemes have involved relatively short term contracts lasting only a few years. Once buyers and sellers become more confident in these schemes rather longer periods may be preferable to avoid sellers changing land uses once PES funding ceases.

It can be difficult to establish the price of ecological services. Some may be set by an international market, such as the price of carbon. But others, such as the price of clean water, may have only a local price. It can take some time to establish just what this price may be. Buyers may not have a realistic idea of how much it costs to deliver a particular service and sellers may not know about national environmental plans or standards that may be driving the behaviour of the buyer. An interesting approach to resolving this is outlined in Box 9.2.

Box 9.2 Establishing the Price of Ecological Services Through Conservation Auctions

Conservation auctions are a cost-effective and transparent means by which ecosystem services can be sold (Stoneham et al. 2003). Conservation auctions work in a setting where there may be several landowners able to act as providers of an environmental service and a single buyer with a limited amount of funds. The buyer calls for sealed bids from landowners to provide that service. This might be to retain existing vegetation, exclude grazing animals, reforest a specified area of bare hills or to revegetate buffer zones along rivers. Each bid is assessed using some form of an 'environmental benefit index' that ranks the expected benefits that the action is likely to generate. The best bid is that where the value of this index, divided by the cost, is highest. This gives an indication of the value-for-money of the bid. A variety of such indices might be developed depending on the environmental factors or services being targeted. Bids are then accepted, starting with the most cost-effective bid, until available funds are exhausted.

These auctions have several advantages for buyers of ecological services. They allow buyers to assess and select bidders able to provide a service at the lowest cost since they provide an incentive for landowners to reveal their real costs of undertaking particular actions that produce desirable environmental outcomes. They also mean priority areas can be targeted and they allow payments to be linked to actual and timed actions. But they also have some advantages for landowners. It is up to them to specify what actions they are prepared to offer and, once the contract is signed, they are in control of the operation. The auctions are also a way of funding improvements in environmental conditions on their own property.

Design principles for the contracts and the auctions are still being developed (Jack et al. 2009). For example, it may be difficult to run sequential auctions if sellers learn what went before and adjust their prices accordingly. Similarly it can be difficult to deal with multiple environmental objectives although there is evidence that this can be done (Gole et al. 2005). It may also be difficult to run such auctions in communities where the idea is unfamiliar or where landholdings are small. Nonetheless, there are features of these auctions that could be attractive to those concerned with forest rehabilitation and the idea deserves to be explored further in a variety of settings.

Some ecological services such as biodiversity conservation and hydrological flows are scale-dependent and require that a minimum area is reforested before any benefit is generated. This could mean that a number of plantation owners must be involved before a service is generated. Likewise, these same services also depend on the spatial distribution of any new forests. Thus a number of small but randomly distributed plantations may generate only modest conservation benefits but the same area planted to form a corridor between two forest remnants may be much more effective. Similarly, random plantings on flat land are unlikely to be as effective in reducing erosion as plantings on steep slopes. Both factors mean that care is needed when deciding where to promote PES schemes to ensure any reforestation will generate the desired outcome.

These spatial considerations aside, areas deserving the highest priority should be those where the functional consequences of reforestation are likely to have the greatest beneficial effect per unit of cost and where reforestation might not otherwise have occurred. This means some landowners are likely to be more important than others. Of course not all landowners (including those in especially critical locations) may wish to reforest their land and it is not difficult to imagine situations where some landowners are willing to engage in a PES scheme for a standard price but others occupying critical positions in the landscape are not. These farmers may demand a much greater payment. Should the buyer accept that some areas are more important than others and pay accordingly or should such landholders be obliged to participate? These landscape issues are discussed further in Chapter 11.

The Role of PES in Improving Livelihoods and Reducing Poverty

Many people living in degraded lands are poor. Are these people likely to be assisted by a PES schemes or is it a mechanism that largely benefits the wealthy? There are three common problems. One is that many poor people do not have legal tenure over their land and, because of this, are unattractive to buyers of ecosystem services. A second is that even those with legal tenure can have trouble benefiting because they may be in a weak position when exposed to negotiations with powerful buyers who are able to manipulate the PES transaction to their own advantage. Finally, many smallholders are unfamiliar with the idea of contracts and may be unable to guarantee they can undertake management of the kind needed to generate particular environmental outcomes because of their limited financial resources.

Taken together, these constraints mean it can be difficult for smallholders to benefit from PES schemes. Nonetheless, evidence is beginning to accumulate to show that PES schemes are beginning to foster reforestation and make a significant contribution to the incomes of poorer farmers in some areas (Wunder et al. 2008; Pagiola et al. 2007). There is also evidence they can sometimes help reinforce land security vis-à-vis neighbours and squatters. And, even when payments are low, some farmers still value PES schemes because of the technical assistance they provide (Huang and Upadhyaya 2007). In all such cases supportive and protective institutions are needed to help build trust and facilitate negotiations between smallholders and the potential purchasers of ecosystem services.

Making PES Schemes Work

Despite the promising indications referred to above PES schemes are still not common in the tropics especially in the Asia-Pacific region. Many schemes are still in their planning stages and rather fewer are in operation. There appear to be a number of pre-conditions that must be fulfilled before a PES scheme can be established. The most obvious of these is that there must be a demand for the service. Or, more precisely, there must be a buyer willing to pay for a service that they may have previously been received at no cost. Such buyers might emerge as deforestation and degradation proceeds and a former environmental service, such as clean water, disappears.

Secondly, there must be a legally recognized seller with whom the buyer can negotiate. Such a seller might be an individual or community with formal land tenure or with recognized rights over an area with planted trees or a regrowth forests. This may be an impediment in many parts of Asia where land tenure arrangements are fluid and uncertain. On the other hand some of the pilot schemes now being trialled are experimenting with tenure or land use rights as payment or reward for ecosystem services (Huang and Upadhyaya 2007). An example of this occurs in Doi Suthep-Pui National Park in northern Thailand where Hmong villagers collaborated in reforestation without payment in the expectation that this would help consolidate their claim to land use rights, and possibly, citizenship (S. Elliott, personal communication, 2009).

A third requirement is that any payments must be sufficient to overcome the opportunity costs involved in forgoing an alternative land use that does not generate a particular service. This is likely to mean PES schemes are only viable when these costs are low because it may be difficult to match the high prices received for, say, a valuable cash crop. As a consequence, PES programs are more likely to succeed on modestly degraded or steeper sites rather than on prime agricultural land (where PES cannot match crop market prices) or severely degraded sites (where the cost of rehabilitation needed to generate the service may be too high to be compensated by a PES scheme). Some landholders will have different opportunity costs than others because of soils, topography or the location of their farms. As noted earlier, this may mean there is not a standard payment for particular ecological services but that this payment varies according to location.

Fourthly, the buyer should receive the ecosystem services they have paid for. But many buyers do not monitor to ensure they receive the service they have paid for. In fact, Wunder et al. (2008) were of the opinion that very few government-funded PES schemes have been terminated because agreed services have not been delivered. By contrast, private buyers have tended to be more insistent that they receive what has been agreed upon. This lack of monitoring may be because it is expensive or because there are few metrics available to measure 'conservation' outcomes but it may also be a function of the fact that the whole PES process is still in its infancy.

A fifth requirement is that some way must be found to dealing with the high transaction costs that can be involved in establishing and managing a PES scheme.

These tend to be high when there are many smallholders acting as suppliers. Costs are also increased when property rights are weak and when the cost of assembling baseline data and monitoring performance is high. Most of those who discuss this issue suggest these costs can be reduced by developing collective agreements involving a group or community rather than dealing with individual landowners.

Many governments will probably continue to decide which forests become protection forests and which areas should be reforested and some governments will continue to offer various kinds of reforestation incentives. But PES-like schemes could be useful devices for assisting the promotion of reforestation when the opportunity cost of following state directives to put more land under forest became too high (Wunder 2005). That is, a PES scheme might be a way of introducing economic incentives at the margin of a state run, land use planning system. And PES schemes could help ensure a stronger linkage between reforestation and the actual delivery of ecological services (since, as noted earlier, not all types of reforestation are equally able to deliver all ecosystem services).

The Carbon Market

One particular ecosystem service provided by plantations that is beginning to attract considerable interest is their capacity to take up and immobilise carbon. Concern about climate change has led to growing interest in a market for carbon. This means that, in principle, a grower could be paid for planting trees to store carbon rather than produce timber. Calculations by Tomich et al. (1997) concerning the profitability of reforesting grasslands in South Kalimantan, Indonesia showed how even modest payments for sequestering carbon could significantly enhance the profitability of tree growing. For a discount rate of 10% the financial profitability of growing *Acacia mangium* on 8-year rotation was US$337 per ha. When the carbon price was US$5 per t and the imputed value of net carbon sequestration was added, the return to growers rose to US$451 per ha. If the carbon price was US$20 per t then the return rose to US$793 per ha.

In practice, of course, the process is a good deal more complicated. One set of complications arise from the need to certify precisely how much carbon is actually absorbed by particular forms of reforestation and whether this truly is a real gain or whether it is being matched by deforestation (and carbon release) going on elsewhere on that owner's land, a process known as 'leakage'. Carbon accounting needs to use an agreed methodology to assess plantation growth rates and verify the size of the carbon stocks acquired by the plantation. There must also be agreement on whether below-ground biomass is included as well as above-ground biomass. Likewise, there must be agreement on whether or not to take account of changes in soil carbon. In some cases it may be small compared with above-ground biomass but in other cases it may be significant. A related issue is to ensure that existing forests are not being felled in order to allow new plantations to be grown on the same sites specifically for carbon sequestration. The present Kyoto protocols, for example, specify that land deforested after 1990 will not be eligible for reforestation.

A second set of complications arise from whether the plantations are 'permanent' in the sense that no harvesting will take place or whether they are temporary and some form of harvesting and replanting will occur. Much depends on and whether the buyers of credits operate in a regulated market or a voluntary market and on the rules these different carbon markets adopt. Owners of permanent plantations would presumably be paid a specific sum on successive occasions depending on how much carbon had been sequestered in their plantations. These payments might track, say, the international carbon price and continue for a defined contract period of 50–100 years. Under these circumstances a mixed-species plantation or an ecologically restored forest would probably be more attractive than a simple monoculture because it is likely to be more resilient over this period. Even so, some allowances might be needed for the possibility of damage (and carbon loss) from fires, storms, insect pests or diseases.

But the situation is likely to be different for owners wishing to periodically harvest timber from their plantations. In this case they would also receive credit for carbon absorbed but would presumably have to pay for carbon emissions caused when they harvested their timber and exited the market. The profitability of participating in the carbon market then depends on the price of carbon and the profitability of the timber they harvest. If the carbon price is increasing over the rotation the cost of purchasing credits to pay for emissions caused by logging might be prohibitively high and overwhelm the timber profits. This cost might be covered by planting more trees on other sites or reforesting the site a second time. Alternatively, it might make more sense to selectively fell only a small number of the best trees and retain the remaining trees as a permanent carbon sink. Again, this situation would favour mixed-species plantations involving high-value trees. If, on the other hand, the carbon price was stable or fell, the grower would have benefited but would then have to decide whether to continue with the carbon market in future rotations. The task for the prospective grower, then, is to balance growth rates, future market prices of forest products and carbon and also interest rates to determine the most profitable rotation length and silvicultural strategy. Large industrial companies might be able to carry out these calculations but most smallholders will not.

A third set of complications concern ways of minimising the transaction costs arising from having large numbers of farmers owning small areas of land. Modelling carried out by Cacho et al. (2008) found there were significant economies of scale and that the viability of carbon sequestration projects could be enhanced by working with community or producer groups rather than individual farmers. This would have the added advantage that the groups could afford consultants to develop financial strategies suitable for small-scale growers to deal with the issues raised above. They concluded payments for carbon sequestration are likely to be especially attractive to landowners in more remote areas because these growers do not have to transport goods to market. The best results were found when degraded land with low opportunity costs and limited existing carbon stocks were reforested.

Finally, there is a need to devise ways to ensure that payments for carbon sequestration actually reach the individual landowners on whose land reforestation is taking place. There is a considerable risk that a significant proportion may be siphoned off by governments, provincial authorities, traders or middlemen leaving only small amounts able to reach growers. If carbon sequestration payments are to act as incentives for reforestation they must, at the very least, cover reforestation costs as well as any costs to the landowner of measuring and certifying the amount of carbon sequestered. Many smallholders will expect these payments to begin being made at an early stage in the life of the plantation even though it may take time for an appreciable stand biomass (and thus carbon store) to accumulate.

In short, carbon trading could have a major impact on the economics of reforestation but there are a number of issues needing to be resolved before it will generate the financial benefits that many currently expect. Thomas et al. (2010) discuss some of these in their review of why so few reforestation or afforestation projects have been able to utilise funding currently provided by the Clean Development Mechanism. It also remains to be seen whether these benefits will be largely captured by large industrial groups or whether smallholders will also benefit. The best way for smallholders to participate in this market might be as members of a consortium or producer group that can advise individual members on how to manage their plantations and that would have sufficient economic power to negotiate more equitable arrangements with other marketplace participants.

Increasing the Incomes Received by Tree-Growers

Some of the factors affecting the profitability of tree-growing and influencing the financial returns to growers have been described above. But what actions or policy settings might improve the returns to smallholders growing trees in a particular location? There are a surprisingly large number of possibilities. Some involve increasing the value of outputs while others involve reducing the cost of inputs. Growers should benefit if:

- *Care is taken with the choice of plantation species*: species choice is critical and many species used in reforestation have been chosen for the wrong reason (e.g. they have been used in the past by neighbours; the seedlings were readily available; they are known to grow well at the site). Better outcomes are more likely when species are chosen that grow well at a site but which also suit a specific market – preferably one that offers relatively high prices. Top down or general prescriptions from government agencies concerning the species to plant in a region may be appropriate if the objective is to simply boost forest cover but may not always be the best way of improving livelihoods.

- *Only high-quality seedlings are used*: poor quality seedlings lead to low survival rates and poor growth. Only seed from reputable sources should be used and seedlings should meet strict quality guidelines whether they are derived from government or privately-owned nurseries. It is a mistake to use seed from just a few parent trees because it is easy to collect or old, pot-bound seedlings because they are cheap.
- *They are able to get reliable technical advice*: tree growing may be a new activity for many landowners and they may need technical advice on how to go about it. This advice may cover site preparation, planting material and ways of boosting productivity and shortening rotation lengths (e.g. by using the correct fertilizers or by thinning to increase final crop tree size) or it may cover marketing. It is also likely to emphasize the importance of good weed control. However, those giving advice should be qualified to do so. Anyonge and Rochetko (2003) give examples where growers have been given conflicting advice by agricultural extension workers (who advised growers to lop trees for fuelwood and fodder) and forestry extension workers (who recommended against lopping but favoured thinning trees with poor form). Specialised technical advice is likely to be needed for those wishing to sell ecosystem services.
- *Plantations are diversified*: a single species grown in a monoculture on short rotations will be attractive and often quite profitable if a good market is nearby. But, where that is not the case, spatial mosaics of monocultures or mixtures of species can diversify incomes and reduce the risk of fluctuating market prices. These plantations can be mixtures of timber trees or mixtures that produce sawlogs and NTFPs including food crops. The latter type of mixture has the advantage of also providing an earlier cash income. Mixed-species plantings are also more are likely to allow growers to more readily enter markets for ecosystem services.
- *Plantation are managed to improve log quality*: higher quality logs attract higher prices. Quality can be boosted by early thinning to remove trees with poor form and by pruning the retained trees to improve timber quality.
- *Plantations are certified by a recognised certification body*: growers able to satisfy the requirements of certification schemes such as that administered by the Forest Stewardship Council may be able to label their timber accordingly and be eligible to receive a higher market price. There are opportunities within some such schemes for small producers.
- *Logs can be taken to the roadside or directly to the market*: a higher return is received because middlemen are avoided. This may be feasible if growers have access to trucks or other transport and know where the market is but it may not be feasible otherwise.
- *There is a variety of potential buyers*: growers dependent on a single market are in a weaker position than those able to negotiate with several potential buyers. In this respect it is important that small rural sawmills and timber processing enterprises are supported by government policies and technical advice on timber processing, seasoning and preservation to ensure there are a number of potential buyers and that timber prices are not determined by a single large entity.

- *A marketing cooperative can be formed with other growers*: growers who produce small volumes of logs at infrequent intervals are likely to obtain lower prices than those who regularly produce larger volumes. The obvious solution is to form grower-cooperatives to increase the negotiating power with traders or mills. This may be especially important for those growing high-value timber species. Such cooperatives might organise transport and also advise growers on other technical matters associated with marketing such as how to measure their trees or logs to ensure they receive a fair return that reflects the real market price. Cooperatives formed to market NTFPs can also be effective and Morris et al. (2004) describe how one such cooperative in northern Laos allowed growers to substantially increase their income by simply introducing scales allowing them to sell their cardamom based on weight.
- *A partnership can be established with an industrial end-user*: by acting as out-growers such smallholders know they have a guaranteed market price and can plan accordingly. Depending on the partnership they may also benefit by having access to initial establishment funds, improved planting material, technical advice and fire protection. Of course the value of the relationship depends on the type of agreement and some industrial groups have taken advantage of their greater negotiating power to form inequitable partnerships.
- *Governments develop a supportive policy environment*: government policies can significantly affect the financial attractiveness of tree-growing by smallholders. Supportive land tenure policies will encourage households to embark on long-term activities such as tree-growing while financial incentives or low-interest loans can reduce the cost of plantation establishment. The return to growers will be increased by the removal of restrictions on harvesting as well as many of the taxes and fees that often litter forest product market chains. Growers will also benefit from a stable policy environment in which existing laws are enforced and the development of illegal taxes or charges is prevented. A supportive government policy will be especially important if smallholders are to participate in any PES market.
- *They acquire market knowledge*: growers often have little knowledge of market prices or of the types of forest products or ecosystem services required by the market. Forest owners with such knowledge will be in a better position to design plantings and obtain a higher price than those without it. Uninformed growers are more at risk of being pressured by middlemen to prematurely sell trees before the reach the most financially rewarding sizes. This improved knowledge of actual market prices might be done by publishing the information in newspapers or broadcasting it on radio and TV in the same way that agricultural crop prices are broadcast.
- *Illegal logging is prevented*: if logs reach the market place and are sold at prices that do not reflect their cost of production (or replacement), then genuine growers will be rapidly squeezed out of the market. Illegal logging not only damages the natural forest but it also reduces the likelihood that landowners will grow trees on degraded land.

In short, these strategies touch on improving production, widening the variety of market opportunities, overcoming market failures and improving grower's knowledge. Most are underpinned by the need for supportive government policy settings. If implemented, will these changes be sufficient to increase the attractiveness of tree-growing to farmers? It will probably depend on the circumstances in which individual farmers find themselves and this forms the basis of the next chapter.

Conclusions

Tree growing is less attractive than many other land uses because there is a considerable delay before the grower receives any return on their initial investment costs. Poorer farmers are obliged to give priority to ensuring food security rather growing timber trees and other longer term crops. Tree growing is also unattractive to many smallholders because it is not a traditional land use activity. On the contrary, many farmers may have spent considerable effort trying to remove trees from their land and have always had access to most of the forest products they need from nearby natural forests. Nonetheless, there is mounting empirical evidence that, in addition to industrial plantation timber companies, farmers from across the Asia-Pacific region are finding tree growing is financially rewarding. In some cases this is because the returns exceed those from alternative land uses. In others it is because a small tree plantation is a way of diversifying the household income or creating a durable but liquid asset to buffer themselves against changes in their financial circumstances.

Broadly speaking, the timber markets can be broken into two categories. The first is the pulpwood or fibreboard market. This is the market targeted by most large industrial tree growers in the region and it can also be an attractive market for smallholders able to form partnerships with these industrial organisations. The second category is the sawlog or plywood market and this can be sub-divided between a market for utility timbers and one for high-value timbers. Trees destined for this market must reach a larger size than those being sold as pulpwood which usually means growing them for longer periods of time and this can be a major disincentive for some landowners.

Not surprisingly, prices for various forest products differ. In particular, pulpwood logs are usually sold at a much lower price than most sawlogs. It is difficult to predict future market prices for either category of timber although there is some evidence of a rising demand for higher quality timbers in the longer term. This difference in price also means that only growers located near export points or pulpmills can normally afford to grow pulpwood timbers while those living in more isolated areas would be better advised to grow higher quality timber able to bear higher transport costs. Most of the sawlogs produced by smallholders will probably be sold into the domestic market but there may be opportunities to sell higher-value timbers in the international market.

There are a variety of silvicultural systems that growers can use and those contemplating growing trees on longer rotations are likely to benefit from using some form of mixed-species planting including those where a non-timber species is established in the plantation understorey. However, it seems likely that the financial returns that smallholders obtain from tree-growing are rarely as much as they could obtain. This is because of an inappropriate species choice or poor plantation management. Likewise many smallholders know little about markets or marketing. A supportive government policy framework that improves silvicultural and marketing advice and removes disincentives such as excessive fees, taxes and other administrative hurdles could change the profitability of tree growing for many smallholders.

Most forest owners have relied on the sale of various forest products to generate an income but there is evidence that in future some owners may be able to also find a market for the sale of the ecosystem services generated by their forests. This market is likely to grow as natural forests decline and there is a diminution of the services previously obtained at no cost. There is also likely to be increasing interest in a market for carbon sequestration. Large industrial plantation owners may be able to participate in such markets rather more easily than smaller growers. Indeed, smallholders may need considerable support and institutional back-up to be able to participate at all. On the other hand, smallholders may be better placed to provide a wider range of ecosystem services than some large industrial growers because of their ability to undertake more varied forms of reforestation.

References

Anyonge C, Rochetko JM (2003) Farm-level timber production: orienting farmers towards the market. Unasylva 54:48–56

Bertomeu M (2006) Financial evaluation of smallholder timber-based agroforestry systems in Claveria, Northern Mindanao, the Philippines. Small-scale For Econ Manage Policy 5:57–81

Bertomeu M (2008) Can smallholder tree farmers help revive the timber industry in deforested tropical countries? A case study from southern Philippines. In: Snelder DJ, Lasco RD (eds) Smallholder tree growing for rural development and environmental services. Springer, Berlin, pp 177–191

Bui HB, Harrison S, Lamb D, Brown S (2005) An evaluation of the small-scale sawmilling and timber processing industry in Northern Vietnam and the need for planting particular indigenous species. Small-scale For Econ Manage Policy 4:85–100

Byron N (1991) Cost benefit analysis and community forestry projects. In: Gilmour DA, Fisher RJ (eds) Villagers, forests and foresters: the philosophy, Process and practice of community forestry in Nepal. Sahayogi Press, Kathmandu, pp 163–180

Cacho O, Hean R, Ginoga K, Wise R, Djaenudin D, Lugina M, Wulan Y, Subarudi LB, van Noordwijk M, Khasanah N (2008) Economic potential of land-use change and forestry for carbon sequestration and poverty reduction. Australian Center for International Agricultural Research, Canberra, Australia

Center for International Economics (1999) Paper, Prices and Politics: an Evaluation of Swedish Support to the Bai Bang Project in Vietnam, Report 99/3. Swedish International Development Cooperation Agency, Stockholm

Fauzi PA, Noor MM (2002) Financial analysis. In: Krishnapillay BA (ed) Manual for forest plantation establishment in malaysia. Forest Research Institute of Malaysia, Kuala Lumpur, Malaysia, pp 137–162

Gole C, Burton M, Williams KJ, Clayton H, Faith DP, White B, Huggett A, Margules C (2005) Auction for landscape recovery. WWF, Sydney [http://www.uwa.edu.au/__data/page/12817/ALR_Final_Report.pdf]

Hansen PK, Sodarak H, Savathvong S (2007) Teak production by shifting cultivators in northern Lao P.D.R. In: Cairns M (ed) Voices from the forest. Resources for the Future, Washington, pp 414–424

Harrison S (2003) Property rights issues in small-scale forestry in the Philippines. Ann Trop Res 25:77–87

Harrison S, Lamb D, Hieu PS (in press) The economics of alternative smallholder silvicultural decisions in northern Vietnam. Small Scale Forestry (in press)

Herbohn J, Harrison S (2005) Improving financial returns to smallholder tree farmers in the Philippines: issues and way forward. In: Suh J, Harrison S, Herbohn J, Mangaoang E, Vanclay J (eds) ACIAR smallholder forestry project ASEM/2003/052 improving financial returns to smallholder tree farmers in the philippines proceedings from the ACIAR project planning workshop held in Ormoc City, the Philippines, The University of Queensland, Ormoc City, the Philippines, 15–17 Feb 2005.

Hines D (1995) Financial viability of smallholder reforestation in Vietnam. United Nations Development Program, Hanoi

Huang M, Upadhyaya SK (2007) Watershed payment for ecological services in Asia. Sustainable Agriculture and Natural Resource Management Collaborative Research Support Program, Working Paper No 06–07, Virginia Polytechnic Institute and State University, Blacksburg

ITTO (2007) Annual Review and Assessment of the World Timber Situation. Document GI-7/07. International Tropical Timbers Organisation, Yokohama

Jack BK, Leimona B, Ferraro PJ (2009) A revealed preference approach to estimating supply curves for ecosystem services: use of auctions to set payments for soil erosion control in Indonesia. Conserv Biol 23:359–367

Le TP, Nguyen VD, Nguyen NQ, Phan LV, Morrison E, Vermeulen S (2004) Making the most of market chains: challenges for small-scale farmers and traders in Upland Vietnam. International Institute for Environment and Development, London

Lee YF (2008) Financial analyses of forest plantations. In: Lee YF, Mohammad A, Chung AYC (eds) A guide to plantation forestry in sabah sabah forest record No 16. Sabah Forestry Department, Sandakan, pp 115–120

Leslie AJ (1987) A second look at natural management systems in tropical mixed forests. Unasylva 39:46–58

Leslie AJ (2005) What will we want from the forests? ITTO Trop Forest Update 15:14–16

Maguire L, Justus J (2008) Why intrinsic value is a poor basis for conservation decisions. Biotropica 58:910–911

Mangaoang E, Herbohn J, Harrison S, Cedamon E (2005) Overcoming problems with tree registration and log transport permits for smallholder tree farmers in Leyte. In: Suh J, Harrison S, Herbohn J, Mangaoang E, Vanclay J (eds) ACIAR smallholder forestry project ASEM/2003/052 improving financial returns to smallholder tree farmers in the philippines proceedings from the ACIAR project planning workshop, Ormoc City, the Philippines, 15–17 Feb 2005

McCauley DJ (2006) Selling out on nature. Nature 443:27–28

McElwee P (2001) Fuelwood harvesting and use in Cam Xuyen District, Ha Tinh Province. Forest Science Institute of Vietnam, Hanoi

Mercer DE, Soussan J (1992) Fuelwood problems and solutions: policy options. In: Sharma N, Rowe R (eds) Managing the world's forests: looking for balance between conservation and development. Kendall/Hunt Publishing, Falls Church, VA, pp 177–214

Midgely S, Blyth M, Mountlamai K, Midgely D, Brown A (2007) Towards improving profitability of teak in integrated smallholder farming systems in Northern Laos. Australian Center for International Agricultural Research, Canberra, Australia

Morris J, Hicks E, Ingles A, Ketphanh S (2004) Linking poverty reduction with forest conservation: case studies from Laos. International Union for the Conservation of Nature, Bangkok, Thailand

Nguyen QT (2004) Agroforestry household system. In: Buurman J, McCarty A, Robertson S (eds) Developing Vietnam: studies in rural and sustainable development. Centre for the Study of Transition and Development, Institute of Social Studies, The Hague, The Netherlands, pp 229–266

Nguyen TP (2009) The role of the government in developing and implementing the policy on payments for forest environmental services in Vietnam. Ministry of Agriculture and Rural Development, Forest Sector Support Partnership Newsletter, pp 5–6

Pagiola S, Ramirez E, Gobbi J, de Haan C, Ibrahim M, Murguerito E, Ruiz JP (2007) Paying for the environmental services of silvopastoral pratices in Nicaragua. Ecol Econ 64:374–385

Pang K (ed) (2009) Prospects for Garahu in Sabah, The current state of plantation forestry in Malaysia: a special focus on Sabah Held 18–20 November 2009. Forestry Department Headquarters, Sandakan

Pasicolan PN, de Haes HAU, Sajise PE (1997) Farm forestry: an alternative to government driven reforestation in the Philippines. For Ecol Manage 99:261–274

Pasicolan PN, Macandog DM (2007) Gmelina boom, farmers' doom: tree growers' risks, coping strategies and options. In: Harrison SR, Bosch A, Herbohn J (eds) Improving the triple botton line from small-scale forestry proceedings of IUFRO 3.08 conference. University of Queensland, Brisbane, Ormoc City, Leyte, the Philippines, pp 313–318

Peters J, Nguyen CT, Nguyen TBT (2009) The pilot payment for forest environmental services in Lan Dong province. Ministry of Agriculture and Rural Development, Forest Sector Support Partnership Newsletter 26:11–13

Roder W, Keoboualapha B, Manivanh V (1995) Teak (*Tectona grandis*), fruit trees and other perennials used by hill farmers of northern Laos. Agroforest Syst 29:47–60

Stoneham G, Chaudhri V, Ha A, Strappazzon L (2003) Auctions for conservation contracts: an empirical examination of Victoria's Bush Tender trial. Austr J Agric Resour Econ 47:477–500

Suyanto S, Leimona B, Permana RP, Chandler FJC (2005) Review of the development environmental services market in Indonesia. World Agroforestry Center, Bogor

Thomas S, Dargusch P, Harrison, Herbohn J (2010) Why are there so few afforestation and reforestation Clean Development Mechanism projects? Land Use Policy 27:880–887

Tomich TP, Kuusipalo J, Menz K, Byron N (1997) Imperata economics and policy. Agroforest Syst 36:233–261

Venn TJ, Harrison SR, Herbohn J (2000) Relative financial performance of Australian and traditional timber species in the Philippines. In: Harrison SR, Herbohn J (eds) Socio-economic evaluation of the potential for Australian tree species. Philippines Australian Center for International Agricultural Research, Canberra, pp 93–110

Wunder S (2005) Payments for ecological services: some nuts and bolts. Center for International Forestry Research, Bogor, Indonesia

Wunder S, Engel S, Pagiola S (2008) Taking stock: a comparative analysis of payments for environmental services programs in developed and developing countries. Ecol Econ 65:834–852

Chapter 10
Reforestation and Farmers

The smallholder may well be the timber producer of the future in many parts of the developing world, and increasingly a provider of high quality tropical timbers. – D.P. Garrity, Director General, World Agroforestry Center (Garrity 2004)

Introduction

In the past many farmers have grown trees on their land using some form of agro forestry. These have been mostly grown for domestic uses. But the recent decline in the area of natural forests has meant the demand for forest products has increased and there is increasing scope for farmers to grow trees and sell timber and other forest products, especially on land that might be marginal for agriculture. The way in which this is done will be different to those earlier forms of smallholder tree-growing and so represents a new form of land use practice.

Most smallholders are not averse to growing new crops or changing their land use practices when a new opportunity arises. One has only to look at the rapidity with which cash crops such as coffee, tea or rubber have spread. And many farmers have responded to degradation by planting trees where they believe it is advantageous for them to do so (Scherr 2000). But tree plantations, and especially those using species grown for timber, are different to agricultural cash crops and they have some significant disadvantages. One is that there can be a long delay between the time an investment is made and the timing of the financial returns. Another is that the cash flow is episodic and there can be long intervals between harvests and, therefore, cash income. Both factors mean the opportunity costs of tree growing can be high. But, as previously discussed, tree plantations offer some advantages not found with other crops; they can provide goods and services, they can utilise marginal lands and they have low maintenance costs once they are established. While tree plantations are unlikely to lift smallholders out of poverty they have the capacity to diversify farm incomes at a relatively low cost and complement other agricultural activities. This means they may be a useful complement to other land use activities on many farms.

D. Lamb, *Regreening the Bare Hills*, World Forests 8,
DOI 10.1007/978-90-481-9870-2_10, © Springer Science+Business Media B.V. 2011

The previous chapter considered some of the financial issues surrounding tree-growing by smallholders. These are important but are not the only factors determining whether or not a landholder engages in reforestation. This chapter discusses some of these other factors that affect the attractiveness of reforestation. It also explores how smallholders might be assisted to make tree-growing part of their farming activities if they choose to do so.

Farmers and the Farming Environment

The choices farmers make about land use practices depend on the particular circumstances in which they find themselves. Ellis (1993) has referred to this as their 'livelihood platform'. This platform includes the assets they have as well as the institutional framework and government policy settings that affect the use that can be made of these assets. Assets include the amount of land (and the form of land tenure they have), the biophysical properties of this land, the amount of labour in the family unit, the household financial resources as well as the supportive infrastructure such as roads. The knowledge farmers have about markets and farming techniques are also important. These various assets are sometimes categorized as Natural Capital, Financial Capital, Physical Capital and Social and Human Capital (Aronson et al. 2007; Fisher et al. 2008; Sayer and Campbell 2004). The value or quality of these resources vary over time so that a farmer's decisions might be made in the context of changing population densities, changing market prices for the goods they produce or seasonal variations in their degree of food self-sufficiency. Like all natural resource managers they must also cope with risks such as storms, droughts, floods, pests and diseases.

Farmers differ in how they perceive their circumstances and how they use the resources and various forms of 'Capital' available to them. Such differences affect their level of interest in tree-planting and, if they choose to engage in reforestation, just how they might go about it. This raises the question – can farmers be categorized or classified in a way that reflects these differences? There are several ways in which this might be done.

Typologies of Farmers Based on Behaviour

One way of classifying farmers is to do so according to their behaviour and the way they use their land resources. This was done by Turkelboom (1999) who worked with an Akha farming community in northern Thailand. He recognized five categories of farmer (Table 10.1). The first he named *Secure Investors*. These were farmers with sufficient rice to last the full year because of the size of their farms (around 2.5 ha), the fertility of their soils or because they could irrigate

Table 10.1 Examples of some farmer typologies

Thailand[a]	Vietnam[b]	Philippines[c]	Australia[d]
Secure investors (13)	Large area – high income (6)	Confident farmers (22)	Traditionalists (6)
Profit maximisers (23)	Large area – moderate income (18)	Disadvantaged farmers (24)	Experienced-comfortable (53)
Diversifiers (19)	Large area-low income (6)	Experienced foresters (18)	Progressive second-generation (7)
Survivors (30)	Small area-high income (8)	Doubtful foresters (13)	High intensity (14)
Dropouts (7)	Small area-moderate income (44)	Well-off households (23)	Retired professional, conservationists and absentees (20)
	Small area-low income (19)	Disadvantaged households (24)	

[a]Turkelboom 1999; [b]Lamb et al. 2006; [c]Emtage 2004, [d]Herbohn et al. 2005
The names used are those given by each author in an attempt to summarize the characteristic of each type. The percent of the community in each class shown in parentheses.

some of their fields. These farmers were willing and able to invest more in their farms to increase production. Some had begun to do so by planting fruit trees. The second category was the *Profit Maximizers.* These farmers had smaller farms (1.5 ha) and were much less self-sufficient in rice. They had chosen to invest land and labour in growing profitable but high-risk cash crops such as cabbages in the hope of making sufficient income to buy their food. In time they hoped to build up enough capital to be able to move into buying more land, hiring labour and moving to a less risky livelihood strategy. The third category was the *Diversifiers*. These, too, had less land than the Secure Investors and were less self-sufficient in rice but chose to achieve food security by having a variety of crops. Most were necessarily focussed on the immediate future and were not as interested in long-term crops like fruit trees. The final two categories included the *Survivors* and the *Dropouts*. Neither group was self-sufficient in rice and farmers in both relied on off-farm income for their livelihoods. The *Survivors* continued to carry out some farming on the own landholdings (about 1 ha in area) while most of the *Dropouts* had lost all their land or no longer engaged in farming because they believed it was not worth the effort.

The value of Turkelboom's typology lies in the way it describes individual farmer's behaviour even though timber tree growing is not attractive to any of these particular farmers simply because of the small areas of land they have access to (mostly <3 ha). Interestingly, this community had established a communal or village forest covering 98 ha despite these differences in attitude towards land use practices. This appears to have been done partly for religious reasons and partly to provide forest resources such as medicines to the village (Fig. 10.1).

Fig. 10.1 A village forest established on degraded grasslands by Akha migrants in northern Thailand. Most of the trees in this forest will not be harvested although members of the community are allowed to collect NTFPs. The forest has religious significance for the community

A Typology of Farmers Based on Resource Limitations

A second approach to classifying farmers was that developed for 210 farm households in northern Vietnam by Lamb and Huynh (2006). Unlike the classification of Turkelboom (1996), this was designed for a situation where land had been specifically allocated to farmers to reforest for production purposes and all farmers had additional land for food production. This process had commenced in the late 1980s at the time of the national land distribution program (MARD 2001). In this situation the issue was not whether reforestation might be an attractive new land-use option for farmers. Rather, the question was what type of reforestation might they adopt?

Two factors thought to be especially influential were the amount of land available to each farmer and the household income (Table 10.1). Landholdings varied considerably and some farmers had up to 30 ha of land although the median area was usually <5 ha. Farmers were categorized as having 'large' areas of land if their holdings exceeded 2 ha and 'small' if they had less (71% of farmers had only small land holdings). There were three levels of household income (well-off, average and poor) based on a self-assessment of household food self-sufficiency. Of the 210 households studied, most (62%) were in the average income category but 25% were classified as poor. Of the six land-income categories the biggest single group were those with small landholdings and moderate incomes (44%).

Farmers in each category had differing views about the merits of investing time and resources into reforestation and about the constraints they faced in doing so. Most of those with large landholdings intended to spend time and money planting more trees within the next 5 years but rather fewer of those with small landholdings planned to do so. None-the-less many of these landowners were also interested in renting land from neighbours to plant more trees.

A Typology of Farmers Based in Their Interest in Reforestation

A third example of a typology was that developed by Emtage (2004) to identify the types of households in an uplands area of the Philippines who might be interested in reforestation and the types of reforestation that might interest them. He identified five groups in a survey covering 195 households (Table 10.1). More than 50% of the households in each group fell below the poverty threshold. Despite this, the first three groups, representing 53% of all households, were interested in further reforestation. The first of these he named the *Confident farmers*. These owned most of their land and grew most of their food. They saw no constraints on continuing to plant more trees. The second group were the *Experienced foresters* who were also able to produce most of their own food. Most had formal land leases but were less dependent on their farms for income because they had other off-farm sources of income. Many had previous experience in using state-owned forest resources and appeared to want to use this experience to grow trees on their own land. The *Doubtful foresters* felt technically capable of carrying out further reforestation but were only able to grow a small proportion of their food and were concerned about the lack of support for tree-growing. In contrast to these potentially interested groups, the *Well-off farmers* were less interested in growing more trees. These households were financially better of than most others but owned limited areas of land and were least dependent on their farms for income. They were more concerned with developing their off-farm income generating activities. Finally, the *Disadvantaged households* were also uninterested but, in this case, the reason was that they owned very small areas of land and had the lowest incomes in the community. They simply could not afford to be involved in tree-growing.

A second typology of farmers based on their interest in reforestation was one developed by (Harrison and Herbohn 2005; Herbohn et al. 2005) in the wet tropics of northern Australia. They recognized six classes of landholders (Table 10.1). These differed in the degree to which they depended on their farms for income, the length of time they had been farming and in their (current) interest in tree-planting. The *Traditionalists* were older farmers with larger land holdings who had a long history of managing their properties (6% of all households). They relied heavily on family labour and derived most of their income from revenue generated by the farm. These were least likely to adopt new crops such as trees. The *Experienced/comfortable* group were also older farmers who were debt free (53% of households). In contrast to the first group, these households made extensive used of outside contractors.

They were ambivalent about tree-planting, perhaps because of their age but appeared willing to discuss the matter. *Progressive Second Generation* farmers (7% of households) had, like the first two, comparatively larger landholdings. However, these farmers were more prepared to consider new land uses and were willing to consider planting trees if the perceived risks could be reduced. *High Intensity* farmers (14% of households) had smaller property sizes and used a high proportion of their land for cropping although many also derived some income from off-farm sources. Most had a shorter farming history but had a strong commercial orientation and sought to maximise the area of their land that was under crops. They were interested in tree-planting but, interestingly, this was for mostly non-commercial reasons. Finally, the *Retired Professionals, Conservationists and Absentee* farmers (20% of households) had a much lower commercial focus because they derived most of their income from off-farm sources. Members of this group were interested in tree-planting for commercial and non-commercial reasons. Because many were willing to try new planting systems they could be seen as community innovators.

These typologies show how misleading it is to view farmers as being all the same. But they are also necessarily simplistic and reflect the circumstances in which they were developed; it may be quite misleading to summarize the attitudes and likely behaviour of farmers using a single term. A particular weakness is that several views or attitudes might be represented within a single household or land-owning group. For example, sons may have a different view than their fathers while women's views may differ from those of their menfolk. This divergence of views is likely to be even greater in situations such as those in many Pacific Islands where land is held by clans containing many families or households. And, finally, people can change their minds as community views change meaning that their perceptions of both constraints on tree-growing and opportunities from tree-growing can also change

Nonetheless, several broad generalisations can be drawn from these four examples. Firstly, there are always likely to be some farmers who will not be interested in any form of reforestation because they have too little land and the opportunity costs of tree-growing are too high. A second group are those who do not think tree-growing is a proper thing to do on good farmland such as the Traditionalists of Herbohn et al. (2005). The numbers of such farmers in any community will vary as older, more conservative farmers retire or as land holdings change although it is not always the older members of a community who are the most conservative. Schuren and Snelder (2008) note how it was the older farmers in their study area in the Philippines who were more predisposed to tree-planting possibly because they could see the need for a cash flow in their old age and because it was less demanding of labour.

Secondly, there are always likely to be other more progressive or entrepreneurial farmers who could be interested in reforesting part of their land if they knew more about it. These farmers are usually those having food security and relatively larger areas of land. Such farmers would like information on costs, profitability and risks and will always be comparing the opportunity costs of reforestation with that of other land use practices. Some of these farmers might have chosen to work off-farm and could be looking for a low-cost way of continuing to use their land. Others might see emerging market opportunities and view forestry as a way of using their poorer quality agricultural land. Such farmers might be the first to take up commercial tree-growing.

Thirdly, there are those looking for opportunities to maximize their income such as the Profit maximisers, Confident farmers or the High Intensity farmers of Table 10.1. Some of these may have land that is marginal for crops and would be quite receptive to tree-growing. Some might already have had some experience in growing trees and are confident of being able to use these within their farms to complement existing activities. Others might be those who obtain most of their income from off-farm activity and who are willing to plant trees on their farm because they will not demand much maintenance once established.

Fourthly, there are often likely to be those who are cautiously interested in some kind of tree-planting because it offers an additional way of diversifying their sources of income and reducing their exposure to risks or because they are progressive farmers interested in testing new crops. These farmers may already be practicing some form of agro forestry and might see plantation mixtures as another version of this. Such farmers would need to be convinced that reforestation carried low risks and had low opportunity costs. Finally, there may be some who are interested in reforestation for largely non-commercial reasons. These may include those interested in 'conservation' or those with a sense of stewardship. But they might be households or communities wanting to protect water resources or medicinal plants on which they are dependent.

The types of silviculture that might be attractive to these different types of potential tree-growers are shown in Table 10.2. Most potential growers will have already achieved food security by either growing their own food or having sufficient income to purchase it. Nonetheless, price levels, price changes, perceived risks and farm size are likely to influence decisions (Godoy 1992). Many farmers interested in reforesting are likely to use simple monocultures because this is the easiest form of silviculture but some, such as those classed as Diversifiers,

Table 10.2 The types of silviculture that might be preferred by different types of farmers (plantation designs are those discussed in Chapters 6 and 7)

No	Plantation design	Type of farmer who might be interested
1	Simple monoculture	Potentially of interest to most types of farmer but especially those with larger land holdings. Location may be influential with those near pulpwood facilities opting to use fast-growing species while those in more distant locations may prefer higher-value timber species
2	Trees and NTFP understorey crop	Poorer farmers and those with small land holdings and requiring income diversity and an early cashflow
3	Trees established beneath a temporary cover crop	Those with larger areas of poor quality land wanting to grow high-value timber trees
4	Tree mixtures – different rotation lengths	Those wishing to diversify income sources and some of the more progressive entrepreneurs
5	Tree mixtures – single long rotation	Progressive entrepreneurs, those such as miners restoring degraded land, conservationists

Experienced foresters or Progressive Second Generation, may be also interested in mixtures and especially in tree and NTFP mixtures. Only farmers with large areas of highly degraded lands are likely to bother with silvicultural systems involving temporary cover crops or nurse crops to improve site conditions although these systems do generate an early cash flow. The more entrepreneurial farmers are likely to be more willing to try various types of tree mixtures and those with a conservational bent may be interested in these as well.

Tree planting is still a novel land use option for many farmers although typologies like these help identify the types of farmers likely to be predisposed towards some kind of reforestation on part of their land. But are there other factors also likely to influence a farmer considering reforestation?

Making Reforestation Attractive to Farmers

Some of the ways in which countries might turn around their patterns of forest loss and begin increasing overall forest cover were discussed in Chapter 2. Explanations included that the changes would follow an environmental Kuznets curve, an 'economic development pathway', a 'forest scarcity pathway' or some kind of government mediated transition. These explanations of changes occurring at a national or macro-scale are not very revealing of the causes of changes in land use practices underway at particular sites or farms.

There have been many studies investigating the circumstances under which farmers adopt new methods or land use practices. Pannell (1999) argues that most farmers will come to any radical new innovation with scepticism, uncertainty, prejudices and preconceptions. Unless they are new to farming they will have tried new ideas in the past and concluded that many have not matched the claims made for them. He suggests at least four conditions must be met before most farmers will change their current practices. The first is that they must be aware of the innovation. Being told about it is rarely sufficient. A rather more powerful introduction is to see the innovation in practice. Field demonstrations that show the benefits of the new land use can be especially persuasive especially if these are on the land of a neighbour with similar field conditions. Secondly, there must be a perception that it is feasible to trial on the farmer's own land. A complex new land use involving unknown species may be less attractive than one that uses well-known species or planting methods. Thirdly, the innovation must be feasible but it must also be seen as carrying a low risk and be sufficiently promising to be worth testing, perhaps in a small-scale trial. For example, the presence of an existing market for a new crop would clearly make it more attractive than if the market is unknown. Finally, it should be clear that the new land use will promote the farmer's interests. The financial costs and profitability of the new land use will obviously be important for many farmers but it is clear from the earlier typologies that other factors involving farmer attitudes and behavioural patterns are often at work as well.

Each of these caveats applies to reforestation which can be thought of as simply a particular form of agricultural innovation. However, there are a number of other socio-economic issues that may also make reforestation more or less attractive to farmers. The combined set of factors is shown in Table 10.3. The first of these concerns whether landowners understand the opportunities that reforestation may offer.

Table 10.3 Some factors influencing the attractiveness of tree-planting to farmers

Factor	Significance
Perception of the opportunities provided by reforestation	Interest in tree-planting is likely to increase as natural forests able to supply goods or services decrease or become more distant
Regulatory environment	Reforestation is less likely in the absence of some form of secure long-term access to land as well as ownership of any trees that are planted. Perceptions that governments may restrict the right to harvest planted trees will be a major dis-incentive to planting. Changing or inconsistent policies only accentuate the problem
Farm area and food security	Reforestation with timber trees is more like when farmers have enough land to ensure food security; those with only small farms may be unable to afford to use land for plantations unless they have off-farm incomes
Land characteristics	Reforestation may be more attractive on lands with steep slopes, less fertile soils or with more stressful climates because these are less attractive for crops
Perceived financial profitability	Tree plantations are more attractive if they are seen to be financially rewarding; this will necessitate growers having access to market knowledge and financial advice. Episodic cashflows may be unattractive to some landowners while long-term out-grower contacts may be attractive to others
Available financial resources	Reforestation is initially expensive; some farmers may need access to cheap loans to establish plantations
Labour available	Tree crops may be more attractive when farm labour is limited since trees require little maintenance once established
Availability of off-farm income	Farmers able to earn off-farm income to purchase food may be more inclined to reforest some or all of their farmland
Silvicultural technologies are understood	Growing timber trees may be seen as a novel and risky land use. Technical advice can help reduce risks and improve the likelihood farmers will benefit although it is always difficult to predict growth rates and specify the productivity of farm plantations
Plantations can be protected	Trees can be lost by theft, fires, grazing or diseases and the risks of these are heightened because of the length of most rotations. Individual farmers will need support from governments and the community to reduce such risks
Attitudes of neighbours	Neighbours can have positive and negative influences. Innovative neighbours can provide examples to be copied but conservative neighbours can also argue against change and diminish the propensity of innovators to take on seemingly risky new land uses

Tree-planting may not seem to be worth carrying out in landscapes where there are many patches of residual forest but may become increasingly attractive, especially to the more progressive and entrepreneurial types of farmers, as these disappear and market prices for timber and other goods increase (Arnold and Dewees 1995; Gilmour and Fisher 1991; Gilmour et al. 1990). A second factor concerns the regulatory environment. Farmers without secure rights are less likely to undertake a long-term venture like plantation establishment because they have no way of ensuring they will benefit from doing so (but recall the exceptions discussed previously in Chapter 3). Likewise limitations on harvesting or selling trees grown on farmers will be a major dis-incentive.

A third group of factors concerns the amount and type of land they have. This influences whether the household has food security and whether they can afford to use some of their land for non-food crops. Farmers with large amounts of land or with large areas of unproductive or relatively inaccessible cleared land are more likely to engage in tree-planting on at least part of this land than those with only small landholdings. Interestingly, Sikor (2001), Predo (2003) and Emtage and Suh (2004) all found farmers with some plots of land distant from their houses or roads were more likely to use these for plantations than land close to their homes or roads. In each of these cases the issue is the opportunity cost of reforestation. What are the agricultural opportunities foregone by planting trees? Might there be some advantages in diversifying income sources especially on land that might be otherwise hard to use?

A fourth group of factors involve financial considerations including the likely profitability of tree-growing, the household's financial resources, the labour available and the extent of off-farm income. These are likely to be crucial for many landholders (Godoy 1992; Predo 2003; Shively 1998). Few farmers will carry out any significant reforestation unless it is seen to be financially attractive and they will only be able to judge this if they understand the markets for forest products (and ecosystem services) and have sound financial advice. The attractiveness of reforestation may be increased by improvements in the price of forest products or services especially if these are increasing *relative* to the market prices of alternative crops. This may happen, virtually overnight, when new roads are built making it easier to transport logs to market. But past disappointments can be very influential. Landholders in Fiji were told that planting *Pinus caribaea* would be like having 'green gold'. In fact few made much money from growing pines and this may be the reason some are now reluctant to start growing *Swietenia macrophylla* even though recent logging in government-owned plantations shows there is a good market for this species (Sai Bulai, personal communication, 2009).

Farmers with limited financial resources and many mouths to feed will find an infrequent cashflow especially unattractive feature of tree-growing. Financial assistance in the form of cheap loans with long payback periods may trigger more planting in some communities although this may be harder to arrange if tenure is uncertain or is shared with a large communal group. This disadvantage might also be overcome using some of the techniques described earlier such as using mixtures or growing annual cashcrops in the plantation understorey. But, trees might also be attractive to such farmers because they could represent a form of long-term savings in locations where banks are absent (Chambers and Leach 1989).

Lastly, the financial attractiveness of tree-growing will depend on the availability of household labour. Once planting is done the maintenance costs of tree growing are usually low. This might make reforestation more financially attractive because it means either a farmer can spend more time working at the most productive parts of their farm or finding off-farm employment and using all their land for tree growing. It might also make reforestation an attractive option for older farmers who are less able to engage in heavy labour.

The final group of factors influencing the attractiveness of tree-growing to farmers are several that are perhaps best linked by the term 'risk'. Landowners are more likely to undertake tree-planting if they understand the technology involved but may be more hesitant if they do not. Their perception of risk is reduced if they receive information from extension workers or community associations or have had some previous tree-growing experience (Emtage and Suh 2004; Pattanyak et al. 2003). Those with previous experience would have a greater understanding of the opportunities provided as well as a greater confidence in their ability to establish and manage their new plantations. There is also risk associated with events such as fire, grazing, theft or diseases and the magnitude of these risks depend on both the ecological as well as the social circumstances in which landowners find themselves. The attitude of neighbours can affect both factors since they may help transmit knowledge and encouragement to sceptics such as the Traditionalist of Herbohn et al. (2005) in Table 10.1 (or do the opposite). But cooperative neigbours can also help reduce the risks of events such as fires, grazing or theft.

These various factors will influence the attitudes and behaviour of some farmers more than others. For example, those described in the previous summary of farmer typologies as being interested in reforestation if they knew more about it will probably be concerned with the perception of opportunities presented by tree-growing, the regulatory environment and the financial opportunities. Those referred to previously as profit maximisers will obviously be especially interested in financial issues while those classed as diversifiers are likely to be more interested in any risks they might face and the types of silvicultural techniques they might use to minimize such risks.

In short, those wishing to promote tree growing amongst landholders need to be sensitive to their different circumstances and attitudes as well as to the variety of concerns these landholders will have about commercial tree-growing. Many, of course, will have already experience in growing trees for largely subsistence purposes but, for others, growing trees as cash crops may be quite different to anything they have done before.

The Transition Away from Traditional Forms of Silviculture

Traditional forms of tree-growing vary a great deal across the Asia-Pacific region. These differ in the numbers of trees planted by each household and in the times over which the trees have been retained before being felled. Some farmers have simply inter-planted trees with other crops or planted a few scattered fruit and nut trees around their houses. In these cases some trees might be felled for fuel after only a short period but fruit trees could be retained for many years. Some farmers have planted scattered trees to improve soil fertility in agricultural fallows as well as to provide fuelwood and building material. The nitrogen-fixer *Casuarina oligdon* has been used in this way in the highlands of Papua New Guinea (Bourke 1997) while Cairns (2007) has described a variety of other systems that have used *Leucaena leucocephala, Alnus nepalensis, Falcataria moluccana, Sesbania* spp. *and Erythrina* spp. Trees in these situations are often felled after 10 years or less. With the advent of cash crops like coffee or tea many farmers have used these same species to provide shade or cover.

Small woodlots or forests have also been planted in traditional communities only weakly connected with the cash economy. These include the species-rich agroforests (discussed in Chapter 5) as well as monocultural plantings of well-known species such as *Melia azadirachta, Manglietia glauca* and species of *Styrax, Santalum* and *Cinnamomum.* Such woodlots have mostly been small and grown for local purposes although some also produced NTFPs for trade and timber trees were sold if the opportunity arose. The silvicultural techniques used by farmers were often relatively simple but these could evolve over time as more became known about a particular species. An especially interesting example from southern China is given in Box 10.1 where the methods being used to manage plantations depended on the length of time farmers had lived in the area and on their level of silvicultural knowledge.

But changes in silvicultural practices are beginning to take place as deforestation has increased and farmers have found themselves having to use more marginal agricultural lands. Pasicolan et al. (1997) describe a number of examples from the Philippines of the way tree planting has become a more prominent land use activity when migrant farmers moved in to farm abandoned lands. In the absence of technical advice these households used well-known tree species they believed would be useful for subsistence or commercial purposes. These included *Leuceana leucocephala, Gliricidi sepium, Gmelina arborea, Pterocarpus indicus* and *Sweitenia macrophylla.* All of these species are already relatively well-known in the Philippines and can be easily sold. There appear to have been two main triggers for reforestation in these cases. One was a perception that there was a good local timber market while the second was the belief among farmers that reforestation would help them acquire some form of land tenure. Walters (2004) also observed the same pattern of so-called 'spontaneous reforestation' in deforested mangrove areas in the Philippines. Again this was driven by the existence a good market for poles to construct fish weirs and a perception that tree-planting would confer land ownership.

In the legally uncertain situations in which these farmers found themselves, reforestation appear to have been started by the more progressive or entrepreneurial individuals in the community. Their enthusiasm carried the rest of the community with them and provided a demonstration of the benefits that others could not ignore. Similar examples of 'spontaneous' forms of reforestation using a variety of species have been reported from other parts of the world where natural forest resources have declined (Fairhead and Leach 1996; Gilmour and Fisher 1991; Holmgren et al. 1994).

Box 10.1 The Evolution of Silvicultural Knowledge in South Western China

Much silvicultural knowledge is acquired through a long period of trial-and-error.

An illustration of this comes from a study carried out on the way traditional communities have grown the important sub-tropical timber tree *Cunninghamia lanceolata* in south western China. This species is currently used in state owned plantations in the region. It is grown in monocultures and there is usually a severe productivity decline in second and third rotations. Many local communities also grow *Cunninghamia* but their approaches differ from those used by government foresters (Chandler 1994, quoted by Menzies and Tapp 2007). But there are also differences between different community groups. These appear to depend on the length of time these groups have been in the area.

The most recent immigrants in the study area are the Ye people. They are of Han lineage and have now been present for five generations or about 125 years. These people grow *Cunninghamia lanceolata* as monocultures using short (<40 year) rotations in much the same manner as the state forestry department. A second group, the Wu, belong to another Han lineage but have lived in the area for 30 generations or 800 years. In their case they use three 35 year rotations of *Cunninghamia lanceolata* but then have a 50 year fallow period during which a which forests containing pines, firs and mixed broad-leaved species usually develop. They then repeat the cycle. This has presumably evolved as a means of avoiding the productivity decline that the more recent Ye migrants are yet to experience. A third traditional group are not Han but belong to the She ethnic minority lineage (Hmong/Miao-Yao language) and are the descendents of the earliest settlers in these mountain areas. They grow *Cunninghamia* in mixture with *Cryptomeria fortunei* and other species on long rotations and saw no need to have a fallow to avoid a productivity decline. In their view, both sets of Han 'newcomers' are using the wrong methods to grow *Cunninghamia*.

> The Wu people have that problem (the need to rotate *Cunninghamia* off the growing site) because they do not understand the soil. The soil is like a person. Turnips are good to eat and the Wu like to eat a lot of turnips. But if they eat only turnips, then very soon they become sick. Pork is also good to eat but if we eat only pork, then very soon we also become sick. If we eat turnips and pork and rice and vegetables together we do not become sick, our health is good and we are happy. The Wu soil is not happy. The Wu give it only *(Cunninghamia)*. They do not give it chumu or dumu (both of the latter names refer to several genera of the Fagaceae) or other trees. Only (*Cunninghamia*) and the soil becomes sick. (*Cunninghamia)* buys a lot of money. The Wu want only money, so the Wu soil becomes sick. (Chandler 1994, quoted by Menzies and Tapp 2007)

The case study illustrat.es how silvicultural practices can evolve as experience accumulates and is shared within a community. But it also demonstrates how difficult it can sometimes be to share knowledge between different communities, especially when that knowledge comes from people who are seen as being somewhat 'backward'.

Reforestation Following Government Assistance

Cases such as these of farmers reforesting marginal or degraded land prompt the question – why is this not taking place more widely? Why has the area of 'degraded' land continued to increase across the region without more people initiating tree planting? Some of the possible reasons were outlined in Table 10.3 and many of these constraints will be overcome as experience grows and the profitability of tree growing makes it a more acceptable on-farm land use. But others will only be overcome by some form of external assistance or by incentives to undertake reforestation.

Governments have usually been the primary source of such assistance motivated by a desire to encourage more sedentary forms of agriculture or to create a new timber resource to replace that lost through deforestation. Governments have taken a variety of approaches although it has often been to 'educate' people about the new technologies and to provide planting material. Most have assumed that there were no other impediments or disincentives and that the regulatory and policy environment was already conducive for tree planting. Information has been communicated by extension teams, radio and TV broadcasts, community or village meetings and meetings of farmer groups, women's groups, schools or religious groups.

The source of the advice provided by governments has often been from research undertaken by the government's own forestry agencies. Much of this research was originally done on the assumption that reforestation would be carried out by the government itself rather than smallholders and that it would take the form of large industrial plantations. The research usually involved species screening trials and studies of how to raise these species in nurseries. Less commonly it may have included studies of growth rates and site preferences of the favoured species.

This sort of information is important because it can take some of the risk out of tree growing (in much the same way that advice from government agricultural research stations do for field crops). On the other hand, not all of this silvicultural research or advice has been always relevant to the needs of many smallholders since their circumstances are different to those of large plantation managers. They may need to target different markets and so need different species. Likewise, they are likely to have more flexible harvesting schedules and so be less concerned about precise rotation lengths than industrial growers. And, lastly, many are likely to be more concerned about risk management than maximising production. Each of these differences generates different species choices and management systems. Some smallholders may be happy to plant the species used for pulpwood or sawlogs used in government plantations and adopt the same silvicultural techniques but others may find these prescriptions at odds with their economic situation.

There is also a danger that government efforts to promote reforestation can be compromised if the way it markets forest produce from its own plantations is not done with care. This is because these plantations, if large, can dominate local timber markets and set log prices. This will help small producers if the price reflects the true costs of production but may disadvantage these growers if the government decides it wants to manipulate the price of timber for other reasons such as reducing the cost of building materials to consumers in urban areas.

The success of government efforts to foster smallholder plantation development has varied across the region. In some countries there is a large and active smallholder plantation program while in others it is much less active because timber tree growing cannot compete with alternative land uses. Some of these national experiences are outlined below.

Smallholder Reforestation in Vietnam

One of the most ambitious examples of a government program designed to foster reforestation by smallholders is that currently underway in Vietnam. This is known as the Five Million Hectare Reforestation Program (5MHRP) and it has been given a high priority by the national and provincial governments. It follows several earlier reforestation projects in Vietnam but, unlike these, relies heavily on reforestation being carried out by smallholders to whom secure land access and use rights has been granted in order to facilitate the process (De Jong et al. 2006; MARD 2001; McElwee 2009). Previous reforestation programs in Vietnam were done entirely through state agencies. They relied heavily on pines, eucalypts and *Acacia* species. Large areas of new forests were planted but the program was rather less successful in terms of the quality of the plantations established, the benefits received by rural people or the value of the environmental outcomes. The new program explicitly seeks to benefit smallholders as well as use a wider range of species and improve environmental outcomes.

Of the target area of five million hectares, three million hectares are to be reforested for production purposes (including two million hectares of timber trees and one million hectares of trees such as rubber and fruit trees) and two million hectares are for protection forests. Of these protection forests, one million hectares will be produced from natural regeneration and the remainder will be planted in critical areas such as degraded watersheds. By 2003, 5 years after the program commenced, around two million hectares had been reforested across the country (de Jong et al. 2006).

A critical element of the program is that smallholders are being allocated land for reforestation as well as for agriculture. This land is provided on long leases with ownership retained by the state. Most households are allocated less than 5 ha of land for reforestation although farmers with poorer quality land in more isolated locations have up to 30 ha. There are also a series of financial incentives to encourage farmers to participate in the program including payments for protecting natural regeneration, grants for establishing protection forests and cheap loans and tax concessions for planting production forests. These loans are sometimes difficult for smallholders to access and may not necessarily be taken up (Nawir et al. 2007a). Planting material is available from government forestry agencies while technical advice from extension teams and in the form of written material, radio and TV broadcasts is also available.

Several years after the program commenced a number of problems have become evident (MARD 2001, de Jong et al. 2006, McElwee 2009). Firstly, the program has been heavily top-down and most farmers have had little opportunity to take part in the planning or design of plantation programs although they are expected to do most of the work and carry much of the risk. In some cases land allocated for reforestation was not unused wasteland but was being used for gathering NTFPs and other purposes by some of the poorer members of the community. As a consequence of the program they have lost access to these resources (McElwee 2009). Secondly, the species used in reforestation have not always reflected the objectives of the program or farmer's interests but have relied rather too heavily on the narrow range of species identified in earlier plantation schemes. This is because there is still too known about the silviculture or site preferences of other species and because it has been difficult to get planting stock of these. Thirdly, many farmers took part in the program with a very poor knowledge about markets or market prices for the timbers they were growing. As discussed in Chapter 9, there is a very good market for sawlogs in many parts of Vietnam but most farmers are unaware of this or have been overly impressed by the fast growth of some of the recommended exotic species. This has not mattered for farmers able to sell pulpwood to local markets but has sometimes been problematic for those in more distant locations.

Finally, the forest land allocation process has been slow and there have often been unclear boundaries between agricultural and forest land. Since most farmers understandably give a higher priority to food production or well-known agricultural cash crops this has sometimes disadvantaged the reforestation program and made the extension program irrelevant. Sikor (2001) described a situation in Son La Province where people had taken over land ahead of the formal land allocation program. Under these circumstances the new, formalised arrangements represented a backward move and many local officials were reluctant to enforce them. Some of the earliest areas allocated to reforestation were along the main rural roads and cash payments for planting these areas were popular. People were happy to take the money and plant the recommended trees but they also planted crops with these trees as a way of retaining cropping land for a few more years. Unsurprisingly, most of the trees died. Meanwhile changes in agricultural technologies and markets enabled villagers to switch from hill rice to wet rice cultivation and from cassava to corn. This resulted in an intensification of agriculture in valleys and much of the agriculture being practiced on the hill areas ceased. This allowed natural tree regeneration to take place in these areas. The net effect was that overall forest cover began to increase even though most of the actual tree plantings failed.

Despite these difficulties the program has been responsible for a substantial increase in forest cover rising from around 25% in the early 1980s (Nguyen and Gilmour 2000) to 40% in 2005 (FAO 2007). It is also clear that many farmers now accept timber tree plantations as being a legitimate rural land use activity. Small woodlots are common in many parts of the country and appear to be contributing to household incomes (Fig. 10.2). Natural regeneration is being protected and there are an increasing number of small private commercial nurseries that sell seedlings to new growers. While many of these produce only eucalypts or *Acacia* some have rather more sophisticated operations suggesting that a more complex range of reforestation options will eventually develop (Fig. 9.5). An illustration of how these national programs affected farmers in a particular community is given below.

Fig. 10.2 Most farm plantations are small and cover only a few hectares

Case study: Promoting tree-growing following land redistribution in the uplands of Bac Kan province, northern Vietnam

Villagers in the Thanh Mai commune live in the mountain area of Bac Kan province in northern Vietnam. Most villagers belong to the Tay or Dao ethnic groups. The area was heavily forested until about 50 years ago but forest cover has since decreased by over 60% because of shifting cultivation, logging and agricultural clearing. Until recently the agricultural system was one based on paddy rice in the valleys with some shifting cultivation on the hills. These largely rice-based systems are now undergoing a dramatic shift to a series of more complex and diversified agro forestry systems (Fatoux et al. 2002).

The changes have been driven by changes in land tenure. Communes were developed in 1960 resulting in private land holdings being amalgamated and collectivised systems of agriculture being developed. This changed in 1982 when some de-collectivisation took place; paddy fields were re-distributed to families in proportion to family size and after 1989 families began to also regain control of their ancestral lands. A further change took place in the 1990s when state-owned 'forest' land in hill areas was also redistributed to individual families. The reason for this latter change was that the Government wanted to phase out shifting cultivation (which is still being practiced) and increase forest cover on these largely degraded hill areas.

As a result of these tenurial changes families have had to change agricultural practices and develop new systems to compensate for the shrinking amounts of agricultural land (largely caused by a reduction in access to hill areas for cropping). The capacity of families to change has been determined by the areas and types of land they now own and by their financial resources. Families with large numbers of dependents (i.e. children or elderly relatives) are usually less well-off and less able to change than those without.

Tree-planting was promoted through a series of national government reforestation initiatives involving the mandatory plantings of certain preferred timber species and some fruit trees. Households have differed in the way they have participated. Families with larger areas of paddy rice per worker have generally been less interested in participating in this than those with smaller areas. Many of the former have preferred to use their financial resources to buy more paddy land. On the other hand, others have been more willing to engage in tree-planting as a way of diversifying their incomes especially farmers with only sufficient water to plant one rice crop each year. The types of reforestation practiced were relatively simple and included monocultural plantations of native fast-growing timber species such as *Manglietia glauca* and various fruit trees.

The level of silvicultural knowledge held by most farmers appears to be low. What knowledge they have seems to have filtered down from previous government reforestation programs or, occasionally, from a farmer participation in a training course. Few farmers have much knowledge about timber markets or what timber species buyers prefer. Interestingly, neighbours tend not to share this knowledge but to keep it for themselves, presumably to obtain some kind of an economic advantage. Fatoux et al. (2002) believe there is considerable scope to enhance farmers' silvicultural knowledge and increase their capacity to market timber perhaps using the existing Farmers' Association. Incomes have tended to increase as a result of these changes with those having more diversified production systems being better off. This has led Fatoux et al. (2002) to hypothesise that most types of farmers in this area will eventually embrace some kind of tree-growing as a means of diversifying and improving household incomes.

Conclusion: secure land tenure, constraints on further forest clearing and a supportive policy environment help make tree-growing attractive to households. But their degree of interest depends on the household's circumstances and not all farmers will necessarily want to participate.

Smallholder Reforestation in the Philippines

A different relationship between government agencies and prospective woodlot owning smallholders has developed in the Philippines. This has evolved through several stages. The government initially sought to carry out reforestation itself on state-owned land, or required logging companies to do so, as natural forests were cleared. Sometimes external contractors (including rural communities) were paid to plant these trees. Few of these plantations succeeded largely because those doing the planting had no long-term interest in the outcome, especially if their payments were delayed. Eventually these various approaches have been replaced by the present system of Community-Based Forest Management Agreements that involve granting tenure to rural communities and assisting them to reforest the land. The target groups are the poorer upland farmers who would otherwise focus on short-term agricultural crops. The government offers these communities land access and use rights for plantation establishment for up to 25 years with the possibility of an extension for another 25 years. In addition, they are awarded contracts to reforest lands and share the profits from doing so with the government. The program also funds a community organiser to coordinate activities and to provide advice on species and silvicultural techniques to use.

The policy framework underlying the program is complex and is still developing. This has discouraged some households from taking part. For example, Emtage (2004) describes some work programs taking up to 4 years before receiving government approval to provide financial assistance to communities. On the other hand, these new approaches appear to have been more successful in establishing forests than the earlier government plantings. An outline of the way these programs work is given by Harrison et al. (2004), Chokkalingam et al. (2006) and Calderon and Nawir (2006).

The granting of land access and use rights has been an important step. Not only has it made tree planting more attractive but it has probably made farmers more willing to seek out advice. The capacity of the government agencies to offer advice is constrained by a fragile financial base which means support to growers can sometimes be abruptly withdrawn. The quality of the advice also varies. Similar silvicultural prescriptions are often given to landholders irrespective of their particular site conditions. Likewise, only a small number of mostly exotic species are recommended to prospective growers (e.g. *Gmelina arborea, Swietenia macrophylla, Leuceana leucocephala, Pterocarpus indicus*) even though some farmers are very interested in growing native species (Santos et al. 2003). These recommended species grow well in a variety of situations but are not necessarily the most financially profitable and there are already examples of the price of *Gmelina* logs being depressed when large volumes of farm-grown timber suddenly appeared on the market (Santos et al. 2003, Calderon and Nawir 2006).

Although the focus of the current government program is on community reforestation some households clearly prefer to operate alone and this has been a trend in recent years. Individualistic behaviour is probably more likely in cases where the community is made up of recent immigrants and lacks the strong social bonds that are commonly found in more traditional communities. It may also reflect some of the differences in the typologies of farmers described earlier in which more entrepreneurial farmers are reluctant to be tied to more cautious or reluctant households. A crucial element in the promotion of reforestation as a new land use is the way in which information is provided to prospective tree-growers. Again, a local case study illustrates some of the difficulties in doing this.

Case study: Developing methods of providing silvicultural advice to farmers on Leyte Island, the Philippines

Improving the silvicultural knowledge of smallholders requires that suppliers of advice have something useful to communicate and that smallholders are interested in hearing about it. This is sometimes more difficult to manage than it might seem. Baynes (2007) describes experiences gained during a specially designed program that sought to provide up-to-date silvicultural knowledge to smallholders on Leyte Island in the Philippines. The island has been extensively deforested and the local sawmilling industry has collapsed. Some farmers are now interested in planting trees on their land but many are not and prefer to grow other well-established crops such as coconuts. Until recently, little silvicultural advice has been available to farmers in a form they can understand or use.

A program was developed to test ways of dealing with this lack of silvicultural information. The program was extensively advertised across the district but the level of interest amongst farmers was only modest. Those that agreed to participate were initially taken on a 1 day field trip that visited existing successful farm plantations in the district so they could meet the owners and discuss their experiences. Participants were then asked if they wished to continue in the program and most agreed to do so. These people were then taken to collect seed from a stand of *Swietenia macrophylla* and taught how to raise seedlings from these in a nursery. Participants were shown how to plan and establish a plantation using seedling material. Follow-up visits were made to those with existing trees on their farms to show them how to prune and thin these trees.

Most of those who took part in the program went on to raise seedlings and establish plantations even though the sites they chose were not always the most suitable ones to use. Few of the farmers fenced these plantations to exclude cattle and it remains to be seen how many seedlings will survive. Most farmers said they planned to grow more trees in future years.

This apparent success has to be qualified. Most of those taking part in the program had some 'unused' land and may not be representative of the farming community as a whole. Further, many of the participants frequently missed mutually agreed appointments with extension staff and clearly regarded the program as a low priority. This may be due to the fact that tree-growing is still not widely recognized as a legitimate land use activity despite the massive amounts of deforestation that have occurred in the Philippines. Or it may reflect the complexities of government regulations governing tree-planting that make it less attractive than it might be. Gilmour (personal communication, 2009) has also noted that politically powerful absentee landowners can sometimes exert considerable authority over smallholders that discourage them from undertaking long-term activities like tree growing even though they may have received nominal land access and use rights from the government.

Conclusion: the presence of an extension service does not necessarily mean its advice will be sought or acted upon. Ways must be found to make information relevant to farmers needs and it may take time for extension services to become trusted.

Smallholder Reforestation in Indonesia

Indonesia has had a long history of reforestation using plantations. Teak was probably introduced to Java in 200 AD for this purpose and has been grown there since that time (Nawir et al. 2007b). The total plantation area was probably never large and it is only since the 1960s that there has been a substantial national effort to reforest degraded lands. Nawir et al. (2007a) argues that the overall success rates of reforestation have been low and have not kept pace with rates of land degradation in different parts of the country.

Communities have not had a large involvement is these reforestation programs until relatively recently. And, even where they have, the organisation has been mostly top-down so that the community had little role in planning the program or in deciding which species to use. Little attention was paid to traditional or informal land rights, or to ecological conditions. The markets the plantations were supposed to supply were often unclear and the role of the community was often simply one of supplying labour to plant seedlings. Many of these reforestation projects have subsequently led to social disturbances and unrest.

The situation changed after 1998 as decision-making became more decentralised and power moved from Jakarta to the provinces. There are now greater efforts to increase community participation in reforestation on lands inside and outside state forest areas. Instead of reforesting to simply increase forest cover and protect watersheds the emphasis has shifted to empowering communities, securing community access to land and raising environmental awareness. It remains to be seen how successful this change in direction will be.

One successful element of the national reforestation effort has been the Farm Forestry program which began in Java in the 1970s. This program aimed to encourage farmers to reforest areas of community land using teak, *Acacia* and a variety of multi-purpose tree species. Initially farmers were encouraged to simply plant teak trees around the edges of their own land. Later, in the 1980s, the focus moved to planting teak on unoccupied barren land. More recently farmers have switched to planting trees on their own land because of the good timber prices being received. Anyonge and Roshetko (2003) report on a study in Lampung Province where farmers with more land, higher incomes and off-farm jobs were especially interested in planting species producing premium quality timber even though these need longer rotations. Those with more limited incomes and land were more inclined to grow species on shorter rotations.

Several factors have made the Farm Forestry program far more successful than most of the other national reforestation schemes in Indonesia. One was the policy settings. Farmers had secure land access and use rights as well as rights to harvest and sell any trees they grew. There was also strong local support from district officials that ensured local institutions were empowered and the rights and responsibilities of government, the Forest Service and people were clearly defined. Technical advice was provided through a network of demonstration plots and village nurseries were established to produce seedlings. And, unlike the other reforestation programs, the Farm Forestry program improved social cohesion and strengthened tenurial security and the capacity of community members to voice the views about other natural resource management issues. Financial returns have been attractive enough to encourage people to use their own money to reforest their land again after the first trees were harvested. Nawir et al. (2007b) suggest farmers in this program have mostly followed relatively simple and well-tried methods but that few have been prepared to innovate or try new approaches of their own. In that sense, the community may still be overly-dependent on government leadership.

Perhaps one of the key outcomes of the Farm Forestry program is that it has increased community awareness about the opportunities offered by tree planting. An example of this occurring in a highly degraded part of Java is outlined in the Case Study below. In this case the community was able to use the extension services offered to take advantage of favourable markets to reforest some highly degraded wastelands.

Case study: Reforestation of highly degraded farmland in the Sewu Hills, central Java

Deforestation in the Sewu Hills region of central Java (Yogakarta Special region) was virtually complete by the middle of the 1800s. Continued population growth and intensive agriculture over the next 100 years led to severe degradation and by 1950 the region was characterized as being in a deep ecological and social crisis (Nibbering 1997, 1999). The population was poor and subsisted on a diet dominated by cassava. Crop yields were low and a number of famines induced by droughts and rat plagues occurred. Forest resources continued to be important for many people and these were obtained from state-owned forest but eventually this source also disappeared. Few people could afford to devote time or land to plant trees because of the need to grow food and, where individuals tried, organized gangs stole their trees. After the 1950s people were leaving the area and the destruction brought about by deforestation seemed irreparable.

National reforestation programs in the 1970s encouraged tree growing by communities and farmers and have continued since then. Planting became attractive because cropping and grazing on the hillsides had declined and the frequency of fires had also decreased. The reforestation programs were top-down but straightforward; seedlings were distributed and cash incentives were provided. Those farmers still living in the area began planting trees on the hills and restricted cropping to the lower valleys. Though it is not reported, most farmers were probably very similar in economic circumstances and attitudes to reforestation given the lengthy period of progressive degradation they had lived through.

The trees they planted were teak (*Tectona grandis*), mahogany (*Swietenia macrophylla*) and *Acacia auriculiformis* for timber while *Cassia timoriensis* and *Sesbania grandiflora* were planted for firewood and fodder. Although failures occurred, the scale of the program enabled farmers to learn how to match individual species to particular sites and how to incorporate tree-growing into their overall farming systems. The program also provided a critical mass of farmers interested in growing trees and allowed them to become self-sufficient in planting material. This large number of farmers engaged in plantings has meant tree theft has not been the problem it once was.

It was found through experimentation that a tree density (around 200 mature trees per hectare) allows shade-tolerant cassava to be grown with the trees and this has increased the attractiveness of tree-growing. Meanwhile the market price for teak has increased throughout this period and made tree-growing a much better option than simply restarting cropping on the previously degraded hillsides. In short, tree growing in what was a highly degraded landscape has added to cash incomes, diversified household economies and provide households with alternative ways of accumulating capital.

Conclusions: reforestation can be especially attractive when land is available and there is a recognized market for forest goods.

Papua New Guinea

Very little timber tree-planting has been carried out by traditional landowners who control most land in Papua New Guinea. The main exception is the use of *Casuarina oligodon* to enrich grass fallows in parts of the highlands (Bourke 1997). This reflects the fact that forest resources are still widely available and that there are limited local markets for small-sized logs from plantations. At present most farmers clearly regard crops such as coffee or cocoa as being far more attractive than timber trees.

As the logging of natural forests has continued the government of Papua New Guinea has begun to take a different view and has sought to develop state-owned timber plantations as a way of sustaining industries in certain strategic locations and to create a permanent plantation forest estate. These plantations have included local species such as *Araucaria cunninghamii* and *Eucalyptus deglupta* as well as exotic species such as teak and pines. It has been difficult for the government (and private timber companies) to do this because few landowners have been willing to sell their land and negotiations over land leasing can be lengthy. But even when land has been purchased or leased, arguments can subsequently develop over who the rightful owners really were. Complicated social structures and over-lapping land claims sometimes mean there are several groups with historical claims to the same piece of land and a large number of groups can be involved in areas of more than a few hundred hectares. These disputes can sometimes lead to violent confrontations. Following independence, land disputes of this kind have led to the demise of the former government-owned teak plantations near Port Moresby. Throughout the country most clan boundaries remain unmapped making land ownership a significant problem for those seeking to identify large contiguous areas suitable for a plantation. The Case Study that follows describes an unusual approach to solving this problem.

Case study: Joint ventures between land-owners and governments: the Fayantina reforestation project in the Eastern Highlands, Papua New Guinea

The highlands of Papua New Guinea represent a rather different situation from the rest of the country since much of the area has been deforested and there are now large areas of relatively un-used grasslands. Most people live some distance from natural forest and, apart from subsistence use, most timber must be imported from other parts of the country. The former colonial government leased land from the traditional highland owners to establish *Pinus patula and Pinus strobus* plantations to supply the local timber market but this practice ceased with independence in 1975. Since then the new government has been unable to lease more land and no land owner has sought to replicate these plantations on their own land despite evidence that the pine trees grew well and clearly represent an important economic asset. The main reasons for not doing so appear to be that people lacked sufficient funds and felt they did not have the necessary silvicultural knowledge. There was also a persistent risk of inter-group disputes that can lead to new plantations being burned or destroyed.

This situation has led to an interesting joint-venture between land owners and the government. It commenced in 1984 and involves grasslands belonging to 13 customary land-owning clans (Garin 2008). Disagreements between the various clans meant it was impossible to finalise a formal legally binding agreement. Instead, planting went ahead using an informal 'gentlemen's agreement' in which landowners provided the land and the government provided finance to carry out the planting and management. The government also undertook to hire clan members for any work needed. In return, the clans undertook to identify land available for reforestation, resolve land disputes amongst themselves and to protect the plantations from disturbances such as wildfire. This informal arrangement has now persisted for 22 years during which time the plantations established under the agreement have expanded to cover 1,800 ha. The success of the arrangement has led to the original clans offering a further 3,500 ha for reforestation while neighbouring clans are now also seeking to join the program. These other clans collectively own another 50,000 ha.

The success of the partnership between the government and landowners has apparently defused some of the land ownership conflicts of the past. But success has also brought some interesting dilemmas. One of these is the need to find ways of delaying harvesting (and thus income generation for landowners) until there is a large enough timber resource to make it a sustainable and profitable operation and not simply a brief boom-and-bust harvest. A second problem is to devise a new financial structure that allows the program to expand since the government does not have the financial resources to cope with the numbers of landowners now wishing to participate. Perhaps some of the future returns from harvesting the first plantations can be used to do this?

Conclusions: it can be difficult to initiate reforestation in socially complex situations where there has been no prior involvement in community or household forestry. In these circumstances governments can have a crucial role in defusing community tensions and helping to initiate reforestation

Solomon Islands

In recent years the Solomon Islands have relied heavily on natural forests to provide export income. However, the rate of logging has been so high that all natural forests will soon have been logged and export incomes will then plummet. The change will be abrupt when it occurs and there are few alternatives presently available to replace this source of government revenue. Many logged-over areas may eventually recover but most will take some time to do so. Raymond and Wooff (2006) suggest the earliest these natural forests will be available for harvesting again might be 2030. Timber plantations offer a means of accelerating the rate at which a new timber resource might be created. The government had previously established some plantations of teak, *Eucalyptus deglupta* and *Gmelina arborea* but subsequently sold these to private timber companies. As in Papua New Guinea, traditional land owning clans control most land in the country and this has made it difficult for these companies to extend these plantations.

Smallholder plantings represent another alternative and both the government and the private timber companies have been seeking to expand these to increase the overall plantation estate (Raymond and Wooff 2006). Their task has been made easier because of the attractive prices currently being received for logs of higher value species. Even a single container load of teak or *Eucalyptus deglupta* logs can be sold profitably on the international market. One of the timber companies has been able to have their plantation products certified as conforming to the standards of the Forest Stewardship Council which has improved prices even more (V. Vigulu, personal communication, 2008).

The current national strategy has several elements. It relies largely on using well-know species with high timber values because these attract the best prices. Teak makes up 80% of plantings and *Swietenia macrophylla* makes up another 10%. The indigenous *Pterocarpus indicus* (rosewood) is also grown. Emphasis is given to working with individual households or families rather than with communities (except for several communities with very strong leadership where community forestry has been successful). No payments of any kind are made and participants must fund their own plantations. This is because past experience in the Solomon Islands has shown that financial assistance encourages people to plant trees for short term gain only and that they cease doing so once the government's money stops. On the other hand, self-funded plantation owners have demonstrated their commitment to establishing and maintaining plantations for the longer term. The program has also attempted to make planting a recurrent annual activity since experience has shown that those who establish high quality plantations gain considerable satisfaction from doing so. But some care is needed to ensure people only plant the number of trees they can tend. Those who over-extend themselves and fail rarely plant again and their failure can have a negative effect on others in the village. The program has been in operation for several years and has now established 6,500 ha of plantation (Gua 2008; Raymond and Wooff 2006). About 10% of households in the country are involved. Evidence to date suggests the program has been remarkably successful although it is also likely that some of the plantings at more isolated locations will be too small and too expensive to market to be economically viable.

After earlier attempts to persuade logging companies to carry out reforestation has been dropped since they have neither the skills nor the interest in doing so. Rather, the emphasis now is on getting them to bequeath a road network that will assist landowners to extract their future logs and transport them to market. But if landowners are to engage in reforestation they need an active extension service to provide advice.

Case Study: Providing silvicultural knowledge at low cost: village based forest extension officers in the Solomon Islands

A national village forestry program requires a good extension service. But recent political events severely damaged the national economy (including an armed insurgency in the late 1990s that effectively toppled the government). It also limited the government's financial and human resources and the capacity of the national forest service to do very much in the field. In the absence of sufficient field staff the government has sought to foster plantation reforestation using village-based Forest Extension Officers. Government foresters have identified potentially interested persons, informed them of market opportunities and provided them with high-quality seed (mostly teak) for which they had to make a modest payment (to ensure the seed were used and not wasted). They were also given a short course in nursery techniques and other simple silvicultural practices and provided with some basic tools such as pruning saws. These extension officers have subsequently gone back to their villages and promoted tree growing amongst their communities. At the end of each year the success of each officer is evaluated to ensure the right people are being supported. Those not taking the program serious are excluded.

The program has been a significant success (Gua 2008). One of the reasons is that the Forest Extension Officers can speak the local language. They also know who is likely to be interested in reforestation and so be receptive to advice and assistance. Now that reforestation is underway the program is likely to move to a new phase in which quality rather than quantity is promoted. Most trees are expected to be grown on a 20–25 year rotation and will need some thinning and pruning to produce high quality logs. It is expected the village-based extension officers will be used to transmit advice on these procedures.

Conclusions: the most trusted sources of information are likely to be knowledgeable people from within the community. Programs that identify and then actively support village-based Forest Extension Officers are a good way of promoting small-scale forestry.

Australia

Most of the cases above have occurred in situations where there was a distinct market acting as an incentive for growers (even though not all farmers were necessarily aware of these market opportunities). But sometimes market situations are less clear and governments may see their role as one of building a resource in order to create conditions in which a new forest products market might develop. An example of this occurred in Queensland in the northern tropics of Australia. Logging in the region's natural rainforests ceased in 1988 when these forests were placed within the national Protected Area system. This appeared to offer a new market niche for plantations of high value timber species to replace those previously supplied by the natural forests.

The national and state governments embarked on a large reforestation scheme using high-value rainforest species in order to provide employment for timber workers who had lost their jobs when logging ceased and to create a new timber resource (Vize et al. 2005). Timber plantations had previously been established in the area but most of these involved the exotic *Pinus caribaea*. The market price of logs from these was not high and few local landowners thought these were an attractive land use. However a number of landowners were attracted to the new program especially because it used high-value native species and was heavily subsidised. The government prepared the sites, provided the seedlings and planted these. Landowners simply had to provide land. Most plantings on private land were only a few hectares in size and the government's hope was that landowners would replicate these. The typology described by Herbohn et al. (2005) was seen as a way of identifying the types and proportions of landowners most likely to be interested in future reforestation. To this end the government and various NGOs provided information, organised field days and helped establish farm forestry groups.

But, in the meantime, the market signal diminished. Although many of the former rainforest timber species previously supplied by natural forests had commanded a high price the market weakened in the absence of continued log sales. This information flowed back to prospective growers who saw little incentive to continue timber plantation establishment for an uncertain market. This created a paradox: without a timber resource there was no market but without a strong market signal it was impossible to persuade growers create the resource. The government could have continued to support the scheme until a sufficient resource was created but decided this would be too costly and terminated the program. Most sawn timber used in the region is now imported from elsewhere although some farmers such as the Absentees and Progressive Second Generation farmers of the Herbohn et al. (2005) typology (Table 10.1) have continued planting.

In this case government assistance was not enough to create a resource big enough to sustain a new timber market although it could be argued that it did help break down some of the anti-tree views held by many families in the rural community who had cleared some of the original forests to establish their farms. And the program also created a much greater knowledge of the silviculture of many of these high value tree species and how to restore multi-species forests on degraded sites for conservation purposes. It is ironic that, since then, there has been increasing interest by private timber companies in reforesting some of these deforested areas. But, rather than using local high value timbers they are mostly planting teak and African mahogany (*Khaya* spp.) for which there are well established external markets.

Conclusions: a declining market price is a major disincentive to tree-planting.

Lao PDR

Tree growing can be a risky investment for some farmers and failure may have serious financial consequences. This was the case in a major reforestation project in Lao PDR funded by the Asian Development Bank (Barney 2008; Lang 2002). The objective of the project was to reforest so-called degraded lands and create an industrial timber resource. The project sought to establish 9,600 ha of eucalypt plantation by loaning funds to large companies (who would plant 7,000 ha) with the remainder being established by small farmers. In fact, few companies expressed any interest in being involved so the emphasis was switched to include more smallholders. Eventually 2,500 households became involved. The project commenced in 1994 (although the actual commencement was delayed until 1997 because of institutional problems) and effectively ceased in 2003. By this time the Bank claimed 7,800 ha had been established and the project had been successful. It is unlikely, in fact, that this area had been achieved. In any case, it was evident that most plantations had very low growth rates or had been abandoned. Shortly after the Bank itself agreed the program had indeed been a failure.

There are a variety of reasons for this outcome (Barney 2008; Lang 2002). A key reason is that vulnerable farmers were persuaded to take on expensive loans for a land use activity about which they knew very little. Most of the farmers were comparatively poor. They grew a single crop on non-irrigated rice each year and many had to find work off-farm to purchase additional food. When it became clear that institutional investors were not going to participate there was pressure on project staff to involve more of these smallholders in order to meet planting targets. These people were persuaded they would benefit from taking out a loan at 7% interest over an 8 year period. None of the farmers had grown trees as a commercial crop and none knew of the likely market or market prices they could expect to receive.

Another problem was the poor supervision of the project. As a result, some of the loan funds were spent on other priorities such as improving housing or buying bicycles. Sometimes fertiliser supplied to farmers was diverted to agricultural crops. Similarly, there was surprisingly little silvicultural assistance from government extension officers. Farmers received seedlings from government nurseries but these often arrived late in the planting season with some coming in September or even October just before the start of the 6-month dry season. Little advice was provided about the need for weed control or fire exclusion and there was no attempt to overcome problems when it became clear that the plantations were not performing as expected.

The scale of the failure became clear in 2005–2006 when attempts were made to begin collecting payments on interest. By then it was evident that most farmers were left with failed plantations and with debts they had no way of repaying. It was also evident that there had been an over-emphasis on meeting planting targets and too little understanding of the vulnerability of farmers to a new land use activity.

Conclusion: an attempt to burden smallholders with a poorly conceived reforestation project failed because they had neither the capacity nor the interest in being involved.

In summary, these various examples and case studies show governments can do much to help smallholders take advantage of the opportunities provided by tree-growing. This assistance can especially help those who might be regarded as 'progressive' landholders looking to take advantage of commercial opportunities or those classified earlier as 'diversifiers' who want to diversify their income sources. As well as facilitating land tenure and providing technical information governments can assist by providing an enabling environment that helps overcome some of the social and economic constraints limiting reforestation. But, it must also be said, that some government actions have had an entirely opposite effect. This is where inappropriate advice has been given concerning species to use or about future markets. This can have serious consequences, especially for poor farmers. The most satisfactory outcomes appear to have occurred when governments have adopted a more participatory approach rather than a simple top-down relationship with farmers.

Reforestation with Assistance from Private Timber Companies

A second kind of partnership is that formed between landowners and a private company. Companies differ from government because they are driven primarily by financial considerations and have little interest in objectives such as improving the national forest cover, increasing employment or improving regional environmental outcomes. When companies form relationships with communities or individual smallholders they largely do so to benefit themselves. The same is true, of course, for the individual smallholders with whom they form these partnerships.

A variety of arrangements have developed including joint ventures, out-growers schemes and crop-sharing schemes (Mayers and Vermeulen 2002; Nawir et al. 2003; Race et al. 2009). In some cases the relationship is relatively informal with the company simply providing advice and a market for growers once their timber is ready for harvest. Such arrangements may be attractive to some growers because they retain overall control although most will still have only one buyer for their trees when these are harvested. On the other hand, they may not be as attractive to companies since they may not know how much timber is being grown or when it will be available. This may not matter so much if the resources controlled by the private landowners are small relative to those controlled by the company.

An informal relationship of this kind has developed at Open Bay on the island of New Britain in Papua New Guinea where the Open Bay Timber Company grows *Eucalyptus deglupta* in its own plantations on a 15 year rotation for plywood

(Yendkoa 2008). These plantations will eventually cover an area of 20,000 ha. Nearby landowners have asked if they could also grow trees for sale to the company and it has agreed. It provides seed and advice and helps clear sites for woodlots but is not involved with plantation management (although the Company does provide the general community with health services and other social infrastructure). Around 250 growers had joined the scheme by 2008 and each is growing 1 or 2 ha of trees. Interestingly, these people could have joined a nearby oil palm scheme but chose to grow timber trees because they believe timber plantations are less socially disruptive to the community than the oil palm plantations. The scale of the contribution made by these household plantations is comparatively small but it does benefit the Company by improving its relationships with the surrounding community.

Most other relationships between companies and household or community growers are rather more formal. In fact Mayers and Vermeulen (2002) argue that there are advantages to both parties if growers can become registered as a separate company since they then share in the mutual rights and controls provided by corporate law.

These agreements usually define which land is available for reforestation and which is to be left for the community to use for other purposes such as for food crops. The agreements usually oblige the company to provide expertise, funds, seedlings and other resources such as fertiliser while obliging the growers to follow certain silvicultural prescriptions and sell the timber to the company at an agreed time for an agreed price. Both parties agree on their separate responsibilities and on how benefits will be shared. Ideally there should be regular meetings to increase the transparency of the arrangements between the company and the growers.

These types of partnerships have advantages for both parties. The grower is able to use land for a purpose they may not otherwise have been able to entertain (because they did not have the knowledge or financial resources to do so). The agreement may also consolidate their claims to ownership of this land. Further, much of the risk in growing a long-term crop like trees is reduced at an early stage and they have some certainty over the price they will eventually get. For its part, the company acquires access to land it would not otherwise get and the possibility of future land use disputes is reduced. It also has self-interested partners able to reduce risks of timber thefts, fires and other disturbance to the plantation resource. Agreements often require companies to provide roads, training and other rural infrastructure. Companies usually bear the costs of harvesting and transport (and may sub-contract this to other community members).

Critics of these company-community partnerships point to the weak negotiating position that smallholders or communities have in their relationships with companies. This can foster a sense of distrust by communities and smallholders (Nawir et al. 2003). The problem may be most acute when growers have no alternative buyers of their trees or alternative uses of their land. However, the improvement of road networks, the arrival of timber merchants or the spread of crops such as rubber or oil palm may change these relationships so it is usually in the long-tem interests of the company to establish equitable partnerships that can withstand changes in crop prices and opportunity costs.

A major international review of partnerships between companies and communities carried out by Mayers and Vermeulen (2002) was unable to find an example of an equitable, efficient and sustainable relation that had persisted over a long time period. However they did conclude there were a number of promising relationships now underway that were delivering benefits to both parties. Such partnerships are encouraged by a favourable government policy environment (e.g. by insisting companies pay attention to the needs of local communities such as their need for food security even when their land rights are not entirely formal) and by having short-term agreements that allow smallholders to leave once a contract is complete. Partnerships can be upset by high transaction costs, misunderstanding and disputes within communities over negotiating targets or the sharing of benefits. Examples of the complexities of some of these relationships are described in a series of case studies described by Calderon and Nawir (2006) in the Philippines and by Nawir et al. (2003), Maturana et al. (2005) and Race et al. (2009) in Indonesia.

Reforestation with Assistance from Non Government Organisations

A third type of partnership is that between private landowners and NGOs (including overseas development aid projects). Some of these have been very successful, especially those where long-term relationships develop (Gilmour and Fisher 1991; Murray and Bannister 2004). Unlike the partnerships involving governments or private companies which have largely on focussed timber production, NGOs have often had a more diverse range of objectives including reforestation to promote conservation objectives as well as timber production and general development. Because of this, a much wider variety of silvicultural approaches have been used including mixed species plantings (Elliott et al. 2006; Goltenboth and Hutter 2004; Tucker et al. 2004). As well providing material or financial assistance NGOs have sometimes also provided legal assistance or acted as advocates to help landowners protect their land tenure or prevent illegal logging in regenerating natural forests.

Many of the partnerships initiated by NGOs are relatively small but, because of this, they have tended to be more participatory and less top-down than those of government agencies or industrial companies. This has the advantage that they are more likely to come to grips with issues of most concern to landholders and can lead to experimentation and testing that is unlikely to occur in the other two partnerships.

Sayer and Campbell (2004) suggest that, on the whole, they have often been more successful and influential than many government projects. On the other hand, small scale projects can be difficult to scale–up and apply to wider areas and sometimes fade once the NGO partners depart.

Are Partnerships Enough? The Role of Incentives

Partnerships and various forms of assistance are important and can help interested smallholders undertake tree-planting. But some argue that incentives are sometimes needed as well. Incentives are policy instruments that increase the comparative advantage of forest plantations thereby increasing the attractiveness of these to landowners. They are most useful when the benefits of doing something are not likely to become evident for some time. A variety of incentives have been used by governments and international agencies to encourage landowners to undertake reforestation. These have included technical advice and free seedlings as well as policy changes that encourage tree planting such as permanent land tenure or long-term leases. Incentives in the form of financial assistance might also be used including grants, cheap loans or tax concessions. Governments usually offer incentives to encourage cautious landowners who are unfamiliar with tree-growing. By doing so they can help build a regional plantation estate and sustain a local timber processing industry. Incentives might also be used to persuade farmers to plant trees on marginal lands, on critical watersheds or to restore forests to provide habitats for certain endangered wildlife.

But incentives have some serious disadvantages. They can be costly to provide and the transaction costs can be high if they are being passed on to many small landowners. They can be wasteful if, for example, subsidised fertiliser is used on agricultural crops rather than on trees or if unwanted tree seedlings are thrown away instead of being planted. And, unless they are strategically targeted, they can have undesirable consequences such as having trees planted in location too distant from markets to be economically viable or being offered in areas where the local agriculture department is already offering incentives for other purposes. Incentives can be used to benefit cronies or garner political favours and it is not always clear just how efficient they have been in achieving certain outcomes (e.g. would reforestation have occurred anyway?). Finally, they may not be enough to make a difference. In the particular case of cheap loans, many farmers may find these unattractive if repayment must be made before the final harvesting is carried out.

Is must be said that the topic is one about which there is considerable debate. Foresters in the Solomon Islands were very firmly of the view that seeds should not be given away free and their insistence on payment does not seem to have hindered planting by villagers. On the contrary, it has been embarked on with enthusiasm. Likewise, many farmers in Vietnam are willing to pay for seedlings (Fig. 9.5). On the other hand, no significant planting was carried out in the highlands of Papua New Guinea until the government formed its partnership with local clans and supplied funds to do so. And, in discussing a long-lasting reforestation project in Haiti, Murray and Bannister (2004) were of the very firm opinion that even a modest charge for seedlings would have largely prevented any tree-planting from taking place.

Enters et al. (2003) distinguish between direct and indirect incentives. The former are items such as free seedlings and cash grants or cheap loans. The later are incentives that enable growers to benefit more from their investment in reforestation. These include:

- Providing infrastructure such as roads that reduce the costs of moving logs to market.
- Providing market information in a form that is accessible and useful to smallholders.

- Carrying out silvicultural research that improves grower's abilities to choose appropriate species for particular sites and increase the growth rates of these species.
- Establishing a stable policy environment in which impediments to tree-growing such as bureaucratic constraints on harvesting planted trees or excessive subsidies in other sectors such as agriculture have been removed.
- Ensuring that environmental and social costs of all land uses are taken into account and not passed on to community as a whole.
- Removing import or export controls that hinder the development of efficient industries and markets.
- Reducing public sector involvement in reforestation that 'crowds out' or discriminates against smallholders (e.g. by establishing artificially low timber prices).

Some forms of direct incentives may be necessary to foster plantation establishment by smallholders but there is a danger that they can foster a degree of over-dependency. This may occur even though the implicit of a project aim is to empower the community and build its capacity to act independently. The same it true of international funds from overseas development agencies. Ultimately reforestation must be driven by local economic circumstances and domestically driven policy initiatives if it is to be successful. Enters et al. (2003) argue that direct incentives should be gradually phased out in favour of indirect or enabling incentives that generate more efficient uses of land and income from tree planting. This means that, from a policy perspective, success has been achieved when there is a diminishing need for continued government support. It is important to recognize that reforestation may not be a competitive land use in all situations and that other land use activities may have a much greater economic or social benefit. Under these circumstances it would be foolish to try to use incentives or subsidies to compete with these alternatives.

Building Socially Resilient Forms of Reforestation

Partnerships and incentives may be necessary to initiate tree-planting by farmers who have never thought of doing this but who have land that could be reforested (i.e. in terms of the earlier typologies they are not those who simply cannot afford to plant trees). Indeed they may be the only way in which the task of reforesting large areas of wastelands and other degraded lands is achieved. But, at some point, landowners should become independent from this outside support and be able to innovate and test alternative ways of establishing and managing plantations in the same way that they manage their agricultural crops.

This issue is related to the earlier discussion about resilience. Then resilience was discussed from a largely ecological viewpoint and was defined as the capacity of any system to absorb disturbances and remain in the same state with essentially the same structure, functioning and feedback mechanisms. When dealing with degraded landscapes the key concern was to find ways of assembling productive new forests able to adjust and adapt to changing circumstances. But resilience has a socio-economic dimension as well. This is because humans form part of many ecological systems and human economic activities affect the way these systems develop. Walker et al. (2004) refers to these systems as social-ecological systems (Box 10.2). Transforming existing social-economic systems and making them more resilient involves changes to both the economic and social sub-systems (Folke et al.

Box 10.2 Transforming Social-Ecological Systems

Researchers concerned with resilience have paid particular attention to the way Social-Ecological Systems can be transformed (Anderies et al. 2006; Berkes et al. 2003; Olsson et al. 2004, 2006). The transformability of these systems depends upon the collective abilities of human actors in the system to influence resilience and manage it. This requires, firstly, that there is some form of monitoring of the ecological, social and economic situation which provides feedback to farmers or managers. And, secondly, it requires that these actors have been persuaded by this feedback that there is a need for the system to be adapted or transformed. Finally, it requires that they have the leadership and institutions able to do so.

Social-Ecological Systems are mostly unpredictable because of their inherent complexity. In such cases the best way of managing change is by developing learning networks. These are self-organised groups that engage in experimentation and testing and then use the knowledge they gain to build the community's adaptive capacity. They are not just concerned with acquiring knowledge but with the practices that knowledge can foster and that will allow their particular Social-Ecological System to adapt to change. Over time the community lessens its dependence on external sources of information such as governments and become more self-sufficient. By doing so the system becomes more resilient. Learning networks can explore new methods of doing things, new policy options and test different ways of sharing knowledge. In effect, they increase social as well as human capital. Olsson et al. (2004) argues that government members of these networks should be able to think of options that may be different to current government policy settings.

Besides requiring leaders to initiate and manage the network there is also a need for members to develop a shared vision about the purpose of the network and a sense of trust. Who decides when to change systems and the types of changes that might be needed by different stakeholder groups? Who controls the implementation process? Lebel et al. (2006) suggest that, ultimately, there also needs to be complementary forms of adaptive governance–by government and non-government bodies – that creates flexibility in institutions so they can nourish and support learning networks and so promote sustainable development.

2003). One way of doing this is by developing networks and institutions able to learn, store and exchange knowledge. A participatory learning network could monitor feedback from reforestation activities and existing plantations and so allow the system to adapt to changing ecological, economic or social conditions. In this way it would become more self-sustaining and independent of outside support.

Learning Networks for Reforestation

Learning networks are cooperative partnerships between landholders, government forestry staff and others with an interest in plantation silviculture. The idea behind a learning network is that government foresters and researchers are not the only sources of silvicultural knowledge and that many farmers have direct experience and insights that can be useful, especially in the early stages of carrying out reforestation in a new area. A learning network might take various forms but it should become a bridge between farmers and researchers, extension officers and government foresters. It should also be structured to enable information to be shared across the network's membership. This means farmers could become partners in learning how to carry out reforestation rather than being mere recipients at the end of the knowledge chain (Sayer and Campbell 2004). The capacity of farmers and project staff to jointly learn from experience and modify their practices appears to be one of the reasons for the success of long-term reforestation projects in Nepal (Gilmour and Fisher 1991) and Haiti (Murray and Bannister 2004).

A local learning network could start by involving some of the more enthusiastic farmers already involved in tree planting. Some of these might have considerable practical experience with tree-growing while others may not. It is likely than many would be representative of the more progressive farmers mentioned in the earlier discussion on farmer typologies. Ideally, these farmers and their farms would be representative of the broader community and of the environments in which trees were being planted. In some cases it might be useful to build on existing community structures to develop the network. But, however it was developed, care should be taken to prevent membership of the network being dominated by a local elite who could redirect it to suit their own particular interests.

In the early stages the network might simply review the experiences of local growers and identify the difficulties they have experienced (Fig. 10.3). It might also explore the problems inhibiting landholders who have not yet engaged in tree planting. This would allow researchers to focus on the most pressing of local problems. Another useful early goal might be to develop a shared 'vision' concerning just what role tree-planting might play both for individual landowners and also for the local region. For example, should growers opt to maximise production using fast-growing monocultures of exotic species or should they seek to build a degree of ecological and economic resilience by using a larger variety of slower growing but higher value species? If the latter, what are the silvicultural questions that growers

Fig. 10.3 The best place to discuss small-scale forestry issues is in the field

might need answering? As plantations become established the network might then become an exchange for information on the performance of different species at various sites as well as experience with pruning and thinning regimes. It could also act as an early warning system for new plantation insect pests and diseases. Others might also be coopted at this stage including owners of private nurseries and small rural sawmills both of whom might be able to contribute specialised knowledge about species or markets.

The role of government foresters and researchers who are members of the network would be to advise farmers of new opportunities. Many farmers have experience growing trees in agro forestry plantings but the variety of species used can sometimes be limited and these species may not necessarily be those needed by future markets. Because of their on-going research activities, researchers and government officers should be in a position to introduce new species, seed sources or silvicultural techniques to farmers as well as knowledge about possible future markets.

The learning network should not be seen as a way of identifying a single 'best' method of reforestation. As the typologies described earlier suggest, communities usually contain people with a variety of interests and aspiration and some farmers may prefer to grow species on short rotations while others may prefer species that need longer rotations. Likewise, some farmers might like to consider several reforestation options and not just one. Learning networks can help ensure that this multiplicity of goals is acknowledged by both extension workers and researchers.

These learning networks can be a means of focussing research on the types of problems faced by farmers and they can also become vehicles by which information is subsequently passed on to other farmers who are not members of the network.

Information may be passed on by word-of-mouth, through radio broadcasts, or field days at plantations owned by network members.

Compared with more traditional forms of farm forestry extension this participatory approach has several advantages:

- It allows problems of concern to growers (and potential growers) to be identified at an early stage. These problems might be biological (e.g. seed shortages, nutrient deficiencies) but they may also include financial problems and bureaucratic issues acting as impediments to farm forestry.
- It increases the capacity of farmers and extension workers to adapt and respond to new situations represented by the marginal and degraded lands sometimes being reforested.
- It uses people with a diversity of backgrounds, experiences and skills meaning that a range of solutions to problems are likely to be aired.
- It enables an examination of the performance of species, plantation designs and management systems in a wide variety of (contrasting?) field situations thereby increasing the knowledge base under-pinning reforestation of degraded lands.
- It helps encourage the use of species and silvicultural systems useful to small-scale growers rather than forcing them to use systems developed for large industrial-scale operations.
- It provides opportunities to examine farm forestry at several scales – site, farm, landscape.
- It enables members of the network to learn how to make the trade-offs that are sometimes necessary (e.g. between production and crop diversity, between the benefits of growing trees and the income from agricultural crops).
- It helps identify demonstration sites on farms that might be used for field days (which may be more credible than field days on government field stations).
- It increases effectiveness of extension officers working with farmers outside the learning network because they become better informed about growers needs and about practical solutions. Ideally, members of the learning network also become *de facto* extension officers because they are likely to be able to communicate effectively with fellow farmers.
- It creates a knowledge bank (a 'social memory') concerning farm forestry that is available to all farmers.
- It creates a cadre of informed practitioners who may find it easier to get support for administrative or policy changes where this becomes necessary.

Networks like these may take some time to develop not least because some government officers may be sceptical of the contribution that uneducated farmers can make (Dove 1992). In some places government researchers are even reluctant to share data with their scientific colleagues (especially those from other agencies or ministries) let alone farmers. This is because information is seen as a valuable commodity and not to be given away lightly, unlike in the western scientific tradition where prestige is gained by sharing new findings. These types of impediments mean learning networks may take time to develop even though their merits seem self-evident. The ways participatory learning networks can be assembled and function have been discussed in some detail by Gilmour and Fisher (1991), Scherr (2000), Folke et al. (2003), Sayer and Campbell (2004).

Monitoring and Evaluating Progress

The rate of progress towards achieving a self-sustaining and resilient farm-forestry movement requires some form of monitoring. Because of their contrasting objectives different stakeholders will be interested in different types of feedback. Extension workers and those interested in promoting farm forestry will be most interested in indicators of progress towards a system that is less dependent on external partners while farmers themselves are likely to be more interested in the immediate benefits they gain by planting trees. Some possible indicators of progress towards a less dependent and resilient form of farm forestry are shown in Table 10.4.

The key question is whether or not farm forestry is being adopted and indicators such as the area of farmland being planted with trees or the proportion of farmers involved in tree-planting are fairly obvious. Two other fundamental questions are whether tree planting is economically rewarding and is it is being accepted as a credible land use? Some of the indicators concerning these questions are quantitative while others are more qualitative but in all cases it is the trends over time that are of most interest rather than the absolute values. Sometimes it may take several years before these trends are clear (e.g. market prices for plantation grown timbers may change only slowly) while in other cases patterns will become evident relatively quickly (e.g. a network of demonstration farms or plantations might be identified at an early stage). The examples in Table 10.4 are not prescriptions of what must be done but are simply indications of the type of data that might be monitored.

Gilmour and Fisher (1991) emphasize the difference between monitoring and evaluation. Monitoring provides information about trends but those doing the monitoring should also look for unintended impacts and consequences. Are the outcomes all positive? Are the economic benefits being shared or are they being captured by an elite segment of the community? Is reforestation affecting local water supplies? Monitoring systems can not be too complex otherwise they become unwieldy. As a way of complementing formal or semi-formal monitoring Gilmour and Fisher (1991) suggest using simple forms of rapid rural appraisal such as small group meetings that allow extension activities and some of these other outcomes of plantings to be regularly evaluated.

An alternative form of evaluation is to use some form of modelling. Modelling can be useful in situations where there are time lags before the outcomes of ecological and economic processes become evident, where the system covers a range of scales or where there is a high degree of unpredictability involved because the socio-economic systems are so complex. Sayer and Campbell (2004) discuss the advantages and disadvantages of modelling in rural settings at some length. They argue that simple 'throw-away' or scoping models can provide some considerable advantages because they help everyone understand the system and where the gaps in knowledge occur. They can also be especially useful in exploring alternative

Table 10.4 Examples of indicators that might be used to assess the success of farm forestry programs

Is reforestation being adopted?	Is reforestation generating economic benefits for farmers?	Is tree-growing being accepted as a credible and worth-while land use practice by farmers?
Overall area of farm plantations is increasing	Prices of timber of different qualities, firewood and NTFPs are stable or increasing	There are farmers and farmer groups discussing tree-growing and using the learning network
Number (and percent) of farmers growing timber trees is increasing	Number of buyers or traders of plantation grown products is increasing	Membership of the learning network and local farm forestry groups is increasing
Number (and percent) of farms with plantations >1 ha is increasing	There are local sawmills and other industries using farm-grown resources	A network of demonstration farm plantations has been identified
Number of plantations grown on long rotations (>10 year) is increasing	Distance over which goods are transported to industrial buyers is decreasing	Farm forestry field days are being held and attracting good crowds
Number of private tree nurseries is increasing because there is a market for seedlings	Seedling prices (including those of the more valuable species) are increasing	Prices of forest products are regularly quoted in the media
Variety of species being grown in these nurseries is increasing	Marketing cooperatives are being formed to sell logs	Farmers are aware of prices for specific forest products
	Administrative procedures regulating harvesting and sale of plantation products are being simplified	Farmer's knowledge about silvicultural practices is increasing
	Payments are being made for the supply of ecosystem services	Farmers are collecting their own seed
		Farmers are seeking improved planting material
		More sophisticated forms of farm forestry are being used (e.g. thinning and pruning is widely practiced). There is interest in quality and not just quantity

scenarios or options when stakeholders have conflicting aspirations (making them a potentially useful tool in the earlier task of the learning network of articulating a 'vision'). The role of modelling in evaluating alternative reforestation scenarios will be discussed further in the next chapter.

Judging Success from a Farmer Perspective

What might individual farmers think of all this? This chapter (like the book itself) has taken the view that reforestation is usually a 'good thing' that can bring benefits to individual landowners as well as to society as a whole (most especially in areas with degraded lands). It is assumed farmers will share this view once they are made aware of these potential benefits. But the development literature is replete with examples of seemingly good ideas being introduced to farmers for the best of reasons and of these ideas being rejected. From a farmer's perspective, change can be risky and it is sometimes difficult for them to judge how much faith they can place in the suggestions being proffered by outsiders who, it must be said, rarely undertake any risk themselves.

So, how might a farmer judge whether the tree planting they have carried out on their land had been successful? Different farmers will have different perspectives (recall the range of attitudes contained in the earlier typologies in Table 10.1) but they might consider things like the expected impact on their future household income or the extent to which reforestation is likely to diversify income sources or whether it has increased the overall value of farm assets. Some farmers might value tree planting not because of timber it produces but because it improves their ability to produce other crops or support animals (because trees stabilised soil surfaces, produced animal feed or mulch); from their viewpoint the best form of reforestation may not necessarily be that which involves planting the most trees. Others might regard tree planting worthwhile because it supports their assertion of land ownership. There is also likely to be a difference in the opinion of men and women. Women are often heavily involved in household decision-making and are likely to be especially concerned with the types of goods produced by different tree species, the opportunities to carry out inter-cropping in plantations or with the implications that farm forestry has on their ability to continue to gather firewood and NTFPs from surrounding lands.

It is interesting that Emtage and Suh (2004) found landholders in their study site in the Philippines who already had some experience with tree-growing were more enthusiastic about establishing further plantations than those who had none. Predo (2003) made a similar observation. But circumstances can change and farmers with trees may decide not to replace them once the first rotation is complete. This may be because they cannot afford the replanting costs or they may prefer to use the land for another more profitable purpose such as a new crop like oil palm or rubber.

Conclusions

Many farmers have had experience planting trees around their houses or in agro forestry situations and so are not entirely unfamiliar with what is involved when land is reforested. On the other hand, rather fewer have experience of growing trees for commercial purposes. They may not be familiar with the most appropriate species to use nor how to manage these to obtain the highest financial return. For many such farmers, commercial tree growing is a novel and risky enterprise, not least because it apparently takes so long for any financial return. But farmers can differ in their attitudes and circumstances. Some will never embrace tree-growing because they have too little land and need all they have to grow food. Others may be quite interested in the possibility of growing trees, especially as regional forest resources decline and the markets for forest products begin to increase. Of these potential growers, some will be looking to maximize household incomes while others may be more interested in diversifying their sources of household income. These differences mean quite differing types of silvicultural advice may be needed for these different types of grower if reforestation is to successfully implemented on a larger scale.

Governments have usually taken the lead role in promoting reforestation. They have mostly done so by providing advice, access to planting material and sometimes financial assistance. The appropriateness and quality of this advice has varied. In most cases the same silvicultural prescriptions have been offered to all landholders.

This has meant that some households and communities have benefited and have embraced tree-growing with a high degree of enthusiasm while others have not because the expected tree growth rates have not materialised or the financial returns were not as great as had been anticipated. Partnerships or relationships between smallholders and companies or NGOs have often been more successsful since small numbers of households are usually involved and the financial objective has been more clearly defined.

In the longer term, new ways of promoting reforestation may be needed. Some have argued in favour of providing direct financial incentives. These may be useful in some special situations but they tend to perpetuate a sense of dependency and a risk that operations will cease if support is later withdrawn. A better approach would be to foster learning networks involving growers, government staff and others with an interest in plantation silviculture that help test new methods of reforestation, share this and that build a more resilient socio-economic system. These networks could ensure landowners, and especially those on marginal lands, become more self-sufficient with respect to silvicultural technologies and that research is directed at the particular problems faced by farmers rather than just those faced by large industrial growers. The learning network then becomes a vehicle for not only increasing and storing knowledge but also distributing it amongst the rural community. In other words, it helps build both social and human capital.

While it is clear that reforestation has the capacity to improve the livelihoods of smallholders it will only do so if there are enough plantings in a district to create an economically attractive resource; small and isolated plantings will rarely be financially viable. And if enough farmers agree to carry out reforestation this may generate some functional changes across the wider landscape. That is, macro-ecological benefits may arise from micro-economic changes. But it all depends on just where in the landscape this reforestation is done. This highlights the fact that individual farmers can have different interests than society as a whole. Farmers are largely interested in profitability and risk management while society is often more interested in the sustainability of land uses, watershed protection and biodiversity conservation. These latter benefits are very dependent on what changes farmers collectively make across the landscape. This is discussed further in the next chapter.

References

Anderies JM, Walker BH, Kinzig AP (2006) Fifteen weddings and a funeral: Case studies and resilience-based management. Ecol Soc 11(1):21, [online] URL: ≤http://www.ecologyandsociety.org/vol11/iss1/art21/≥

Anyonge C, Roshetko JM (2003) Farm-level timber production: Orienting farmers towards the market. Unasylva 54:48–56

Arnold JEM, Dewees PA (1995) Tree management in farmer strategies: Responses to agricultural intensification. Oxford University Press, Oxford

Aronson J, Milton SJ, Blignaut JN (2007) Restoring natural capital: Science, business, and practice. Island Press, Washington, DC

Barney K (2008) Local vulnerability, project risk and intractable debt: The politics of smallholder eucalypt promotion in Salavane Province, Southern Laos. In: Snelder DJ, Lasco RD (eds) Smallholder tree growing for rural development and environmental services. Springer, Dordrecht, pp 263–286

Baynes J (2007) Evaluating the effectiveness of a small-scale forest extension program on Leyte island, the Philippines. In: Harrison S, Bosch A, Herbohn J (eds) Improving the triple bottom line: Returns from small-scale forestry; Proceedings of the IUFRO 3.08 Conference, Ormoc City, Leyte, The Philippines, pp 17–32

Berkes F, Colding J, Folke C (2003) Navigating social-ecological systems: Building resilience for complexity and change. Cambridge University Press, Cambridge

Bourke RM (1997) Management of fallow species composition with tree planting in Papua New Guinea. Research School for Pacific and Asian Studies, The Australian National University, Canberra

Cairns M (ed) (2007) Voices from the forest. Resources for the Future, Washington, DC

Calderon MM, Nawir AA (2006) An evaluation of the feasibility and benefits of forest partnerships to develop tree plantations: Case studies in the Philippines. Center for International Forestry Research, Bogor

Chambers R, Leach M (1989) Trees as savings and security for the rural poor. World Dev 17:329–342

Chokkalingam U, Carandang AP, Pulhin JM, Lasco RD, Peras RJJ, Toma T (2006) One century of forest rehabilitation in the Philippines: Approaches, outcomes and lessons. Center for International Forestry Research, Bogor

De Jong W, Sam DD, Hung TV (2006) Forest rehabilitation in Vietnam: History, realities and future. Center for International Forestry Research, Bogor

Dove MR (1992) Foresters' beliefs about farmers: A priority for social science research in social forestry. Agroforest Syst 17:13–41

Elliott S, Blakesley D, Maxwell JF, Doust S, Suwannaratana S (2006) How to plant a forest: The principles and practice of restoring tropical forests. Biology Department, University of Chiang Mai, Chiang Mai

Ellis F (1993) Peasant economics: Farm households and agrarian development. Cambridge University Press, Cambridge

Emtage N (2004) Typologies of landholders in Leyte, Philippines and the implications for development of policies for smallholder and community forestry. In: Baumgartner DM (ed) Proceedings of human dimensions of family, farm and community forestry international symposium, Washington State University Extension, Washington State University, pp 81–88

Emtage N, Suh J (2004) Socio-economic factors affecting smallholder tree planting and management intentions in Leyte Province, Philippines. Small-Scale Forest Econ Manag Policy 3:257–270

Enters T, Durst P, Brown C (2003) What does it take? The role of incentives in forest plantation development in the Asia-Pacific region. Unasylva 54:11–18

Fairhead J, Leach M (1996) Misreading the african landscape: Society and ecology in a forest – savanna mosaic. Cambridge University Press, Cambridge

FAO (2007) State of the World's Forests 2007. Food and Agriculture Organisation of the United Nations, Rome

Fatoux C, Castella J-C, Zeiss M, Pham HM (2002) From rice farmer to agroforester within a decade. The impact of Doi moi on agricultural diversification in a mountainous commune of Cho Moi District, Bac Kan Province, Vietnam. In: Castella JC, Quang DD (eds) Doi Moi in the mountains land use changes and farmers livelihood strategies in Bac Kan Province, Vietnam. The Agricultural Publishing House, Hanoi, pp 73–97

Fisher RJ, Maginnis S, Jackson W, Barrow E, Jeanrenaus S (2008) Linking conservation and poverty reduction: Landscapes, people and power. Earthscan, London
Folke C, Colding J, Berkes F (2003) Synthesis: Building resilience and adaptive capacity in social-ecological systems. In: Folke C, Colding J, Berkes F (eds) Navigating social-ecological systems: Building resilience for complexity and change. Cambridge University Press, Cambridge, pp 352–384
Garin P (2008) Fayantina Afforestation Project in Eastern Highlands Province, Papua New Guinea. Capacity building on restoration, management and rehabilitation of degraded forests and deforested land in the Pacific – a regional seminar for improved practices enhancing forest functions. Land Resources Division, Secretariat of the Pacific Community, Nadi, Fiji Islands
Garrity DP (2004) Agroforestry and the achievement of the millennium development goals. Agroforest Syst 61:5–17
Gilmour DA, Fisher RJ (1991) Villagers, forests and foresters. Sahayogi Press, Kathmandu
Gilmour DA, King GC, Applegate GC, Mohns B (1990) Silviculture of plantation forest in central Nepal to maximise community benefit. Forest Ecol Manag 32:173–186
Godoy RA (1992) Determinants of smallholder commercial tree cultivation. World Dev 20:713–725
Goltenboth F, Hutter C-P (2004) New options for land rehabilitation and landscape ecology in Southeast Asia by 'rainforestation farming'. J Nat Conserv 12:181–189
Gua B (2008) Solomon Islands Country Report. Capacity building on restoration, management and rehabilitation of degraded forests and deforested land in the Pacific – a regional seminar for improved practices enhancing forest functions. Land Resources Division, Secretariat of the Pacific Community, Nadi, Fiji Islands
Harrison S, Emtage N, Nasayao E (2004) Past and present forestry support programs in the Philippines, and lessons for the future. Small-Scale Forest Econ Manage Policy 3:303–317
Harrison S, Herbohn J (2005) Relationship between farm size and reforestation activity: Evidence from Queensland studies. Small-Scale Forest Econ Manag Policy 4:471–484
Herbohn J, Emtage N, Harrison S, Smorfitt D, Slaughter G (2005) The importance of considering social issues in reforestation schemes. In: Erskine P, Lamb D, Bristow M (eds) Reforestation in the topics and subtropics of Australia using rainforest tree species. Rural Industries Research and Development Corporation, Canberra, pp 224–244, https://rirdc.infoservices.com.au/items/05-087
Holmgren P, Masakha EJ, Sjoholm H (1994) Not all African land is being degraded – a recent survey of trees on farms in Kenya reveals increasing forest resources. Ambio 23:390–395
Lamb D, Huynh DN (2006) Mixed species plantations of high-value trees for timber production and enhanced community services in Vietnam and Australia (unpublished report FST2000/003). Australian Center for International Agricultural Research, Canberra
Lang C (2002) The pulp invasion: The International Pulp and Paper Industry in the Mekong region. World Rainforest Movement, Montevideo
Lebel L, Anderies JM, Campbell B, Folke C, Hatfield-Dodds S, Hughes TP, Wilson J (2006) Governance and the capacity to manage resilience in regional social-ecological systems. Ecol Soc 11(1):19, [online] URL:http://www.ecologyandsociety.org/vol11/iss1/art19/
MARD (2001) Five Million Hectare Reforestation Program Partnership: Synthesis report. International Cooperation Department, Ministry of Agriculture and Rural Development, Hanoi
Maturana J, Hosgood N, Suhartanto AA (2005) Moving towards company-community partnerships: Elements to take into account for fast-wood plantation companies in Indonesia. CIFOR Working Paper No 29, Center for International Forestry Research, Bogor
Mayers J, Vermeulen S (2002) Company-community forestry partnerships; from raw deals to mutual gains? International Institute for Environment and Development, London
McElwee P (2009) Regforesting 'bare hills' in Vietnam: Social and environmental consequences of the 5 million hectare reforestation program. Ambio 38:325–333
Menzies N, Tapp N (2007) Fallow management in the borderlands of southwest China: The case of *Cunninghamia lanceolata*. In: Cairns M (ed) Voices from the forest: Integrating indigenous knowledge into sustainable upland farming. Resources for the Future, Washington, DC, pp 425–434

Murray GF, Bannister ME (2004) Peasants, agroforesters and anthropologists: A 20-year venture in income generating trees and hedgerows in Haiti. Agroforest Syst 61–62:383–397

Nawir AA, Anyonge C, Race D, Vermeulen S (2003) Towards equitable partnerships between corporate and smallholder partners: Relating partnerships to social, economic and environmental indicators: workshop synthesis. Food and Agriculture Organisation of the United Nations, Rome

Nawir AA, Kassa H, Sandewall M, Dore D, Cambell B, Ohlsson B, Bekele M (2007a) Stimulating smallholder tree planting – lessons from Africa and Asia. Unasylva 58:11–18

Nawir AA, Murniat, Rumboko L (2007b) Forest rehabilitation in Indonesia: Where to after three decades? Center for International Forestry Research, Bogor

Nguyen VS, Gilmour DG (2000) Forest rehabilitation policy and practice in Vietnam. In: Anon (ed) Forest rehabilitation policy and practice in Vietnam: Proceedings of a national workshop, Hoa Binh, International Union for Conservation of Nature, Hanoi, pp 4–34

Nibbering JV (1997) Upland cultivation and soil conservation in limestone regions of Java's south coast. In: Boomgaard P, Colombijn F, Henley D (eds) Paper landscapes: Explorations in the environmental history of Indonesia. KITLV Press, Leiden, pp 153–184

Nibbering JW (1999) Tree planting on deforested farmlands, Sewu Hills, Java, Indonesia: Impact of economic and institutional changes. Agroforest Syst 46:65–82

Olsson P, Folke C, Berkes F (2004) Adaptive co-management for building resilience in social-ecological systems. Environ Manag 34:75–90

Olsson P, Gunderson LH, Carpenter SR, Ryan P, Lebel L, Folke C, Holling CS (2006) Shooting the rapids: Navigating transitions to adaptive governance of social-ecological systems. Ecol Soc 11, URL: http://www.ecologyandsociety.org/vol11/iss11/art18/

Pannell DJ (1999) Social and economic challenges in the development of complex farming systems. Agroforest Syst 45:395–411

Pasicolan PN, de Haes HAU, Sajise PE (1997) Farm forestry: An alternative to government driven reforestation in the Philippines. Forest Ecol Manag 99:261–274

Pattanyak S, Mercer DE, Sills E, Yang J (2003) Taking stock of agrofrestry adoption studies. Agroforest Syst 57:173–186

Predo CD (2003) What motivates farmers? Tree growing and land use decisions in the grasslands of Claveria, Philippines. Research Report 2003-RR7. Economy and Environment Program for Southeast Asia International Development Research Center, Singapore

Race D, Bisjoe AR, Hakim R, Hayati N, Julmansyah KA, Kuniawan KP, Nawir AA, Nurhaedah PDU, Purwanti R, Rohadi D, Stewart H, Sumirat B, Suwarno A (2009) Partnerships for involving small-scale growers in commercial forestry: Lessons from Australia and Indonesia. Int Forestry Rev 11:88–96

Raymond DH, Wooff WG (2006) Small-scale forest plantations are the key to the future of the Solomon Islands forest industry. Int Forestry Rev 8:222–228

Santos F, Bertomeu M, Vega B, Mangaong E, Stark M, Bullecer R (2003) Local knowledge on indigenous trees: towards expanding options for smallholder timber plantating and improved farm forestry in the Philippine uplands. In: Sim HC, Appanah S, Durst P (eds) Bring back the forests: Policies and practices for degraded lands and forests FAO. Bangkok, Kuala Lumpur, pp 75–84

Sayer JA, Campbell B (2004) The science of sustainable development: local livelihoods and the global environment. Cambridge University Press, Cambridge

Scherr S (2000) A downward spiral? Research evidence on the relationship between poverty and natural resource degradation. Food Policy 25:479–498

Schuren SHG, Snelder DJ (2008) Tree-growing on farms in northeast Luzon (The Philippines): smallholders' motivations and other determinants for adopting agroforestry systems. In: Snelder DJ, Lasco RD (eds) Smallholder tree growing for rural development and environmental services. Springer, Netherlands, pp 75–97

Shively G (1998) Economic policies and the environment: The case of tree planting on low-income farms in the Philippines. Env Dev Econ 3:83–104

Sikor T (2001) The allocation of forestry land in Vietnam: Did it cause the expansion of trees in the north west? Forest Policy Econ 2:1–11

Tucker N, Wardell-Johnson G, Catterall CP, Kanowski J (2004) Agroforestry and biodiversity: Improving conservation outcomes in tropical northeast Australia. In: Schroth G, da Fonseca GAB, Harvey CA, Gascon C, Vasconcelas HL, Izac A-M (eds) Agroforestry and biodiversity conservation in tropical landscapes, Island Press, Washington, DC, pp 431–452

Turkelboom F (1999) On-farm diagnosis of steepland erosion in northern Thailand. Dissertation Faculteit Landbouwkundige en Toegepaste Biologische Wetenschappen, Katholike Universiteit Leuven, Leuven

Vize S, Killin D, Sexton G (2005) The community rainforest reforestation program and other farm forestry programs based around the utilisation of rainforest and tropical species. In: Erskine P, Lamb D, Bristow M (eds) Reforestation in the tropics and subtropics of Australia using rainforest tree species. Rural Industries Research and Development Corporation, Canberra, pp 7–22, https://rirdc.infoservices.com.au/items/05-087; accessed 20 September 2010

Walker B, Holling CS, Carpenter SR, Kinzing A (2004) Resilience, adaptability and transformability in social-ecological systems. Ecol Soc 9: 5 [online] URL: http://ecologyand society.org/vol9/iss2/art5

Walters BB (2004) Local management of mangrove forests in the Philippines: Successful conservation or efficient resource exploitation? Hum Ecol 32:177–195

Yendkoa F (2008) A case study of the reforestation project in the Open Bay Area in East New Britain Province of Papua New Guinea. Capacity building on restoration, management and rehabilitation of degraded forests and deforested land in the pacific – a regional seminar for improved practices enhancing forest functions. Land Resources Division, Secretariat of the Pacific Community, Fiji

Chapter 11
Reforestation at a Landscape Scale

> Much of the current confusion and distress surrounding environmental issues can be traced to decisions that were never consciously made, but simply resulted from a series of small decisions. Consider, for example, the loss of coastal wetlands on the east coast of the United States between 1950 and 1970. No one purposely planned to destroy almost 50% of the existing marshland along the coasts of Connecticut and Massachusetts. In fact, if the public had been asked whether coastal wetlands should be preserved or converted to some other use, preservation would probably have been supported. However, through hundreds of little decisions and the conversion of hundreds of small tracts of marshland, a major decision in favour of extensive wetlands conversion was made without ever addressing the issue directly.
>
> (Odum 1982, p. 728)

Introduction

Reforestation must be undertaken on a large scale if the adverse effects of deforestation and degradation on biodiversity conservation and ecosystem functioning are to be overcome. But the success of reforestation at a landscape scale depends on how decisions are made. Reforestation by individual farmers may increase the supply of goods and begin to affect erosion and regional hydrological processes. These efforts might also influence the ability of species to move across the landscape. The increased forest cover may even prevent certain species from becoming regionally extinct. However, the effectiveness of any of this reforestation depends on where these reforested areas are located.

Most farmers, understandably, focus on their own land and pay much less attention to the broader regional context. This means the landscape evolves through a series of small ad hoc decisions or what Odum (1982) referred to as the tyranny of small decisions. If reforestation is done in this way it is likely to produce some local benefits but, in most cases, the collective result will be sub-optimal because it is the result of unconnected events rather than being carefully designed and integrated. Nor do these individual decisions make any use of our current knowledge about landscape ecology. Better planning could allow better outcomes (and build greater resilience) for the

D. Lamb, *Regreening the Bare Hills*, World Forests 8,
DOI 10.1007/978-90-481-9870-2_11, © Springer Science+Business Media B.V. 2011

same expenditure of effort and resources. But planning done by government officers drawing lines on maps without knowledge of the actual field conditions or the opinions of landholders will usually generate sub-optimal outcomes as well. The alternative to these un-planned or over-planned alternatives is a more consultative approach that has become known as Forest Landscape Restoration.

Maginnis and Jackson (2007) have defined Forest Landscape Restoration as being 'a process that aims to regain ecological integrity and enhance human well-being in deforested or degraded forested landscapes'. It is not an attempt to restore the original 'pristine' forest cover but to strengthen functionality and resilience of the social-ecological systems present. It might be appropriate to label any kind of large-scale as forest landscape restoration but the most effective form is where reforestation is both purposeful and strategic. It should be concerned with how much reforestation takes place but, perhaps more importantly, with the location and type of reforestation that is employed.

Another distinguishing feature of Forest Landscape Restoration is that it seeks to strike a balance between conservation and production by enabling stakeholders with differing views to negotiate trade-offs at a landscape scale. Forest Landscape Restoration is therefore intimately concerned with the construction of new landscape mosaics. But, equally importantly, it is concerned with the way that communities can work together to develop a shared vision of these future landscapes.

This chapter describes how Forest Landscape Restoration might be undertaken. It takes for granted that governments rarely have sufficient funds to enable reforestation to be carried out on a scale that solves the many environmental and conservation problems created by previous land clearing. It also assumes land users have de facto if not de jure land tenure so that reforestation is something they might contemplate carrying out. The task, then, is to find ways of undertaking reforestation that satisfy the financial needs of these individual landowners and that, at the same time, also help solve some of these regional conservation problems. The chapter begins with a discussion of the nature of landscape mosaics and some design principles for enhancing resilience at a landscape scale. It then addresses two sets of issues. The first deals with the ecological questions of how much reforestation is necessary, where it should be located and what types of new forests should be established at these sites. Then, secondly, it considers how these questions might be resolved in practice and how any forest restoration program might be carried out at a landscape scale.

The Nature of Landscape Mosaics

The term 'landscape' can be interpreted quite differently by different people and some of these views are discussed in Box 11.1. All would agree that landscapes are a heterogeneous mosaic of separate components. The spatial distribution of these is influenced by the underlying geology and soils as well as by the topography and drainage patterns. One can think of an ecological mosaic and a socio-economic mosaic (Fig. 11.1). The ecological mosaic is made up of a

Box 11.1 Definition of Landscapes

The term 'landscape' has been given a wide variety of meanings. Some have indicated a spatial extent while others have not. Thus Schroth et al. (2004) have defined a landscape as a mosaic of ecosystems or habitats present over a kilometre-wide area while ITTO (2002) defines a landscape as simply a cluster of interacting ecosystem types without mentioning the spatial scale. Fisher et al. (2008) recognized landscapes as contiguous area, intermediate in size between an 'ecoregion' and a 'site' with a specific set of ecological, cultural and socio-economic characteristics distinct from it neighbours. Lindenmayer and Fischer (2006) note the definition may depend on the context in which it is being used. From a human perspective it is usually seen as being an area covering 100s to 1,000s of hectares but from a conservation biology perspective it is a function of the scales over which a particular species moves and how this species perceives its environment. It is also possible to think of a number of 'landscapes' that overlap with each other such as a land-use landscape, a cultural landscape, an aesthetic landscape, an economic landscape or a conservation landscape. Perhaps the most useful way of thinking of landscapes is not so much as a planning unit but as the scale at which it is necessary to intervene if one is to balance trade-offs and optimize conservation and livelihood benefits (Sayer and Boedhihartono, in press).

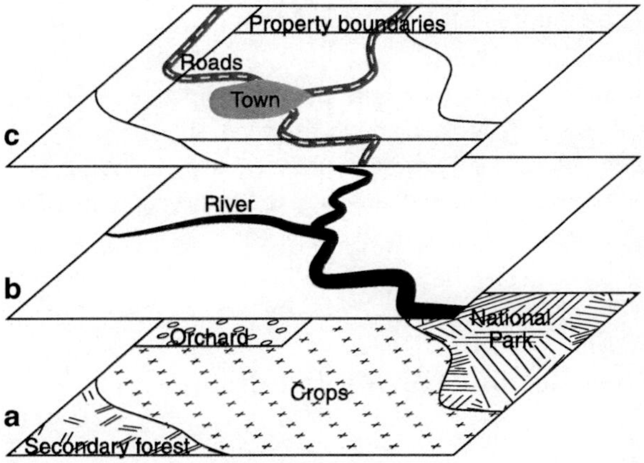

Fig. 11.1 Landscapes are composed of ecological and socio-economic mosaics. Each of these can be dis-aggregated into different layers each with its own spatial pattern. Layer (**a**) shows vegetation patterns, layer (**b**) shows drainage systems and layer (**c**) shows location of towns, property boundaries and roads

patchwork of different ecosystems scattered across the landscape. Some of the vegetation present may be represented by natural forests including undisturbed primary forest (i.e. late successional habitats) as well as fallow and secondary regrowth (i.e. early successional habitats). Other areas in the mosaic could include annual and perennial agricultural crops, agroforests and timber plantations. Some of the forests might be protected in National Parks while others may be subject to logging. The spatial heterogeneity will vary with some landscapes being highly fragmented and having small and isolated patches of forest while others may still have large forest areas all of which are relatively close together. This ecological mosaic may change over time as forests are cleared for agriculture, crops change or as farmland is abandoned and regrowth forests become established. Likewise, wildlife will move across the landscape depending on the changing availability of habitats and food resources.

The ecological mosaic is matched by a socio-economic mosaic that reflects the scale and organisation of past land use activities as well as the distribution of roads and markets. Part of the landscape may be owned by state forestry or conservation agencies with other areas being owned or controlled by corporations, traditional community groups or individual households. Some of the non-government owners will have legal title to the land they are using but others may have only de facto ownership. There may be disputes over land ownership and over the location of property boundaries. The amounts of land each manager or user controls will vary with individual households usually owning only small areas while wealthier households, communities and other owners controlling much bigger areas. All of these land users will differ in the way they manage their lands and the ecological mosaic will reflect these choices. Some decisions will reflect traditional community practices while other will be a consequence of recent agricultural market changes (e.g. attractive prices for palm oil generate new oil palm plantations) while other areas, such as those with long-lived crops like fruit orchards, will be the legacy of choices made many years earlier. As was the case with the ecological mosaic, these patterns are not static. Populations and cropping practices fluctuate as migrants arrive or younger people leave and move to urban areas. Likewise, new roads or changes in market prices for particular products will alter the attractiveness of different land use practices. These changes in the socio-economic mosaic will feed through to and alter the ecological mosaic.

Ecological Processes in Evolving Landscapes

The transition from a forest to an agricultural landscape disrupts most ecological processes. Since the transition may take place over a number of years these processes may not reach a new quasi-equilibrium for some time and their present condition will depend on the types of vegetation communities that are developing and their spatial heterogeneity. In the context of forest landscape restoration, the most important ecological processes affected by deforestation are those associated with the maintenance of biodiversity, the regulation of hydrological flows and with watershed protection.

Biodiversity

Most agricultural landscapes support fewer species than the forests they replace. This is because any remnant patches of natural forests are now scattered across an agricultural matrix that may be hostile to some forest dwelling species. The species most disadvantaged by forest clearings are the forest specialists, particularly those large species of wildlife with extensive home ranges (their demise being accelerated by higher hunting pressures in rural landscapes). Changes in the relative abundances of frugivores, insectivores and nectivores are also common (Tscharntke et al. 2008). Species favoured by clearing are mostly habitat generalists and open country species.

It is not useful to classify agricultural landscapes as either 'forest' or 'agriculture' because many also contain various forms of woody vegetation such as orchards or agroforests in addition to areas of undisturbed natural forest. Agroforests in particular can be biologically diverse and structurally complex and an agricultural landscape with scattered patches of agroforests could allow considerable wildlife movement compared with one dominated by, say, a grain crop. Nonetheless, reviews by Scales and Marsden (2008) and Bhagwat et al. (2008) have found agroforestry systems often sustain only 60% of the species richness of natural forest systems. This proportion varies widely depending on the nature of the agroforest and the distances to intact forest.

But in the present context, perhaps the more interesting comparisons are not simple two-way choices between forest and agroforests but those that explore how much biodiversity is able to persist in rural landscapes that include a variety of land uses as well as patches of residual forest. Thus Sodhi et al. (2005) studied the birds present in landscapes in Sulawesi in Indonesia that included mixed rural agricultural ecosystems (villages, gardens and scattered forest remnants) as well as primary forests, 40 year old secondary forests and plantations containing clove trees (*Syzygium aromaticum*), coffee, bananas and maize. These secondary forests and mixed agricultural areas contained 82% and 76% respectively of the forest bird species while the mixed plantations had only 32%. They concluded that some forest birds could be found in all areas where there was at least 20% native tree cover and that there was potential to further enhance the conservation of these forest species with appropriate management.

Rather lower proportions of forest bird species were observed in a mixed rural landscape in Peninsular Malaysia (Peh et al. 2006). In this case the landscape contained three distinct types of vegetation, namely orchards of fruit trees, rubber plantations and oil palm plantations. Surveys found the mixed rural landscape had only 28–32% of the forest birds present in nearby intact forest. They concluded these lower numbers might be explained by the fact that none of these plant communities were structurally complex and they lacked large canopy trees. They also questioned how many birds in the agricultural landscape still depended on the nearby forest for survival.

This issue was explicitly addressed in a study carried out in Costa Rica by Sekercioglu et al. (2007). They observed that 75% of Costa Rica's land birds are able to persist in deforested agricultural areas provided some canopy trees and forest patches remain. In a detailed radio-tracking study of three species that differed in their vulnerability to deforestation, they found these were all able to persist and breed in agricultural landscape without the need to commute back to larger areas of intact forest. This was because of the remnant trees, riparian strips and small areas of forest that remained. The areas with trees in their study site represented only 11% of land cover but this appeared to be sufficient for the particular forest-dependent bird species studied.

For those contemplating the restoration of degraded landscapes these are encouraging results. They suggest rural landscapes containing at least some structurally complex vegetation can support a significant proportion of forest bird species. On the other hand, the few studies carried out to date provide only limited guidance about the spatial layout of vegetation that will be necessary to achieve this. It is also clear that a great deal more needs to be known about the specific habitat requirements of different types of wildlife and feeding guilds as well as about their capacity to move across and reproduce in rural landscapes.

Hydrology, Sedimentation and Watershed Protection in Landscape Mosaics

Some of the effects of deforestation on hydrological flows and soil erosion have been discussed in earlier chapters. The key principles are reasonably well established, namely that evapo-transpiration is decreased, that run-off is increased and that deforestation may lead to persistent erosion and sedimentation. But the extent of the hydrological changes depends on the scale at which deforestation occurs and on the type of vegetation that replaces the former tree cover. In most cases, the impact of a small clearing is unlikely to be detected at a larger regional level. The impact of large scale changes depend on the types of new ecosystems that replace the forests and on their spatial distribution. Guo et al. (2000) describe a study carried out with in the upper watershed of the Yangtze River in Hubei province of China where 90 vegetation-soil-slope complexes were identified. Each of these had its own hydrological response. They were able to use maps of the units and their particular hydrological responses to assess their collective impact on downstream hydro-electric power generation and to plan compensation for those forest owners whose future logging activities might need to be regulated for the sake of improving hydrological efficiency.

In the case of erosion and sedimentation, the type and extent of the new ecosystem replacing the original forests as well as their location also matters. Extensive grasslands may have low rates of erosion and the quality of water draining from these may be higher than from landscapes containing a greater variety of vegetation communities. Erosion from agricultural croplands can be high but also depends on the degree of disruption involved when crops are harvested and replanted and the length of time before a new vegetative cover is established. The extent of cultivation and weed control is also important. The greatest soil losses occur on steeper slopes and Turkelboom (1999) and Sidle et al. (2006) have outlined the levels of risk associated with different types of crops in these circumstances. Most erosion is likely to occur when lateral water movement is able to take place over long distances. In this case larger volumes of water accumulate and can mobilise soil. Tracks, ditches and furrows within agricultural areas are especially prone to generating erosion in this way. Cropping systems and spatial pattern of vegetation that intercept overland flow or which act as filters along streams can reduce sedimentation and the loss of nutrients applied to crops as fertilisers. Deep rooted trees are also likely to reduce the incidence of landslides in agricultural landscapes (Sidle et al. 2006).

In summary, a variety of ecological processes are disrupted and functional changes occur when forests are replaced by agricultural landscapes. These changes affect the diversity of biota able to persist in the new landscape as well as the spatial movement of water, soils and nutrients. In some cases a relative stable new system may develop but in other cases a period of continued instability may ensure leading to ongoing losses of biodiversity or increasing soil erosion.

Building Resilience at the Landscape Scale

The need to ensure new forests have some degree of resilience has been discussed in previous chapters. The focus then was on building resilience at a site level and one of the dilemmas was in knowing how to balance the immediate needs of those with small farms and low incomes against longer-term concerns about the stability of social-ecological systems. Forest landscape restoration offers a way of at least partially resolving this problem (for the community if not for each farmer) since trade-offs are much easier to make at a landscape scale than at individual farms.

From a biological viewpoint the principles for building resilience in agricultural landscapes are reasonably clear:

- Protect residual forest areas from further clearing, where-ever possible, to conserve the overall biodiversity (recognizing that this will depend on the level of deforestation that has occurred and on the need for land to ensure food security).
- Reforest buffer areas to increase the size of smaller forest fragments to reduce the risk of them being damaged and to protect 'core' forest needed for species that are forest specialists.
- Use reforestation to enhance connectivity within the landscape. This might be done by creating corridors linking existing forest patches or by planting discrete new forest areas that break up an otherwise deforested agricultural matrix and provide stepping stones for biota to assist them move across the landscape.
- Use reforestation to stabilise exposed slopes and prevent soil or nutrient movement.
- Increase the overall number of species functional types used in reforestation even though only one or a few species might be planted at any single site. That is,

increase the collective or gamma diversity across the landscape while tolerating, if necessary, a low alpha diversity at individual sites.

These principles point to the fact that some areas of the landscape are more important than others. This means some land might be used largely for production provided other areas can be used for these other purposes. By doing so the landscape retains more of its biota, its functionality and its capacity to adjust to future changes.

Economic resilience is enhanced by increasing the variety of goods and services produced by reforestation and the variety of markets into which these can be sold. The most obvious way of doing this is to increase the species used and the types of plantations in which these are grown. But would not every farmer seek to produce only the most saleable product? Many will indeed do this, especially those close to a major mill or factory. However, household circumstances differ and others might choose to other species because they are too distant from that key market or because they are uneasy about relying on just one (unreliable?) buyer. Resilience can also be enhanced by taking advantage of these different approaches or pre-dispositions by encouraging businesses and markets with which such growers might form economic relationships. Examples might include small rural sawmills, furniture factories, handicraft suppliers and sellers of honey, traditional medicines and fuelwood. All of these businesses operate at an ideal scale for small growers and can be widely distributed across the landscape and close to growers rather than being restricted to a few distant urban areas. The usual impediments to the development of these businesses are discriminatory regulations and taxation systems as well as poor business skills. There are a variety of methods and policies that can help overcome most of these impediments (Molnar et al. 2007). Such assistance might be especially useful during the transition from a forest industry that is based on large sawmills handling large logs from natural forests to a new industry based on plantations and dominated by much smaller log sizes and NTFPs.

Social resilience is enhanced when communities develop a capacity for self-organisation and adaptive management. This necessarily develops at a landscape scale because this is the scale at which social practices operate and evolve. The learning networks described previously in Chapter 10 are one example of the way in which the self-organisational capacities of farmers could be developed. But forest landscape restoration involves many other community groups. These include members of the business community interested in buying forest products, other farmers who don't grow trees and, quite probably, a large number of external stakeholders. These might be users (and potential buyers) of ecological services such as water supply bodies, farm irrigators and conservation groups. All of these are likely to have some degree of interest in the way reforestation is undertaken. Resilience is likely to be enhanced when they can take part in the discussion concerning where and how reforestation will be carried out. The nature of these decision-making processes will be discussed in more detail below.

In short, landscapes offer ways of enhancing the resilience of social-ecological systems that complement those operating a site level. Some of these occur because of the trade-offs that can be made more easily within the landscape mosaic while others are a consequence of processes that only operate at this larger landscape scale. Together these can create a mutually re-enforcing feedback system that also generates socially preferred outcomes.

How Much Reforestation?

One of the primary tasks of Forest Landscape Restoration is to improve overall ecosystem integrity thereby enhancing the functional effectiveness of the communities present. The most obvious question is how much reforestation – natural or planted – is needed to do this? Of course this will partly depend on how much forest still remains in the landscape and on the quality of that forest. In this respect different wildlife species will regard the quality of these forests rather differently than humans. But the answer also depends on whether the responses to reforestation develop monotonically or whether there is a certain minimum area that must be reforested to achieve certain outcomes. The notion of thresholds is difficult because they are characterised by non-linear dynamics and by multiple-factor controls that operate at a diverse range of spatial and temporal scales (Groffman et al. 2006). From the viewpoint of someone planning reforestation at a landscape scale the problem is made even more difficult because of the need to reconcile ecological imperatives with economic realities; reforestation that improves the connectivity between two remnant patches of forest may have minimal effect on improving the livelihoods of people living in that particular area. If this is indeed the case it may be difficult to carry out.

How Much Reforestation is Needed to Improve Biodiversity Conservation?

Most ecological research to date has considered the reverse of this question – how is biodiversity affected by habitat loss? Numerous studies have consistently shown that deforestation results in a proportional loss of forest species. But once the residual area falls below 20–30% cover these studies suggest the spatial pattern and fragment size also become important (Andren 1994). Similar results have been reported by Flather and Bevers (2002). It is unlikely this represents a general ecological threshold since there are considerable differences between species and responses to habitat loss and the populations of some species are affected by deforestation well before the forest cover falls to 30% (Fahrig 2001). Some species are also able to use agricultural matrix thereby diminishing the importance of the forest and non-forest difference. It is usually assumed these species must be habitat generalists but, as already noted, empirical evidence from studies in Costa Rica and Sulawesi has shown that even some forest-dependent species can be supported in spatially complex agricultural landscapes including some with less than 20% forest cover (Sekercioglu et al. 2007; Sodhi et al. 2005).

What, then, are the implications for reforestation? The species-area relationship suggests that 'more' forest is better and that a landscape with larger areas of forests will contain more species than the same one without. While it might be difficult to reach a forest cover of 50% that Soule and Terborgh (1999) argued is necessary to prevent further species extinctions, even modest amounts of reforestation should have some significant conservation benefits. It is not possible to predict how many species will be recovered if a certain amount of reforestation occurs. Nor is it usually possible to specify which particular species may be able to recolonise a particular site: it depends on the quality of the new habitats being created and on the requirements and tolerances of these species although we know that forest interior species mostly require large forest areas. Given all these uncertainties, perhaps the most realistic objective for biodiversity conservation may be to simply focus on reducing the adverse effects of fragmentation and the small size of residual forest fragments by using new plantings to increase connectivity within the landscape and, by protecting small fragments with buffer zones, to increase the viability of the species and communities they contain. Both activities will increase the overall area but the focus becomes the location and quality of reforestation rather than the size of the reforestation effort.

How Does Increasing Reforestation Area Affect Hydrology and Watershed Protection?

There is widespread empirical evidence showing that increasing the proportion of a watershed that is reforested will reduce the amount of run-off (van Dijk and Keenan 2007). This is largely because of increasing amounts of evapo-transpiration. The change is an incremental one and does not involve a threshold although there is some evidence that reforestation of a cleared land must exceed 15% of the area before an effect is strongly evident (van Dijk and Keenan 2007). The actual rate at which run-off is reduced in a particular watershed will depend on the types of species used for reforestation and on their ages. An area may be reforested over several years and so include trees with a variety of age classes and involve plantation with differing rotations. These differences will affect the magnitude of the overall hydrological changes that occur at particular times.

In some locations reforestation will have significant consequences for down stream water users, especially where agriculture is expanding and there is already competition amongst farmers for water. As seen earlier, plantations may also reduce water flows in the dry season when the demand from irrigators can be highest. But ultimately, the question of whether extensive reforestation generates too many hydrological disadvantages becomes a question of scale. It may be possible to reforest a high proportion of small watersheds but it is less likely that all of a large watershed will be reforested so that, in many cases, the overall impact of reforestation on regional water supplies may be modest.

The situation is less clear in the case of the relationship between reforestation area and soil erosion because of the interactions between rainfall intensity, slope and soil conditions. In addition, species other than trees are often quite effective in preventing erosion and mixtures of trees and shrubs are likely to be more effective than simple tree monocultures. In areas with moderate slopes the area (or proportion) of forest cover is likely to be less important than the spatial location of this cover. On the other hand, a complete vegetative cover may be necessary to prevent erosion in steep landscapes and Turkelboom (1999) recommended this should occur on cleared land with slopes exceeding 60%. Whisenant (1999) also argued that 'repair programs should address the largest scale at which the process damage occurred' since large scale problems are not adequately repaired at smaller scales.

How Much Reforestation is Needed to Generate Socio-Economic Benefits?

Putting aside the matter of opportunity costs, large forest areas are commercially more attractive for industries using forest products than small areas although the magnitude of the benefit also depends on the types of species used and the location of the reforested area with respect to markets. Those with small plantations are only able to offer small volumes of goods at infrequent intervals putting them at a substantial disadvantage in most market places. This might not matter so much if there are many small plantations and growers are able to form a marketing cooperative. But if the overall plantation area is itself small then there may be no market at all. This is what happened when logging of natural rainforests in north eastern Australia ceased in the late 1980s. Some landowners assumed the price for high quality cabinet timbers would soar as supply declined and began planting trees. But, in the meantime, a lack of logs meant that local prices for these timbers collapsed. Once this became known it reduced the financial attractiveness of growing these species in plantations.

The problem is not only one of having enough plantation area to form a commercial resource but also of being able to develop a harvesting schedule that permits the operation to be sustainable. The risk is that a small plantation might be harvested in a single operation. The most common way of avoiding this is to ensure that plantation contains a range of age classes and that harvesting operations are carefully regulated. If the primary markets are local sawmills that are largely dependent on these resources then the plantation area must also be big enough to sustain their annual demands. These are the types of problems facing forest managers in the highlands of Papua New Guinea seeking to persuade growers to keep enlarging their plantations while delaying the start of harvesting (Chapter 10).

Size also matters when secondary forests are being managed for subsistence or commercial purposes since harvesting rates must match the productive capacity of the forest. In many cases this productive capacity is low, especially in young forests, meaning that small areas of secondary forest can be easily degraded if harvesting rates are not strictly controlled.

In summary, there is no clear answer to the question of how much reforestation should be done since it depends on the extent of deforestation, the conservation and other environmental threats, the opportunity costs of reforestation and on the state of the landscape mosaic. In principle, large areas are usually better for both conservation and commercial reasons. On the other hand, changes in water yields from a particular watershed are more affected by the proportion that is reforested than by the actual area covered. In practice, it may be difficult to develop many new, large forest areas in landscapes with numerous small landholders although it may be possible to increase the overall forest area.

Where to Undertake Reforestation

Though total forest area is important, the location of any new forest is also important. This is because large but poorly sited reforestation blocks may not be as effective for conservation purposes or even as economically valuable as smaller but better-placed plantings. On the other hand, it may be better to begin work with opportunities provided by the landscape such as eroding hills or exposed areas unsuitable for agriculture. These are sites that everyone can agree need reforesting. Once the process of reforestation begins and is seen to be successful it may be easier to then extend it to more contentious sites.

Where to Reforest to Improve Biodiversity Conservation?

There are two possible ways of answering this question. One would be to identify locations where a new area of forest will help conserve existing biota, especially those most vulnerable to extinction. A second approach would be to increase the connectivity between forest remnants to allow species to move across the landscape once more.

One way of using the first approach would be to reforest areas around patches of natural forest since these often retain significant numbers of species including those that are vulnerable to local extinction (Table 11.1). New forests at these locations can act as buffer zones which protect the remaining forest from further disturbances. Once the new forests mature they effectively increase the overall forest area and thus the conservation value of the forest patch.

Some remnant forest patches are more critical than others and there are at least four criteria that might be used to evaluate them. One would be based on the species they currently support; thus a recently formed remnant that still contains a comparatively large number of species would normally deserve a higher priority than a degraded patch that had been isolated for a longer time or a patch of young secondary forest that had fewer species. Similarly, a remnant containing vulnerable or endangered species (or forest type) would normally deserve a higher priority than a remnant without such species.

Table 11.1 Priority locations for reforestation to achieve particular outcomes

Type of outcome needed	Location in landscape
Improved biodiversity conservation	Buffer strips around residual forest patches, especially those containing endangered species or threatened by fire or grazing
	Patches of residual forest linked by corridors and stepping stones including forests on altitudinal or latitudinal gradients
	Riparian zones (especially in seasonally dry areas)
	Strips in fire prone areas to act as breaks
	In areas of extensive grassland where natural regeneration is unlikely
	Any areas within the non-forest matrix
Reduction in erosion and sedimentation	Lands with steep slopes
	Actively eroding lands
	Riparian strips or belts
	Areas with high-intensity rainfall and likely to be more affected by erosion
	Areas with young secondary forest
	Coastal protection zones
Change in hydrological functioning	Lands with compacted soils and poor infiltration capacity
	Waterlogged areas (induced by prior deforestation?)
	Recharge areas (when there is a need to reduce groundwater levels to control salinity)
Livelihood improvements	At sites that are marginal for agriculture (low soil fertility, steeper slopes)
	Sites close to roads or markets (to reduce transport costs)
	Sites distant from roads (to lower opportunity costs)
	In clusters with other growers to generate economies of scale

A second way of evaluating remnant forest patches would to use the size; a small patch might be seen as more deserving of a buffer strip than a larger patch because species-area relationships suggest a small increase in area will lead to a greater increase in its capacity to conserve biodiversity than would the same amount of restoration adjoining a large patch (Fig. 11.2). But not all small remnants deserve being prioritised in this way and there is probably a size below which the remnant might be seen as being simply too small to have any conservation value (<10 ha?) unless it could form part of a corridor between several larger remnants. Likewise, some larger patches containing especially vulnerable biota may be in particular need of surrounding buffer zones to protect them from future disturbances. When increasing the size of a remnant it is probably better to aim for compactness where this is possible to reduce the length of the perimeter and, thus, problems associated with the ‘edge effect’.

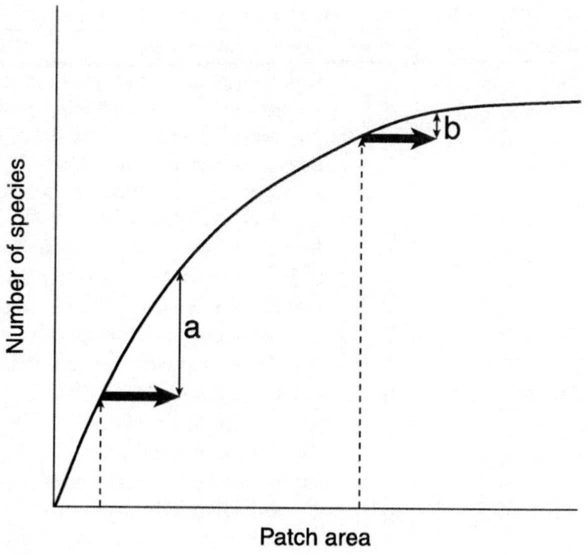

Fig. 11.2 The species-area relationship suggests more species can be conserved by building buffer zones and enlarging small remnant forest patches (**a**) than when a buffer zone of the same area is established around a large patch (**b**). This assumes colonists can reach the new forests and ignores the conservation status of these species. In fact the smaller remnant may only provide habitats for generalists and that a buffer zone around the larger remnant of natural forest safeguards species with larger home ranges and more specialized habitat requirements

A third way of evaluating patches is to take account of their location. A patch of remnant forest in the middle of an extensive agricultural landscape and distant from intact forest might be considered as a repository of the 'living dead' containing species that are reproductively isolated from the main population centres in intact forest areas and therefore doomed to extinction. Alternatively, it might be seen as a key location upon which a network of corridors or stepping stones might be built in order to foster landscape heterogeneity and facilitate species movement across the landscape. This judgement depends on future plans. If an extensive reforestation program is unlikely then it might be best to abandon the site and choose alternative locations for reforestation. Alternatively, if future nearby reforestation activity is likely, it could be useful to immediately enlarge the remnant to protect it from further disturbances and to safeguard the existing biota.

Finally, the risk that an existing remnant will be being logged or cleared in the near future might be an important determinant of reforestation priorities. Even a simple monocultural plantation surrounding a remnant of natural forest might be enough to protect it from those who would otherwise regard it as unclaimed or under-utilized land.

A second location within a landscape deserving a high priority for reforestation would be areas between natural forest remnants, especially in landscapes where the agricultural matrix represents a barrier to species movement. Reforestation that created corridors or stepping stones between these forest patches would help improve connectivity and allow species movement between meta-populations or enable species to recolonise parts of the landscape from which they had disappeared. These corridors arranged along altitudinal (i.e. ridgetop to gulley) or latitudinal gradients could also help species adapt to changes induced by global warming. Plantings along drainage lines might be seen as another form of corridor and would be especially important in seasonally dry areas where rivers are often centre of biological activity. Such plantings can sometimes support wildlife movement within a surprisingly short period. Jansen (2005) found a number of forest-dependent birds had moved into restored forest that formed a corridor between two rainforest patches in north Queensland within 3 years of it being established. The study suggested it was likely to allow movement of many forest species once the corridor forest matured. Other possible locations for reforestation corridors include areas around fire-prone sites where the new forests could act as 'green breaks' to limit the spread of fires. These corridors and stepping stones are obviously at the expense of spatial compactness that was described earlier as something to aim for when enlarging existing remnants. That is, there is a compactness-versus-connectivity trade-off.

All of these locations are places where reforestation would have to be carried out by planting seedlings. Places where natural regeneration is already underway should deserve some special consideration since the cost of this form of reforestation is relatively low (and mostly takes the form of protecting the site from further disturbances) and because these secondary forests can often contain large numbers of species.

Sites that might be given a lower reforestation priority are those where it would be difficult to exclude recurrent disturbances such as fires or grazing animals or that are likely to be regularly invaded by weeds because there is a large nearby source of these that cannot be eradicated.

Where to Reforest to Improving Ecosystem Functioning?

Much reforestation is carried out to prevent erosion and protect watersheds (Table 11.1). Erosion is usually accelerated in deforested areas because there is more overland flow which transports soils (and nutrients) across the landscape and into waterways. Areas with steep slopes and existing gulley or sheet erosion are obvious targets. Reforestation of such areas using block plantings or bands of trees along contours will intercept overland water flows and help trap sediment and any areas where natural regeneration is taking place are obvious priority areas for protection. Areas with high-intensity rainfall might be especially targeted. Agricultural lands are a common source of stream sediments and reforestation of riparian areas along streams will help limit the amounts of sediment and nutrients reaching waterways (van Noordwijk et al. 2007). Such riparian plantings should be carefully designed and evidence from McKergow et al. (2004) suggests that unless these strips are wide (>15 m) or have dense understories they may not be very effective. Sediment may be temporarily trapped but then remobilised during subsequent rainfall events.

Sometimes reforestation is carried out to change hydrological flows (Table 11.1). One way of doing this is to increase the infiltration capacity of the topsoil. Many degraded lands have compacted or eroded topsoils with limited infiltration capacities. Reforestation can increase the topsoil infiltration capacity by producing litter, adding organic matter to the soils and improving the diversity of soil fauna. As a consequence, more rainfall enters the soil and less is lost from the site as overland flow. Guo and Gan (2002) and Guo et al. (2003) describe a situation in China where strategically targeted areas in a large watershed were identified for reforestation in order to maintain river flows in the late dry season for hydro-electric power generation. They used a GIS to map the location of 90 vegetation-soil-slope units across the watershed. Many of these were represented by degraded forests, shrublands or grasses. They were able to identify those units where water retention was likely to be greatest if they were reforested.

But reforestation also increases water loss through evapotranspiration. This might be a desirable outcome if the site has become waterlogged because of prior land uses but may be an undesirable outcome if it is important to maintain the overall water yield. The impact of reforestation varies in different parts of a watershed. It is known that soils tend to become deeper when moving from catchment ridges to lower slopes. These soils acquire and store water from surface run-off and groundwater movement. Because of this, planting trees on lower slopes has a greater negative effect on streamflow than planting the same area on upper slope positions, at least in locations where water is seasonally limited (van Dijk and Keenan 2007). But beyond such generalisations, the significance of the spatial location of plantations for run-off is not fully understood and much of our present understanding comes from hydrological models rather than empirical evidence. A great deal of this work has been done to find ways of controlling salinity by lowering groundwater levels or reducing groundwater recharge (Stirzaker et al. 1999; Vertessy et al. 2003). Van Dijk et al. (2007) found carefully targeted planting locations could be up to seven times as effective as random plantings in changing water tables and affecting stream salinity. They noted the magnitude of the change was dependent on the scale of reforestation and was rather less at smaller scales

But how should one rank these alternatives? Should one tackle the most degraded site first or the easiest sites? In some situations it may make sense to begin work at the latter since a much larger area of degraded can be treated for the same amount of money and there is a greater chance of quickly generating beneficial outcomes. And success breeds success: once reforestation is shown to be effective it may be easier to generate resources or community support to treat the more difficult areas. On the other hand, some highly degraded areas such as eroding hillsides or old minesites may demand early attention because of the amount of sediments or toxic leachates they generate.

Where to Reforest to Improve Livelihoods?

The best locations for commercial tree plantations are those with gentle topography and good soils and close to roads or markets (Table 11.1). These, of course, are also the best locations for most agricultural crops so the opportunity costs of reforestation at such locations may be high. In practice, most plantations are usually established at areas with poorer soils or on steep lands that are distant from roads or dwellings. From an economic perspective tree-planting then becomes a useful complementary land use activity because it can make better use of sites that are marginal for agriculture. But, even in these situations, some farmers may still regard tree-growing as an inferior land use and prefer crops or other land uses. The main exception to this pattern of relegating reforestation to more marginal areas occurs at sites close to a large timber market such as a paper mill. In such cases many farmers will find it advantageous to grow trees as well as food crops irrespective of the agricultural merits of the land.

Plantations grown in steeper and more distant areas are subject to higher harvesting and transport costs. The financial viability of these plantations may be marginal unless the timbers or NTFPs produced are especially valuable. The problem is even greater when there are only one or two small plantations in the area. On the other hand, a cluster of plantations may transform the economics because there is a larger overall resource of different ages which allows for a regular harvest. Clustering of smallholder plantations can have other advantages as well such as improving opportunities for better fire protection and fostering tree grower cooperatives to assist in marketing.

In summary, the areas where reforestation should be carried out depend on the location of degraded areas within the landscape mosaic and the present distribution of intact forest. Some areas may deserve priority because of their particular conservation significance or because they deal with a severe erosion problem. Other locations may have economic advantages. But there may be some locations where reforestation can achieve multiple benefits. For example, reforestation of steep areas may protect watersheds, provide buffers around remnant forest or create corridors between forest patches and have still low opportunity costs for farmers. One of the tasks for those undertaking forest landscape restoration is to look out for complementarities like these. Some areas may be difficult to reforest and should not be given a high priority because the costs of doing so become too high. These include areas those where repeated disturbances are hard prevent, where land disputes are common or where the costs of reforestation are high in comparison with the benefits generated.

What Types of Reforestation at Particular Locations?

Four main types of reforestation have been described in previous chapters. These included plantation monocultures, mixed species plantings, Ecological Restoration and secondary forest regrowth. Each differed in their ability to produce commercially valuable products as well as in their capacities to conserve biodiversity, protect watersheds or provide other ecological services. Each is also likely to appeal to a different group of stakeholders with landowners and commercial investors such as corporate forest growers being especially interested in types of reforestation that generate income while external stakeholders might be more interested in forms that reduce stream sedimentation or improve wildlife habitats. The two most likely to be used to cover large areas are plantation monocultures (because they are attractive to large industrial enterprises) and natural regrowth (because it has a low cost).

There appear to be three ways of deciding the type of reforestation to be used at particular locations. One approach might be to identify those parts of the landscape mosaic where the choice is constrained by site conditions. For example, a simple plantation containing a single tree species tolerant of the particular site conditions might be the only way of reforesting a badly degraded site with infertile soils. Likewise, a mixed species plantation involving deep-rooted trees and a ground cover might be the best way of reforesting an eroding and unstable slope. This approach fixes ecological problems but may not be attractive in landscapes that are not severely degraded.

A second approach might be to use forms of reforestation at particular sites that maximise certain benefits. For example, a plantation monoculture using fast-growing pulpwood species may be highly profitable in areas immediately surrounding a pulpwood mill but be much less so in more distant locations. Likewise, Ecological Restoration that formed a corridor between two patches of secondary forest may help ensure the survival of an endangered species in one area but have little impact when planted as an isolated new forest in the midst of farmland. This approach means different types of reforestation are used in different locations. It increases landscape heterogeneity but necessitates spatial trade-offs across the landscape and requires individual land owners to agree with these trade-offs.

A third approach might be to seek out landowners interested in diversifying income sources rather than simply maximising income. Some of these communities and households are likely to be already interested in tree-planting and especially in forms of reforestation yielding multiple benefits. They might be those with land distant from their dwellings or those owning land in steeper areas where the opportunity costs of reforestation are lower than in flatter areas. Reforestation might be carried out using mixed species plantings involving high-value species or it might it might be possible to take advantage of secondary forest regrowth. As with the second approach, landscape heterogeneity is increased although the types and spatial distribution of forests resulting from this approach might be different.

In fact, it is unlikely that any one of these approaches would be followed exclusively. What seems far more probably is that all three approaches might be used in different parts of a landscape and that much will depend on the attitudes and circumstances of individual landholders. The issue, then, becomes one of reconciling these different approaches to achieve the best overall outcome?

Planning Forest Landscape Restoration

This review of how much, where and what type of reforestation might be done highlights the fact that reforestation at a landscape scale involves trade-offs and that some kind of planning is needed if reforestation is to be more than simply a series of small, disconnected and ad hoc decisions by individual landowners or managers. But it is difficult to be prescriptive about this process because of the wide variety of situations that exist within what was referred to earlier as the socio-economic mosaic. Landowners or managers are the primary stakeholders in the planning process but there is another group of secondary stakeholders living outside the area who see themselves as having a legitimate interest in the outcome of any decision-making process. These external stakeholders include downstream water users, conservation groups, townspeople who use some of the area for recreation and state agencies who see themselves as representing society as a whole. Many of these might wish to alter the types of decisions being made by farmers and have them plant more trees or plant particular types of trees using alternative planting designs. The task, then, is to find trade-offs that reconcile these various objectives and that, as far as possible, optimise the outcomes. The more stakeholders involved then the more difficult this process will be.

Top-Down or Bottom-Up Planning?

In many countries land use planning is a largely top-down process carried out by technical specialists working for a government agency and following prescriptions or guidelines. Sometimes this can work well but the history of many large-scale rural planning operations show these are often fraught with problems. Institutional failures have been common across the region and, in many cases, seemingly elegant plans have left rural communities in desperate situations (Scott 1998). The Indonesian Mega-rice project described in Chapter 2 is a good example.

Despite these problems, top-down approaches still remain attractive because they seemingly allow planners to have an over-view of the problem enabling them to make any trade-offs more efficiently than individual stakeholders. Governments and some conservation groups would find this especially appealing. In recent years some quite sophisticated computer-based planning tools have been developed to take advantage of the of large conservation data bases that are becoming available (e.g. Chetkiewicz et al. 2006; Drielsma et al. 2007; Hargrove et al. 2005; Millspaugh and Thompson 2009; Thomson et al. 2009). These tools are often used to target the needs of just one or two species of interest or, perhaps, a group such as birds. Planning is carried out with the biological attributes and habitat needs of these species or this group in mind (Dobson et al. 1999). This approach addresses the current threat to these species and seeks to correct this within the existing biological and socio-economic environment by targeting areas to improve landscape linkages. Other modelling tools have been developed to simulate historical landscapes, project future landscapes or explore the consequences of various land use alternatives including reforestation (Mladenoff 2004; Polasky et al. 2008; Wimberley 2007).

These approaches may be efficient in finding an optimal biological solution but they are often difficult to implement and, at least in many tropical countries, are sometimes politically unrealistic (Chomitz et al. 2006; Knight 2006; Knight et al. 2006). There are at least six problems in using this essentially technical and top-down approach. Firstly, there are usually a variety of species of interest threatened by deforestation and fragmentation and it is rare to know much about the biological attributes of these or of their habitat preferences. In most situations it is necessary to promote forms of landscape reforestation that are likely to benefit many species and not just one. These species may have quite contrasting habitat requirements. Secondly, the process makes debateable assumptions about trade-offs between different environmental benefits. Reforestation for biodiversity conservation may be important but so too is reforestation for watershed protection purposes. Reforestation may often improve both but sometimes a choice must be made. Thirdly, from a political point of view, the task is not one of minimizing the cost of achieving a certain outcome but of determining what kinds of environmental benefits can be achieved with the funds available and under the constraints imposed by the views and aspirations of stakeholders. Fourthly, the optimisation process focuses on where to intervene but not on how to induce landowners to comply; it relies on either compulsion (which is politically costly), compensation (which is costly) or on universal cooperation (which may not be forthcoming). Fifthly, it is an overly static process with a short time perspective that pays little attention to the often dynamic ecological and economic circumstances prevailing in these landscapes. Finally, most of these approaches depend on large data bases, modelling expertise and institutions able to undertake this work and these are rare in most tropical regions.

Experience suggests a more consultative planning process is needed involving more than governments or conservation NGOs. Forest Landscape Restoration may have to begin as a top-down process because many stakeholders will be unaware of the broader context and not know of the scale of the problems that past deforestation or land degradation has created. Nor will they have access to tools or resources to develop the planning process. But, once these other stakeholders become involved, they are likely to generate ideas and initiatives. And when this happens the process is neither top-down nor bottom-up but a mixture of both. Shaping the new landscape then becomes a matter of negotiation as much as one of planning.

A variety of tools are being developed to assist the planning-negotiation process. At one end of the spectrum are the sophisticated computer-based tools referred to earlier (Chetkiewicz et al. 2006; Drielsma et al. 2007; Polasky et al. 2008). These may have a role to play in some situations but, by themselves, are rarely sufficient to drive the process. At the other end of the spectrum is a more informal process that tries to take account of biological goals but is a more consultative process that Sayer et al. (2008) describe as 'muddling through'. By this they mean a sometimes messy process that tries to help build human capacities and foster institutions in order to improve forest landscape restoration outcomes at a local level. This might not seem to offer the 'purposeful and strategic' planning referred to earlier as being necessary for Forest Landscape Restoration. But, in fact, it does because it aims to build a consensus among stakeholders around a plausible set of compromises and helps set in place a process of adaptive management which allows goals to be modified over time in the light of the feedback

Table 11.2 Circumstances favouring use of alternative approaches to planning forest landscape restoration

Attribute	Optimising computer models best when	'Muddling through' best when
Numbers of stakeholder	Lower	Higher
Funds for incentive payments and compensation	Ample	Limited or unavailable
Scientific knowledge, data bases and maps	Substantial	Patchy
Strength of institutions and legal frameworks	Stronger	Weaker
Formal land tenure	Widespread	Less frequent

received. The approach is described in a little more detail below. The circumstances where these two contrasting approaches – modelling and 'muddling through' – might work best are shown in Table 11.2. Many situations, if not most, probably fall somewhere between these two extremes.

Steps in Planning Reforestation at a Landscape Scale

Experience from a wide variety of field situation suggests there are a number of elements in a participatory planning process aimed at fostering Forest Landscape Restoration (Gardner et al. 2009; Gilmour and Fisher 1991; Hobley 1996; Reitbergen-McCracken et al. 2007; Sayer et al. 2007; Shepherd 2004). A synthesis of these is outlined below. A facilitator is usually needed to organise and manage the process, share information, resolve disputes and help ensure agreements are maintained over time. Government agencies or conservation NGOs are often the most common promoters of Forest Landscape Restoration and could also facilitate the planning process although they might run the risk of being seen to have a conflict of interest. In such cases a more independent facilitator or 'honest broker' will be needed.

The sequence below is outlined as if it is a linear process. In fact, feedback links between the various stages may be common and it is likely to be a rather more messy process than it appears here. Some of the possible feedbacks are shown in Fig. 11.3. The process of operationalising conservation planning, together with other approaches that might be used, is discussed in more detail by Sayer and Campbell (2004) and Knight et al. (2006). It is also important to recognize that some landscape-scale reforestation is only part of a large region planning program. One example of this is given in Box 11.2 which describes such a regional program in Fiji.

Stage 1: Develop a Landscape View of the Problem

The first stage is to develop a landscape view of the problem. This means assembling information and maps about the present landscape mosaic including the location and status of residual forests, especially those of high conservation value, and areas that probably should be reforested including sites with significant erosion problems. Information about the plant an animal biota present and their conservation status might also be gathered. This information will come from scientific reports and government sources but also 'traditional' knowledge held by landholders and long-term residents. Only in special cases will there have been a systematic conservation assessment. It is inevitable, therefore, that this knowledge will be imperfect and, in particular, it is probable that information on the diversity of species present, let alone their population dynamics or conservation status, will be incomplete.

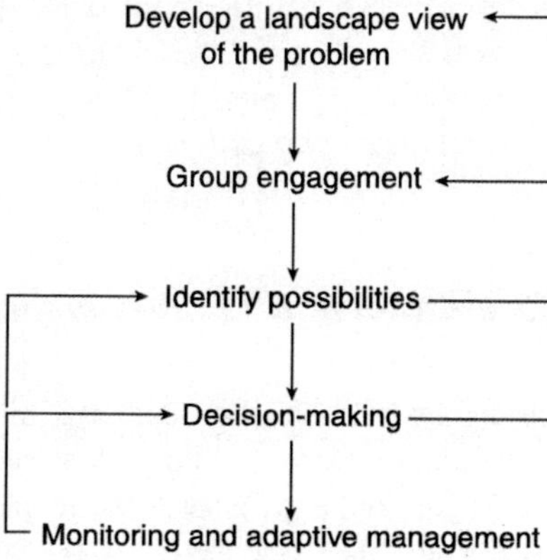

Fig. 11.3 Steps in planning reforestation at a landscape scale

Box 11.2 Forest Landscape Restoration as Part of a Larger Land Use Planning Program in Fiji

The Drawa area on the Fijian island of Vanua Levu has eleven traditional land owning communities and covers 6,300 ha. In recent years changes have occurred in these societies as the roles of traditional authorities have broken down and more families have challenged the rights of former customary leaders to regulate their activities and decide their land use practices. The situation has been further complicated by land ownership disputes and by increasing absenteeism. Some people have begun to leave the area seeking outside employment because of the limited economic opportunities available. During this period of social change, unregulated forest clearing by farmers seeking to develop cash crops has increased. All these events have caused a gradual loss of forest cover and enhanced rates of erosion.

A land use planning program has been carried out to improve agricultural productivity and incomes, safeguard exist forest areas and reforest some of the degraded lands (Fung 2008). The program is being undertaken using a

(continued)

Box 11.2 (continued)

process of intimate, participatory land use planning that links community groups and government staff. It involves multiple consultations as well as building the capacity of the community to be involved and eventually work independently of outside agencies. The intent of the plan is to ensure farmers make informed decisions about the agricultural lands they use, to develop new tree crops, including timber trees, and to improve the capacity of the landscape to sustain a future ecotourism industry. The plan forms part of a larger project involving the development of management plans for harvesting the natural forests in the area.

The process began with consultations being carried out across the community to gather views about how this might be achieved. Several community-base groups were involved including the Drawa Landowners Association and a Land Use Working Group. In the meantime the government assembled representatives from a variety of government agencies including agriculture, forestry and environment bodies as well as organizations like the National Food and Nutrition Center and the Department of Women to coordinate its activities using a land use planning group.

Following training, community-based teams then carried out surveys of the areas to assess current land use practices, document sites of ecological, historic or cultural interest and to settle land ownership boundaries. Participatory rural appraisals were undertaken to document current livelihood practices, community organisational structures and traditional knowledge and skills. Government staff also assembled maps of ownership boundaries, soils and forest types.

These various data bases were used by the Land Use Working Group to define areas with good agricultural land, areas that might be used for ecotourism and areas that needed to be reforested. A digital map was prepared showing where these various land use proposals would be located. These maps were then taken around to the stakeholders for verification and confirmation. Once an agreement was reached Memorandum of Understanding were signed with the relevant landowners concerning areas of conservation forest that are to remain undisturbed. The land use plan is being implemented by villagers under the supervision of the Land Use Working Group who also assist with advice and training. Monitoring is to be carried out by the Landowners Association of Drawa.

The land use plan is a pragmatic response to a series of economic and ecological problems and to an emerging social problem. Rather than being a top-down operation it has evolved as an intimate relationship between government technical experts and villagers. The planning process has created an agreed land use plan and its implementation is being overseen and monitored by a significant local body with the political strength to do so. The process has also left the community with new skills that will help them implement the plan and fine-tune it whenever this proves necessary thereby making them less dependent on outside bodies.

Information should also be collected on cadastral boundaries and current land use practices. For example, is deforestation continuing and, if so, what is the subsequent land use that is planned? As well, an effort should be made to understand the history of land use in the region and the current drivers of ecological and economic change. Are the prices of certain crops rising? Are tourist numbers increasing? Again, it may be difficult to obtain some of this information but stakeholders will be in a better position to make decisions if they have a common understanding of the conservation and economic circumstances in which they find themselves. Much of this information will have to be collated by government land use planners, scientists or outside consultants.

Stage 2: Group Engagement

The second stage is to identify all the key stakeholders with an interest in the area and the changes that reforestation might bring. Stakeholders differ in their degree of dependency and in the extent to which they will bear costs or share benefits. They also differ in their willingness or capacity to embrace changes and undertake reforestation. Those living within the landscape would be expected to be more wary than those living outside because they are the ones being asked to make the greatest immediate contribution. These participants are more likely to be willing to engage in some form of forest landscape restoration if they can see the need for collective action and believe they will benefit from being involved. Some of the factors likely to increase the willingness of landholders to participate are summarised in Table 11.3. Some of these concern environmental and economic attributes of the landscape while others are to do with attributes of the participants themselves. Particular efforts should be made to involve those living in parts of the landscape where reforestation is most needed and where it might be contentious. Of course not all such stakeholders may wish to be involved, especially those happy with the status quo. It may be that such people can be drawn into the process at a later stage through the use of incentives or compensation (see below). There may be land ownership or boundary disputes in some landscapes and it is at this stage of the process that these must be resolved. It may be especially difficult to accommodate the wishes of larger landowners or corporate bodies because of their commercial power and political influence.

The identification of stakeholders is only the first step in the process of engagement. Ways must also be found of ensuring all stakeholders are able to attend meetings, if they so wish (recognizing that participation comes at the expense of their other activities), and become included in the planning process. Unless this is done discussions will be dominated by the more articulate or more politically powerful stakeholders or speakers. People can also be disadvantaged by their ethnicity or class background and by their gender.

Sometimes the numbers of stakeholders are simply too large to be manageable. In these cases representatives of the various stakeholder groups will have to be selected. Such representatives should be able to attend successive meetings so that some continuity is maintained between meetings. Local participation and

Table 11.3 Factors likely to increase the extent to which landholders engage in forest landscape restoration (After Ostrom 1990)

Environmental and economic attributes favouring some reforestation	Attributes of the participants likely to favour forest landscape reforestation
Spatial extent of area: not so great that there is not a recognition of their inter-dependence and hence that they have a community of interest	Common understanding: participants have a shared knowledge of the needs for reforestation and the benefits that forest landscape restoration could bring
Sites: not so degraded that it would be too expensive to reforest critical areas	Discount rate: participants have a sufficiently low discount rate in relation to future benefits from reforestation
Existing plantations: existing plantations support the idea that tree-growing is feasible	Distribution of interests: wealthy and poor participants all perceive they will benefit from participation
Markets: there is a need and perhaps even a market for the environmental services and goods provided by reforestation	Trust: participants trust each other to keep to agreements
	Prior organizational experience: participants have some experience in working with others in local associations or groups

multi-stakeholder engagement is often easier said than done and Hobley (1996), German and Taye (2008) and Sayer et al. (2008) describes some methods that have been used to achieve representative participation and manage meetings.

Stage 3: Identify Possibilities

The primary task of the stakeholder meetings is to canvas alternative viewpoints and then arrive at some kind of consensus or 'vision' concerning the future landscape. Stakeholders will usually come to a meeting with differing understandings not only of the existing situation but also of the ecological and economic possibilities open to them. This means that the first step must involve a process by which these views are shared. This will include making available the information collected earlier about the existing landscape mosaic and its history as well as information about future economic and conservation possibilities. In addition, stakeholders will need to be advised about the ecological consequences and economic opportunities surrounding questions of how much, where and what type of reforestation might be carried out.

Participants will commonly have contrasting view about reforestation. Farmers will value reforestation according to the extent it will improve livelihoods (with women often having different priorities than men) while conservation NGOs will place a high value on the type of reforestation most able to conserve biodiversity. Government staff will usually give priority to land uses and forms of reforestation that will contribute to national economic goals.

Stage 4: Decision-Making and Priority Setting

This is clearly the most difficult stage. It can be difficult to comprehend the advantages and disadvantages of a variety of proposed land use changes, especially in spatially complex landscapes and a number of different tools have been developed to assist stakeholders evaluate alternatives. Maps, simple 3D models made of paperboard and plaster and other forms of visualisation are common methods of illustrating alternatives. Other ways might involve computer models to compare different scenarios, role playing games or economic analyses. In recent years a variety of market-based instruments have also been developed. Some of these different approaches are discussed in more detail in the following section.

The intent of this decision-making stage is to make the consequences of each scenario as transparent as possible to all stakeholders and the purpose of these tools is to share information in a way that is meaningful to participants. This means the type of tool to use will depend on the stakeholders present; simple maps may be sufficient in some cases but more complex tools may be needed in others. An example of how a complex tool that produced simple maps was used to resolve a somewhat tense situation in one area in northern Thailand is given in Box 11.3.

In some cases it is possible to reach an agreement relatively quickly in which case the discussion then moves to which areas should be tackled first, the practicalities of timetables, arranging finances and finding enough seedlings of the preferred species to plant. But reforestation schemes that bring benefits to the wider community sometimes do so to the disadvantage of individual landholders who own land in certain strategic locations. Trade-offs then become necessary with the aim of maximizing collective gains and minimizing individual losses. Trade-offs and compromise can be made more palatable if some form of (mostly financial) compensation is available to those who see themselves as bearing most of the burden but sharing few of the benefits of reforestation. The advantages and disadvantages of various kinds of incentives for reforestation have been discussed in Chapter 10. It was argued then that incentives that took the form of infra-structure such as roads or information about market were usually preferable to cash payments. These might not be always sufficient in the present circumstances and some more tangible and immediate forms of recompense may be necessary. Goldman et al. (2007) suggest alternatives including 'cooperation payments' for those in particularly important locations and the formation of 'ecosystem service districts' to reduce the transaction costs when payments for ecological services arising from reforestation are being paid. A particularly important incentive might be the offer of formal or conditional tenure to land users who happen to be, in a strictly legal sense, unlawful land users.

Not all disagreements can be resolved, even when compensation is offered, and sometimes compromises must be struck that might seem to be sub-optimal to many stakeholders. Sayer and Boedhihartono (in press) concede that, as conservationists, they find it easier to work with stakeholders who share at least part of their vision for a future landscape. But Forest Landscape Restoration is necessarily a long-term process and change may have to be incremental and depend on the availability of financial resources (Lamb et al. 2005). Indeed, this may be a distinct advantage,

Box 11.3 Evaluating Alternative Reforestation Scenarios

A traditional Hmong community living in the Upper Mai Sa valley in northern Thailand found it self enveloped by the Doi Suthep-Pui National Park when this was established in 1981. The villagers initially practiced shifting cultivation but, over time, have changed to more sedentary forms of agriculture. Most now grow vegetables and fruit trees. The villagers have neither Thai citizenship nor legal land tenure. Because of this they have had an acrimonious relationship with National Park managers who see them as illegal occupants destroying the conservation values of the Park.

A meeting was held to resolve these differences and to plant a reforestation program that would cover some of the deforested lands (D. Pullar, personal communication). On Day 1 facilitators from the University of Chiang Mai met with National Park staff to determine their view of the problems and to seek ideas about a way forward. On Day 2 a similar meeting was held with representatives of the villagers to seek their views. On Day 3 the two groups were brought together. The Head of the National Park described what he saw as the problem and how the villager's livelihoods might be met in future. A representative of the villagers then gave their perspective on the problems they faced and on a way forward. The facilitators then helped the two groups to link these views and develop some shared goals. This included having the groups acknowledge (a) that forest conservation was something that both groups supported and that some cleared areas should be reforested to protect water supplies and (b) that villagers could continue to practice agriculture on some of the land currently being used but that their future economic opportunities lay with tourism and employment outside the Park.

With this common understanding some prospective locations for reforestation within the Park were identified. The merits of these alternatives were compared using a Scenario Analysis tool which generated maps (derived from detailed satellite imagery) of the various proposals and where areas of current agricultural use and high conservation value could be easily seen (Pullar and Lamb in press). Everyone then visited all these sites in the field for further discussion. A final reforestation plan was then negotiated.

Two factors in particular appeared to help make the process successful. One was that the facilitators were well-known to both parties and had worked in the area for many years. Secondly, there were detailed maps showing exactly what each group had proposed. It was important that these could developed in time to be taken into the field on Day 3 where they gave participants confidence that they understood they trade-offs being made.

rather than a disadvantage, especially when plantation species are being tested and silvicultural systems are being developed in degraded landscapes with uncertain markets. Based on experiences emerging during the implementation of the large and top-down Sloped Land Conversion Program in south west China, Weyerhaeuser et al. (2005) explicitly recommend that pilot projects be undertaken before any large-scale reforestation program is implemented even when there is general support for reforestation. In any case, new reforestation opportunities may arise in future, particularly if populations move to urban areas (the forest transition described earlier as the 'economic development pathway') or because changes in the market prices of forest goods and ecological services (the 'forest scarcity pathway').

Stage 5: Monitoring and Adaptive Management

Implicit in any agreement is that certain outcomes should be achieved with a defined period if the group's landscape 'vision' is to be achieved. These targets should then become the basis for a monitoring program that uses a small number of easily measurable indicators that are able to reflect the changes occurring across the landscape and that show progress towards the goals or, alternatively, show when a change in goals or management procedures might be necessary. Gilmour (2007) discusses the particular role of adaptive management in Forest Landscape Restoration and views the process as one of an 'action learning cycle' in which managers use monitoring to observe and reflect on what happens when their plans are implemented and then draw lessons from this. Monitoring can also provide a form of quality control needed when payments are being made for an ecological service that reforestation is meant to provide.

Several sets of indicators are needed. The first set should assess progress in implementing the agreed reforestation program. Is reforestation being carried out according to the agreed timetable? Are seedling survival rates acceptable? The other sets of indicators might assess whether this reforestation is achieving the program's objectives. These might monitor changes to biodiversity and ecological functioning, improvements in the livelihoods of the various social groups present and developments in institutional and governance arrangements concerning land management. Some possible indicators are shown in Table 11.4. Many of these indicators will have a simple positive or negative (or a 'yes' or 'no') value. This type of monitoring program could be complemented by a simple network of permanently marked photo-points that provide a visual record of changes. Perhaps the ultimate indicator of success is how committed the stakeholders remain to the Forest Landscape Restoration process. Some other possible indicators are considered by the Landscape Measures Resource Center (http://www.landscapemeasures.org).

The design of any monitoring program poses a number of dilemmas. One concerns which parts of the landscape mosaic to assess? Should monitoring largely deal with the most recently reforested areas or should it cover the landscape as a whole? The recently reforested areas are those where changes will develop most rapidly but it is usually the collective impact that is of most interest which means the whole landscape must be monitored. Nonetheless, some locations within this landscape will obviously be more critical than others. Such points should be

Table 11.4 Potential indicators for monitoring the implementation as well as the outcomes of a Forest Landscape Restoration program

Goal	Potential indicator
Reforestation plan implemented	The planned reforestation is being carried out
	All proposed new forest locations are being planted
	Seedlings are being tended and protected from weeds and pests
	Survival and growth rates of all tree species acceptable
Biodiversity and functioning	There is no further deforestation of residual natural forests
	Wildfires are being excluded
	There is improved connectivity between forest patches
	Populations of endemic flora and wildlife are being maintained (or are increasing)
	Weeds and pest not spreading
	Key species are using corridors or new forest areas to move across the landscape
	Reforestation has reduced erosion and water quality is improving
Livelihoods	Food production is stable or improving
	Household incomes are improving
	Land prices are stable or improving
	Reforested areas are generating income (from goods or services)
	Prices of forest products (eg. firewood, timber, NTFPs) are stable
	There is increased employment from tourism
Institutions and governance	All stakeholders remain supportive
	The planning group continues and its decisions are being implemented
	Incentive or compensation schemes remain effective
	Dispute resolution mechanisms in place
	State agencies are effective and supportive
	The need for external support and incentive payments is decreasing

marked in the field and mapped so the same location can be used over time. A second question concerns the frequency of monitoring. More frequent assessments mean changes will be picked up quickly but the process will then be more expensive than a less frequent assessment. Most of the indicators in Table 11.4 can probably be satisfactorily reviewed annually but the indicators concerned with biodiversity are likely to be different.

Monitoring changes in the size and distribution of plant and animal populations will always be difficult and require quite specific assessment procedures and sampling protocols if they are to be done well. The problem is that biologist with the skills to undertake these types of studies are usually in short supply in most tropical regions. Of those present, many will be associated with universities or NGOs and will be unavailable for long-term monitoring. In most circumstances the best option may be to simply monitor the development of structurally-complex forests containing a variety of vegetative life forms and food trees and use these as biodiversity proxies. The assumption would be that the development of structural complexity will reflect improvements that favour biota with more specialised habitat requirements. This may be sufficient if the task is to encourage the recovery of biodiversity in general but may not be so useful if the focus is on restoring populations of particular endangered species.

The task of monitoring is not a trivial one. Forest Landscape Restoration takes time and it may take some years before benefits are observed. There must be a mechanism in place to provide the funding and other resources to carry out monitoring over this time and interpret the data as it is collected. The best approach may for it to be done by a government agency under the supervision of a sub-committee of the stakeholder group but there may be alternative mechanisms. A balance has to be struck between developing an elaborate monitoring system that cannot be sustained because it is too costly and having one where monitoring is so superficial as to be worthless in detecting changes. In the end it is probably better to use a few, carefully chosen indicators than a larger and more comprehensive set. But the over-riding test of any program must be whether or not it gives feedback enabling repairs and a process of adaptive management to develop (Gilmour 2007; Sayer et al. 2007).

The advantage of a monitoring program which allows stakeholders to adaptively manage the reforestation program is obvious. But there can be an additional benefit in circumstances where landholders begin to have second thoughts about their continued participation. Monitoring can provide evidence of the changes that are indeed underway and also of collective progress towards the mutually agreed vision. Affirmation of continuing progress may be critical in maintaining the participation of key landowners in what will normally be a long-term program.

Approaches and Decision-Support Tools for Forest Landscape Restoration

Forest Landscape Restoration is likely to be easiest when there are a relatively few stakeholders and an even smaller number of landowners. The task is more difficult when there are many stakeholders with alternative views about how the landscape mosaic should be transformed. A variety of tools are being developed to assist decision-making. Some of these have a strong spatial element while others are more concerned with helping landowners and communities adjust to changing ecological and economic environments and improve their adaptive capacity (Sayer and Campbell 2004; Van Noordwijk et al. 2001). Those outlined below are simply an indication of some of the approaches being used.

Visualisation

Discussions about alternative reforestation plans are invariably helped by maps showing which parts of the landscape might be planted and how much could be done at each site. Different stakeholders will usually wish to develop their own set of maps. The advantage of these maps is they make clear who is most likely to be affected by any particular proposal. They can also help clarify ownership boundaries. Such maps can take a variety of forms ranging from GIS print-outs to satellite imagery (e.g. via Google Earth). Sometimes simple 3-D landscape models can be useful as well (Fig. 11.4).

A particular type of visualisation referred to here as 'rich maps' has been described by Boedhihartono and Barrow (2008) and Sayer and Boedhihartonon (in press). They get small groups drawn from the larger consultative group to produce hand-drawn maps of their land. The scale of these should be specified (e.g. 1:50,000 or 1:100,000) so that the maps refect the landscape as a whole rather than the area of particular interest of an individual. Each group is asked to draw the landscape and its current uses as they perceive it. They are then asked to draw alternative visions about how they would like the landscape to change. Different visions emerge depending on the composition of the group and maps produced by men's and women's groups can be quite different. The artistic skills of the map-makers vary and in some cases it can be useful to scan the images and manipulate them on a computer. These 'rich maps' then become the basis for discussions by the larger group (Fig. 11.5).

Another form of visualisation is the use of imagery produced with the aid of a computer. These images can take several forms (Pettit et al. 2008). The simplest are digital pictures of scenery showing the landscape before and after reforestation. The intent of these is to help the viewer to appreciate what the changes might mean.

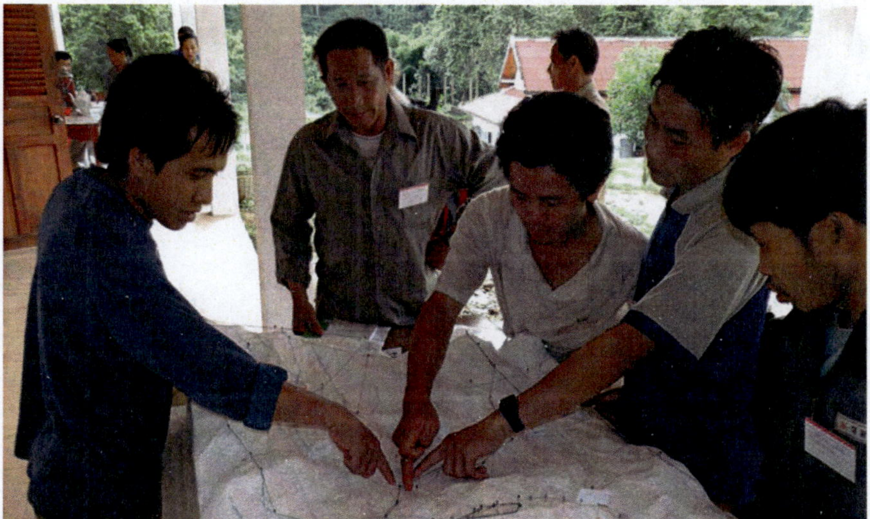

Fig. 11.4 A 3-D plaster landscape model being used in Laos PDR to discuss land use options (Photo: Jean-Christophe Castella)

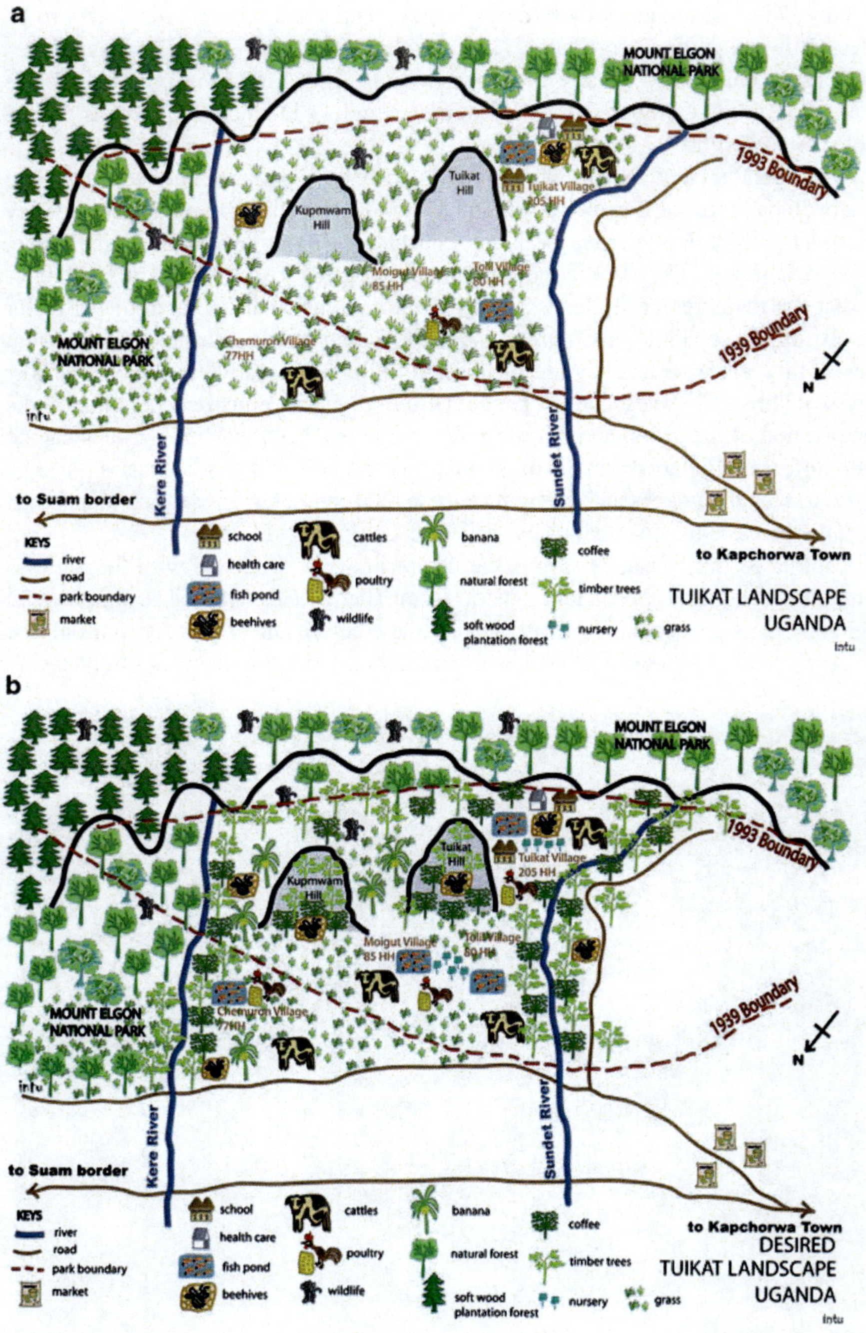

Fig. 11.5 Rich maps drawn by smallholders showing (**a**) their perceptions of key biophysical and socio-economic attributes and (**b**) where reforestation might be carried out. The *dotted lines* show the former boundary of the forest in 1939 and in 1993 (Source: Itu Boedhihartono)

More sophisticated techniques allow the viewer to examine the image from a variety of perspectives. Though technically impressive, it is not clear that these presently offer many advantages over simpler technologies and they can be quite costly to produce.

Scenario Analysis

Different visions of the future are sometimes referred to as scenarios. The maps or images referred to above might describe alternative scenarios with various stakeholders putting forth their own scenario for consideration by the group as a whole. They can be very useful in cases where farm sizes are small or tenure is insecure. But maps alone are sometimes not enough for considered judgements to be formed or trade-offs made. In some situations an idea of the functional consequences of each reforestation scenario and its implications for livelihoods is also needed.

One approach to analysing the advantages and disadvantages of the alternative scenarios proposed by different stakeholders has been outlined by Pullar and Lamb (in press). In this case, each alternative landscape scenario is mapped using a GIS and is then various attributes are scored using simple metrics or indicators. Stakeholders can compare the merits of alternative scenarios by comparing the maps and these scores.

The worth of the system clearly depends on the attributes used to assess each scenario and on the scoring system. The attributes currently used include:

1. Physical landscape changes: a variety of metrics that show the total area of new forest, improvements in the degree of forest connectivity, the length of forest edge, the area of 'core' forest that is free of edge effects, etc.
2. Biodiversity: assessed using an index that takes account of the type of new forest, its area and its distance from another area of forest. For example, in comparison with natural forest, a small monocultural plantation might be given a value of 0.1 while a large mixed-species plantation might receive a value of 0.7. The biodiversity score for that forest is the product of the area and the relevant index value. The biodiversity value of that scenario is the sum of the scores for the various reforestation areas.
3. Watershed protection: again, this is assessed using an index that takes account of slope and the type of reforestation. The index would be highest on agricultural cropland on steep slopes but lowest where there is natural secondary forest on gentle slopes.
4. Commercial value of reforestation: assessed using an index based on a ratio of the future value of the forest and its present agricultural value. Different types of reforestation have different commercial values and while the agricultural value of land will depend on its fertility and other attributes. A productive forest on poor quality agricultural land will have a higher ratio than, say, regrowth forest on productive agricultural forest.

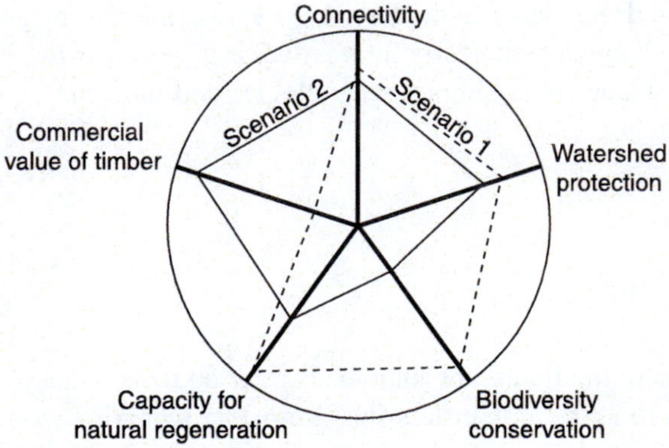

Fig. 11.6 A radar diagram showing the value of five attributes in each of two alternative reforestation scenarios (Scenario 1 = *dotted line* and Scenario 2 = *full line*). Each attribute is measured on a linear scale ranging from zero in the center to 100% at the end of each axis. In this case Scenario 1 scores well for biodiversity conservation but less well for timber productivity

5. Capacity for natural regeneration: assessed using an index reflecting the degree of past agricultural use and the presence/absence of weeds.

Qualitative indices like these lack precision but are useful in situations where quantitative data are unavailable. Once a reforestation scenario has been agreed upon these attributes are assessed and the scores standardized so they can be compared. The results are then plotted on a radar diagram (Fig. 11.6). Stakeholders can then assess the merits of each alternative scenario.

The tool has been constructed using a variety of attributes and indicators although it is expected that users will only select four to five of these depending on their interests and the points of contention. It has also been developed in a way that new indicators can be added and the scoring system can be modified if users wish to suit local circumstances. For example, the biodiversity index can be modified according to local judgements about the biodiversity value of different forms or ages of reforestation. The tool is available at www.gpem.uq.edu/cser-tools.

Although the tool purports to show the likely consequences of alternative scenarios the value of the tool is likely to be used more as a means of promoting discussion about the biophysical and economic implications of the different options. Box 11.3 gives an example of how the tool has been used in practice.

Simple Models

Marjokorpi and Otsamo (2006) describe a simple model used by a large reforestation company to prioritize areas for reforestation in a degraded landscape in West Kalimantan so that efforts could be concentrated in areas with the greatest economic and cultural value to local people. The model assessed the regeneration potential of buffer zones around patches of man-made and residual forest. Once these areas were identified the company could then locate areas for it own industrial pulpwood plantations. These plantation lands are rented from local farmers on a 45 year lease and this income, together with that from employment in the plantations, reduces the need for people to carry out as much forest clearance as before.

The landscape in which they worked contains fields of upland rice, *Imperata* grasslands, fallow lands with woody regrowth arising from shifting cultivation and patches of residual natural forest. There were also areas of culturally important man-made agroforests producing fruits, nuts and resins or rubber (in 'jungle rubber' forests) as well as crops like oil palm. Natural forests covered about 20% of the area while man-made forests were usually small (mostly <5 ha) and covered about 2%. Forest regeneration in buffer zones around these culturally important forest patches would protect them from damage, reduce edge effects and increase connectivity between forest patches. The task was to locate forest patches where this objective could be most efficiently achieved.

The area was mapped and a GIS was used to create a range of buffer zones of different widths around each culturally important forest patch. The wider the width the larger the area of landscape affected by rehabilitation; with a buffer zone width of 40 m the area of forest patches doubled and about half of the forest patches became interconnected.

The value of each buffer zone for rehabilitation was assessed using a weighed scoring system. Four classes of vegetation were recognized within each buffer zone: forest (excluding man-made forest), natural regrowth, *Imperata* grasslands and other vegetation types such as active swidden cultivation areas. A Rehabilitation Index (R) was calculated for each forest patch or group of patches as follows:

$$R(x_1, x_2, x_3, x_4, A) = (4x_1 + 3x_2 + 2x_3 + x_4) + A$$

where x_1 = percent of buffer zone area covered by natural residual forest, x_2 = percent of buffer zone area covered by regrowth forest, x_3 = percent of buffer zone area covered by grasslands and x_4 = percent of buffer zone area covered by other vegetation types. A = the area (ha) of cultural important or man-made forests within the fragment area. The subjective weights given to the different vegetation types reflect their regeneration capacity and need for external inputs to assist regeneration.

The more natural or man-made forest present in an area then the higher the index. Sites where buffer zones have a high proportion of advanced regrowth are obviously preferable to those with a high proportion of grassland since the former will recover if simply protected while the latter may require active planting.

A single numerical score for each buffer zone could be calculated by standardizing the percent cover and area measures by taking account of means and standard deviations to create a single index value. This allowed a map to be produced showing areas of the landscape with the highest Rehabilitation Index and the greatest degree of connectivity for a given buffer zone width. Final land use plans were developed in consultation with the Dayak landowners.

Role Playing Games

Boissau et al. (2004) and Castella et al. (2005) developed an agent-based, spatial computational model to explore land use changes by communities living in the highlands on northern Vietnam. The model sought to link (a) farmer strategies and decision-making processes, (b) the institutions and policies that regulate resource access and use and (c) the biophysical and socioeconomic environments in which the communities found themselves. Though not explicitly directed at Forest Landscape Restoration the approach could be adapted for this purpose.

The views of farmers about land use decisions were explored using a role playing game. Representatives from the community played the game using a set of small cubes lying on the game board or grid. Each cube represented a unit of land and different faces of the cube had a different colour to indicate alternative land uses (paddy rice, orchard, timber trees, forest regrowth, etc.). Participants developed a virtual landscape resembling their own village on the grid using these cubes to show current land uses. They then drew cards that described their virtual families and the resources available to them (e.g. composition of families, number of paddy fields, buffaloes, etc.). Players then had to manage these resources to feed their virtual families.

Before the game started the costs and benefits of alternative activities were firstly agreed upon amongst the players. For example, the yields from certain crops, the labour needs to carry out certain activities, how much material might be gathered from natural or regrowth forests, etc. Players could then manage their lands by changing the land uses. A change in land use would be reflected by a change in colours displayed by the cubes spread across the board. Thus, a farmer might open up fallow land for cropping, turn an upland rice crop into fallow or plant a tree crop. At the end of the game a player's income could be assessed and payments made in virtual rice credits or they might be forced to borrow funds to remain in the game. Up to six repetitions of the game could be played in a day. By the end of the day enough data was usually collected to enable the agreed rules, the games and sequence of decisions to be captured on a computer-based GIS. This meant players could review their day's decisions in a short time period. Castella et al. (2005) subsequently went on to build models that allowed them to use local management rules to simulate farmer behaviour across the broader landscape and tested these against actual land use changes.

The process showed itself to be a powerful tool by which to observe the actions of farmer-players and the ways in which they made decisions when confronted by different situations. But, more importantly, it enabled a discussion about land use choices and decision-making in a way that is usually not possible in simple question-and-answer interviews. The intent of the game was not necessarily to replicate real-life situations and the rules of the game were deliberately kept open. As a consequence of this, players sometimes explored choices the organisers had not thought of or which the players might have found to be too risky to carry out in real life. The game also brought out the contrasting strategies of different players (e.g. repetition, imitation, innovation and cooperation) which became the starting point for further discussions. Although time consuming for the players, there is clearly scope for using games like this by modifying resource bases, land allocation patterns or market prices for forest products to explore attitudes to various Forest Landscape Restoration initiatives.

Cost Effectiveness

Assuming all other factors are equal, is it better to reforest a badly degraded site or site that is only moderately degraded? Restoring the badly degraded site might generate the largest functional benefit but reforesting the moderately degraded site might be much cheaper meaning that larger areas can be treated for the same amount of money. A similar dilemma confronts those trying to decide how to allocate resources to conserve threatened species. Which threatened species deserves support when funds are limited? Based on work by Joseph et al. (2008) a simple way of deciding which of two degraded sites should be treated is to assess the cost effectiveness of the reforestation effort. Cost effectiveness (CE) can be calculated as:

$$CE = (B \times P)/C$$

where B is the overall 'benefit' generated by reforesting a particular area, P is the probability of reforestation being successful and C is the discounted cost of reforesting the site. A simple approach might be to ask managers to estimate the likely value of B (in comparison with the status quo) on, say, a 1–100 scale (with no real benefit being 1 and a significant improvement being 100). This subjective judgement of benefit might be based on the estimated overall value of the reforested site or on the way reforestation would improve, say, current levels of erosion. They could also rate the risks involved in reforesting the site and give a probability of reforestation success based on past experience. Thus a site with an infertile soil and invasive weeds would be more risky than a site with few biophysical constraints. Reforestation costs could also be estimated on the basis of past work. An example of these types of calculations is given in Table 11.5. In this case reforestation at location 1 generates a greater benefit than at location 2 but the cost is higher and the probability of success is lower. The data suggest it would be rather more cost effective to reforest at location 2 than location 1.

The metric is necessarily simple and assumes reforestation will generate a similar type of response at both sites. But there are ways in which it might be developed. For example, the benefit B might be broken into a watershed protection

Table 11.5 Assessing the relative cost effectiveness of alternative reforestation locations

Location	Present degree of degradation	Expected benefit	Probability of reforestation success	Discounted cost of reforestation	Cost effectiveness (× 100)
1	High	90	0.3	2,000	1.3
2	Low	50	0.9	500	9.0

benefit (Bw), a conservation benefit (Bc) and a (timber) production benefit (Bp). Separate scores might be given to each component based on the likely consequences of reforestation. In addition, the importance of each might be weighed (W_n) to reflect its importance to land managers. The overall value of B would then be the sum of these:

$$B = (Bw \times W_1) + (Bc \times W_2) + (Bp \times W_3)$$

This value for B would be inserted in to earlier cost effectiveness equation. Cost effectiveness is not, of course, the only criteria that might be used to evaluate priorities but it could be a useful tool that might be combined with others.

Market-Based Instruments

Most of the approaches and tools described above are designed to help stakeholders negotiate an agreement amongst themselves. In some cases it may be necessary to also provide financial incentives such as the PES schemes discussed in Chapter 9 or some form of compensation using external funds to reach such an agreement. An alternative approach is to use these same funds to devise market instruments that guide landowners towards preferred solutions. In this case there is not a negotiation process amongst stakeholders but individual landholders decide whether to become engaged in reforestation depending on their perception of the benefit they will receive from doing so. The process is based on the reverse auctions also described in Chapter 9.

An example of how this might work is given by Chomitz et al. (2006) who explored the use of a reverse auction to conserve regrowth forests in the Atlantic forest area of Brazil. The region is one where landscapes contain a mixture of land uses and where residual forest and secondary forests tend to occur on land that is least attractive for agriculture. Chomitz et al. (2006) carried out a simulation of the process whereby a hypothetical government agency with a fixed budget invites bids from landowners specifying the amount and quality of existing forest cover on their land and the minimum one-time payment that would induce them to put the property onto a permanent conservation easement. The agency then rates the quality of the submissions based on the amount of forest and its capacity to recover. Contracts were awarded to landowners whose bids provide the best quality forest for the least cost. Certain rules are needed to promote the desired conservation outcomes. Thus, preference was given to those offering patches of forest above a certain area. Likewise, areas where the distances between fragments are small are preferred over those where they are greater.

The obvious question is whether a voluntary and uncoordinated approach like this can generate a reforestation program that is as functionally effective as one designed by planners or generated by negotiation between stakeholders? The result of the simulation suggested it could be and that a significant amount of connectivity could be achieved. This connectivity increased as the funds per hectare increased. Even better outcomes might be achieved by zoning the landscape to solicit bids from priority zones (or exclude them from areas where regrowth would not occur). Parkhurst and Shogren (2007) argued that benefits would also accrue if farmers were paid an 'agglomeration bonus' when land offered adjoined a preferred area (e.g. intact natural forest, a riverine strip, the habitat of a key wildlife species). Chomitz et al. (2006) concluded the cost of the process would be substantially less than if a simple fixed price compensation payment was made. Besides being cost efficient, transparent and simple they argued this process could be politically more acceptable than some of the negotiated outcomes described above.

Evidence supporting these results came from an actual field trial in the tropical woodlands of northern Australia (Windle et al. 2009). In this case the intent was to develop a regional conservation corridor using existing woodland on farmer's properties. None of the farmers were familiar with the auction process so time was spent describing not only the arrangements for the auction and the design of the contracts but also the environmental index used to evaluate the competing bids. The environmental index had three components: the connectivity outcomes provided by the bid (weight of 44%), the biodiversity value of the forest (weight of 33%) and the present ecological condition of the sites (weight of 22%).

Unlike the process used by Chomitz et al. (2006), three bidding rounds were carried out. The purpose of this was to allow bidders to modify their bids in light of new information they received about the geographic location of other farmer's bids. This generated greater coordination while retaining the competitive element (later bids by individual farmers tended to have a lower price than their initial bids). The process was successful in generating a considerable degree of landscape connectivity at a relatively low cost. The final corridor covered an area of 85,000 ha and over 70% of the bids (making up 77% of the total bid area) were part of a group that formed a distinct corridor or landscape linkage with only single or part-property gaps. These examples were based on using existing regrowth or forest to form a corridor but, in principle, there is no reason why it should not also work by having landowners bid to carry out reforestation.

These market-based mechanisms will not be suitable for all situations. Firstly, there obviously needs to be a source of funds. These may come from a government agency seeking to achieve a particular outcome or an NGO such as downstream water users or a conservation organisation. Secondly, there also needs to be a sufficiently large pool of bidders to make the system work but not all farmers will be comfortable with the idea of auctions and bidding. Nor may the approach work with many poor farmers who have only small landholdings. Finally, there needs to be a relatively sophisticated financial institution able to manage and monitor the program that participants feel they can trust. These constraints probably limit the number of situations in which the technique can be used. But, in places where these limitations do not apply, there may be some considerable merit in the idea.

Conclusion

Any large-scale reforestation of degraded landscapes is likely to improve, directly or indirectly, the livelihoods of people and communities living within them. It is also likely to have many conservation benefits. But livelihoods and conservation outcomes will both be enhanced when there is some degree of coordination and planning. The question, then, is not whether coordination and planning is necessary but how this might be done. The more stakeholders involved, the more complex the process. When faced with planning large and complicated rural settlement schemes government planners are often tempted to adopt a largely top-down approach. Such approaches are least likely to work in degraded landscapes with many poor farmers. On the other hand, a process that is largely bottom-up has problems as well. This is because many smallholders may unaware of the larger ecological or economic drivers of change and the options that may be open to them. Because of this there is increasing agreement that Forest Landscape Restoration should be coordinated through some form of negotiation rather than using just one or other of these approaches.

Conservationists often emphasize how little is known about tropical ecosystems and biota. But enough is now known to suggest guidelines concerning how much forest cover should be present, where reforestation should be done and what types of reforestation should be carried out at particular locations to achieve particular outcomes. There is little doubt that considerable conservation and other environmental benefits could be achieved if these guidelines could be implemented.

There are also usually clear indications about where and how reforestation should be carried out to get the best economic outcomes. But there are also uncertainties. Just what markets will be most valuable in, say, 20 years time? Will they be markets for goods or for ecosystem services? What will be the implications for the types of forest established and the location of these forests? Different stakeholders will have different points of view about these questions. The fundamental problem for those interested in restoring forests at a landscape scale is, then, to find ways of reconciling these sometimes contrasting viewpoints.

The spatial extent of landscapes means trade-off should be easier to make at this scale than is the case at a single site. But they will still be difficult, especially when certain locations are valuable for both conservation and commercial purposes. There are a variety of tools becoming available to assist stakeholders make decisions. These don't provide blue prints or recipes – there are no recipes – and there can be a role for both 'muddling through' and sophisticated computer modelling.

There are a number of situations where Forest Landscape Restoration may be difficult to undertake. These are areas where the market price of certain agricultural crops are rising sharply (so the opportunity costs of reforestation are high) or where many farmers do not have tenure (because their traditional ownership claims are not recognized or because they are recent migrants). It may also be difficult in badly degraded landscapes where the costs of reforestation are high or where land disputes are unresolved. In such cases external stakeholders who are often important beneficiaries of Forest Landscape Restoration have an especially important role to play. If they wish to increase reforestation at particular sites they will have to help provide the incentives or compensation to landholders for the opportunities forgone when reforestation is carried out or to settle disputes. For those without formal ownership some form of tenure may be the most appropriate incentive.

Land use planning is widely practiced but there are, as yet, too few examples of large-scale Forest Landscape Restoration. We do not know if agreements amongst stakeholders will hold in the face of future changes in their economic circumstances. What, for example, might be the consequence of a 20% increase in the price of, say, coffee (or of timber) on reforestation activities? Nor do we know which species will be favoured as restoration proceeds (e.g. at what point will top order carnivores benefit?) or whether it is possible to improve the conservation status of other vulnerable species through specially targeted reforestation activities. Forest Landscape Restoration must therefore be, above all else, an adaptive process through which participants learn and adapt as the spatial mosaics change.

Forest Landscape Restoration is a step along the way towards undertaking reforestation at a scale that matches the rate at which degraded lands are being created. But it is not enough. Both site and landscape-based interventions need a policy framework and national institutions to encourage and support them. These are discussed in the next chapter.

References

Andren H (1994) Effects of habitat fragmentation on birds and mammals in landscapes with different proportions of suitable habitat – a review. Oikos 71:355–366

Bhagwat SA, Willis KJ, Birks HJB, Whittaker RJ (2008) Agroforestry: A refuge for tropical biodiversity? Trends Ecol Evol 23:261–267

Boedhihartono I, Barrow E (2008) Baseline photography and participatory drawing in East Africa. Arborvitae special; learning from landscapes. International Union for the Conservation of Nature, Gland, Switzerland

Boissau S, Anh HL, Castella JC (2004) The SAMBA role play game in northern Vietnam: An innovative approach to participatory natural resource management. Mt Res Dev 24:101–105

Castella JC, Trung TN, Boissau S (2005). Participatory simulation of land-use changes in the northern mountains of Vietnam: The combined use of an agent-based model, a role-playing game, and a geographic information system. Ecol Soc 10, http://www.ecologyandsociety.org/vol10/iss11/art27/

Chetkiewicz C, St. Clair CC, Boyce MS (2006) Corridors for conservation: Integrating pattern and process. Ann Rev Ecol Evol Syst 37:317–342

Chomitz KM, da Fonseca GAB, Alger K, Stoms DM, Honzak M, Landau EC, Thomas TS, Thomas WW, Davis F (2006). Viable reserve networks arise from individual landholder responses to conservation incentives. Ecol Soc 11, http://www.ecologyandsociety.org/vol11/iss2/art40/

Dobson A, Jensen DB, Knight RL, Gatewood S, Mills L, Boyd-Heger D, Mills LS, Soule ME (1999) Corridors: Reconnecting fragmented landscapes. In: Soule ME, Terborgh J (eds) Continental conservation: Scientific foundations of regional reserve networks. Island Press, Washington, DC, pp 129–170

Drielsma M, Manion G, Ferrier S (2007) The spatial links tool: Automated mapping of habitat linkages in variegated landscapes. Ecol Modell 200:403–411

Fahrig L (2001) How much habitat is enough? Biol Conserv 100:65–74

Fisher RJ, Maginnis S, Jackson W, Barrow E, Jeanrenaud S (2008) Linking conservation and poverty reduction: Landscapes people and power. Earthscan, London

Flather CH, Bevers M (2002) Patchy reaction-diffusion and population abundance: The relative importance of habitat abundance and arrangement. Am Nat 159:40–56

Fung C (2008) Rural land use planning of the drawa model area: A participatory and integrated approach. Capacity building on restoration, management and rehabilitation of degraded forests and degraded lands in the Pacific. Regional Seminar for improved practices enhancing forest functions, 28–31 October, 2008, Land Resources Division, Secretariat of the Pacific Community, Fiji

Gardner TA, Barlow J, Chazdon R, Ewers RM, Harvey C, Peres CA, Sodhi NS (2009) Prospects for tropical biodiversity in a human-modified world. Ecol Lett 12:561–582

German L, Taye H (2008) A framework for evaluating effectiveness and inclusiveness of collective action in watershed management. J Int Dev 20:99–116

Gilmour DA (2007) Applying and adaptive management approach in FLR. In: Rietbergen-McCracken J, Maginnis S, Sarre A (eds) The forest landscape restoration handbook. Earthscan, London, pp 29–37

Gilmour DA, Fisher RJ (1991) Villagers, forests and foresters. Sahayogi Press, Kathmandu

Goldman RL, Thompson BH, Daily GC (2007) Institutional incentives for managing the landscape: Inducing cooperation for the production of ecosystem services. Ecol Econ 64:333–343

Groffman P, Baron J, Blett T, Gold A, Goodman I, Gunderson L, Levinson B, Palmer M, Paerl H, Peterson G, Poff N, Rejeski D, Reynolds J, Turner M, Weathers K, Wiens J (2006) Ecological thresholds: The key to successful environmental management or an important concept with no practical application? Ecosystems 9:1–13

Guo ZW, Gan YL (2002) Ecosystem function for water retention and forest ecosystem conservation in a watershed of the Yangtze River. Biodivers Conserv 11:599–614

Guo ZW, Gan YL, Li YM (2003) Spatial pattern of ecosystem function and ecosystem conservation. Environ Manag 32:682–692

Guo ZW, Xiao XM, Li DM (2000) An assessment of ecosystem services: Water flow regulation and hydroelectric power production. Ecol Appl 10:925–936

Hargrove WW, Hoffman FM, Efroymson RA (2005) A practical map-analysis tool for detecting potential dispersal corridors. Landscape Ecol 20:361–373

Hobley M (1996) Participatory forestry: The process of change in India and Nepal overseas development institute. Overseas Development Institute, London

ITTO (2002) ITTO guidelines for the restoration, management and rehabilitation of degraded and secondary tropical forests. ITTO Policy Development Series No 13, International Tropical Timbers Organization, Yokohama

Jansen A (2005) Avian use of restoration plantings along a creek linking rainforest patches on the Atherton Tableland, North Queensland. Restor Ecol 13:275–283

Joseph LN, Maloney RF, Possingham HP (2008) Optimal allocation of resources among threatened species: A project prioritization protocol. Conserv Biol 23:328–338

Knight AT (2006) Failing but learning: Writing the wrongs after Redford and Taber. Conserv Biol 20:1312–1314

Knight AT, Cowling RM, Campbell BM (2006) An operational model for implementing conservation action. Conserv Biol 20:408–419

Lamb D, Erskine P, Parrotta J (2005) Restoration of degraded tropical landscapes. Science 310:1628–1632

Lindenmayer DB, Fischer J (2006) Habitat fragmentation and landscape change. CSIRO Publishing, Collingwood

Maginnis S, Jackson W (2007) What is FLR and how does it differ from current approaches? In: Rietbergen-McCracken J, Maginnis S, Sarre A (eds) The forest landscape restoration handbook. Earthscan, London, pp 5–20

Marjokorpi A, Otsamo R (2006) Prioritization of target areas for rehabilitation: A case study from West Kalimantan, Indonesia. Restor Ecol 14:662–673

McKergow L, Prosser IP, Grayson R, Heiner D (2004) Performance of grass and rainforest riparian buffers in the wet tropics, far North Queensland. 2. Water quality. Aust J Soil Res 42:485–498

Millspaugh J, Thompson FR (2009) Models for planning wildlife conservation in large landscapes. Academic, Burlington, NJ

Mladenoff DJ (2004) LANDIS and forest landscape models. Ecol Modell 180:7–19

Molnar A, Liddle M, Bracer C, Khare A, White A, Bull J (2007) Community based forest enterprises: Their status and potential in tropical countries. International Tropical Timbers Organization and Forest Trends, Yokohama

Odum WE (1982) Environmental degradation and the tyranny of small decisions. Bioscience 32:728–729

Ostrom E (1990) Governing the commons. Cambridge University Press, Cambridge

Parkhurst GM, Shogren JF (2007) Spatial incentives to coordinate contiguous habitat. Ecol Econ 64:344–355

Peh KS, Sodhi NS, De Jong J, Sekercioglu CH, Yap CAM, Lim SLH (2006) Conservation value of degraded habitats for forest birds in southern Peninsular Malaysia. Divers Distrib 12:572–581

Pettit C, Cartwright W, Bishop I, Lowell K, Pullar D, Duncan D (2008) Landscape analysis and visualisation: Spatial models for natural resource management and planning. Springer-Verlag, Berlin

Polasky S, Nelson E, Camm J, Csuti B, Fackler P, Lonsdorf E, Montgomery C, White D, Arthur J, Garber-Yonts B, Haight R, Kagan J, Starfield A, Tobalske C (2008) Where to put things? Spatial land management to sustain biodiversity and economic returns. Biol Conser 141(6):1505–1524

Pullar D, Lamb D (in press) Prioritising forest restoration to improve landscape function using GIOS scenario modelling; an example for linkage restoration. In: Stanturf JA, Madsen P, Lamb D (eds) A goal-oriented approach to forest landscape restoration. Springer, Dordrecht

Reitbergen-McCracken J, Maginnis S, Sarre A (2007) The forest landscape restoration handbook. Earthscan, London

Sayer J, Boedhihartono I (in press) Developing and tracking restoration scenarios. In: Stanturf JA, Madsen P, Lamb D (eds) Forest landscape restoration: Integrating natural and social sciences. Springer, Dordrecht

Sayer J, Bull G, Elliott C (2008) Mediating forest transitions; grand design or muddling through. Conserv Soc 6:320–327

Sayer J, Campbell B, Petheram L, Aldrich M, Perez MR, Endamana D, Dongmo ZLN, Defo L, Mariki S, Doggart N, Burgess N (2007) Assessing environment and development outcomes in conservation landscapes. Biodivers Conserv 16:2677–2694

Sayer JA, Campbell B (2004) The science of sustainable development: Local livelihoods and the global environment. Cambridge University Press, Cambridge

Scales B, Marsden S (2008) Biodiversity in small-scale tropical agroforests: A review of species and diversity abundance shifts and the factors influencing them. Environ Conserv 35:160–172

Schroth G, da Fonseca GAB, Harvey C, Vasconcelas HL, Gascon C, Izac A-M (2004) Introduction: The role of agroforestry in biodiversity conservation in tropical landscapes. In: Schroth G, da Fonseca GAB, Harvey C, Vasconcelas HL, Gascon C, Izac A-M (eds) Agroforestry and biodiversity conservation in tropical landscapes. Island Press, Washington, DC, pp 1–14

Scott JC (1998) Seeing like a state: How certain schemes to improve the human condition have failed. Yale University Press, New Haven and London

Sekercioglu CH, Loarie SR, Brenes FO, Ehrlich PR, Daily GC (2007) Persistence of forest birds in the Costa Rican agricultural countryside. Conserv Biol 21:482–494

Shepherd G (2004) The ecosystem approach. The International Union for the Conservation of Nature, Gland/Cambridge

Sidle RC, Ziegler AD, Negishi JN, Nik AR, Siew R, Turkelboom F (2006) Erosion processes in steep terrain-truths, myths, and uncertainties related to forest management in South East Asia. For Ecol Manage 224:199–225

Sodhi NS, Koh LP, Prawiradilaga DM, Tinulele I, Putra DD, Tan THT (2005) Land use and conservation value for forest birds in Central Sulawesi (Indonesia). Biol Conserv 122:547–558

Soule ME, Terborgh J (1999) The policy and science of regional conservation. In: Soule ME, Terborgh J (eds) Continental conservation: Scientific foundations of regional reserve networks. Island Press, Washington, DC, pp 1–17

Stirzaker RJ, Cook FJ, Knight JH (1999) Where to plant trees on cropping land for control of dryland salinity: Some approximate solutions. Agric Water Manage 39:115–133

Thomson J, Moilanen AJ, Vesk PA, Bennett AF, MacNally R (2009) Where and when to revegetate: A quantitative method for scheduling landscape reconstruction. Ecol Appl 19:817–828

Tscharntke T, Sekercioglu CH, Dietsch TV, Sodhi NS, Hoehn P, Tylianakis JM (2008) Landscape constraints on functional diversity of birds and insects in tropical agroecosystems. Ecology 89:944–951

Turkelboom F (1999) On-farm diagnosis of steepland erosion in northern Thailand. Dissertation Faculteit Landbouwkundige en Toegepaste Biologische Wetenschappen, Katholike Universiteit Leuven, Leuven

van Dijk A, Keenan RJ (2007) Planted forests and water in perspective. For Ecol Manage 251:1–9

Van Dijk AIJM, Hairsine PB, Pena Arancibia J, Dowling TI (2007) Reforestation, water availability and stream salinity: A multi-scale analysis in the Murray-Darling Basin, Australia. For Ecol Manage 251:94–109

van Noordwijk M, Agus F, Verbist B, Hairah K, Tomich T (2007) Watershed management. In: Scherr SJ, McNeely JA (eds) Farming with nature: The science and practice of ecoagriculture. Island Press, Washington, DC, pp 191–212

Van Noordwijk M, Verbist B, Vicent G, Tomich TP (2001) Simulation models that help us to understand local actions and its consequences for global concerns in a forest margin landscape. International Centre for Research in Agroforestry, Bogor

Vertessy RA, Zhang L, Dawes WR (2003) Plantations, river flows and river salinity. Austr Forestry 66:55–61

Weyerhaeuser H, Wilkes A, Kahrl F (2005) Local impacts and responses to regional forest conservation and rehabilitation programs in China's Northwest Yunnan province. Agric Syst 85:234–253

Whisenant SG (1999) Repairing damaged wildlands: A process-oriented, landscape scale approach. Cambridge University Press, Cambridge

Wimberley MC (2007) Understanding landscapes through spatial modelling. Proceedings of the IUFRO conference on forest landscape restoration, Seoul, Korea, 14–19 May 2007, pp 70–72

Windle J, Rolfe J, McCosker J, Lingard A (2009) A conservation auction for landscape linkage in the southern desert uplands, Queensland. Rangeland J 31:127–135

Chapter 12
Developing Institutional Support for Large-Scale Reforestation

> As long as the task environment of an institution remains repetitive, stable, and predictable, a set of fixed routines may prove exceptionally efficient. In most economies and in human affairs generally, this is seldom the case, and such routines are likely to be counter productive once the environment changes appreciably.
>
> (Scott 1998, p. 354)

Introduction

Earlier chapters described the gradual loss of natural forests and the increase in the area of under-used former agricultural land across the Asia-Pacific region. The growing interest in different forms of reforestation has also been described. This means forestry practices are changing. Many new groups, in addition to state forestry agencies, are becoming involved in reforestation including private companies and smallholders. There are also other stakeholders with an interest in what is done because they are users of the goods and services being generated by the newly-established forests. This diversity of participants means the policies and practices that served in the past are becoming increasingly out of date. But what types of changes are needed to deal with the new circumstances? This chapter is concerned with the institutions and policies needed if reforestation to be undertaken on a national scale and at a rate that matches the rate at which forest and land degradation is occurring.

Ostrom (2005) has defined institutions as being the prescriptions or rules that humans use to organise all forms of repetitive and structural interactions. That is, they can be seen as the arrangements developed to coordinate collective action. Institutions include such things as traditional local customs, industrial codes of practice as well as national arrangements used to organise or regulate industries or social interactions. A large variety of informal or formal institutions have evolved to enable collective management of natural resources. Some of the principles emerging from the study of these provide useful insights into the types of institutions needed to encourage reforestation. But, before considering these policy and institutional

D. Lamb, *Regreening the Bare Hills*, World Forests 8,
DOI 10.1007/978-90-481-9870-2_12, © Springer Science+Business Media B.V. 2011

issues further, it is useful to firstly consider the ecological and socio-economic context in which future reforestation is likely to occur.

The Future Context?

If the rate of deforestation and degradation experienced in recent years does not decline then the environmental and socio-economic problems described earlier will continue and the rate at which they develop will possibly accelerate. However there are a number of other changes underway that will be equally important. Some of these are likely to strongly influence future opportunities for reforestation. They include:

Population Growth and the Need for Greater Food Production

By 2000, the world's population had reached 6.1 billion and it is projected to reach 9.2 billion by 2050 (United Nations 2007). This represents a 50% increase and means that food production will also have to increase quite substantially to meet this future demand. The world's capacity to produce and distribute food has been under some stress in recent years. This has been caused by a rising demand, but also because of competition for land for other crops like biofuels, rubber and oil palm. Concerns about food production are increasing. In 2008, the market price index for rice rose by a staggering 270% during the course of that year. It subsequently fell but, in mid 2009, was still 50% above the level present only a year or so earlier (Economist May 30, 2009). The situation was deemed serious enough for some rice-exporting countries such as Vietnam to cease overseas sales during this period to ensure food security at home. New agricultural technologies may help increase the productivity of existing farmland but, in the decades ahead, it is also likely that more natural forest and marginal land will have to be used for food production. This could include quite unsuitable areas such as steep hills or sites with infertile soils. Based on past experience, farming on some of these marginal sites will fail and they will be abandoned. All of this means there may be less land available for reforestation in future and that the quality of much of the land that is available will be poor.

Urbanisation

The world's population is increasing, but the distribution of this population is also changing. Many rural people are leaving the country for urban areas and cities have recently grown to contain more than half the world's people. Global rural populations are projected to reach a maximum of 3.46 billion people by 2020 and then fall to 2.79 billion by 2050 (United Nations 2007). Within Southeast Asia, rural populations reached a peak of 313 million in 2000 and have declined since then. They are

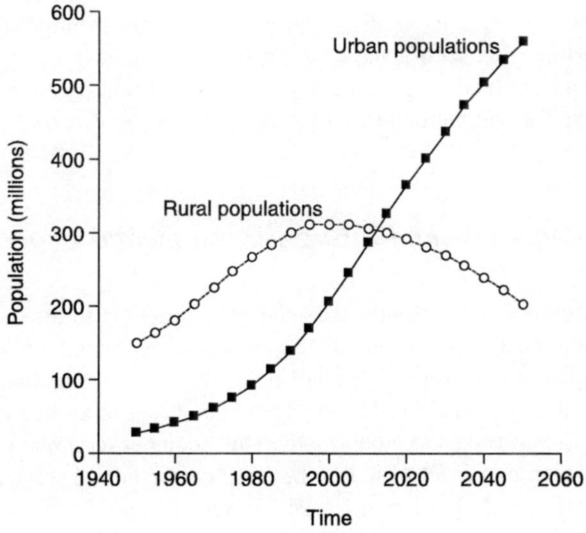

Fig. 12.1 Projected changes in rural and urban populations in Southeast Asia between 1950 and 2050 (United Nations 2007)

projected to fall to 205 million by 2050 (Fig. 12.1). By this time rural populations will represent only 27% of the total Southeast Asian population (down from 84% one hundred years earlier) Not surprisingly, there are large differences across the region. In Indonesia rural populations are expected to fall to 48% of their peak over this time frame while those in Vietnam are expected to only fall to 77% of their peak. In the Pacific, some urban areas will grow but most islands will have relatively stable rural populations.

The prospect of urban drift and a decline in rural populations led Rudel et al. (2005) to suggest this could be one trigger for reforestation (the 'economic development pathway'). There are various ways in which this could happen. One way is by the amalgamation of farms. Rigg (2006) argues that rural lives and livelihoods are being increasingly de-linked from farms. Farming is becoming seen by many as a low status occupation and one in which it is difficult to improve household incomes, especially when landholdings are small. In his view, the best way of reducing rural poverty would be to help poor farmers leave their farms, especially those using more marginal lands. This would mean small farms were amalgamated and that fewer (wealthier) farmers managed larger farms. Under these circumstances there could be greater opportunities in these larger farms for an improved partitioning of land uses with crops grown on better land and trees on more marginal land. In this case, the understanding these farmers have about the opportunities offered by trees-planting will be crucial.

An alternative outcome might be something like what is now occurring in parts of Peninsular Malaysia where urbanisation appears to be leading to the development of significant areas of abandoned land (Jomo et al. 2004; Kato 1994). How this land might eventually be used probably depends on its spatial distribution. Large contiguous blocks could be attractive to industrial groups interested in large agricultural or forestry enterprises. But such groups would probably be less interested in small fragmented patches of land. Might natural regeneration occur on these areas? Might they be attractive for commercial tree-growing by absentee owners?

A Rising Middle Class and Rising Environmental Concerns

A number of countries in the Southeast Asian region belong to the so-called 'tiger economies' with most experiencing long periods of sustained growth in recent years. This has led to a decline in (though certainly not an eradication of) poverty and the growth of a middle class. The change is already leading to a rise in the per capita consumption of food and other goods. This, together with rising populations, will limit opportunities for reforestation because many unused wastelands will be needed for new forms of agriculture, especially for meat production.

On the other hand, there is likely to be an increased public concern about environmental matters and conservation groups are now being found in many tropical countries (Koop and Tole 2001; Steinberg 2005). There are already signs that governments are also becoming more interested in environmental issues and conservation. For example, protection forests form a significant component of the national Five Million Hectare Reforestation Program in Vietnam (MARD 2001; Ohlsson et al. 2005). In recent years community groups in Australia and elsewhere in the region have been engaged in non-commercial forest restoration plantings (Elliott et al. 2000; Erskine et al. 2005). These changing social attitudes are already affecting the ways some timber companies design plantations, causing them to think more about the landscapes in which these plantations are established (Cyranoski 2007; Wooff 2009). There is also increased interest in timber produced in plantations that have been certified to have reached certain environmental standards and the possibility of consumer boycotts of timber produced by plantation companies that do not reach these standards (Laurance 2008).

New Markets for Forest Products and Ecosystem Services

Until recently, timber has been readily available throughout the region at relatively low prices because of largely unconstrained logging but this is likely to change. As the areas of unlogged natural forests decline, alternative supplies will have to be found. Projections are difficult to make and depend on assumptions about population growth and per capita consumption as well as global patterns of growth. Countries such as China and India that were once not heavily involved in the global market are now becoming major consumers of imported forest products.

A recent detailed study by Whiteman and Jonsson (2009) took account of the declining production from natural forests and foreshadowed a demand in the

Asia-Pacific region (including China, India, Japan and Korea) for industrial roundwood production (sawlogs plus pulpwood). They estimate it will rise from 86 million cubic meter in 2005 to 294 million cubic meter in 2030. The sawlog component was estimated to grow from 45 million cubic meter in 2005 to 233 million cubic meter by 2030. Projections based on recent plantings of fast-growing species using short rotations suggested the overall timber demand could be met from local plantations but that much of this production would come in the form of pulpwood and not sawlog timbers. While some pulpwood quality timbers can be reconstituted into other products such as panel boards, there could be a significant deficit in sawlog timbers in the region and a substantial surplus of pulpwood logs by 2030. On the other hand, there could be a compensatory rise in the market for biofuels that might use this surplus.

This leaves open the market for higher value, high-quality 'cabinet' or decorative timbers. Some of these timbers may continue to come from natural forests in the region or elsewhere (e.g. Latin America). Likewise, some are being replaced by bamboos or new technologies that are leading to improvements in the durability, stability and aesthetics of some of the utility timbers (including a product purporting to be a substitute for teak). But there appears to also be a potential future niche market for plantation-grown timbers of these species provided the log quality is high enough and a regular supply can be assured. Well-known plantation species such as teak (*Tectona grandis*), mahogany (*Swietenia macrophylla*.) and rosewood (*Pterocarpus* spp.) already command attractive prices. But, given the biological diversity of the region's forests, there should be opportunities for other high quality species as well, especially those coming from certified plantations. Efforts will be needed by growers to identify where these specialty timber markets are located and to form a relationship with them.

In the view of Leslie (2005), however, the real market opportunities of the future will lie in the in the provision of ecosystem services and the production of NTFPs. He estimated the value of ecosystem services provided by forests would more than double between 2010 and 2040. If he is right, this re-enforces the need for silvicultural systems that are able to generate both high-value cabinet wood timbers and ecosystem services rather than those producing just utility timbers or pulpwood.

Climate Change

A changing global climate will have profound consequences for both the remaining tropical forests and for the prospects for reforesting some of the lands that have been deforested. Although there is a growing scientific consensus about the nature of the threat the impacts at particular locations are far less clear. But some things appear certain including the fact that there are likely to be changes in temperatures as well as in the amount and seasonality of rainfall in different parts of the Asia-Pacific region. The median projections from current climate models for the Southeast Asian region predict temperatures will rise by 2.5°C and annual rainfall by 7%.

In the case of rainfall, the increase could be up to 15% although dry seasons will be more severe (IPCC 2007). There is also likely to be an increase in extreme weather conditions including heatwaves and intense precipitation and a 10–20% increase in tropical cyclone intensities.

These changes, together with associated changes in fire regimes, will have consequences for most tropical ecosystems and many species will be forced to move up altitudinal or along latitudinal gradients. Others may retreat to niche refugia (Colwell et al. 2008; Williams et al. 2003). This will cause a reshuffling of communities because of differences in the sensitivities of species to environmental changes and the migration capacities of some species may not be able to accommodate the rates of change that are forecasted (Svenning and Condit 2008). A study of the distribution of *Eucalyptus* species in Australia is illustrative of the problem. Hughes et al. (1996) found 41% had natural distributions spanning areas with less than a 2°C difference in mean annual temperature (and 25% of species where the variation was less than 1°C). Likewise, 23% of species have ranges in mean annual rainfall that span less than 20% variation. Although the actual environmental tolerances of some species will be larger than the climatic envelopes they currently occupy, these relationships suggest many eucalypts (including eucalypts growing in plantations outside Australia) will be significantly affected by climate change. In the case of the tropical lowland flora, Corlett (2009) estimated the dispersal distances needed to compensate for climatic changes over the next century will exceed 100 km and this seems impossibly large for most plant species.

Climate changes of this magnitude will also affect human livelihood by changing the availability of water resources, the levels of agricultural productivity at particular sites and the activities of pests and diseases affecting agricultural crops and human populations. The net effect could be major changes in the geographic location of current food-producing areas and human populations. Poor rural communities are likely to be especially vulnerable to these changes.

There are two sets of responses to climate change and both will have consequences for reforestation policies. One set of responses are concerned with trying to avoid the changes by stabilising and then reducing atmospheric carbon levels. In the forestry context, this means reducing deforestation and degradation of natural forests and, at the same time, reforesting degraded lands to sequester more carbon. The international REDD+ (reduction in deforestation and degradation plus reforestation) schemes that pay forest and land owners for the carbon they currently store or will sequester in future are still being developed. If they do emerge they will make natural forest protection and reforestation more attractive land use options than they are at present (see previous discussion in Chapter 9). However, the workings of an international scheme and the future market price for carbon are still unclear. In the case of the reforestation options, might the high transaction costs involved in dealing with large numbers of small farmers exclude them from a scheme or will some regional body represent them? What will be the financial consequences in the carbon market of pulpwood plantations grown on short rotations and sawlog plantations grown on long rotations? Will soil carbon also be involved? These questions will be resolved over time. On the other hand, it is likely that, in the immediate future, reforestation will continue to be driven by domestic policies such as land tenure arrangements and local institutions even though the prospects of a carbon market may sometimes tip the balance in favour of tree-planting.

The second approach to climate change assumes some changes are inevitable and seeks ways of accommodating or adjusting to these changes. This is likely to involve changes in both the location of various agricultural activities as well as in the crop species used at particular sites. Likewise the availability of land for reforestation may also change and some currently favoured tree plantation species may not be suited to some of the sites available for reforestation in the future. Reforestation may become more important in areas subject to heavier rain and erosion but less attractive in areas where water resources are limited. In short, it is still too early to say just how reforestation will be affected by climate change although the changes it will make are likely to be profound.

To summarise: there are a number of trends underway that will affect the amount and type of reforestation carried out in future. Some, such as urbanisation, may increase the opportunities for reforestation while others, such as rising populations, will have the opposite effect. Both patterns could occur simultaneously but in different parts of the region. Some trends, such as a rising interest in environmental matters and the changes associated with global warming also point to the fact that a wider variety of reforestation methods will be needed in the future. Of course there will be other changes as well such as increasing oil prices, changes in the price of fertilisers, rising demands for water and those arising from the process referred to as globalisation that are already affecting patterns of global trade. These trends are complex, sometimes contradictory and often hard to understand (Kates and Parris 2003). In many cases, the problems they create will be ones that few people will have had much experience in solving and future ecological and economic 'surprises' are likely. Any future forms of reforestation will have to be resilient and capable of adapting to these changes.

Undertaking Reforestation in the Future

Current patterns of reforestation are caused by an amalgam of market forces and government policy settings. The evidence suggests the present combination has been unable to promote reforestation at the rate needed or in the places where it is most required. Some new policy settings and institutional arrangements are needed to deal with the backlog of degraded lands needing to be reforested and to cope with the degree of uncertainty that is likely to be encountered in future. There are several specific questions that must be resolved:

- How to resolve conflicts over future land use (especially with agriculture) and how to identify areas where reforestation should be carried out in future to protect watersheds, conserve biodiversity or accommodate the changes induced by global warming?

- How to develop non-financial incentives to make large-scale reforestation an attractive land use alternative?
- How to develop – and promote – new forms of reforestation able to supply the goods and ecosystem services required in future as well as methods for reforesting particularly degraded areas?
- How to access capital and develop financial mechanisms that make investing in long-term forestry ventures attractive and that also reward those supplying ecosystem services?

The Role of Markets

Markets can be important drivers of reforestation and create incentives for innovation and entrepreneurship. Both the 'economic development pathway' and the 'forest scarcity pathway' identified by Rudel et al. (2005) are ways in which different types of market activity can increase forest cover over large areas. For reasons outlined earlier, some combination of the two may be especially powerful in future. People are more inclined to carry out activities such as tree planting if they know that they, or their families, will benefit from doing so. This means that those seeking to increase forest cover must be aware of the formidable role of markets in synthesising information and changing people's behaviour. This role is especially important given the variety of uncertainties described above.

Despite their undoubted power, markets have some significant limitations. Firstly, unregulated markets do not always generate ecologically satisfactory outcomes. Markets can easily value goods such as timber, pulpwood, oil palm or real estate. But are mostly unable to put a market price on biodiversity or clean water until after the system passes some kind of degradation threshold by which time it may be difficult and expensive to cross back. Nor are markets very good at sending signals that induce people to rehabilitate severely degraded lands. Sometimes special financial instruments can be developed to deal with these types of problems. But there will always be situations where markets cannot be relied upon and governments must step in and demand, or initiate, appropriate forms of reforestation on behalf of the community.

Secondly, markets are affected by geography. In isolated locations such as remote highland areas or isolated Pacific islands there may be no market because of high transport costs. In other cases, the high cost of transport can mean the only market is for more valuable products such as specialty timbers. The geographical distribution of market links also matters. Foster and Rosenzweig (2003) found there may be little relationship between the local demand for forest products and forest cover when economies are open and buyers can purchase these products from a wide variety of sources. This means deforestation can continue without generating an incentive for local landowners to carry out any reforestation. The situation may be quite different, however, if economies are more closed, or when local growers do supply much of the local demand.

Thirdly, markets may not work well when it is difficult to respond quickly to a market signal. Reforestation takes time and it may be some years before goods or services can be produced to supply a market previously supplied by natural forests. Clever landowners able to monitor the situation might be able to anticipate such future market signals and begin reforesting to supply this market. But, in most local situations, there are few landowners with a sufficient overview of what is occurring for this to happen.

Finally, unregulated markets do not always produce socially acceptable outcomes and in some cases they can cause extreme hardship to some segments of society. Obvious examples are when sudden price changes devalue the worth of a smallholder's crop or plantation or when there is only a single buyer able to dominate the market and set prices that suit themselves.

The Role of Governments

Governments have a uniquely important role in facilitating reforestation (Table 12.1). In the past, government agencies often carried out reforestation to compensate for the loss of natural forests by creating timber resources to provide rural employment or to protect certain environments such as coastal zones or mountain areas. But, in recent years, most governments have been less active in reforestation because of shortages of land and financial resources. Some have established wholly-owned entities to establish and manage plantations with largely commercial objectives (e.g. Peruntani in Indonesia, State Forest Enterprises in Vietnam and the Forest Industry Organization in Thailand). Some have sought to promote reforestation (and avoid further degradation) by devising new policies. Among the most important of these are the provision of land tenure and legal frameworks governing ownership rights. Some have also reviewed policies that discourage reforestation including tax regimes and cumbersome administrative procedures and have provided financial incentives of various kinds to make reforestation more attractive. Many have also sought to enhance the economic benefits of reforestation by creating supporting infrastructure such as roads to get goods to markets and by reducing the complexity of market chains and the costs of marketing.

Much future reforestation will probably be carried out to provide certain ecosystem services rather than just timber production. In such cases it will be the location of reforestation within a landscape rather than just the area covered that is important. Governments are uniquely equipped to take a broader perspective than most other land managers and identify where reforestation should be carried out for the national benefit. These locations might be eroding mountain areas, sensitive coastal zones or areas of high conservation significance. Governments can offer incentives or subsidies to ensure such sites are treated and legitimize tree-growing as a profitable land use activity. Governments can also help reduce some of the risk from reforestation by initiating silvicultural research and by developing extension services to share this knowledge. More often than not, governments have provided

Table 12.1 Participants in reforestation and some of the advantages they have in being able to promote reforestation as well as their disadvantages

Participants	Advantages	Disadvantages
Governments	Can identify where reforestation should be carried out in the national interest	Often have limited understanding of local issues or constraints faced by smallholders
	Can provide technical information to growers and undertake long-term silvicultural research	Silvicultural prescriptions not always relevant to social or economic circumstances of growers
	Can provide policy and governance framework (e.g. land tenure, legal system, tax system, enforce regulations)	Often undertake narrowly focussed research
	Can improve operation of markets (e.g. costs of externalities recognized, share market information, reduce costs of marketing)	Sometimes unable to implement policies because of limited on-ground capacity
	Can provide supporting infrastructure (e.g. roads)	Sometimes impervious to advice or feedback
	Have a convening and organising capacity (e.g. to arrange PES including carbon markets, monitor water quality or occurrence of pests and diseases)	Sometimes captured by special interests
Private companies	Can assemble funds and expertise to reforest large areas	Use limited variety of species
	Often technically and financially efficient	Mostly prefer pulpwood plantations, less interested in long rotations or more complex forms of silviculture
		Uninterested in degraded sites, limited interest in biodiversity conservation
		Usually prefer short financial investment periods
NGOs	Can influence public awareness of importance of reforestation	Sometimes work at only small spatial scale
	Able to work closely and build trust with communities, farmers and local champions	Sometimes their funding means they can only work for short periods
	Are often sources of technical expertise and funds	May have narrow interests (e.g. only conservation outcomes)
	Often linked to international networks providing access to new ideas	Their number and diversity often makes coordination between NGOs and with governments difficult

(continued)

Table 12.1 (continued)

Participants	Advantages	Disadvantages
Community	Have capacity to enforce protection of forest areas	May not have a single viewpoint; possibility of internal conflict over land use objectives or need for reforestation
	Can develop suitable rules governing usage of natural forests	Limited knowledge of commercial plantation silviculture techniques
	Hold traditional ecological knowledge about area and its species	Limited knowledge of markets
	Sometimes able to foster restoration plantings (for cultural or religious purposes)	Lack national perspective
	Often have a convening and organising capacity	
Private smallholders	Often have considerable agroforestry expertise	Many (but not all) have less technical expertise
	More inclined to use variety of species and plantation designs	Unlikely to be able to tackle most degraded sites
		Limited interest in biodiversity conservation
		Often unable to tolerate risk
		Lack national perspective

the institutional memory that safeguards much of this formal scientific silvicultural knowledge (but see Box 12.1 for an example of the risks inherent in there being only one knowledge node). Finally, governments have a special convening and organising capacity that enables them to link all those with a stake in forests to undertake national or regional reforestation initiatives such as developing systems of payment for ecosystem services. In short, large-scale reforestation is likely to be difficult without some degree of active government support.

But it is easy to over-estimate the role of governments and lose sight of the fact that many of the policies they have adopted in the past have been inappropriate. Top-down approaches that pay insufficient account of local practices or realities often run into problems and examples of this can be seen across the Asia-Pacific region. Sometimes local communities filter (or ignore) central government directives or add rules that have been generated by local institutions. Vietnam, for example, has a large national reforestation program and its forest cover is increasing (MARD 2001). However, a number of authors have argued that some reforestation in Vietnam has happened despite, rather than because of, the government's policies (Clement and Amezaga 2008; Fahlen 2002; Sikor 2001; Sowerwine 2004). Sometimes land supposed to be planted with trees was far more valuable to local farmers as agricultural land and its reforestation could have caused unnecessary stress. In such cases communities have found ways of circumventing government directives and achieving plausible outcomes although different to those intended.

Box 12.1 Losing Knowledge

It is commonly assumed that knowledge steady accumulates and that, speaking collectively, we know more now than we did last year. In fact, this is not always the case and a striking example from the Solomon Islands illustrates the point. Between 1998 and 2003 a period of disturbances and lawlessness occurred in the Solomon Islands. This resulted in the disintegration of government activities and the breakdown of an effective civil service. One national government agency badly affected by these events was the government's own agriculture department. In the absence of staff, rain, mould and termites entered government offices in the capital Honiara and effectively destroyed many of the department's records. These included most of the department's agronomic and soil databases as well as records of field trials and reports carried out by international aid consultants (Barry Evans, personal communication, 2008). Some of this information could be retrieved from district offices away from the capital but much else had to be sought by consulting former staff, including expatriate and former colonial-era staff no longer living in the country. In effect, an attempt had to be made to create a retrospective network of knowledge-holders.

Elsewhere, knowledge has been lost when key staff retired, organisations have been restructured, fires have burned government buildings or computer disks have been inadvertently wiped cleaned. Erskine et al. (2005, p. 270) describe several cases from northern Australia where the organisational restructuring of government departments showed seemingly impregnable 'institutional memories' were much less secure than was imagined.

Nor have governments always been very innovative in either the silvicultural techniques they have promoted or the financial instruments they have used. Thus they have often recommended 'safe' silvicultural approaches that have been successful elsewhere rather than exploring silvicultural systems that might suit the environmental (or economic) circumstances at a particular site. For example, in recent years, most government research has concentrated on the same small number of 'fashionable' exotic species and looked for ways to increase the productivity of plantations using these rather than exploring a wider variety of options. Sometimes the farmers taking this advice have been disappointed with the outcome (the dangers of a single, top-down prescription have already been noted in the case of the rice planting problem and the Islamic calendar described in Box 2.1). Some governments have also provided financial support for reforestation although this has often been given to favoured companies rather than being used more generally for the national interest.

The Role of Plantation Timber Companies

In recent years private timber companies have reforested some large areas and it is likely they will continue to do so in future. Some have done this without government support but many have received substantial financial incentives or taxation concessions from governments to attract them to do so (Enters et al. 2003; STCP 2009). The strong commercial imperative of these companies has meant that some have developed highly efficient plantation operations (although, it must be said, others have not and there are many examples of company plantations that have failed).

Most industrial plantations have been established on grassland or secondary regrowth although some have replaced natural forests (and some companies have logged natural forests in order, they say, to establish plantations but have then failed to do so). Few companies have used highly degraded sites or those with infertile soils unless these disadvantages could be compensated for by gentle topography and good locations close to transport (apart from those mining companies legally obliged to do so). Most industrial timber plantations are grown on a short rotation for woodchips, pulpwood or for veneer and only a few companies have invested in sawlog plantings. Because of this, the range of species used has been comparatively small. In future these companies are likely to continue with this type of reforestation. Those that have undertaken any research have largely confined their activities to improving the productivity of their chosen species and few have explored the role their plantations might have in generating ecosystem services although a carbon market could dramatically change this.

However, this pattern is changing. In recent years some timber plantation companies are beginning to pay more attention to the landscapes in which they establish their plantations and try to embed them within a matrix of secondary forest regrowth on steeper lands and along water courses (Cyranoski 2007; Wooff 2009). More are making more effort to ensure harmonious and beneficial relationships with local communities (e.g. Marjokorpi and Otsamo 2006). These are significant changes and, together with moves towards the certification of plantation timbers, mean that companies could play an important future role in helping to reforest cleared lands in ways that provide some environmental and social services in addition to goods. Mining companies, too, are increasingly involved in research to develop better ways of rehabilitating former mine sites.

The extent to which private companies will continue to invest in timber plantations depends on the availability of land, how their funding sources perceive the future profitability of these investments and also on the economic and political risks involved. The willingness of governments to continue offering incentive payments or tax concessions so these companies can continue tree-planting is likely to remain crucially important (STCP 2009). It is possible that some companies could switch to an alternative crop such as oil palm or other land uses if this was deemed to be more financially attractive than timber tree plantations.

The Role of Non-Government Organisations

In this context, the term NGO is used in rather broadly and taken to include local and international conservation organisations, international development assistance groups, university researchers as well as organisations like farmer's associations. This diversity makes it hard to generalise about the role of NGOs but it is clear they have been very influential in raising public awareness about environmental issues and of the need for reforestation (Table 12.1).

NGOs usually have different priorities than governments and have often sought to change government policies. For example, while recognizing the need to improve rural livelihoods, some place a much higher priority than governments on the need for reforestation to protect biodiversity. These differences mean they have often promoted a wider range of silvicultural options than used by most government agencies or timber companies. They have also often taken a lead role in improving information flows and promoting new financial instruments such as payments for ecosystem services.

Most NGOs have taken a more participatory approach than governments and have often worked in smaller, more intimate relationships with local partners. On the other hand, some NGOs have also been able to work across several scales and have formed relationships with national governments, international bodies and corporate entities. This has allowed them to bring in new ideas and techniques from comparable situations in other countries. However, NGOs have some disadvantages. Some have limited budgets and can only tackle short-term projects; this is a severe disadvantage when working on reforestation. And, where there are large numbers of NGOs operating, each with differing objectives, capabilities and time frames it can be difficult to coordinate their activities. This can reduce the overall impact of their potential contributions.

The Role of Households and Communities

Most of the large reforestation efforts in the past have been carried out by government agencies and timber companies. But one of the themes of this book is that smallholders and communities are an unrecognized or under-valued group of potential growers. When given the opportunity, communities and households have often been very effective at protecting small patches of residual forests, including new secondary forests, and enforcing rules governing access and usage (Table 12.1). In addition, households and communities have also established some very large areas of plantation even though individual plantings are usually small (Table 3.3). The scale of these contributions are substantially under-estimated because the individual areas involved are mostly small and because several countries, apparently, do not collect any statistics on smallholder plantings.

Households are primarily interested in reforestation because of the direct livelihood benefits it can bring but their plantings often generate a wider range of conservation benefits and add heterogeneity to rural landscapes. This is not because they are more selfless than governments or private timber companies. Instead, it is because they are growing trees for a wider range of purposes and are prepared to use a more varied set of silvicultural methods. Not all of these methods are ideal from a commercial point of view and many smallholders are poorly informed about markets for their forest products (let alone any ecosystem services they generate). Nor do they necessarily know the species or silvicultural methods most likely to allow them to take advantage of future markets. This is why the learning networks described in Chapter 10 are so important. Farmer's views and aspirations change over time as new markets appear and new opportunities arise and their future interest in reforestation, and as well as their capacity to undertake it, will be heavily influenced by government policies.

In summary, there are a number of groups interested in reforestation and able to make a contribution in the future but no one of these various players – governments, companies, NGOs or smallholders – has all the skills needed to address the four questions raised earlier (i.e. how to undertake better land use planning, devise incentives for reforestation, develop new forms of reforestation and devise new financial mechanisms to encourage reforestation?). On the other hand, there is considerable complementarity between these groups and also the potential for some considerable synergies between them. The best outcome would be if they could be somehow be brought together. But how might this be done?

New Institutional Settings to Encourage Reforestation

When social-ecological systems are complex and face an uncertain future then the form of governance required is one that is able to deal with this. Large-scale and centralised forms of governance are unlikely to be adequate. In recent years, there has been considerable interest in how communities have developed ways of managing common property resources and so avoided the 'tragedy of the commons' (Ostrom 2005). Plantation forests, and especially those established by smallholders, are not common property resources like fisheries or water used for irrigation. Nonetheless, the extensive literature concerning how communities have evolved methods to manage common property resources provides a lens through which to examine the institutions that might be developed to foster reforestation on a larger scale. Despite the obvious differences, there are also some similarities between the situations faced by managers of these natural resources and those establishing plantations. Firstly, both managers of common property resources and tree-growers are concerned with restoring or re-creating a resource. Many common property resource institutions and management regimes only develop after it becomes obvious that the resource is being over-exploited. Likewise, those carrying out reforestation are usually prompted to do so when natural forests are unable to supply the goods or services they once did.

Secondly, in both cases, participants depend on collective action to generate benefits. Tree-growers can establish trees, whether or not their neighbours do so, but they may depend on their neighbour's plantations to create a sufficient resource to make their own trees commercially valuable. Without a regular and predictable supply of a product, a single, small plantation may be virtually worthless. Likewise, a single small plantation may be unable to provide ecosystem services such as watershed protection or biodiversity conservation. In short, the supply of both goods and services are scale-dependent meaning that collective action is needed to generate a benefit.

Thirdly, both sets of participants, consciously or unconsciously, also depend on collective action to manage the new resources and maintain them over time. Markets depend on reliable supplies of goods and services. When plantation owners withdraw from the market (e.g. by felling trees and not replanting) they may diminish the economic and ecological value of the remaining resource. Note that those withdrawing at an early stage do not suffer because they sell before the costs of their withdrawal become evident. In this sense, these people are the plantation equivalent of common property resource free-riders (Ostrom 1990).

Several design principles have emerged from studies of the institutions used to manage common property resources (Ostrom 1990). Some of these are concerned with rules governing access to existing common property resources and with penalties for infringing these access rules. But there are also several that appear to bear directly on the task of promoting reforestation. These are:

1. Collective choice arrangements: those affected by policies and rules should have a role in formulating these.
2. Congruence between policies and local conditions: policies and rules developed at a national level should be consistent with local conditions.
3. Nested enterprises: when the resources (or plantations) are parts of larger systems, then rules and procedures developed at one level (e.g. a district or province) must complement or be integrated with those at another (e.g. national).
4. Conflict resolution: participants should have low-cost access to local arenas to resolve conflicts among themselves or between participants and the government.

Some of these principles are reflected in the various forms of 'participatory', 'joint forest management' or 'community-based forest management' that have developed in recent years (Fisher 1995; Petheram et al. 2004). But how might these principles be used to encourage reforestation at a national scale?

A System of Cooperative Advisory Groups

Perhaps the strongest of these lessons is the first one concerning the importance of having stakeholders involved in the development of policies that affect them. Once stated, the advantages seem self-evident; those most intimately involved in the day-to-day business of reforestation are likely to be better equipped to identify problems (or opportunities) than those in remote government offices. On the other hand, there is undoubtedly also a need for a central government body to retain an overview and be able to coordinate activities. In any case, few government Forestry Departments are likely to give up their dominant role in establishing and administering policy.

One solution would be to develop a form of governance that brought the various parties together. A way of doing this could be to have a system of cooperative advisory groups. These groups could include representatives from private plantation companies, farmers' groups, NGOs and other government agencies such as agricultural and conservation departments. In addition, the groups might involve timber industry representatives, users of the ecological services provided by forests and non-government specialists such as resource economists, land use planners and conservation biologists. The purpose would be to link all those interested in reforestation and allow them to share views on the effectiveness of current policies and how impediments to further reforestation could be overcome.

Some of the specific issues to be addressed might be:

- The policies and practices needed to prevent further forest and land degradation.
- Ways of developing more participatory forms of land use planning (e.g. Where might reforestation rather than agriculture be a priority? How to respond to climate change? How to balance national needs for, say, watershed protection against the livelihood needs of individual landowners? How to use abandoned wastelands?).
- The policies needed to protect secondary regrowth or encourage plantings by both smallholders and industrial companies (e.g. land tenure, property rights, the role of incentives, removal of perverse subsidies, taxation issues, micro-finance, silvicultural needs of different types of farmers).
- How changes in reforestation practice might be monitored and evaluated over time so that alterations could be made to policies in response to changing circumstances (e.g. Are rural livelihoods being improved by reforestation? Are equitable arrangements being used in out-grower schemes? Are ecosystem services being produced in national or provincial reforestation programs to the extent that governments assume?).
- Ways of liaising with the learning networks of Chapter 10 to ensure coordinated national research efforts are carried out on common problems.
- The role of markets and market-based instruments in funding reforestation programs (e.g. business models for smallholders; the ways in which local industries able to process forest products might be supported, how reforestation can tackle poverty and improve livelihoods?).
- The identification and development of new markets for forest products as well as ecosystem services (e.g. ways of improving financial benefits to forest growers, ways of supporting small rural industries using forest products, ways of entering carbon markets and using these to benefit smallholders as well as large companies,

location of reforestation areas for ecotourism, encouraging smallholders to achieve certification of their plantations).

- How to ensure that legal ambiguities are avoided and that national, provincial or local reforestation policies are harmonised. Likewise, ensuring that reforestation policies of forestry agencies do not conflict with land use policies of other government agencies.

This is a formidable list and it would be sensible to have a network of sub-groups specialising in different topics rather than a single body. Alternatively, separate groups located in different geographic areas might able to give feedback on draft national policies by exploring local views on how these policies might work in practice. All groups would be coordinated by, and responsible to, a central advisory body on which each of the primary stakeholders was represented.

Those studying the way different kinds of institutional arrangement work have often concluded that diversity is beneficial. Having separate groups working on the same issue often results in a variety of solutions that are all worth exploring and testing. Bodin and Norberg (2005) have demonstrated how loosely organised networks are more likely to be successful than a tightly organised one because they are likely to generate multiple solutions. Similarly, Ostrom (2008) concluded that there are considerable advantages in having seemingly redundant design teams when the problem being worked on is such there is a high probability of errors being made.

The groups or network then ensures that information sharing becomes a bottom-up as well as a top-down process. Because information is shared across the network it avoids the problem of having a single centralised repository of knowledge (see Box 12.1). The matter of how these collaborative groups might be established and the advantages of this type of approach is discussed in rather more detail by Sayer and Campbell (2004), Anderies et al. (2004), Olsson et al. (2004, 2006), Lebel et al. (2006) and Berkes (2007).

The work done by a set of advisory bodies such as these differs from that carried out by the silvicultural learning networks described in Chapter 10 by being broader in scope and involving more than just growers and researchers. But, in a sense, the idea is simply an enlargement of the same vision in which greater benefits and a more resilient form of adaptive management can be developed when those with a stake in the outcome of reforestation are involved in establishing policies governments use to promote it.

One of the other features of the institutions emerging over time to manage common property resources is that they usually establish penalties to prevent rules being circumvented (to avoid what is referred to as the 'free-rider' problem). Penalties to prevent free-riders are not as relevant in the present situation. In fact, one of the purposes of the advisory groups is to spread knowledge about management systems as widely as possible. There are advantages in collective action to coordinating sales of products emerging from plantations but it would be extremely difficult to insist on this being done and, in any case, would probably be counter-productive. A far better approach would be for the advisory network to encourage producers to see this as being in their own self-interest.

Problems in Implementing Change

These are hardly radical proposals but most governments jealously guard their powers and are reluctant to share them with others. This is especially true of the relationships between Provincial and National governments. And forestry departments are often especially resistant to change. Muthoo (2009) notes that forest authorities are sometimes among the oldest, largest and most powerful land management agencies in many countries. Their long tradition has facilitated an administrative sense of mission that perpetuates established norms and traditions and makes them resistant to external pressures. Unlike farmers who are usually pragmatists and often open to new ideas, many bureaucrats are not always receptive to alternative ways of establishing forests or managing lands. Even when changes are agreed to at forestry headquarters, local officials are often reluctant to implement them in the field. This may be because of professional egotism, a genuine belief that the changes will lead to the overturning of national land use practices that they support or sometimes, it must be said, a belief that the devolution of power will curtail their ability to manipulate events for personal financial gain (Edmunds and Wollenberg 2001). Both governments and their staff can often find ways of slowing or preventing change by having lengthy approval processes or burdensome administrative procedures.

Problems can also occur in on the non-government side when trying to create advisory groups such as:

- It can be difficult to identify non-government participants who are truly representative of other stakeholder interests and who are prepared to be accountable to their constituents. It may be possible to use traditional community leaders but communities made up of recent migrants rarely have the coherence of traditional communities while the institutions and leaders once present in traditional villages are sometimes overwhelmed. New organisations like growers associations could provide representatives but these are not always present or active.
- Not all parties may wish to be involved. For example, some larger corporations may prefer having one-on-one discussions with governments when matters of particular interest to them arise. Others may not think it worth the effort.
- The process can be captured by local elites who find ways of manipulating the process to suit their own purposes. Similarly the process can become corrupted, especially if large amounts of external funds (e.g. aid funds) are involved.
- Differences in the views of, say, powerful industrial corporations and small local growers that make collaboration or agreement difficult. For example, small growers may see large corporations being competitors rather than partners. Likewise, some participants will expect their views to carry more weight than their fellow member. Ways will need to be developed to take account of these differences if the group is to work together effectively. Some principles that might increase the effectiveness of the consultation process are outlined in Box 12.2.

Box 12.2 Consultative Principles

Alexander (2009) suggested a set of principles that might allow effective consultative networks develop. They were developed for forest conservation practices in general rather than reforestation in particular but are still relevant. They need to be discussed, and agreed to, by stakeholders if they are to be useful.

1. Engage a broad range of stakeholders
   These should represent civil society, industry and various levels of government. An issue to be resolved is how representatives are chosen to ensure the each sector is given a voice.
2. Institute reliable operating structures and processes
   These are needed to ensure meetings are planned and effective. Representatives need to feel their voices will be heard.
3. Practice transparency
   The results of the group's work need to be shared so all those with an interest in the topics discussed are able to take advantage of the outcomes.
4. Use effective communication channels
   Different forms of communication may be needed for different stakeholders.
5. Foster a focus on interests and not positions or personalities
   The group will work best if vested interests are put aside and the group works for the common good.
6. Allow for independent verification
   It may be useful to have outside entities or consultants check findings or conclusions.
7. Be responsive to all concerns
   The group may not be able to tackle or resolve all the concerns that it is asked to address. Care should be taken to ensure the needs of smallholders are given as much attention as those of industry.
8. Make use of existing networks
   Existing growers, industries or marketing networks should be utilised to share knowledge, communicate concerns and solicit feedback and advice.
9. Undertake capacity building
   Build the capacity of stakeholders to use new information and knowledge and to benefit from the group's work.
10. Undertake periodic reviews
    Periodic reviews are useful to ensure the purposes and objectives of stakeholders continue to be met as ecological, social and economic circumstances unfold.

The history of devolution of natural forest management in the Asian region has been extensively reviewed by Edmunds and Wollenberg (2001) and they conclude that many of the outcomes have been disappointing. Muthoo (2009) also gives an equally dispiriting account. Might it be different when the objective is to create new forests rather than manage existing ones? One answer is that it may be difficult to expect government forestry agencies to devolve authority or explore more participatory forms of governance in the reforestation sector in times of rapid economic growth, where there is a scramble by investors to acquire land or where there is intense inter-agency competition within government circles for resources and influence. These are occasions when trust can be in short supply and deceptive or selfish behaviour is more common. Under such circumstances governance often worsens and democracy sometimes malfunctions (Collier 2007).

On the other hand, promising changes have appeared when, after repeated failures, it becomes clear that something different is needed. This was the case in the Philippines where it was evident that reforestation would only occur if a completely new set of institutional arrangements were developed (Chokkalingam et al. 2006). Collaborative arrangements also seem to work best when there are at least several ‘champions’ of the idea who are prepared to work to make them succeed. One collaborative forestry network that has worked well is one developed in Fiji to carry out land use planning for reforestation in the Drawa forest area. This was briefly described in Box 11.2. In this case, the champion was the German development agency GTZ and the real test of its usefulness will come when this external champion (and the funds it provides) departs. Given that most land in Fiji is owned by traditional communities, among whom there is likely to be much greater levels of trust, participatory forms of planning and forest development should work well there.

Despite setbacks of the type described by Edmunds and Wollenberg (2001) and Muthoo (2009), there appears to be a general trend towards devolution across the region and sufficient promising examples are emerging to show that collaboration and networking amongst forestry stakeholders is not only possible but can also be beneficial (Gilmour and Fisher 1991; Hobley 1996; Magno 2001; Sayer and Campbell 2004).

Revisiting Resilience

The development of an appropriate institutional framework is the final step in building resilient social-ecological systems at deforested or degraded sites. The topic of resilience is one discussed at several stages throughout this book and the variety of future uncertainties outlined at the commencement of this chapter means it is appropriate to return to the idea once more in order to synthesize these different discussions.

Recall that there are three elements of resilience within social-ecological systems, namely an ecological component, an economic component and a social component (Table 12.2). Making reforestation ecologically resilient involves ensuring the new ecosystems have sufficient biodiversity such that one functional species type can be supplemented or replaced by another when conditions change. This might be achieved at a particular site by growing mixed-species stands, or it might be achieved across a landscape by creating a mosaic of simple monocultures each involving a different species. There is no way of knowing just how much diversity is needed given the variety of changes that may occur in future. This means there should be some kind of a monitoring system in place to track the changes that occur and monitor the capacity of the ecological system to adapt itself to these changes. Signs that the new forests cannot adjust to the changes being experienced and are being pushed towards some kind of a threshold would be the trigger for management changes.

The second element is the economic one. Land managers with tenure are likely to consider reforestation as long as they think it will produce a satisfactory economic outcome. These benefits may be from the goods being sold or from payments for ecosystem services they provide to external bodies. The greater the

Table 12.2 Components of a resilient system of reforestation

Ecological	Economic	Social
Diversity of species used (differing in their functionality and in the economic products they generate)	New forests able to produce a variety of goods	Land tenure and property rights provided
These grown in polycultures	New forests able to generate ecosystem services	Supportive institutions including stakeholder advisory networks established
Spatially heterogeneous landscapes developed	Diverse markets available for goods and services	Learning networks for growers established
Ecological monitoring systems developed	Market information available to land managers	Growers associations and marketing cooperatives developed
	Growers can easily access markets	Equitable systems of governance including legal systems developed
	Economic monitoring system developed	
	Supportive policy environment for rural forest industries (e.g. sawmills, furniture factories)	

variety of markets available to growers, the more resilient the system is likely to be. But declining prices for goods or services, a sharp reduction in the number of buyers, increased costs of bringing goods to markets or increasingly complex administrative systems all act as warning signals to growers that something might be amiss. Again, a monitoring system that provides an early warning of adverse changes is essential so growers can develop responses in management practices or even in land use before it is too late.

Lastly, the system must be socially resilient. A socially resilient system is one where growers are aware of ecological and economic feedback and are continually engaged in testing and refining their practices. This means they can adapt to change and modify their silvicultural and economic practices if it becomes necessary to do so. This kind of responsiveness is encouraged by a supportive institutional framework and equitable forms of governance. It is also encouraged by a financial environment where there is access to funds and loans that can help when crises arise and changes are necessary.

These principles may be satisfying for theoretical ecologists, but how appealing might they be for smallholders and other stakeholders interested in reforestation? The most difficult component of resilience to promote will undoubtedly be the ecological elements. Despite the diversity of species used in traditional agricultural systems there appears to be a relentless move towards monocultures. Part of the dilemma is that ecologists cannot specify how much diversity is needed to generate a certain quanta of resilience. If a grower asks, should I use three species or six or ten, an ecologist can only admit that they are not sure; it depends on the nature of the future changes that may occur. In some cases three species might be sufficient while in others it will not. The dilemma may resolve itself in the sense that large industrial growers will mostly continue to grow monocultures but many will do so by embedding these monocultures within a diverse landscape mosaic of regenerating secondary forests. At the same time, many smallholders will continue following their conservative inclinations and grow a variety of trees species not because of the ecological virtues of doing so but because of the economic benefits they receive from maintaining this diversity. And again, their farm and plantation will be just one in a landscape mosaic that includes woodlots, home gardens, secondary forest and residual stands of primary forests. That is, some ecological resilience will be achieved at individual sites but it might be more easily achieved at the landscape scale.

Building economic resilience may also be difficult. Most farmers are inherently cautious and may be interested in producing several income streams as a way of insuring themselves against unexpected change. This means they should be receptive to the principle of building economically-resilient forms of reforestation even though commercial tree-growing is likely to be a new land use activity for most of them. But it depends on circumstances. Those acting as out-growers and having a long-term contract with an industry partner might not see any purpose in having more than one product stream because they believe their contract provides them with a form of insurance against change. Others selling a single product into a currently profitable market might have a similar view. In both cases they would diminish their income if they diversified their products. On the other hand, diversification might be easier to promote when growers live in more isolated areas or where market conditions are less predictable. Given the future uncertainty of the market for commodity timbers described earlier and the increasing interest in ecosystem services, the best way of promoting economic resilience will be to educate growers into the nature of the past and likely future markets for forest products and services. In the meantime, efforts should be made to maintain the existing rural industries that consume the goods produced by forest growers and continue support for research into ways these markets might be diversified and new niche markets encouraged.

Perhaps the easiest component of resilience to promote will be the social elements. Many growers are already members of farmers groups and are likely to be very willing to join grower cooperatives to share knowledge or help them market their produce. The advantages are self-evident and there are few costs apart from the time invested. There seems no reason why this could not be enhanced by such devices as the Learning Networks (Chapter 10) and the Cooperative Advisory Group system of this chapter. Some of these social interactions will take place without government support or involvement. Others may need government assistance to begin even though the role of government may eventually decline.

The promotion of resilience will be a long-term business. Commercially-oriented reforestation is still a new land use activity and the majority of plantations were only established after 1970. Many are still in their first rotation. This means that most growers are yet to experience the full range of ecological risks including pests and diseases, fires, storms or droughts. Similarly, the economic conditions under which they were established may not continue into the future. It is likely, therefore, that at least some present silvicultural practices will prove to be unsustainable in the longer term. Roberts (2009) has argued that much of the world's food production system is already monolithic and brittle. There is a danger that plantation silviculture could fall into the same high-risk trap unless deliberate steps are taken to avoid it.

Conclusions

Large scale reforestation needs some coordination. Some reforestation will occur when market conditions attract landholders to plant commercial tree crops. It can also occur when regrowth develops on abandoned land. But neither of these pathways will necessarily lead to reforestation of areas most in need of reforestation such as eroding watersheds or areas important for the conservation of biodiversity. Nor will they necessarily lead to any improvement in the livelihoods of rural smallholders who could benefit from growing trees on part of their land. If reforestation is to improve environmental outcomes, or rural livelihoods, then some form of government involvement is usually necessary in order to coordinate and perhaps guide reforestation. The question is, just how big a role should governments play? In the past, much reforestation was actually carried out by government forestry agencies and these usually took a largely top-down approach when dealing with others interested in planting trees.

Thus, they tried to specify the species to be planted and the way the plantation should be managed. While this approach had some successes (as well as some failures), it is unlikely to be very successful in the future. The variety of potential forest growers has increased and these have a much wider range of objectives than past government forestry agencies. In addition, the environmental and economic conditions of the future seem rather less predictable. This means that past forms of reforestation may be quite inappropriate.

Studies of the arrangements used by other long-lasting natural resource management groups suggest the best way of evolving new institutions able to foster reforestation are those that build on the capacity of groups to organise themselves. Governments have a critical role to play in helping this to occur by encouraging the formation of networks of advisory groups involving representatives from smallholder groups, industry and NGOs. These groups could help identify the impediments and then cooperate with governments to develop new policies that foster reforestation at both a local and national scale. The development of supportive institutions able to formulate and guide appropriate reforestation policies represents the final elements in a framework of factors likely to lead to the creation of resilient new forest ecosystems.

References

Alexander SS (2009) Ground rules. Arborvitae 39:14–15

Anderies JM, Janssen MA, Ostrom E (2004) A framework to analyze the robustness of social-ecological systems from an institutional perspective. Ecol Soc 9, http://www.ecologyandsociety.org/vol9/iss1/art18/

Berkes F (2007) Community-based conservation in a globalised world. Proc Natl Acad Sci USA 104:15188–15193

Bodin O, Norberg J (2005) Information network topologies for enhanced local adaptive management. Environ Manage 35:175–193

Chokkalingam U, Carandang AP, Pulhin JM, Lasco RD, Peras RJJ, Toma T (2006) One century of forest rehabilitation in the Philippines: Approaches, outcomes and lessons. Center for International Forestry Research, Bogor

Clement F, Amezaga JM (2008) Linking reforestation policies with land use change in northern Vietnam: Why local factors matter. Geoforum 39:265–277

Collier P (2007) The bottom billion: Why the poorest countries are failing and what can be done about it? Oxford University Press, Oxford

Colwell RK, Brehm G, Cardelus CL, Gilman AC, Longino JT (2008) Global warming, elevational range shifts, and lowland biotic attrition in the wet tropics. Science 322:258–261

Corlett R (2009) Seed dispersal distances and plant migration potential in tropical East Asia. Biotropica 41:592–598

Cyranoski D (2007) Biodiversity: Logging: the new conservation. Nature 446:608–610

Edmunds D, Wollenberg E (2001) Historical perspectives on forest policy change in Asia. Environ His 6:190–212

Elliott S, Kerby J, Blakesley D, Hardwick K, Woods K, Anusarnsunthorn V (eds) (2000) Forest restoration for wildlife conservation Proceedings of a Workshop on international tropical timbers organisation and forest restoration research unit, Chiang Mai University, Chiang Mai, 30 Jan to 4 Feb 2000

Enters T, Durst P, Brown C (2003) What does it take? The role of incentives in forest plantation development in the Asia-Pacific region. Unasylva 54:11–18

Erskine P, Lamb D, Bristow M (2005) Reforestation in the tropics and subtropics of Australia using rainforest tree species. Rural Industries Research and Development Corporation, Canberra, https://rirdc.infoservices.com.au/downloads/05–087.pdf; accessed 20 September 2010

Fahlen A (2002) Mixed tree-vegetative barrier designs: Experiences from project works in northern Vietnam. Land Degrad Dev 13:307–329

Fisher RJ (1995) Collaborative management of forests for conservation and development. International Union for the Conservation of Nature, Gland

Foster AD, Rosenzweig MR (2003) Economic growth and the rise of forests. Quart J Econ 118:601–637

Gilmour DA, Fisher RJ (1991) Villagers, forests and foresters. Sahayogi Press, Kathmandu

Hobley M (1996) Participatory forestry: The process of change in India and Nepal. Overseas Development Institute, London

Hughes L, Causey EM, Westoby M (1996) Climatic range sizes of *Eucalyptus* species in relation to future climatic change. Global Ecol Biogeogr 5:23–29

IPCC (2007) Climate change 2007: The physical science basis. Cambridge University Press, New York

Jomo KS, Chang YT, Khoo KJ (2004) Deforesting Malaysia: The political economy and social ecology of agricultural expansion and commercial logging. Zed Books in Association with United Nations Research Institute for Social Development, London and New York

Kates RW, Parris TM (2003) Long-term trends and a sustainability transition. Proc Natl Acad Sci USA 100:8062–8067

Kato T (1994) The emergence of abandoned paddy fields in Negera Sembilan, Malaysia. Southe Asian Stud 32:145–172

Koop G, Tole L (2001) Deforestation, distribution and development. Global Environ Change 11:193–202

Laurance WF (2008) Changing realities for tropical forest managers. Trop Forest Update 18:6–8

Lebel L, Anderies JM, Campbell B, Folke C, Hatfield-Dodds S, Hughes TP, Wilson J (2006) Governance and the capacity to manage resilience in regional social-ecological systems. Ecol Soc 11(1):19, http://www.ecologyandsociety.org/vol11/iss1/art19/

Leslie AJ (2005) What will we want from the forests? ITTO Trop Forest Update 15:14–16

Magno F (2001) Forest devolution and social capital: State-civil society relations in the Philippines. Environ Hist 6:264–286

MARD (2001) Five million hectare reforestation program partnership: Synthesis report. International Cooperation Department, Ministry of Agriculture and Rural Development, Hanoi

Marjokorpi A, Otsamo R (2006) Prioritization of target areas for rehabilitation: A case study from West Kalimantan, Indonesia. Restor Ecol 14:662–673

Muthoo M (2009) Are forestry institutions failing to adapt? In: Leslie RN (ed) The future of forests in Asia and the Pacific: The outlook for 2020; RAP publication 2009/03. Food and Agriculture Organization of the United Nations, Bangkok

Ohlsson B, Sandewall M, Sandewall RK, Nguyen HP (2005) Government plans and farmers intentions: A study on forest land use planning in Vietnam. Ambio 34:248–255

Olsson P, Folke C, Berkes F (2004) Adaptive co-management for building resilience in social-ecological systems. Environ Manage 34:75–90

Olsson P, Gunderson LH, Carpenter SR, Ryan P, Lebel L, Folke C, Holling CS (2006) Shooting the rapids: Navigating transitions to adaptive governance of social-ecological systems. Ecol Soc 11: http://www.ecologyandsociety.org/vol11/iss11/art18/

Ostrom E (1990) Governing the commons. Cambridge University Press, Cambridge

Ostrom E (2005) Understanding institutional diversity. Princeton University Press, Princeton and Oxford

Ostrom E (2008) Institutions and the environment. Econ Aff 28:24–31

Petheram RJ, Stephen P, Gilmour D (2004) Collaborative forest management: A review. Austr Forestry 67:137–146

Roberts P (2009) The end of food: The coming world crisis in the world food industry. Bloomsbury Publishing, London

Rigg J (2006) Land, farming, livelihoods, and poverty: Rethinking the links in the rural south. World Dev 34:180–202

Rudel TK, Coomes OT, Moran E, Achard F, Angelsen A, Xu JC, Lambin E (2005) Forest transitions: Towards a global understanding of land use change. Global Environ Change 15:23–31

Sayer JA, Campbell B (2004) The science of sustainable development: Local livelihoods and the global environment. Cambridge University Press, Cambridge

Scott JC (1998) Seeing like a state: How certain schemes to improve the human condition have failed. Yale University Press, New Haven and London

Sikor T (2001) The allocation of forestry land in Vietnam: Did it cause the expansion of trees in the north west? Forest Policy Econ 2:1–11

Sowerwine JC (2004) Territorialisation and the politics of highland landscapes in Vietnam: Negotiating property relations in policy, meaning and practice. Conserv Soc 2:97–136

STCP (2009) Encouraging industrial forest plantations in the tropics: Report of a global study ITTO technical series No 33. International tropical timbers organization, Yokohama

Steinberg P (2005) From public concern to policy effectiveness: Civic conservation in developing countries. J Int Wildlife Law Policy 8:341–365

Svenning JC, Condit R (2008) Biodiversity in a warmer world. Science 322:206–207

United Nations (2007) The world urbanisation prospects: The 2007 revision population database. United Nations Department of Economic and Social Affairs, Population Division, Washington, DC

Whiteman A, Jonsson R (2009) Trends and outlook for forest product markets in Asia and the Pacific. In: Leslie RN (ed) The future of forests in Asia and the Pacific: Outlook for 2020. Food and agriculture organization of the United Nations, Bangkok, pp 178–198, RAP Publication 2009/03

Williams SE, Bolitho EE, Fox S (2003) Climate change in Australian tropical rainforests: An impending environmental catastrophe. Proc R Soc Lond B Biol Sci 270:1887–1892

Wooff W (2009) Sabah forest industries experiences in plantation forestry. Conference on the current state of plantation forestry in Malaysia: A special focus on Sabah, Forestry Department Headquarters, Sandakan, 18–20 Nov 2009

Chapter 13
The Way Forward

> If this (viz; the study and conservation of biodiversity) is not done, future ages will certainly look back upon us as a people so immersed in the pursuit of wealth as to be blind to higher considerations. They will charge us with having culpably allowed the destruction of some of those records of Creation which we had in our power to preserve: and while professing to regard every living thing as the direct handiwork of a Creator, yet, with strange inconsistency, seeing many of them perish irrecoverably from the face of the earth, uncared for and unknown.
>
> Wallace (1863, p. 234)

Introduction

The last 100 years has seen a massive assault on the world's tropical forests. Global statistics on the scale of deforestation and land degradation are surprisingly imprecise but there is no doubt whatsoever of the need for some reforestation to overcome the environmental problems that deforestation and land degradation have unleashed. Though much deforestation was triggered by a search for agricultural land the process has been hugely wasteful and has generated large tracts of underused or abandoned land. At the same time, deforestation has caused a widespread loss of biodiversity, increased rates of soil erosion, large increases in the emission of greenhouse gases and the persistence of poverty amongst many of the communities living in these landscapes. Perhaps most worrying of all, there is no clear sign that many of these negative trends are slowing. In the meantime, human populations and population densities are continuing to rise.

The question posed at the commencement of this book was how reforestation might be carried out to restore sufficient biodiversity to re-establish ecosystem functioning on these lands and, as well, to do this in a way that improved human livelihoods. Earlier chapters emphasised there are a variety of silvicultural techniques that might be used. These techniques differ in four important respects. Firstly, they differ in whether reforestation is achieved by relying on natural regeneration or whether it is carried out by planting seeds or seedlings. Secondly, they differ in the types of species used with some plantings relying on fast-growing trees

D. Lamb, *Regreening the Bare Hills*, World Forests 8,
DOI 10.1007/978-90-481-9870-2_13,

grown on short rotations while others use slower-growing species grown on longer rotations. Thirdly, they differ in the type or design of plantations with some growing trees in simple monocultures and some growing trees in more complicated designs involving mixtures of species. And, finally, the various techniques differ in their capacity to satisfy different objectives; some are most suited for timber production, others are more suited to improving conservation and environmental outcomes while some can satisfy, at least in part, both objectives.

The greatest conservation and environmental benefits are usually generated by either natural regeneration or forms of planting involving native species and more complex planting designs. A plantation monoculture, for example, may have greater conservation benefits than regularly burned grassland but regrowth forests or mixed-species plantations on the same site are likely to generate substantially more benefits than these monocultures. However, the effectiveness of any planting for conservation or environmental protection also depends on the landscape context. For example, plantations established on cleared lands around existing patches of natural forest or those that improve the connectivity between several patches will usually generate more benefits than small, isolated plantings. Likewise, plantings on steep slopes will usually prevent more soil movement than those using the same design on flat land.

Although there are forms of reforestation able to generate improved conservation outcomes nobody should be under the impression these are necessarily capable of restoring all of the diversity that was once present. This may happen in some locations such as where areas adjoining natural forests are reforested. But in other cases the new forests, including those established using the techniques described here as Ecological Restoration, may only acquire a sub-set of the original biota. Large wildlife species with extensive home ranges are most at risk of not being restored. This is either because these species are now (locally) extinct, because it is impossible to re-create sufficiently large contiguous areas of suitable habitat or because the complex mutualisms and trophic relationships that once supported them are simply too difficult to re-assemble. From a conservation point of view, it is far better to protect the original forests than trying to restore them. The problem, of course, is that it is not always possible to achieve this and, that in some areas, it is too late..

Most forms of reforestation also have the capacity to improve livelihoods although tree planting alone will not lift poor households out of poverty. In particular, tree planting may not benefit households with limited amounts of land or income. Nor may it be as easy for those without land tenure to benefit as readily as those with tenure. On the other hand, households with larger land holdings, those with some marginal land unsuitable for agriculture and those with some land but deriving most of their income from off-farm employment may find tree-growing is a very beneficial land use activity. Not only does it offer a way of diversifying income sources but it can build a capital asset and a buffer against hard times. The best form of reforestation to use depends on local circumstances. Fast-growing species established in monocultural plantations may be best in some situations but may prove to be a poor choice in others. It depends on the market for the products produced and the capacity of the grower to transport their products to these markets. It also depends on the time horizon of the grower and whether they can afford to take a long-term view or must deal with more immediate concerns.

One commonly voiced disadvantage of tree plantations is that they provide only an episodic income. This disadvantage certainly holds for those growing trees in monocultures although there are obvious differences between plantations grown on short or long rotations. But some silvicultural systems such as plantations involving mixtures of trees and understorey crops can generate a more frequent and regular income. Likewise, even those growing monocultures may be able to arrange their plantations in an age sequence in order to have a regular income stream from an annual harvest. Finally, plantations may be highly complementary with other farm activities since the labour costs are generally low once a plantation is established. This means the labour opportunity costs are also low. In short, many farmers are likely to find tree-growing can improve their incomes provided they choose a type of plantation that matches their ecological and economic circumstances.

It is important to acknowledge that livelihoods can sometimes be adversely affected by reforestation. One situation where this might occur is if people are forced from their traditional lands by governments who do not recognise these ownership claims and award these lands to large private plantation companies. Another is when extensive reforestation affects water flows to downstream farmers. Both can be serious problems but are caused by the way reforestation is managed rather than reforestation *per se.*

Perhaps the key question in all of this is whether the full range of these potentially advantageous forms of reforestation will ever be used or whether growers will continue to rely on the relatively small number of species and silvicultural systems that have been in vogue for the last several decades? There seems little doubt that most industrial growers will continue with the methods they currently use, at least for the foreseeable future. These plantations achieve their objective and meet the current market demand. Many present and future small landholders will also use the same methodologies for the same reason and will profit from doing so. But there are signs from across the Asia-Pacific region that other farmers are interested in using alternative methods of reforestation (Chokkalingam et al. 2006; Erskine et al. 2005; Fatoux et al. 2002; Nawir et al. 2007; Nibbering 1999; Pasicolan et al. 1997; Santos et al. 2003).

One change that may favour these other, more diverse, silvicultural systems is the development of global markets for ecosystem services. In recent years the loss of timber from natural forests and the process of globalisation have been two of the major drivers of reforestation dictating where it is carried out and the type of reforestation that is used. But emerging markets for ecosystem services could change these patterns and have a very large effect on both the rates of reforestation and on the types of reforestation implemented. The REDD+ process, for example, could have a very large impact on reforestation depending on how it is structured. The scale of any enhanced planting arising from a global carbon offsets market could be very large. Zomer et al. (2008) estimated there could be 750 million hectares around the world that are biophysically suitable and which meet the current Clean Development Mechanism rules for afforestation and reforestation. Within Australia, Lawson et al. (2008) calculate that under a carbon pollution reduction scheme, a carbon price of AUD 29/tCO_2e (around US$25/t$CO_2$e) would make reforestation preferable to current land uses across 21.8 million hectares of cleared agricultural land. Planting trees to offset carbon emissions would not face the location constraints faced by current tree-growers who must transport their goods to markets. This means reforestation could also be carried out on degraded lands in more isolated areas.

A carbon market could also alter the type and extent of future reforestation. Many forest established to sequester carbon will not be harvested. This means fast-growing timber species are not necessarily the most appropriate choice. On the contrary, it would be more prudent to use native species and develop species-rich and resilient forests better able to deal with some of the biological hazards that could develop over the next century. These types of reforestation obviously offer opportunities to generate better watershed protection and biodiversity outcomes than monocultural plantings. It is not clear, at present, whether naturally occurring secondary forests will be eligible to participate in the carbon market but, if so, this would substantially increase the likelihood that such forests will be protected rather than being cleared for some other purpose. In short, a carbon market could potentially have a significant impact on reducing both the extent and the consequences of forest and land degradation in the tropics.

A rising concern about environmental protection is already persuading a number of governments to promote the establishment of forests over sometimes very large areas for what may be loosely referred to as 'protection' or conservation reasons rather than for timber production. Some of these are on state-owned land and others are being encouraged on privately managed land. These include the Wet Tropics Tree Planting Scheme in Australia (Erskine et al. 2005), the Riparian Forest Restoration Project in the Atlantic Forest region of Brazil (Rodrigues et al. 2009; Wuethrich 2007); the Sloping Land Program or Grain for Green program in China (Li 2004; Lui et al. 2008; Morell 2008; Stone 2009; Uchida et al. 2005), the national reforestation program in Korea (Lee and Suh 2005; Tak et al. 2007) and the Five Million Hectare Reforestation Program in Vietnam which specifically includes two million hectares of protection forests (MARD 2001; McElwee 2009; Ohlsson et al. 2005).

In some cases, these protection or conservation forests have used simple monocultural plantations involving the same species as used in nearby production forests. Presumably this was because it was cheaper to do this and because too little is known about other silvicultural possibilities. But, increasingly, many of these plantings are beginning to use indigenous species and more complex planting designs. The more this happens, the more likely they are to achieve their ecological objectives. Large programs such as these face daunting ecological, social and bureaucratic problems (e.g. Uchida et al. 2005; McElwee 2009). A process of adaptive management as well as considerable patience will be necessary if they are to succeed.

Alternative Visions of the Future

When trying to envisage what might be the outcome of different institutional arrangements, policies or market conditions it can be useful to imagine alternative visions or scenarios. Three such scenarios are outlined here that span a range from pessimistic to optimistic.

Scenario 1: A Gloomy Outcome

In this case current policy failures are not recognized, social attitudes remain unchanged and new silvicultural opportunities are passed up. Attempts to establish participatory institutions to develop new policies fail because government agencies and their staff are complacent and unwilling to make real changes.

Deforestation and land degradation continue because the government is unable to combat unregulated and illegal logging of natural forests. This is driven by a continuing demand for cheap timber and by ineffectual implementation of regulations designed to prevent forest loss and degradation. Such logging undercuts any attempts to create financially viable new timber plantations because it maintains a low timber price. Despite the efforts of forestry agencies to prevent it, large areas of both primary and secondary forest are cleared without any real attempt at land-use planning or any understanding of the likely consequences this clearing might have. Biodiversity is lost and erosion and stream sedimentation become more common. Crops established on some of these deforested areas fail because the lands are marginal for agriculture. Wildfires burn through abandoned lands and large areas of grassland are created. Smallholders are inhibited from investing resources in overcoming degradation because of a lack of legal standing or clear property rights.

Some reforestation does occur but most government agencies see reforestation as a lower priority than clearing natural forest to increase the area of agricultural lands. Where reforestation is done it is almost entirely carried out by industrial growers granted cheap land and enticed by generous government subsidies. All of these plantations involve fast-growing exotic species grown for pulpwood. Some of these plantations are regularly burned by fires lit by traditional land owners who have been forced to move. Most of these plantations use clonal planting material of a few exotic species. As a result landscapes are simplified even more and resilience declines further. Some households obtain contracts to supply timber to these companies but it is a buyer's market and they receive only modest prices. Few are able to improve their standard of living. Only limited amounts of reforestation are undertaken by other smallholders and those that do use the same methods used by industrial growers. Many of these plantations fail and, of those that survive, most provide only limited environmental benefits. Growers receive only modest returns from their plantations because of low prices and bureaucratic impediments in the marketing chain. In short, forest and land degradation continues and reforestation makes little contribution to improving the livelihoods of rural people or conserving a threatened biological heritage.

Scenario 2: A Modest Improvement

In this case the scope of the problems generated by deforestation and land degradation are recognised. But the basic institutional arrangements and silvicultural options remain unchanged. Greater emphasis is given to regulating logging in designated production forests to ensure Codes of Practice are enforced. More emphasis is given to harvesting a small number of high quality specialty timbers rather than all commercial species. As a result less damage occurs in natural forests. Attempts are also made to prevent damage inside National Parks by forcibly removing people living there. Most of these are either unsuccessful or result in large numbers of landless people moving to urban squatter settlements. Improvements are made in land use planning and some modest successes are achieved. Further agricultural clearing is only allowed on land with gentle slopes and suitable soils. Most of the subsequent agricultural development appears to be successful. A land redistribution program begins to allocate land to households but operates very slowly and is troubled by accusations of favouritism; much of the best land is granted to a political and financial elite.

Large industrial plantation companies are given generous incentives to undertake reforestation although no particular emphasis is given to using degraded lands. As a result, many areas with advanced secondary forest having significant biodiversity value are cleared and planted. Most of these new plantations continue to use fast-growing species to produce pulpwood. A number of out-growers schemes develop and rules ensure that the contracts are fair to both parties. Additional encouragement is given to smallholders to engage in reforestation primarily by providing technical information but few of the new staff have much practical experience. Much of the advice they provide reflects traditional practices and concerns species that grow quickly and not necessarily species producing goods with a high market value. Growers are initially excited about the growth of their trees but then disappointed with the financial outcomes. The rate of reforestation by smallholders remains modest. Nonetheless, previously cleared landscapes begin to acquire patches of new forest although the composition of these forests is very similar. There are modest improvements in watershed protection and in the populations of some wildlife species though most of these are habitat generalists and are relatively common species.

Scenario 3: A Conservational Outcome

New participatory arrangements develop allowing more devolved forms of governance and policy development to occur. Government forestry agencies change from being management bodies to facilitators and from being largely concerned with production to being more sensitive to, and engaged in, conservation and environmental protection. Protection of forests designated as production forests and the enforcement of Codes of Practice allows the development of sustainable logging practices. Increased efforts are made to identify ecosystems not represented in the protected area network and to protect existing National Parks. People living within Parks are found land outside and are given assistance to move. Greater emphasis is given to land use planning with the consequence that inappropriate deforestation ceases and secondary forests in particular areas are able to recover. Land tenure and property rights are granted to all land users. Some owners (or their children) sell land to neighbours and migrate to urban areas allowing farms to consolidate and average farm size to increase.

A series of advisory groups made up of various stakeholders are established by the government to advise on policies to encourage reforestation. Tree-planting is more widely adopted, especially at sites with soils that are marginal for agriculture. Landholders participate in Learning Networks to exchange silvicultural information. Marketing cooperatives are formed to assist growers with market research and log sales. Monoculture plantations of fast-growing species continue to be established but a much wider variety of other species, including indigenous species, are used as well. These include mixed-species plantations grown on long rotations. Growers engage in innovation and experimentation in collaboration with government researchers. Subsidies are offered by the government to encourage the reforestation of degraded lands and other locations in the landscape needed to improve functional outcomes. Landscapes become spatially heterogeneous and the loss of regional biodiversity ceases although many species move across the landscape in response to climate changes. The populations of some previously threatened species begin to recover. Efforts are made to devise market-based tools to encourage reforestation. A market for various ecosystem services begins and the government establishes an agency to facilitate this and allow smallholders to benefit. Funds generated by payments for ecosystem services are used to undertake more reforestation including ecological restoration. Newly restored forests become attractive targets for eco-tourism.

These are obviously just fictional examples from a very large number of possibilities. Scenario 3 might evolve from Scenario 1 but different countries will follow different paths. Indeed various scenarios are likely to occur in different parts of the same country.

Some Things We Still Need to Know

How do we avoid Scenario 1 and move in the direction of Scenario 3? A good deal is already known about how reforestation might be done and simply applying this existing knowledge (including much traditional knowledge held by local communities) would have a considerable benefit. But not all the technical problems have been resolved and more needs to be known to widen the variety of silvicultural options that are available and to understand their ecological consequences. Nor are the unresolved problems all silvicultural or ecological in nature. Indeed, some of the most difficult are concerned with socio-economic matters. Again, a good deal is known about possible solutions and considerable progress could be made if policies and institutions allowed this existing knowledge to be used more widely. But, again, there are other issues needing to be explored as well if societies are to become more innovative and independent of top-down advice. In short, what is needed is a program of research into socio-ecological systems.

A number of ecological and socio-economic questions are listed below representing issues that have emerged earlier in this book. Many are inter-related. Such lists are necessarily subjective and the intent is to simply stimulate and provoke the reader since actual research priorities will always vary according to location and circumstance. Elliott (2000) and Chazdon et al. (2009) offer lists of other outstanding research questions for Southeast Asia and Meso-American respectively while Gardner et al. (2009) have suggested a conservation research planning framework within which many of these questions could fit.

Ten Ecological Questions

1. *How can reforestation programs make more use of indigenous tree species?*
   A relatively small number of species are currently used in reforestation. What species will be attractive in future market places? How can seed of these species be collected and stored? Can they be easily grown in nurseries?

2. *What are the site preferences of these species?*
   Conditions at many deforested sites are such that species that once grew there can no longer do so. Where can they be grown now? What is their productivity at these sites? Which species need some early shade (how much and for how long)? What are their nutritional requirements? Where should they not be grown?

3. *How can complementary species able to grow in multi-species plantations be identified?*
   Mixed-species plantings have certain ecological and commercial advantages. But random assemblages of species will rarely be effective. Are there characteristics of species (shade tolerance? growth rate? crown architecture?) enabling judgements to be made about their likely complementarity with other species? What types of species might complement each other and which types will not?

4. *How much secondary forest is present in various ages or degradation classes?*
   There are large areas of secondary forest in most countries within the Asia-Pacific region. These have resulted from regeneration after cleared land is abandoned or left after logging. How much of this forest is present across the Asia-Pacific region in various age classes? What is the conservation status of these different types or age classes of forest? What is their like future commercial or conservation value?

5. *How well are different types of planted forests able to generate the ecosystem services previously supplied by natural forests?*
   Not all planted forests are the same, especially in terms of their environmental impacts. How well are these different forests able to supply various ecosystem services including habitats for wildlife and watershed protection? How is this affected by the age of these forests or by their spatial distribution or their landscape context?

6. *What role might reforestation have in maintaining the populations of species usually found in forest interiors?*
   Certain habitat specialists and top-order predators require large areas of relatively undisturbed forest. These species become increasingly vulnerable as natural forests shrink. What forms of reforestation might help conserve such species in the modified landscapes of the future?

7. *How does the type, scale and spatial arrangement of forest patches affect the ability of biota to persist and move across an otherwise agricultural landscape?*

   Certain forest wildlife can move across heterogenous landscapes containing at least some woody vegetation. Some can also reproduce in these landscapes away from natural forest. Which are these species and what types of forests and landscape patterns do they require? Which forest wildlife species will not persist or breed in such landscapes?

8. *How can tropical forests be ecologically restored?*

   Ecological Restoration might be an appropriate objective in some locations. How can this be most effectively be carried out? What is the role of species with different functional traits? Does the assembly sequence matter or do alternative pathways eventually coalesce around a limited number of endpoints? What is the role of facilitators or framework species? Is it easier to restore the less-complex tropical forests found on isolated (Pacific) islands than more complex forests found on (Asian) mainlands?

9. *How to increase the ecological resilience of plantations at both local and landscape scales?*

   There are guidelines for improving ecosystem resilience but not much experience in actually applying these to reforestation programs. How much functional redundancy should plantation managers build into their silvicultural designs or landscape mosaics? What trade-offs are needed to enhance economic resilience?

10. *What are the design principles for reforesting degraded landscape in the face of climate change?*

    Planted forests have the capacity to sequester significant amounts of carbon and systems of payment are likely to become available to growers whose plantings provide this service. Which species should be used in particular locations? How can the sequestration of soil carbon be enhanced? Where should reforestation be carried out to maximise carbon sequestration? Where should it be done to enable tropical biota to adapt to climate change?

Ten Socio-Economic Questions

1. *What plantation designs might improve livelihoods as well as generate conservation benefits?*

   Reforesting degraded land has the capacity to generate financial as well as conservation benefits. Some landowners are happy to choose just one of these but increasing numbers are interested in doing both. What are the trade-offs that must be made? How do these vary with forest age? How are these influenced by landscape context?

2. *How can secondary forests be managed to improve livelihoods as well as generate conservation benefits?*

   Secondary forests are usually capable of restoring forest cover relatively cheaply. But many are cleared because they are thought to be valueless. What mechanisms or policy settings could help retain and protect secondary forests?

3. *How can farmers be helped to choose the most appropriate tree species and type of plantation to suite their circumstances*?
   Many farmers prefer to use fast-growing exotic species such as *Eucalyptus* or *Acacia*. These may be the best choices in some situations but not others. How are farmers to make the most appropriate species choices to suit their circumstances?
4. *How can the spread of silvicultural information be hastened? How can research findings about conservation biology be implemented?*
   Many farmers would carry out reforestation if they were more familiar with the technology. Managers of National Parks would also use more suitable forms of reforestation to rehabilitate degraded areas within their parks if they knew how to do so. How can existing knowledge be shared? How can it be communicated in a way that makes sense to the people who might use it?
5. *How can plantation owners maximise the benefits they receive from their investment?* How might they learn of market opportunities? How could they receive better prices for the products they sell?
6. *How can a system of payments for ecological services be established?*
   There is often an imbalance between providers and beneficiaries of the ecological services provided by reforestation. How might providers of ecological services be paid? What type of system is needed to avoid high transaction costs when there are many growers with small farms?
7. *How can more resilient forms of reforestation be promoted?*
   Farmers with low incomes are often attracted to simple types of reforestation involving species such as *Eucalyptus, Acacia* or teak. Such plantations are often very productive and financially rewarding. Why should they do anything else? What are the benefits to a single household of diversity or landscape heterogeneity? How can more resilient forms of reforestation be promoted?
8. *How can Forest Landscape Restoration be promoted?*
   Strategic interventions in the landscape mean reforestation activities are more effective than random acts of tree planting. But what is the best way of coordinating forest landscape restoration? Should government agencies always have the coordinating role? How might private land owners be compensated for undertaking activities that primarily benefit the wider community?
9. *What role might different forms of incentives play in encouraging reforestation and conservation practices by farmers?*
   Reforestation may not always be an attractive land use practice and some farmers may need an incentive of some kind if they are to plant trees on their land or to use particular species or types of planting designs. What are the most cost-effective incentives (from a government viewpoint)? Are there alternative forms of financing that might make long-term crops like trees more attractive?
10. *How can social-ecological systems be helped to adapt to climate change?*
   The effects of climate change will be widespread but the impacts at particular locations are still uncertain. What forms of ecological and socio-economic

monitoring will be needed to enable land managers to adapt in a timely fashion? What are the key issues that they will need to take into account when making changes to the ways they manage their land and forest resources?

Finally

Given recent history it is easy to be pessimistic about the future of the world's tropical forests. The warning given by Wallace more than 100 years ago and quoted at the head of this chapter is still relevant. We still have only a rudimentary knowledge of the biota and ecological systems that make up the tropical world and much of what we know is based on a comparatively small sub-set of species (Gardner et al. 2009). Unregulated logging and deforestation persist and the rates of deforestation are not much less than they were a few decades ago. It is hardly surprising that the enormous areas of degraded lands and forest continue to increase.

Nonetheless there are grounds for some cautious optimism. More people living in the Asia-Pacific region are becoming concerned about these events and are seeking improvements (Franzel et al. 2004; Steinberg 2005). There are now a variety of silvicultural tools available to assist reforesting degraded areas and there is evidence that more landowners and governments are interested in using these. Many countries are now trying to increase their forest cover and there are numerous cases of local successes from across the region. One country (Vietnam) has already under gone the 'forest transition' and it's net forest cover has begun to increase.

But there are still many things to be done if these first positive signs are to truly make a difference. There are four tasks of over-riding importance. Firstly, it is important to increase the overall forest cover but it is also important to increase the quality of these new forests. This means improvements in the quality of the goods they produce and improvements in the environmental benefits they are able to provide. Ways must be found to use a much greater variety of genotypes, species and types of plantings when reforestation is carried out to avoid the current trends towards increased biological simplification and landscape homogenization that are now underway. In this respect it is important that efforts to encourage the regeneration of secondary forests be continued. But all these changes must also be accompanied by a reduction in the clearing and degradation of natural forests. An increase in reforestation at the expense of natural forests is a very poor trade-off.

The second task is to identify areas where reforestation should be given a special a priority. Funds and resources are limited. Just where are the degradation 'hotspots' and the areas most in need of being reforested? Where should incentives or subsidies be directed? Which of these areas can be reforested using plantation monocultures and which might need some form of Rehabilitation or Ecological Restoration?

A third task is to reduce the current dependency on imported technologies and to develop networks of people within the region who have the capacity to actively explore and test new methods of reforestation. These methods should improve the economic and ecological benefits of reforestation as well as ensure that the new forests will be resilient in the face of future changes, especially climatic change. Both these things will be easier to do if they are supported by more appropriate institutions and policies than have been common in the past. If so, forest growers may feel more comfortable about taking a longer-term view rather than being obliged to always have a short-term perspective.

Finally, there is a need to promote improved standards of governance if degradation is to be constrained and reforestation is to become more widespread. Various forms of cooperative activity have been discussed that could contribute to this. These include the Learning Networks for local capacity building (Chapter 10), methods to facilitate collaborative forms of Forest Landscape Restoration (Chapter 11) and national Advisory Group networks to bring together all those interested in encouraging reforestation (Chapter 12). But some wider issues are also involved and many of these are well known. It was 200 years ago, for example, when Thomas Malthus first wrote his famous tract that recognized the importance of property rights, the rule of law and representative government:

> The first grand requisite to the growth of prudential habits is the perfect security of property; and the next perhaps is that respectability and importance, which are given to the lower classes by equal laws, and the possession of some influence in the framing of them. The more excellent therefore is the government, the more does it tend to generate the prudence and elevation of sentiment, by which alone in the present state of our being, poverty can be avoided
>
> (Malthus 1890, p. 479).

Governance and reforestation are inter-linked. Property rights, a fair legal system, transparency, accountability and a proper institutional framework are all elements in a system of governance likely to reassure landowners they can afford to invest time and resources in reforestation. But the reverse is also true; good reforestation practices that enhance the experiences, skills and capacities of communities and their members improves, and also has the effect of re-enforcing, good social and democratic practices. Fortunately, there are grounds for being optimistic that both the rates of reforestation and the standards of governance will continue to improve across the Asia-Pacific region in future.

References

Chazdon RL, Harvey CA, Komar O, Griffith DM, Ferguson BG, Martinez-Ramos M, Morales H, Nigh R, Soto-Pinto L, van Breugel M, Philpott SM (2009) Beyond reserves: A research agenda for conserving biodiversity in human-modified tropical landscapes. Biotropica 41:142–153

Chokkalingam U, Carandang AP, Pulhin JM, Lasco RD, Peras RJJ, Toma T (2006) One century of forest rehabilitation in the Philippines: approaches, outcomes and lessons. Center for International Forestry Research, Bogor

Elliott S (2000) The Chiang Mai research agenda for the restoration of degraded forest lands for wildlife conservation in Southeast Asia. In: Blakesley D, Anusarnsunthorn V, Kerby J, Navakitbumrung P, Kuarak C, Zangkum S, Hardwick K, Elliott S (eds) Forest restoration for wildlife conservation. International Tropical Timbers Organisation and Forest Restoration Research Unit, Chiang Mai University, Chiang Mai, pp 383–411

Erskine P, Lamb D, Bristow M (2005) Reforestation in the tropics and subtropics of Australia using rainforest tree species. Rural Industries Research and Development Corporation, Canberra, https://rirdc.infoservices.com.au/items/05–087; accessed 20 September 2010
Fatoux C, Castella J-C, Zeiss M, Pham HM (2002) From rice farmer to agroforester within a decade. The impact of Doi Moi on agricultural diversification in a mountainous commune of Cho Moi District, Bac Kan Province, Vietnam. In: Castella JC, Quang DD (eds) Doi Moi in the mountains land use changes and farmers livelihood strategies in Bac Kan Province, Vietnam. The Agricultural Publishing House, Hanoi, pp 73–97
Franzel S, Denning GL, Lilleso JPB, Mercado AR (2004) Scaling up the impact of agroforestry: Lessons from three sites in Africa and Asia. Agroforest Syst 61:329–344
Gardner TA, Barlow J, Chazdon R, Ewers RM, Harvey CA, Peres CA, Sodhi NS (2009) Prospects for tropical forest biodiversity in a human-modified world. Ecol Lett 12:561–582
Lawson K, Burns K, Low K, Heyhoe E, Ahammad H (2008) Analysing the economic potential of forestry for carbon sequestration under alternative carbon price paths. Australian Bureau of Agricultural and Resource Economics, Canberra
Lee DK, Suh SJ (2005) Forest restoration and rehabilitation in Republic of Korea. In: Stanturf JA, Madsen P (eds) Restoration of boreal and temperatre forests. CRC Press, Boca Raton, FL, pp 383–396
Li W (2004) Degradation and restoration of forest ecosystems in China. For Ecol Manage 201:33–41
Lui J, Li S, Ouyang Z, Tam C, Chen X (2008) Ecological and socioeconomic effects of China's policies for ecosystem services. Proc Natl Acad Sci 105:9477–9482
Malthus T (1890) An essay on the principle of population, 6th edn. Ward Lock & Co., London
MARD (2001) Five million hectare reforestation program partnership: synthesis report. International Cooperation Department, Hanoi
McElwee P (2009) Regforesting 'bare hills' in Vietnam: social and environmental consequences of the 5 million hectare reforestation program. Ambio 38:325–333
Morell V (2008) Let 1000 forests bloom. Science 320:1442–1443
Nawir AA, Murniati, Rumboko L (2007) Forest rehabilitation in Indonesia: Where to after three Decades? Center for International Forestry Research, Bogor
Nibbering JW (1999) Tree planting on deforested farmlands, Sewu Hills, Java, Indonesia: Impact of economic and institutional changes. Agroforest Syst 46:65–82
Ohlsson B, Sandewall M, Sandewall RK, Nguyen HP (2005) Government plans and farmers intentions: A study on forest land use planning in Vietnam. Ambio 34:248–255
Pasicolan PN, de Haes HAU, Sajise PE (1997) Farm forestry: An alternative to government driven reforestation in the Philippines. For Ecol Manage 99:261–274
Rodrigues RR, Lima R, Gandolfi S, Nave AG (2009) On the restoration of high diversity forests: 30 years of experience in the Brazilian Atlantic Forest. Biol Conserv 142:1242–1251
Santos F, Bertomeu M, Vega B, Mangaong E, Stark M, Bullecer R (2003) Local knowledge on indigenous trees: Towards expanding options for smallholder timber plantating and improved farm forestry in the Philippine uplands. In: Sim HC, Appanah S, Durst P (eds) Bring back the forests: Policies and practices for degraded lands and forests FAO. FAO, Bangkok; Kuala Lumpur, pp 75–84
Steinberg P (2005) From public concern to policy effectiveness: civic conservation in developing countries. J Int Wildlife Law Policy 8:341–365
Stone R (2009) Nursing China's ailing forests back to health. Science 325:556–558
Tak K, Chun Y, Wood PM (2007) The South Korean forest dilemma. Int Forestry Rev 9:548–557
Uchida E, Xu JC, Rozelle S (2005) Grain for green: Cost effectiveness and the sustainability of China's conservation set-aside program. Land Econ 81:247–264
Wallace AR (1863) On the physical geography of the Malay Archipelago. J R Geographic Soc (London) 33:217–234
Wuethrich B (2007) Reconstructing Brazil's Atlantic rainforest. Science 315:1070–1072
Zomer RJ, Trabucco A, Bossio DA, Verchot LV (2008) Climate change mitigation: A spatial analysis of global land suitability for clean development mechanism afforestation and reforestation. Agric Ecosyst Environ 126:67–80

Glossary

Adaptive management:	A systematic process for continually adjusting policies and practices by evaluating and learning from the outcome of previously used policies and practices. Each management action is viewed as an experiment designed to test hypotheses and probe the system as a way of learning about the system.
Afforestation:	Tree planting on sites that have not had trees for more than 50 years.
Agroforest:	A regrowth or secondary forest enriched by planting trees or other plants having subsistence or commercial value.
Agroforestry:	A collective name for land use practices in which woody perennials are deliberately integrated with crops and/or animals on the same land management unit. The integration can be either in a spatial mixture or in a temporal sequence.
Biodiversity:	The variety of genes, species and ecosystems present in an area.
Canopy layer:	The leafy crowns of the taller trees in a forest.
Community forestry:	A form of forestry where there is some element of community participation in management and some commitment to improved or secure provision of at least some forest products to rural people living in or near forests.
Deforestation:	When natural forests are replaced by non-forest land uses or the tree cover falls below 10%.
Degraded forest:	A natural forest that has its structure, biomass or composition temporarily or permanently changed by human activities in a way that lowers its capacity to provide goods or services (see also Box 1.3).
Degraded land:	A reduction in the productivity capacity of land caused by changes in soil fertility, erosion, weeds or recurrent fires due to inappropriate human activities (see also Box 1.3).

Direct seeding:	Direct applications of forest seed to help re-establishment a forest.
Disturbance:	Any event that alters the structure, composition or functioning of a forest.
Ecological Restoration:	The process of assisting the recovery of an ecosystem that has been damaged, degraded or destroyed (see also Box 4.1). As used here the term is applied to forms of reforestation intended to lead to the re-establishment of ecosystems resembling those existing prior to a disturbance.
Enrichment planting:	Where plants (usually tree species) are planted within an existing forest to increase its commercial or conservation value.
Ecosystem services:	The benefits that people derive from ecosystems. These include services such as clean water, stable hill slopes, habitats for biodiversity, pollination, biogeochemical cycles, carbon storage as well as cultural or non-material benefits.
Forest Landscape Restoration:	A process that aims to restore ecological integrity and enhance human well-being in deforested or degraded forest landscapes. Rather than returning forests to their original 'pristine' condition Forest Landscape Restoration aims to strengthen the resilience and functionality of the forest landscape to keep future forest management options open.
Forest transition:	The changeover within a country or region from having a net loss in forest cover to having a net gain in forest cover.
Functional type:	Species having a similar ecological function within an ecosystem. These may be classified on the basis of phylogeny, life form, resource use, response to a defined perturbation or role in ecosystem function.
Institutions:	The prescriptions or rules humans use to organise all forms of repetitive and structural interactions. They are the arrangements enabling collective action.
Landscape:	The spatial scale at which it is necessary to intervene if one is to balance trade-offs and optimise conservation and livelihood benefits in a particular area (see Box 11.1).
Mixed-species plantations:	Plantations involving two or more species. These may be all trees or mixtures of trees and understorey crops.

Monoculture plantations: Even-aged plantations established using a single species. Over time such plantations may acquire a species-rich understorey. In old (>60 years) plantations some of these species may even grow up to join the canopy layer at which time the plantation may be indistinguishable from a mixed species plantation or, in extreme cases, a natural forest.

Monocyclic silvicultural system: A management system in natural forests where all trees in patch are harvested at a single time enabling seedlings already present on the forest floor to grow in the open. It results in a new even-aged forest. The length of time before the next harvest depends on the growth rate of trees but is in the order of 60–100 years. This means the cutting cycle or rotation matches the age of the oldest trees.

Natural regeneration: The re-establishment of native trees and other plants by self-sown seed or by vegetative regrowth.

Pioneer species: Species normally found in early successional stages because of their tolerance of exposed conditions and their capacity to be widely dispersed. Most are short-lived but many can form large, dormant soil seed stores. Germination of these seed can be triggered by soil or forest disturbances.

Plantation: Forests established by planting or seeding, usually for a commercial purpose or to provide some environmental service. May use exotic or indigenous species. Often, but not always, initially established at a regular spacing. See also Monocultural plantations and mixed-species plantations.

Polycyclic silvicultural system: A management system in which only trees with a diameter greater than a prescribed size are felled. Those remaining are expected to continue growing and exceed the cutting limit. A second harvest takes place when a sufficient volume has accumulated. This time period (known as the cutting cycle) will be much less than the age of the older trees.

Poverty: A pronounced deprivation of well-being caused by a lack of assets or capital, a sense of powerlessness and increased vulnerability to natural or economic crises.

Primary forest: Undisturbed and ecologically mature forest that has reached an advanced successional stage. Sometimes referred to as old-growth forest.

Protected areas:	A clearly defined geographical space, recognised, dedicated and managed, through legal or other effective means, to achieve the long-term conservation of nature with associated ecosystem services and cultural values. The IUCN classification includes National Parks, Nature Reserves, Wilderness Areas, National Monuments, Species Management Areas and areas where some harvesting operations are allowed such as state-owned production forests.
Reforestation:	An all-embracing term used to cover the re-establishment of forests on cleared land that has had forest within the previous 50 years by natural regeneration, direct seeding or planting, irrespective of the number of species used. Its use here includes Monocultural Plantations, Rehabilitation plantings and Ecological Restoration (see Box 4.1).
Rehabilitation:	The establishment of new forests made up of some, but not necessarily all of the original species. Rehabilitated forests may include exotic species (see Box 4.1).
Resilience:	The capacity of a system to absorb disturbances and remain in the same state with essentially the same structure, functioning and feedback mechanisms. It has ecological, social and economic components. Resilience is a property of social-ecological systems. A resilient system can rebuild itself where necessary.
Restoration:	Often used interchangeably with the term 'reforestation' to describe the re-establishment of forests irrespective of the methods used or the objectives of those carrying it out. Not to be confused with 'Ecological Restoration' which is used here to describe situations where the intention is to return the ecosystem to a state approaching its pre-disturbance condition (see Box 4.2).
Rotation:	The period over which a plantation is grown. It may be <10 years for pulpwood plantations that use fast-growing species to >30 years for some plantations grown to produce sawlogs.
Secondary forest:	Forests regenerating after all forms of severe disturbances including poorly managed logging and agricultural clearings as well as natural disturbances such as landslips and wildfires.

Shifting cultivation (swidden agriculture): An agricultural system in which plots of land are used temporarily and then abandoned while the farmers moves to another location. There are many forms of shifting cultivation but the system commonly involves clearing forest, using the land for a single cropping period (often involving a variety of species that are grown together) and then leaving the land fallow for 10 years or more to exclude weeds and restore fertility before repeating the cycle.

Silviculture: The art of producing and tending a natural forest or plantation. Silvicultural practice involves the treatments that may be applied to a natural forest or plantation to maintain or enhance their utility for any purpose. A silvicultural system is a program for treating a stand during a whole rotation.

Smallholder: A farmer owning, or able to use, a small area of land.

Socio-ecological system: Complex integrated system of people and their natural environments.

Stakeholder: A person, group or organisation with a stake in an organisation because they can be affected by that organisation's policies, objectives or actions.

Tenure: Rights of access to land and use of the resources found on that land. These rights usually prescribe the individuals or communities that can use various resources, the conditions under which they can do so and the length of time over which they have control.

Understorey: The tree seedlings and other plants found growing on the forest floor beneath the forest canopy.

Wildlings: Naturally occurring seedlings found growing on forest floor and collected to plant at another site.

Index

G

O

P